Monoclonal Antibodies

A Practical Approach

Edited by

Philip Shepherd

Guy's, King's and St Thomas' Schools of
Medicine, Peter Gorer Department of
Immunobiology, Guy's Hospital, New Guy's
House, London Bridge, London SE1 9RT, U.K.

and

Christopher Dean

Section of Immunology, Institute of Cancer
Research, McElwain Laboratories, 15 Cotswold
Road, Belmont, Sutton, Surrey SM2 5NG, U.K.

OXFORD

UNIVERSITY PRESS

OXFORD
UNIVERSITY PRESS

Great Clarendon Street, Oxford OX2 6DP

Oxford University Press is a department of the University of Oxford.
It furthers the University's objective of excellence in research, scholarship,
and education by publishing worldwide in

Oxford New York

Athens Auckland Bangkok .Bogotá Buenos Aires Calcutta Cape Town
Chennai Dar es Salaam Delhi Florence Hong Kong Istanbul Karachi
Kuala Lumpur Madrid Melbourne Mexico City Mumbai Nairobi Paris
São Paulo Singapore Taipei Tokyo Toronto Warsaw

with associated companies in Berlin Ibadan

Oxford is a registered trade mark of Oxford University Press in the UK and in
certain other countries

Published in the United States by Oxford University Press Inc., New York

British Library Cataloguing in Publication Data
Data available

Library of Congress Cataloging in Publication Data

Monoclonal antibodies : a practical approach / edited by Philip S. Shepherd,
Christopher J. Dean.
    p. ; cm.—(The practical approach series)
Includes bibliographical references and index.
ISBN 0-19-963723-7 (hbk.: alk. paper)—ISBN 0-19-963722-9 (pbk.: alk.
paper)
1. Monoclonal antibodies—Laboratory manuals. I. Shepherd, Philip S.
II. Dean, Christopher J. III. Series.
[DNLM: 1. Antibodies, Monoclonal. 2. Immunologic Techniques. QW
575.5.A6 M7455 2000]
QR186.85 M65646 2000   616.07'98—dc21   99-059426
1 3 5 7 9 10 8 6 4 2

ISBN 0-19-963723-7 (Hbk.)
ISBN 0-19-963722-9 (Pbk.)

Typeset in Swift by Footnote Graphics, Warminster, Wilts
Printed in Great Britain on acid-free paper
by The Bath Press (Avon) Ltd

Preface

This volume presented in the Practical Approach Series deals with the preparation, testing and derivation of monoclonal antibodies as well as with some of their applications. As with the previous book on antibodies in this series, edited by David Catty, we have tried to maintain a similar format in making it a collection of bench protocols, even if some of the techniques require specialised apparatus not normally found in a routine molecular immunology laboratory.

The first part includes chapters dealing with the preparation of rodent and human monoclonal antibodies by the standard somatic hybridisation technique, as well as by recombinant techniques, including the use of phage libraries and ARM display. Leading on from their preparation there are chapters dealing with small and large scale production, including production in plants together with purification techniques and methods for labelling them with radionuclides and non-radioisotopic compounds such as enzymes or fluorochromes and colloidal gold for light and electron microscopy. The second half covers their application, with chapters on immunoblotting, enzyme linked immunoassays, immuno-fluorescence, immunocytochemical staining of cells and FACS analysis. Also included are chapters on the clinical applications of monoclonal antibodies in rheumatoid arthritis, tissue typing and the management of organ graft rejection, the detection of chemically modified DNA in lymphocytes from patients undergoing chemotherapy and analysis of clinical haematological samples in transplantation for malignancy.

New immunological techniques incorporating tried and tested methodologies are of interest to established Immunologists as well as those entering the field for the first time. One or two of the chapters will be of interest to more specialised workers, e.g. those using confocal microscopy, but still of interest to those investigators whose laboratories lack such specialised apparatus.

The authors have been given a free rein to cover their topic in the manner they chose. This has led to some repetition which we as editors have not removed so as to keep the understanding and continuity within and between chapters such that the reader does not have to keep referring to the other chapters to follow the subject under presentation.

We are deeply indebted to all the authors for their hard work and dedication in writing the chapters, and to Dora Paterson and Jeremy Cridland for their secretarial and computer assistance.

Both of us have learnt a lot about new antibody production techniques while editing this book and we hope this will be the same experience for all whom refer to the finished volume.

P. Shepherd and C. Dean

Contents

10 Non-radioactive antibody probes   *237*

G. Brian Wisdom

11 Immunogold probes for light and electron microscopy   *247*

Paul Monaghan and David Robertson

12 Characterization of cellular antigens using monoclonal antibodies   *265*

Gillian Hynes

CONTENTS

17 FACS analysis of clinical haematological samples in transplantation for cancer *371*

Barbara C. Millar

18 Immunocytochemical staining of cells and tissues for diagnostic applications *391*

Andrew R. Dodson and John P. Sloane

CONTENTS

Protocol list

Abbreviations

Ab	antibody
ABC	avidin biotin complex
Ag	antigen
AP	alkaline phosphatase
APAAP	alkaline phosphatase anti-alkaline phosphatase
APES	3-aminopropyltriethoxysilane
ARM	antibody–ribosome–mRNA complex
ATG	anti-thymocyte globulin
BFC	bifunctional chelating agents
B-LCL	B-lymphoblastoid cell lines
BM	bone marrow
BSA	bovine serum albumin
CD	cluster of differentiation
CDR	complementary determining region
CPMV	cowpea mosaic virus
CV	column volumes
DAB	3,3′-diaminobenzidine
DIG	digoxigenin
DMEM	Dulbecco's modified Eagle's medium
DMSO	dimethyl sulfoxide
DSG	deoxyspergualin
DTPA	diethylenetriaminepentaacetic acid
EBV	Epstein–Barr virus
ECF	enhanced chemifluorescence
ECL	enhanced chemiluminescence
EGF	epidermal growth factor
ELIFA	enzyme-linked filtration assay
ELISA	enzyme-linked immunosorbent assay
EM	electron microscope
FACS	fluorescence activated cell sorter
FCA	Freund's complete adjuvant
FCS	fetal calf serum
FIA	Freund's incomplete adjuvant

ABBREVIATIONS

FITC	fluorescein isothiocyanate
FSC	forward scatter
GR-FL	green fluorescence
H	heavy chain
HAT	hypoxanthine, aminopterin, thymine
HGPRT	hypoxanthene-guanine phosphoribosyl transferase
HIC	hydrophobic interaction chromatography
HRP	horseradish peroxidase
HSA	human serum albumin
HTAR	high temperature antigen retrieval
IFM	immunofluorescence microscopy
Ig	immunoglobulin
IMAC	immobilized metal affinity chromatography
KLH	keyhole limpet haemocyanin
L	light chain
LM	light microscope
LPS	lipopolysaccharide (*E. coli*)
mAb	monoclonal antibody
MHC	major histocompatibility complex
MNC	mononuclear cell
m.o.i.	multiplicity of infection
OD	optical density
OR-FL	orange fluorescence
OVA	ovalbumin
PAGE	polyacrylamide gel electrophoresis
PB	peripheral blood
PBMC	peripheral blood mononuclear cells
PBS	phosphate-buffered saline
PBSC	peripheral blood stem cell
PCR	polymerase chain reaction
PE	phycoerythrin
PEG	polyethylene glycol
p.f.u.	plaque forming units
PHA	phytohaemagglutinin
PI	propidium iodide
PLT	progressive lowering of temperature
PMA	phorbol myristic acetate
PMSF	phenylmethylsulfonyl fluoride
PS	polystyrene
PVC	polyvinyl chloride
QE	quantum efficiency
RA	rheumatoid arthritis
RD-FL	red fluorescence
Rh	rhesus
RHS	right-hand scatter
RIA	radioimmunoassay

RT	room temperature (20°C)
SC	secretory component
SDS–PAGE	sodium dodecyl sulfate–polyacrylamide gel electrophoresis
SEC	size exclusion chromatography
sFv	single chain Fv
SPA	scintillation proximity assay
TBS	Tris-buffered saline
TMV	tobacco mosaic virus
TNF	tumour necrosis factor
TRITC	tetramethylrhodamine-5-isothiocyanate
V	variable region

Chapter 1

Preparation of rodent monoclonal antibodies by *in vitro* somatic hybridization

Christopher Dean* and Philip Shepherd[†]

*Section of Immunology, Institute of Cancer Research, McElwain Laboratories, 15 Cotswold Road, Belmont, Sutton, Surrey SM2 5NG, U.K.
[†]Guy's, King's and St. Thomas' Schools of Medicine, Peter Gorer Department of Immunobiology, Guy's Hospital, New Guy's House, London Bridge, London SE1 9RT, U.K.

1 Introduction and strategy

Like all great discoveries the development of hybridoma technology by Cesar Milstein and Georges Köhler was essentially a simple procedure namely, fuse an antibody-producing B cell with a myeloma cell line lacking the DNA salvage pathway to generate a continuously growing hybridoma producing a single antibody specificity. In this way the genes coding for the specific heavy and light chains are captured from the B cell and expressed by the hybridoma. The ability to generate monoclonal antibodies (mAbs) with precise antigenic specificities from the spleens or lymph nodes of immune rodents has revolutionized our approaches to biomedical science. Not only have these antibodies opened the doors behind which many gene products would have lain hidden but also they have provided us with exciting new approaches for the diagnosis and treatment of disease.

Before embarking on the generation of monoclonal antibodies it is important to be sure that a monoclonal antibody really is required. In some instances such a reagent is not essential and a specific high affinity polyclonal rabbit or sheep antiserum suitably absorbed will suffice. This will depend largely on the use to which the antibodies will be put. For example for the development of certain radioimmunoassays an absorbed 'monospecific' polyclonal antibody preparation of high affinity for the antigen is often preferred. Indeed one of the disadvantages of rodent monoclonal antibodies can be their relatively low affinity for antigen.

In many countries the use of animals for research purposes is strictly controlled and only licensed personnel experienced in small animal surgery are able to carry out the required immunization procedures. Also, the protocols used to achieve immunity are themselves subject to strict control so it is worth considering how essential a mAb reagent is for the purpose in hand or if a high affinity polyclonal antiserum would suffice. Nevertheless, for the precise identification of antigenic epitopes or for diagnostic and therapeutic applications mAbs are essential reagents. In the two decades since the publication of Köhler

1

and Milstein's findings, somatic cell fusion has been the most widely used procedure for the generation of mAbs using either immune rodent spleen cells or lymphoid cells from other immunized species including EBV-transformed human peripheral blood cells.

It is worth remembering that about 80% of the antibodies developed *in vivo* bind to conformational epitopes whereas only 20% recognize sequential determinants. Antibodies to the latter will be essential for the detection of denatured proteins in, for example, immunoblots or formalin-fixed paraffin embedded sections while the former will bind to native macromolecules and may exhibit biological activity. Thus, when rodents are used as the source of immune B cells the response will depend critically not only on the route of immunization and the adjuvant, but also on the immunogen (peptide, protein, whole cell, etc.) used and the number of immunizations given. Compared to rabbits, mice and rats respond poorly to peptide immunogens and it is often not possible to prepare good mAbs from such animals. This problem can be circumvented by giving an initial challenge with the protein of origin then boosting with peptide, but our experience is that proteins either native or denatured constitute far better immunogens than short peptides (20 amino acids or less). Recombinant proteins prepared in *E. coli* as fusion proteins should, where possible, be cleaved from their carriers (e.g. β-galactosidase) which are often highly immunogenic making selection of the specific hybridomas difficult.

Perhaps the most important aspect of monoclonal antibody production is the screening protocol used to identify the antibody of choice either from hybridoma culture supernatants or from recombinant libraries of antibody genes. For example, these vary from binding to antigen coated plastic (e.g. 96-well microtitre plates) to direct binding to target cells or to inhibition of binding of a ligand to its receptor. An important point to remember is that the conformation of peptides and even large proteins (> 100 kDa) bound to plastic can be influenced by charges on the plastic surface. We have found that up to half the antibodies identified by binding to HIV gp120 in this way did not bind to the native protein. Whatever method is chosen for the initial screening further tests will be required to ensure that the chosen monoclonal antibody has the required specificity.

In this chapter we describe the preparation and selection of rodent hybridomas secreting monoclonal antibodies (mAbs) based on procedures that we use routinely and discuss those factors which influence the success of hybridoma production. The basic protocols used to generate hybridomas and select those producing specific antibodies are summarized in *Figure 1* and detailed below. Essentially, the most important requirements are to:

(a) Generate specific immune B lymphocytes.

(b) Fuse them successfully with a continuously growing myeloma cell line.

(c) Identify the antibodies that are sought in culture supernatants.

(d) Isolate and clone the specific hybridomas.

Figure 1: please see plate section between pages 226–227.

2 Choice of host for immunization and myeloma for cell fusion

Most of the monoclonal antibodies that have been produced to date have been generated using mice of the BALB/C strain because the fusion partners (myeloma cell lines) have been developed from plasmacytomas induced in this strain by the intraperitoneal injection of mineral oil. However, rats of the LOU/wsl strain develop ileocaecal plasmacytomas with a high incidence and two of these have been developed for hybridoma production (*Table 1*). These myelomas have been selected for loss of the enzyme hypoxanthene-guanine phosphoribosyl transferase (HGPRT) and, in consequence, are unable to utilize the salvage pathway for DNA synthesis in the presence of the folic acid synthesis inhibitor aminopterin. Normal lymphocytes do not proliferate in culture but, because they contain the gene for HGPRT, the hybridomas formed on fusion with the myeloma will grow if the medium containing aminopterin (A) is supplemented with hypoxanthine and thymine (HAT selection medium).

While it is common to use the same host as the source of immune lymphoid cells this is not essential and the mouse or rat strain can be chosen on the basis of their suitability. For example some rat strains have high levels of natural antibodies to DNA which have precluded their use for the generation of anti-

Table 1. Some myeloma and heteromyeloma cell lines used for the preparation of hybridomas from different species

Myeloma	Origin	Ig expression	Reference
Mouse			
P3-X63/Ag8	BALB/C mouse	IgG1 (κ)	1
NS1/1.Ag4.1	BALB/C mouse	κ chain (non-secreted)	1
X63/Ag8.653	BALB/C mouse	None	1
Sp2/0	Sp2 mouse hybridoma	None	1
NS0/1	NS1/1.Ag4.1	None	1
Rat			
Y3-Ag 1.2.3.	Lou/wsl rat myeloma	κ chain	1
YB2/3.0 Ag 20	Hybrid YB2/3 (Lou × AO)	None	1
IR984F	Lou/wsl rat myeloma	None	2
Ovine[a]			
IC6.3a6T.1D7	Sheep–NS0 heterohybridoma	None	3
Rabbit			
240E 1-1-2	Rabbit plasmacytoma	None	4
Human			
LICR-LON-HMy-2	Human plasma cell leukaemia	–	5
Bovine[a]			
NS0	Calf–NS0 heterohybridomas	–	6
Equine[a]			
NS0	Horse–NS0 heterohybridomas	–	7

[a] Lymphoid cells fused directly with the mouse NS0 myeloma.

bodies to adducts formed with DNA-reactive drugs such a cis-platinum or melphalan. By immunizing rats containing low levels of endogenous antibodies to DNA, mAbs specifically recognizing the DNA adducts were readily produced (8). Furthermore, heterohybridomas can be prepared where the lymphoid partner can be derived from a different species although the hybrids are often unstable compared to their mouse and rat counterparts. However, several of these heterohybridomas have been used successfully as fusion partners in other species (*Table 1*). Indeed, the preparation of high affinity sheep monoclonals has been successful using this route. In general, the lower cost and ease of handling of the smaller rodents has made them the first choice of researchers as the source of immune lymphocytes.

3 Choice of immunogen

It is not essential to use purified antigen for immunization, although it can make a difference when poorly immunogenic molecules are involved. Whole cells and partially purified cell extracts can be used as well as recombinant proteins produced in bacteria, yeast, or insect cells. If antibodies against carbohydrate determinants are desired it should be remembered that recombinant glycoproteins produced in bacteria will not be glycosylated, and that recombinant human glycoproteins produced in insect or Chinese hamster ovary cells will be glycosylated according to the host cell sequences. Recombinant proteins are frequently expressed in bacteria as fusion products with bacterial proteins such a β-galactosidase. The latter are highly immunogenic and antibodies against the bacterial product can predominate unless the protein is cleaved and purified before immunization. Antibodies that bind to sequential determinants are essential reagents for identifying proteins in formalin-fixed, paraffin embedded sections and immunoblots but, when native proteins are used for immunization, only some 20% of the antibodies generated will be of this type. Careful screening of culture supernatants will be required to identify these antibodies. Although peptides based on the predicted coding sequences of cloned cDNAs have been used to prepare rabbit antibodies against sequential determinants on the full-length protein, mice and rats respond poorly to peptide immunogens compared to rabbits. Peptide immunogens are usually ineffective in generating antibodies against conformational determinants, and if these are required then properly folded proteins or protein fragments must be used. The conformation of many transmembrane proteins changes when they are isolated from the cell surface and their biological activity may change. Live cells overexpressing such cell surface molecules, e.g. the receptor for epidermal growth factor (9), the related product of the c-*erb*B-2 proto-oncogene(10), or the human thyrotropin receptor (11) have proved to be excellent immunogens for producing specific anti-receptor antibodies. Where radiolabelled ligand is available it is a simple matter to identify antibodies which modify ligand–receptor interaction using cell monolayers expressing the receptor as target (see *Protocol 12*).

4 Preparation of antigen for immunization

4.1 Soluble antigens

When soluble proteins or carbohydrates are used, they are mixed 1:1 with Freund's adjuvant to give a stable emulsion, e.g. by vigorous vortexing—a sample dropped from a Pasteur pipette onto a water surface should contract into a droplet that remains stable for at least one hour (*Protocol 1*). Freund's adjuvant is an oil (croton oil, which is a tumour promoter) containing mycobacteria (Freund's complete adjuvant; FCA) that acts as a slow release agent that prevents rapid dispersion of the soluble immunogen and elicits a strong cellular infiltrate of neutrophils and macrophages at the site of injection. The use of adjuvant substantially improves the antibody response. Subsequent injections are made using an emulsion made with Freund's incomplete adjuvant (FIA) which lacks the mycobacteria. Other adjuvants that have been used and are less toxic to the animals include precipitates made with aluminium hydroxide or specific precipitating antibodies. Immunoprecipitates made in agar gels can be cut out and homogenized directly for injection.

4.2 Antigens expressed on live cells

Whole cells, either grown in suspension or, when adherent, removed by treatment with EDTA or trypsin, are resuspended in PBS or serum-free medium and injected directly without the addition of an adjuvant.

Protocol 1

Preparation of antigens for immunization

Equipment and reagents

- Stoppered LP3 tubes, Bijou bottles, or 30 ml Universals
- Vortex mixer
- Phosphate-buffered saline (PBS): dissolve in water 1.15 g Na_2HPO_4, 0.2 g KH_2PO_4, 0.2 g KCl, 8.0 g NaCl and make up to one litre (the pH should be 7.4)
- Dulbecco's modified Eagle's medium (DMEM): containing glucose (1 g/litre), bicarbonate (3.7 g/litre), glutamine (4 × 10^{-3} M), penicillin (50 U/ml), streptomycin (50 μg/ml), and neomycin (100 μg/ml); stored at 4–6 °C, and used within two weeks of preparation
- Peptides conjugated to carrier protein (see *Protocol 2*) and dissolved in PBS
- Soluble proteins or other macromolecules dissolved in PBS to give 0.1–4.0 mg/ml

- Soluble recombinant protein extracted from cells or supernatants of eukaryotic cells (e.g. Chinese hamster ovary cells or insect cells expressing recombinant baculovirus) or bacteria (such as *E. coli* harbouring plasmids or recombinant viruses) and dissolved in PBS
- Protein separated electrophoretically in sodium dodecyl sulfate-containing polyacrylamide gels (SDS–PAGE) and eluted from gel slices into PBS
- Cultured cells grown in suspension or as monolayers in flasks, then detached by treatment with 0.1% trypsin-EDTA, and resuspended in PBS or serum-free DMEM
- Freund's complete adjuvant (FCA)
- Freund's incomplete adjuvant (FIA)

Method

1. Mix proteins, peptide conjugates, or eluates from polyacrylamide gels in PBS 1:1 with adjuvant (FCA for the first immunization, FIA for subsequent immunizations) in a capped plastic tube (LP3, Bijou, or 30 ml Universal) by vortexing until a stable emulsion is formed.

2. Check that phase separation does not occur on standing at 4°C for > 1 h. Alternatively, allow a drop of emulsion to fall from a Pasteur pipette onto a water surface; the drop should contract, remain as a droplet, and not disperse.

3. Suspend live cells in PBS or DMEM at 5×10^6 to 10^7 cells/ml.

4.3 Plasmid DNA

Immune responses can be elicited in rodents by direct transfer of DNA, e.g. by injecting an expression plasmid containing the specific gene directly into muscle or other tissue. Some of the muscle cells will express the specific gene and the recombinant protein may activate the host immune system. Alternatively, naked DNA or gold particles coated with low concentrations of the plasmid DNA can be transferred into the tissue using a pneumatic gun or by particle bombardment. These procedures circumvent the need to express the proteins *in vitro* and details of their application are given in refs 12 and 13.

Protocol 2

Conjugation of peptides to carriers

Equipment and reagents

- PPD kit (tuberculin-purified protein derivative) (Cambridge Research Biochemicals, Ltd.)
- Keyhole limpet haemocyanin (KLH), ovalbumin (OVA), or bovine serum albumin (BSA) at 20 mg/ml in PBS
- Glutaraldehyde (specially purified grade 1, Sigma): 25% solution in distilled water
- 1 M glycine–HCl pH 6.6

Method

1. Mix the peptide and protein carrier (KLH, OVA, or BSA) in a 1:1 ratio, e.g. pipette 250 μl of each into a 5 ml glass beaker on a magnetic stirrer. Small fleas can be made from pieces of paper clip sealed in polythene tubing by heating. Add 5 μl of 25% glutaraldehyde and stir for 15 min at room temperature.

2. Block excess glutaraldehyde by adding 100 μl of 1 M glycine and stirring for a further 15 min.

3. Use directly, or dialyse overnight against PBS and store at −20°C.

Protocol 2 continued

4. PPD kit: **Read the instructions supplied with the kit very carefully**. Inhalation of the ether-dried tuberculin PPD is dangerous for tuberculin-sensitive people to handle. Follow specific instructions to couple 2 mg of peptide to 10 mg of PPD and, after dialysis store at $-20\,°C$.

4.4 Peptides

Peptides that contain 20–30 amino acids are often poorly or non-immunogenic because they lack sequences recognized by T helper cells, which are essential for generating an antibody response. For this reason, peptides are conjugated to larger proteins that contain many T cell reactive epitopes, such as keyhole limpet haemocyanin or ovalbumin. Conjugation of the peptide to the protein is done either with glutaraldehyde, which will link the peptide in random conformations (*Protocol 2*) or by disulfide linkage, where the peptide has been synthesized with a terminal cysteine. In this case, the peptide will be attached via one end to the carrier protein and this directional orientation will restrict its immunogenicity. Many of the antibodies produced will be directed against the carrier protein and it is important therefore when screening for specific antibodies to use either the peptide alone as target or to conjugate it to another protein unrelated to that used for immunization.

5 Route of immunization

5.1 Generation of immune spleen cells

To obtain high titre specific antibodies in serum, it is usual to immunize mice or rats at three or four sites subcutaneously or intramuscularly together with a challenge intraperitoneally (*Protocol 3*). The immunization is repeated at two to four week intervals until bleeds taken either from the tail vein (mice) or jugular (rats) show good titres of specific antibody in the serum. Immunization in this way will stimulate the production of specific antibody-producing B cells and their precursors in the local lymph nodes and spleen. Mature antibody-secreting cells fuse poorly with myelomas and it is the committed precursors that are required for hybridoma production. For this reason the animals are given a final challenge with antigen two or three days before the spleen or lymph node cells are harvested.

5.2 Immunization via the Peyer's patches of rats

In rats, cells from immune lymph nodes are an excellent source of specific B cells and we use routinely cells taken from the mesenteric nodes of animals immunized via the Peyer's patches. As with spleen cells, the animals are re-challenged via the Peyer's patches three days before removing the mesenteric nodes for fusion. Using this route for immunization we have successfully prepared rat × rat IgA-secreting hybridomas following a short immunization schedule whereas hyperimmunization yielded high affinity IgG antibodies (14).

Protocol 3

Generation of immune spleen or lymph node cells[a]

Equipment and reagents

- 1 ml tuberculin syringe
- BALB/C mice aged 6–8 weeks, rats of any strain aged 10–12 weeks
- Live cells in PBS or DMEM (see *Protocol 1*)
- Soluble proteins or conjugated peptides emulsified in FCA or FIA (see *Protocols 1 and 2*)

Method

1. Anaesthetize the animals and take a blood sample from the tail vein (mice) or jugular (rats) into a capped 0.5 ml or 1.5 ml microcentrifuge tube to act as a pre-immune sample. Allow to clot, centrifuge (1500 g), remove serum, and store at −20 °C.

2. For fusions that will use spleen cells, immunize each animal at five sites (4 × s.c. and 1 × i.p.) with a total of 10^7 cells in PBS or DMEM, or with a total of 50–500 µg of soluble antigen in FCA. Test bleed 14 days later and re-immunize using the same protocol but with antigen in FIA. Test bleed and re-immunize at two to four weekly intervals until sera are positive for antibodies to the antigen (see *Protocols 11–13*). Three days before the fusions are done, re-challenge the animals i.v. with the antigen in PBS alone.

3. For fusions that will use the mesenteric nodes of rats, the antigens are injected into the Peyer's patches that lie along the small intestine. The surgical procedure is described in *Protocol 4*.

[a] The use of animals for experimental purposes is under strict control in many countries, and licences are necessary before surgical procedures can be performed.

Protocol 4

Immunization of rats via the Peyer's patches

Equipment and reagents

- 1 ml syringe
- Blood collection tubes
- Soluble antigen in PBS, emulsified with an equal volume of Freund's adjuvant
- Live cells suspended in DMEM
- Sterile 0.9% NaCl

Method

1. Anaesthetize rats, bleed from the jugular (pre-immune control), and then open the abdomen along the central line.

2. Carefully extend the small intestine and locate the 8–16 Peyer's patches (strain dependent) that lie along the peritoneal wall of the small gut.

3. Take up the antigen-containing samples into a 1 ml syringe using a 27 gauge needle and inject between 10–15 μl into every other Peyer's patch to give a total dose of between 0.05–0.1 ml/animal.

4. To prevent the formation of adhesions, place 2 ml of sterile 0.9% NaCl in the peritoneal cavity and close the abdomen.

5. Two to four weeks later, test bleed the rats and re-challenge intraperitoneally with antigen in FIA or cells in DMEM.

6. Six to eight weeks after the initial challenge, anaesthetize the rats, test bleed, open the abdomen parallel to the initial incision, and re-immunize using the unchallenged Peyer's patches as recipient for antigen. Three days later kill the animals, remove the mesenteric nodes by blunt dissection, and use for cell fusion.

6 Growth of myeloma cell lines

A number of mouse and rat myeloma cell lines are currently in use and we have experience in the use of two of the mouse cell lines (NS1/1 Ag4.1 and Sp2/0) and the rat myeloma Y3Ag1.2.3 which we use routinely to generate rat × rat hybridomas. The mouse myelomas are grown in flasks as static cultures diluting with fresh medium to maintain them in exponential growth. In static culture the rat Y3 myeloma grows as an adherent cell line and will not fuse with lymphocytes. For this reason it is essential to maintain it as an exponentially growing single cell suspension in spinner culture (*Protocol 5*). Spinner cultures can be maintained and used for up to one month by daily fourfold dilution with fresh medium.

Protocol 5

Growth of myeloma cell lines

Equipment and reagents
- 25 cm^2 or 75 cm^2 flasks
- 200 ml spinner flask (Bellco)
- Mouse NS1 or Sp2/0 myeloma cell line
- Y3 rat myeloma
- DMEM–20% FCS

Method

1. Grow mouse myelomas in static flasks or spinner culture and keep growing exponentially in DMEM containing 10% or 20% FCS (dilute to 2–3 × 10^5 cells/ml the day before fusion).

2. The rat myeloma Y3 has to be grown in spinner culture to fuse well. Seven to ten days before cells are required, about 5 × 10^6 cells stored frozen in liquid nitrogen as 1 ml aliquots in freezer medium are thawed quickly at 37°C, diluted with 10 ml

Protocol 5 continued

of DMEM–10% FCS, centrifuged (500 g for 2 min), then resuspended in 100 ml of the same medium and placed in a 200 ml Bellco spinner flask. Stand for two days at 37°C to allow a monolayer of dividing cells to form on base of vessel, then place on magnetic stirrer running at about 160 r.p.m.

3. The Y3 myeloma has a generation time of about 10–12 h and exponentially growing cultures require feeding daily by fourfold dilution with fresh medium.

7 Preparation of cells for fusion

Myeloma cells are harvested by centrifugation, washed in serum-free DMEM, and resuspended at 10^7 cells/ml. Cells from the spleens or mesenteric nodes of immune animals are harvested by disaggregating the lymphoid tissue by passage through a fine stainless steel mesh (tea strainer) using a sterile spatula (*Protocol 6*). Spleens from immune mice generally yield about 10^8 cells in total whereas immune spleens and mesenteric nodes from one rat can yield up to 4 $\times 10^8$ cells (excess cells can be frozen in liquid nitrogen for subsequent fusion). The latter often give excellent yields of hybridomas probably because the freezing mixture contains dimethyl sulfoxide which is known to assist fusion of cell membranes.

Protocol 6

Preparation of immune lymphocytes and myeloma cells for fusion

Equipment and reagents

- Sterile, fine mesh stainless steel tea strainer and spoon-headed spatula
- Sterile 6 cm Petri dish
- 30 ml plastic Universal
- Two 50 ml sterile, capped centrifuge tubes

- Dulbecco's modified Eagle's medium (DMEM)
- Mice or rats taken three days after final immunization
- Exponentially growing mouse or rat myeloma cell line

Method

1. Centrifuge exponentially growing mouse or rat myeloma cells in 50 ml aliquots for 3 min at 400 g, wash twice by resuspension in serum-free DMEM, count in a haemocytometer, and resuspend in this medium to 1–2 $\times 10^7$ cells/ml.

2. Kill immune animals by cervical dislocation or CO_2 inhalation, test bleed, and open the abdominal cavity. Remove spleens or mesenteric lymph nodes by blunt dissection.

3. Disaggregate spleens or nodes by forcing through a fine stainless steel mesh into 10 ml of serum-free DMEM using a spoon-head spatula (dipped in ethanol and flamed to sterilize it).

Protocol 6 continued

4. Centrifuge cells for 5 min at 400 g, wash twice in serum-free DMEM, and resuspend in 10 ml of the same medium.

5. Count viable lymphoid cells in haemocytometer. Spleens from immune mice yield about 10^8 cells, from rats $3–5 \times 10^8$ cells, and the mesenteric nodes of rats, up to 2×10^8 cells.

8 Cell fusion

A number of protocols have been described for the generation of mouse hybridomas and workers have their own particular protocol. We have used a standard procedure for the fusion of mouse and rat myelomas with considerable success over the last twenty years and this is described in *Protocol 7*. Basically, 10^8 lymphocytes are mixed with 2×10^7 mouse myeloma cells or 5×10^7 Y3 cells in a round-bottomed tube and pelleted by centrifugation. Then the cells are fused by the addition of 1 ml of 50% PEG 1500 and plated into HAT selection medium to allow the growth of the hybridomas generated.

Protocol 7

Hybridoma formation

Equipment and reagents

- Sterile capped 10 ml tube
- 24-well or 96-well plates (Nunc)
- Dulbecco's modified Eagle's medium (DMEM)
- Fetal calf serum (FCS): inactivated by heating for 45 min at 56°C and tested for ability to support the growth of hybridomas
- PEG solution: weigh 50 g of polyethylene glycol (1500 molecular weight) into a capped 200 ml bottle, add 1 ml of water, and autoclave for 30 min at 120°C. Cool to about 70°C, then add 50 ml of DMEM, mix, and, after cooling to ambient temperature, adjust the pH to about 7.2 with NaOH (mixture should be coloured orange). Store as 1 ml aliquots at −20°C.
- HAT selection medium: prepare 100 × HT by dissolving 136 mg hypoxanthine and 38.75 mg thymine in 100 ml of 0.02 M NaOH pre-warmed to 60°C. Cool, filter sterilize, and store at −20°C in 2 ml

aliquots. Prepare 100 × A by dissolving 1.9 mg aminopterin in 100 ml of 0.01 M NaOH, then filter sterilize, and store in 2 ml aliquots at −20°C. Prepare HAT medium by adding 2 ml of HT and 2 ml of A to 200 ml of DMEM containing 20% FCS.

- HT medium: add 1 ml of HT to 100 ml of DMEM containing 10% FCS
- Feeder cells for fusion cultures and cloning of hybridomas (essential for fusions using rat myelomas): rat fibroblast cell lines derived from the xiphisternae of various strains are suspended in DMEM and irradiated with about 30 Gy (3000 rad) of X- or gamma rays, then frozen in 95% FCS, 5% DMSO, and stored in liquid nitrogen as aliquots of 5×10^6 cells. Use one aliquot for each 200 ml of HAT or HT medium. Alternatively, use thymocytes from spleen donors.
- Freezer medium: freshly prepared 5% dimethyl sulfoxide, 95% FCS

Protocol 7 continued

Method

1. Mix 10^8 viable lymphocytes with 2×10^7 mouse myeloma cells or 5×10^7 rat myeloma cells in a 10 ml capped centrifuge tube and centrifuge for 3 min at 400 g.

2. Pour off the supernatant, drain carefully with a Pasteur pipette, then release the cell pellet by gently tapping the tube on the bench.

3. Stir 1 ml of PEG solution, pre-warmed to 37 °C, into the pellet over a period of 1 min. Continue mixing for a further minute by gently rocking the tube.

4. Dilute the fusion mixture with DMEM (2 ml over a period of 2 min and then 5 ml over 1 min).

5. Centrifuge for 3 min at 400 g, then resuspend the cells in 200 ml of HAT selection medium (containing feeder cells where necessary), and plate 2 ml aliquots into four 24-well plates or 200 μl aliquots into ten 96-well plates. Incubate at 37 °C in 5% CO_2.

6. Examine plates 7–14 days later for the presence of hybridomas and screen for the presence of specific antibodies (see *Protocols 10–12*).

9 Screening hybridoma culture supernatants for specific antibody

The fusion cultures are examined for the growth of hybridoma colonies from seven days onwards. Mouse hybridomas tend to grow initially as discrete colonies (sphaeroids) whereas hybridomas produced with the Y3 myeloma are quite diffuse and not easy to recognize. With further growth the Y3 hybridomas form discrete colonies that can be quite firmly attached to the plastic. Screening is normally carried out initially at 10–14 days post-fusion by removing samples of about 200 μl of supernatant.

The importance of the initial screening assay cannot be overemphasized because it is on the basis of the result that the hybridoma colony is picked, expanded, and cloned. Many different procedures have been used to screen hybridoma culture supernatants from direct binding to cells or tissues using fluorescent, radiolabelled (see *Protocol 9*), enzyme-linked, or otherwise tagged second antibodies (anti-mouse, anti-rat, anti-sheep, etc.) to detect bound monoclonal antibody, to functional assays, e.g. inhibition of binding of ligand to its receptor. The important feature of the assay(s) used is that it is quick, reliable, and specific. Most hybridomas grow rapidly with a generation time of 10–12 hours (rat) to 15–24 hours (mouse) so that long-term assays are precluded. Assays that determine binding to antigen, whether present on/in cells, to proteins or peptides coated onto plastic multiwell plates or pins or to tissue sections are to be preferred to biological assays that cannot be performed in one day.

9.1 Antigen coated multiwell plates

Tests that can be carried out on 96-well plates coated with antigen either as protein or as cell monolayers are ideal for these initial screens. Bound anti-

Incubate hybridoma
supernatants with
antigen-coated plates
or cell monolayers

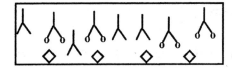

Wash away unbound
non-specific antibodies

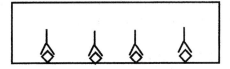

Detect bound antibodies
with 125-Iodine labelled
or enzyme labelled
secondary antibodies

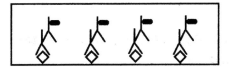

◇   Specific antigen   ⋏   Specific antibody

⋏   Non-specific antibody   ⊢ 125-Iodine or enzyme-
                                  labelled second antibody

Figure 2 Screening for specific antibody using the indirect method.

bodies can then be detected by RIA or ELISA using second antibodies specific for mouse or rat immunoglobulins radiolabelled with [125]iodine or conjugated to alkaline phosphatase or other enzyme for ELISA (see Chapters 10 and 11 for detailed protocols). A simple assay using multiwell plates coated with soluble antigen or a confluent monolayer of adherent cells is illustrated in *Figure 2*.

For a first screen we use antibodies to mouse or rat F(ab')$_2$ since these reagents will detect all antibody classes and subclasses. With a successful fusion the wells will contain more than one hybridoma colony and in some cases a number of colonies will be present. To determine if one or more of the colonies is secreting specific antibody, colonies are picked under the microscope using a Pasteur pipette and plated in DMEM–20% FCS into individual wells of a 24-well plate. The cells are allowed to grow to near confluence before re-testing for the presence of specific antibody. Additional tests are performed at this time to confirm the specificity of the selected antibodies.

Protocol 8

Preparation of antigen coated plates for screening culture supernatants

Equipment and reagents

- 96-well flat-well polystyrene (PS) or polyvinyl chloride (PVC) flexiplates
- PBSA: PBS containing 0.05% NaN_3 (Note: sodium azide should be handled with care; it is an inhibitor of cytochrome oxidase and is highly toxic and mutagenic—aqueous solutions release HN_3 at 37°C)
- PBST: PBSA containing 0.4% Tween 20
- PBS–BSA: PBSA containing 0.5% BSA

- PBSM: PBSA containing 3% skimmed milk powder; centrifuge or filter through Whatman No. 1 paper to remove undissolved solids (cheaper than BSA and just as effective)
- PCB: plate coating buffer pH 8.2, containing 0.01 M Na_2HPO_4/KH_2PO_4 and 0.14 M NaCl
- Soluble antigen: protein, recombinant protein, conjugated peptide

Method

1. Dissolve the antigen at 1 μg/ml in PCB.
2. Coat plates by incubation with 50 μl of antigen/well for 2 h at 37°C or overnight at 4–6°C.
3. Block remaining reactive sites on the wells by incubation for 2 h at 37°C with 200 μl/well of PBS–BSA or PBSM.
4. Wash plates with PBST before use. In many cases the plates can be stored at 4–6°C for several weeks (fill wells with PBST) but check for antibody-binding capacity before use.

Protocol 9

Labelling of proteins with [125]iodine[a]

Equipment and reagents

- Pharmacia PD-10 column containing Sephadex G25: equilibrated before use with UB and pre-treated with 100 μl of FCS to block sites that bind protein non-specifically
- LP2 tubes for collection of samples
- Lead pots for storage of radiolabelled antibodies
- Purified monoclonal antibodies or secondary reagents: e.g. sheep, rabbit, or goat antibodies to mouse or rat F(ab)$_2$ (either prepared in-house or purchased from Amersham Pharmacia, Sigma, Serotech, or other supplier) dissolved in PBS[b]

- Gamma counter
- 1.5 ml polypropylene microcentrifuge tubes coated with 10 μg of Iodogen (1,3,4,6-tetrachloro-3-6-diphenylglycouril, Pierce Chemical Co.) by evaporation, under a stream of nitrogen from a 100 μg/ml solution in methylene chloride; store tubes at 4°C in the dark
- Carrier-free [125]iodine, radioactive concentration 100 mCi/ml (e.g. code IMS-30, Amersham Pharmacia Biotech)
- UB: 100 mM phosphate buffer (Na_2HPO_4/KH_2PO_4) pH 7.4, containing 0.5 M NaCl and 0.02% NaN_3

Method

All procedures should be carried out in a Class 1 fume hood according to local radiation safety regulations.

1. Add 50 μg of purified antibody in 0.1 ml of PBS to an Iodogen coated tube followed by 500 μCi of ^{125}I (e.g. 5 μl of IMS-30). Cap, mix immediately by 'flicking', and keep on ice with occasional shaking.

2. After 5 min transfer the contents of the tube with a polythene, capillary-ended Pasteur pipette to a prepared Sephadex G25 column, then wash in and elute with 0.5 ml aliquots of UB.

3. Collect 0.5 ml fractions by hand, count 10 μl aliquots, and pool the fractions containing the first peak of radioactivity.

4. Store at 4°C in a lidded lead pot (e.g. iodine-125 container).

[a] The protocol is based on that of Fraker and Speck (15).

[b] If the antibodies have been prepared by affinity chromatography using chaotropic ions, such as KSCN to elute the bound antibodies, it is important to ensure that these ions are removed by dialysis or gel filtration since they interfere with the iodination procedure.

Protocol 10

Assay using antigen coated PS or PVC coated multiwell plates

Equipment and reagents

- Gamma counter and/or multiwell plate reader (Titertek)
- Antigen coated PS or PVC plates
- Second antibody (anti-mouse or anti-rat F(ab')$_2$ labelled with ^{125}iodine (see *Protocol* 9) or conjugated to alkaline phosphatase
- PBST
- PBSM
- 100 mM diethanolamine pH 9.5 containing 500 mM NaCl
- *p*-Nitrophenyl phosphate substrate

Method

1. Transfer 200 μl samples of culture supernatant from each well of the hybridoma cultures into a 96-well plate.

2. Transfer 50 μl aliquots of the supernatants to duplicate wells of the antigen coated PVC flexiplates for RIA or PS flat-bottomed wells for ELISA and incubate at ambient temperature for 1 h.

3. 'Flick' off supernatant and wash four times with 200 μl PBST/well.

4. Add 50 μl of ^{125}I-labelled second antibody (10^5 c.p.m./50 μl in PBSM), or for ELISA 50 μl of second antibody conjugated to alkaline phosphatase, and incubate for 1 h at ambient temperature.

Protocol 10 continued

5. Discard supernatant, being careful to dispose of the radiolabelled material according to local safety regulations, and wash plates four times with 200 μl PBST/well. Cut out individual wells from PVC plates and count in gamma counter.

6. For the ELISA wash plates with PBS and then twice with 10 mM diethanolamine pH 9.5 containing 500 mM NaCl. Add 50 μl of *p*-nitrophenyl phosphate substrate to each well, incubate for 10–30 min, then stop reaction by addition of 50 μl of 100 mM EDTA/well, and read at 405 nm.

9.2 Live or fixed cells

Live cells, which overexpress the protein/glycoprotein of interest, can form excellent targets for antibody selection when grown either in suspension or as adherent monolayers in multiwell plates. For example, tumour cells that overexpress growth factor receptors, or Chinese hamster ovary cells expressing recombinant human proteins, or insect cells expressing baculovirus constructs. Also, such cell lines can be used to target intracellular proteins after fixation and permeabilization of the cells.

Protocol 11

Assay using live or fixed adherent cells grown in multiwell plates

Equipment and reagents

- Ice bath
- LP2 tubes
- Gamma counter or multiwell plate counter
- DMEM–5% FCS
- 4% paraformaldehyde in PBS
- Permeabilizing solution: 0.05% Triton X-100 in PBSA containing 10^{-3} M phenylmethylsulfonyl fluoride (PMSF)
- Methanol pre-cooled to $-20\,°C$
- Alkaline Sarkosyl: 1% sodium dodecyl sarcosinate in 0.5 M NaOH
- Confluent monolayers of cell lines expressing cell surface antigen: for example, tumour cell lines overexpressing

a transmembrane protein, e.g. the receptor for EGF or the product of the *c-erb*B-2 proto-oncogene. Or rodent or insect cell lines expressing a recombinant protein, e.g. CHO or 3T3 cells transfected with genes for human transmembrane proteins or adherent cell lines expressing high cytoplasmic levels of the specific antigen that can be accessed following fixation and permeabilization.

- Confluent monolayers of control cell lines known not to express specific protein
- Second antibody (anti-mouse or anti-rat F(ab′)$_2$ labelled with ^{125}iodine (see *Protocol 9*) or conjugated to alkaline phosphatase (see Chapter 11 for method)

Method

1. Wash cell monolayer with DMEM–5% FCS to remove non-adherent/dead cells then proceed as for antigens bound to PVC plates (see *Protocol 10*) but using DMEM–5%

Protocol 11 continued

FCS for all diluents and washings. If the effect of antibody binding on the behaviour of the membrane protein is unknown, all incubations should be carried out at 4°C (float plates on ice bath and pre-cool diluents). Monolayers of cells vary widely in their adhesion to plastic, and also, they may round up after prolonged incubation at 4°C because of depolymerization of microtubules.

2. When RIA is used, after the final wash lyse the cells by incubating for 15 min at ambient temperature with 200 μl/well of alkaline Sarkosyl, and transfer lysates to LP2 tubes to determine the amount of ^{125}I present. Background should be about 50–200 c.p.m. and positive wells at least five times this value.

3. When carrying out ELISAs on live cells it is not necessary to lyse the cells after the substrate has been added and the coloured product formed can be read directly from the multiwell plate.

4. To access cytoplasmic proteins it is necessary to fix and permeabilize the cell monolayers before treatment with culture supernatant. Wash cell monolayer in ice-cold PBS then either:

   (a) Incubate for 10 min at −20°C in pre-cooled methanol, then wash in DMEM-5% FCS, and use directly.

   (b) Alternatively, fix for 30 min at 4°C in 4% paraformaldehyde in PBS, wash in ice-cold PBS, then incubate for 30 min in permeabilizing solution, wash in DMEM-5% FCS, and use as for live cells. This method can be particularly useful when handling cells which are poorly attached to the plastic wells.

9.3 Ligand binding assays

Antibodies that bind to growth factor and other cell surface receptors have proved useful in determining the function of these molecules and in some cases have proven to be useful diagnostic and therapeutic reagents. Where the specific ligand is available, e.g. epidermal growth factor (EGF), and is not affected by radiolabelling or other conjugation procedure, specific antibodies that interfere with ligand binding can be selected readily on live cells.

Protocol 12

Ligand-binding assays for anti-receptor antibodies

Equipment and reagents

- Ice bath
- Gamma counter
- DMEM-5% FCS
- Multiwell plates with confluent monolayers of live cells expressing specific receptor (e.g. human tumour cell line overexpressing the receptor for EGF)

- [^{125}I]EGF (10–100 μCi/μg), either labelled in-house (*Protocol 9*) or purchased ready labelled (e.g. Amersham Pharmacia)
- Alkaline Sarkosyl

Protocol 12 continued

Method

1. Wash cell monolayers with DMEM–5% FCS.

2. Mix culture supernatant or medium control with an equal volume of $[^{125}I]$EGF, add 50 µl aliquots (10^5 c.p.m.) in triplicate to the wells, and incubate 1 h on ice.

3. Wash the cells four times in ice-cold DMEM–5% FCS, then lyse with 200 µl alkaline Sarkosyl/well, and determine the c.p.m. bound. Antibodies which block ligand binding can reduce the binding of radiolabel to background.

10 Cloning of hybridomas

Cells from positive wells are grown up, samples frozen in liquid nitrogen, and the cells cloned twice by limiting dilution (plate out < 50 cells into a 96-well plate containing feeder cells if necessary and select cells from six wells that contain only single colonies). Make sure that samples of the first and second clonings are frozen in liquid nitrogen (*Protocol 14*). When the final selection has been made, be ruthless in disposing of unwanted stocks!

Protocol 13

Cloning of hybridomas

Equipment and reagents

- DMEM–20% FCS containing HT and either thymocytes (mouse) or rat fibroblasts (rat or mouse) as feeders
- Exponentially growing hybridoma cultures

Method

1. Centrifuge cells from at least two wells of a 24-well plate that contain confluent monolayers of hybridoma cells.

2. Count the cells and then dilute to give < 50 hybridoma cells in 20 ml of medium containing feeder cells.

3. Examine the plates seven to ten days later and screen those wells containing only single colonies.

4. Pick cells from positive wells into 24-well plates, expand, and freeze in liquid nitrogen (*Protocol 14*).

5. Re-clone the best antibody-producing clones.

Protocol 14

Long-term storage of cells in liquid nitrogen

Equipment and reagents

- 2 ml plastic ampoules (Nunc) for storage of cells in liquid nitrogen
- Controlled freezing equipment or $-70\,°C$ freezer
- Polystyrene boxes to hold Nunc cryotubes
- Freezing mixture: 5% DMSO, 95% FCS
- Hybridoma cells spun down from culture supernatant

Method

1. Resuspend cells in freezing mixture to give about 10^6 cells/ml.
2. Transfer 1 ml aliquots to properly labelled Nunc cryotubes, cap, and place in controlled freezing equipment or in polystyrene boxes and transfer to $-70\,°C$ freezer and leave overnight.
3. Transfer vials of frozen cells to liquid nitrogen storage tanks and record location.

11 Characterization and use of the antibodies obtained

Where more than one hybridoma has been obtained, it is important to assess the relative usefulness of the different antibodies. For example, affinity, i.e. the strength with which it binds antigen, could be important, particularly if the antibody is to be used in radioimmunoassays to estimate antigen concentration. Secondly, the antibody may have important effector functions when used *in vivo*, i.e. to activate complement and recruit host immune effector cells by binding Fc receptors, factors that could affect biological usefulness. These properties are critically dependent on the isotype (class or subclass) of the antibody (*Protocol 15*), e.g. rat IgG2b antibodies are particularly good activators of human effector cells and the complement cascade. Thirdly, the antibody may bind to sequential amino acids on the antigen, antibodies of this type have proved to be ideal for immunoblotting of denatured proteins and for binding to cellular determinants in formalin-fixed, paraffin embedded tissue sections. Some 80% of antibodies generated against biological antigens bind to conformational determinants. While these antibodies are of little use to the pathologist who works with formalin-fixed sections, they will bind to frozen sections and live cells and they often have important biological activities. For the molecular or cell biologist who is interested in using mAbs to determine the cellular localization and function of novel proteins, all of the criteria discussed previously will apply. Indeed, the use of confocal microscopy plus the ability to inject antibodies or other proteins into individual cells has demonstrated the essential function of a number of cellular proteins while the facility of being able to immunoprecipitate target antigens from detergent extracts of unlabelled or metabolically labelled cells and to subject them to electrophoretic analysis (*Protocol 17*) has facilitated determination of their size, extent of phosphorylation, etc.

11.1 Isotyping of antibodies

Antibodies can be separated into different classes (IgM, IgG, IgA, and IgE) and subclasses (IgG1, IgG2a, IgG2b, and so on) on the basis of the structure of their heavy chains. This can be achieved using antisera directed against epitopes on the heavy chains that are specific for the particular isotype. The different classes have different properties and this may be important in determining the usefulness of the antibodies for example in diagnostic and therapeutic applications.

Protocol 15

Isotyping of monoclonal antibodies

Equipment and reagents

- 96-well PVC plates
- Monoclonal or polyclonal antibodies specific for mouse or rat heavy chain isotypes
- PCB pH 8.0
- PBSM
- PBST
- ^{125}I-labelled antibodies to mouse or rat F(ab')$_2$

Method

1. Coat plates with monoclonal or polyclonal antibodies (5 μg/ml of PCB) using the method described in *Protocol 8*.

2. Block with PBSM.

3. Add 50 μl of test mAb to each well and incubate for 1–4 h at room temperature or overnight at 4°C.

4. Wash three times with PBSM.

5. Add 50 μl/well (10^5 c.p.m. in PBSM) of ^{125}I-labelled antibodies to mouse or rat F(ab')$_2$ and incubate for 1 h at room temperature.

6. Wash three times with PBST and determine the amount of radioactivity bound. Include mAbs of known isotype as controls.

7. Occasionally, a mAb will show reactivity with more than one anti-isotype antibody in this capture assay. In this case, try immunoprecipitation in agarose gels (Ouchterlony procedure, ref. 16). Because a crosslinked lattice must be formed between several different epitopes on the antigen and several antibodies to give an immunoprecipitate the Ouchterlony procedure gives few false positives.

11.2 Epitope reactivity

Where several antibodies have been isolated that bind to the same antigen it is important to determine if they bind to the same or different epitopes on the target antigen. This can be done by labelling one of the antibodies and determining which of the other antibodies compete for its binding to specific antigen (*Protocol 16*). By labelling each antibody in turn their epitope reactivity can be determined.

Protocol 16

Competitive radioimmunoassays

Equipment and reagents

- Gamma counter
- Live cells grown in multiwell plates
- Antigen coated PVC plates
- DMEM–5% FCS
- PBSM

- PBST
- Purified monoclonal antibody labelled with [125]iodine
- Hybridoma culture supernatants or other purified antibodies

Method

1. Make doubling dilutions of culture supernatant or purified antibody starting at 20 μg/ml in DMEM–5% FCS (live cells) or PBSM (antigen coated plates).

2. Add to each well 50 μl ^{125}I-labelled specific antibody containing $2–4 \times 10^4$ c.p.m. in DMEM–5% FCS or in PBSM.

3. Transfer 50 μl aliquots of the mixtures to the cells or antigen coated PVC multiwell plate and incubate for 1 h at 4°C (live cells) or at ambient temperature (PVC plates).

4. Wash plates three times with DMEM–5% FCS or PBST then determine [125]iodine bound. Competing antibodies can reduce the binding of the radiolabelled target to background.

Analysis in this way of a number of rat antibodies directed against the human receptor for EGF showed that they fell into one of four epitope clusters (A–D). The antibodies in clusters C and D strongly inhibited the binding of ligands (EGF and TGFα) to the receptor, blocked ligand-induced phosphorylation, and prevented the growth *in vitro* and *in vivo* of the EGFR overexpressing tumour cell line HN5 (17). Antibodies in clusters A and B were much less effective inhibitors of cell growth and ligand binding indeed one antibody (ICR9, cluster A) enhanced ligand binding. These data illustrate the importance of epitope mapping when considering the significance of the biological activity of certain antibodies.

11.3 Specific immunoprecipitation

Immunoprecipitates prepared from metabolically radiolabelled cells can be used not only to assess the molecular size of a given protein but also to determine if it is phosphorylated or is part of a multiple structure. Antibody-containing culture supernatants from cloned hybridomas can be used for immunoprecipitation by first binding the mAbs to a bead support coated with anti-Fab antibodies, then incubating the specific beads with extracts of the cells made with non-ionic detergents such as Triton X-100 or Nonidet 40.

Analyses in single and two-dimensional gels (isoelectric point versus molecular weight) have proved to be essential for the analysis of cellular proteins and detailed protocols for the analysis of novel cellular antigens using monoclonal antibodies are given in Chapter 12.

Protocol 17

Immunoprecipitation of specific antigen

Equipment and reagents

- Equipment for electrophoresis in sodium dodecyl sulfate-containing polyacrylamide gels
- Equipment for either drying gels onto Whatman 3MM paper or transferring (Western blot) onto nitrocellulose or PVDF membranes (Amersham Pharmacia Biotech) and autoradiographing at $-70\,^{\circ}$C with an enhancing screen
- Antibody-containing culture supernatant ($> 1\ \mu$g/ml) or purified antibody

- Protein A (mouse antibodies) or specific anti-mouse or anti-rat F(ab')$_2$ covalently linked to Sepharose 4B or similar bead support
- Extracts of cells, labelled with [^{35}S]methionine, prepared by solubilization in PBSA containing 0.5% Triton X-100 and 10^{-3} M PMSF

Method

1. Label cellular proteins by incubating cultures in 25 cm^2 flasks for 4–24 h with 100–500 μCi [^{35}S]methionine in methionine-free medium.

2. Wash cells three times in complete medium then incubate for a further hour in the same medium.

3. Wash the cells in ice-cold PBSA containing 10^{-3} M PMSF then lyse the cells with the minimum volume (1–2 ml/25 cm^2 flask) of PBSA containing 10^{-3} M PMSF and 0.5% Triton X-100 by incubating for 30 min on ice.

4. Transfer lysate to a centrifuge tube and spin at 30 000 g for 30 min at 4 $^{\circ}$C to remove cell debris.

5. Prepare specific antibody by incubating overnight at 4 $^{\circ}$C 1 ml of hybridoma culture supernatant or purified antibody with 100 μl of packed beads of Protein A–Sepharose or anti-mouse or anti-rat F(ab')$_2$ linked to Sepharose 4B.

6. Incubate 1 ml of cell lysate ($c.$ 3 $\times$ 10^6 cells) with 100 μl of packed beads overnight at 4 $^{\circ}$C.

7. Wash the beads three times with lysis buffer, pelleting the beads by centrifugation.

8. Elute the antigen (+ mAb) by heating beads for 5 min at 95 $^{\circ}$C with an equal volume of SDS sample buffer and run on an SDS-containing 10% polyacrylamide reducing gel. Run pre-stained markers on these gels because they will transfer to blots and assist in determining the size of the proteins.

9. Either dry the gel onto Whatman 3MM paper and autoradiograph with an enhancing screen, or electrolytically transfer the proteins onto cellulose nitrate filters (18), dry, and autoradiograph.

References

1. Galfré, G. and Milstein, C. (1981). In *Methods in enzymology* (ed. J. Langone and H. Van Vunakis), Vol. 73, p. 3. Academic Press, New York.
2. Bazin, H. (1982). In *Protides of the biological fluids, 29th colloquium* (ed. H. Peeters), p. 615. Pergamon, New York.
3. Flynn, J. N., Harkiss, G. D., and Hopkins, J. (1989). *J. Immunol. Methods*, **121**, 237.
4. Spieker-Polet, H., Sethupathi, P., Yam, P.-C., and Knight, K. L. (1995). *Proc. Natl. Acad. Sci. USA*, **92**, 9348.
5. Edwards, P. A. W., Smith, C. M., Neville, A. M., and O'Hare, M. J. (1982). *Eur. J. Immunol.*, **12**, 641.
6. Anderson, D. V., Tucker, E. M., Powell, J. R., and Porter, P. (1987). *Vet. Immunol. Immunopathol.*, **15**, 223.
7. Richards, C. M., Aucken, H. A., Tucker, E. M., Hannant, D., Mumford, J. A., and Powell, J. R. (1992). *Vet. Immunol. Immunopathol.*, **33**, 129.
8. Tilby, M. J., Styles, J. M., and Dean, C. J. (1987). *Cancer Res.*, **47**, 1542.
9. Modjtahedi, H., Styles, J. M., and Dean, C. J. (1993). *Br. J. Cancer*, **67**, 247.
10. Styles, J. M., Harrison, S., Gusterson, B. A., and Dean, C. J. (1990). *Int. J. Cancer*, **45**, 320.
11. Johnstone, A. P., Cridland, J. C., DaCosta, C. R., Harfst, E., and Shepherd, P. S. (1994). *Mol. Cell. Endocrinol.*, **105**, R1.
12. Tang, D., DeVit, M., and Johnson, S. (1992). *Nature*, **356**, 152.
13. Vahlsing, H., Yangkauckas, M., Sawday, M., Gromkowski, S., and Manthorpe, M. (1994). *J. Immunol. Methods*, **175**, 11.
14. Dean, C. J., Styles, J. M., Gyure, L. A., Peppard, J., Hobbs, S. M., and Hall, J. G. (1984). *Clin. Exp. Immunol.*, **57**, 358.
15. Fraker, P. J. and Speck, J. C. (1978). *Biochem. Biophys. Res. Commun.*, **80**, 849.
16. Ouchterlony, O. and Nilsson, L. A. (1986). In *Handbook of experimental immunology*, 4th edn (ed. D. M. Weir), p. 32.1. Blackwell Scientific, Oxford.
17. Modjtahedi, H. and Dean, C. J. (1994). *Int. J. Oncol.*, **4**, 277.
18. Towbin, H., Staehlin, Y., and Gordon, J. (1979). *Proc. Natl. Acad. Sci. USA*, **76**, 4350.

Chapter 2

Preparation of recombinant antibodies from immune rodent spleens and the design of their humanization by CDR grafting

Olivier J. P. Léger* and José W. Saldanha[†]
*Centre d'Immunologie pierre Fabre, 5 Avenue Napoleon III–BP497, 74164 Saint-Julien en Genevois, Cedex, France
[†]Division of Mathematical Biology, NIMR, The Ridgeway, Mill Hill, London NW7 1AA, U.K.

1 Introduction

The biotechnological generation of high affinity monoclonal antibodies has traditionally involved the production of hybridomas from spleen cells of immunized animals. Recent advances in recombinant technology, using phage display technology, have opened a new route for the generation of antibodies. Instead of immortalizing B cells for production of monoclonal antibodies, the antibody heavy (H) and light (L) chain V genes are immortalized by gene technology. The IgG mRNA of B cells of immunized animals can be used as a source of variable region (V) gene libraries. Murine VH and VL genes are amplified using PCR and used to construct libraries of millions of recombinant single chain Fvs (scFvs). ScFv antibodies consist of the variable regions of the light and heavy chains of antibodies linked via a short peptide spacer (1). Repertoires of scFv are cloned into the gene encoding the minor coat protein (gene III) of filamentous bacteriophage, thus creating a large library of phage, each displaying an individual heavy and light chain combination. The fusion protein created, consisting of the antibody fragment at the N-terminus of the coat protein, is incorporated into the phage particle (a phage antibody). Each recombinant phage genome contains the DNA encoding the specific antibody displayed on its surface, allowing the phage particle carrying antibody gene to be selected directly using the binding properties of the expressed protein.

The phage system replaces all steps after immunization and isolation of spleen cells with simple, rapid procedures utilizing DNA manipulation and bacteria. The time required from isolation of spleen cells to selected stable clones is reduced from several months to several weeks. In the process, many millions, or billions of potential binding molecules may be surveyed using large phage libraries, compared with only several thousand molecules screened using a traditional hybridoma fusion. This method gives access to more, and sometimes better, antibodies than when working with hybridomas. For example, from an immune murine phage antibody library, Chester and colleagues identified an anti-CEA antibody with an affinity substantially higher than ever obtained with the conventional hybridoma technology (2).

There are several important advantages associated with the production of recombinant antibodies. First, the inherent advantage of phage display technology is its direct link of DNA sequence to protein function (3). It includes a powerful enrichment strategy that allows rapid and simple selection of reagents with desired properties and also facilitates the rapid sequencing of the corresponding genes to identify the number of unique antibodies. Secondly, the scFv fragments can be produced in large quantities in mammalian cells, yeast, bacteria, or plants. Thirdly, the genes for the scFvs can be manipulated to introduce mutations that increase antibody affinity into the picomolar range (4). Therefore the production of a library of scFv antibodies represents an important advance toward the development of recombinant antibody-based therapies. Ultimately, the VH and VL genes can be subcloned to construct complete IgG molecules (the use of human constant regions will yield chimeric antibodies less immunogenic than murine monoclonals) and complementarity determining region (CDR) grafted to design and construct humanized antibodies in order to further reduce the immunogenicity for therapy in humans (see Appendix to this chapter for the design of humanized antibodies). More recently, transgenic mice carrying human immunoglobulin gene loci in germline configuration were proposed to generate therapeutic antibodies (HuMAb-Mouse™ technology, Medarex Inc. and XenoMouse™ technology, Abgenix Inc.). High affinity human antibodies (as high as 10^{10} power) can be isolated from these hyperimmunized transgenic mice either by traditional hybridoma technology or through construction of combinatorial libraries.

2 Preparation of mouse spleen

In most cases, the spleen is the best choice for antigen-specific lymphocyte isolation. The immunization regime used to immunize mice is identical in every respect to that used for conventional hybridoma production. In particular about ten days after each injection, test bleeds should be carried out to test for the presence of specific antibodies and only animals giving the best antiserum should be selected. Do not expect to recover high affinity antibodies from phage antibody libraries that are not found in the test serum. Three to five days before harvesting the spleen for RNA extraction, the immunized mouse is given a final boost. This is best achieved by an intravenous injection done concurrently with an intraperitoneal injection.

Protocol 1

Extraction of mouse spleen

Equipment and reagents

- One pair of scissors and forceps
- Hyperimmunized mice (see Chapter 1)
- 95% (v/v) ethanol
- TRI REAGENT™ (see *Protocol 2*)

Method

1. Sacrifice mouse. This can be done efficiently and humanely by cervical dislocation.

2. Dip entire mouse briefly in 95% (v/v) ethanol then lay the mouse down on its right side (as you look down its back towards the head).

3. Using forceps, grip a small fold of the skin and lift away from the body a small distance (not enough to move the mouse).

4. Make a short, shallow incision. Separate the skin by grasping opposing sides with fingers and pulling apart in opposite directions. Be sure to remove edges from the area surrounding the spleen, which will be visible through the peritoneum.[a] Dip forceps and scissors in 95% (v/v) ethanol.

5. Lift the peritoneum gently with forceps, without grasping any internal organs, and make an incision. Continue cutting, holding the peritoneal layer up and slicing parallel to the spleen, until the spleen has been entirely uncovered. Dip forceps and scissors in ethanol.

6. Lift the spleen with forceps gently (it will tear) away from the surrounding tissues and carefully cut from any tissue.

7. Place the spleen in a 100 mm tissue culture dish and trim off and discard any contaminating tissue from the spleen.[b] Homogenize the spleen in 10 ml of TRI REAGENT™ (see *Protocol 2*).[c,d]

[a] The spleen will be an oblong, reddish brown organ, usually resting tail-side of the ribs and sometimes extending upwards below the spine on the mouse's left side.

[b] The spleen can be stored in RNA*later*™ solution (Ambion, Inc., 7020) at 4 °C for up to a month, before RNA extraction.

[c] The volume of the tissue should not exceed 10% of the volume of TRI REAGENT™.

[d] Alternatively, rather than using the whole spleen directly, splenocytes can first be prepared from spleen by using standard procedures described for preparation of hybridoma production (see Chapter 1) before total RNA extraction. A spleen from an immunized mouse contains approx. 5×10^7 to 2×10^8 lymphocytes. For total RNA extraction, substitute 10^8 lymphocytes for a spleen and simply add 10 ml of TRI REAGENT™ (1 ml of reagent for 10^7 cells) to the sterile PBS-washed cell pellet. Allow incubation at room temperature for 5 min for lysis, and proceed to *Protocol 2*, step 2.

3 Isolation of total RNA from spleen

Because we are dealing with a relatively large number of cells ($1-3 \times 10^8$ spleno-cytes), we recommend first isolating total RNA from the entire spleen, followed by mRNA isolation (see *Protocol 3*).

Protocol 2

Cytoplasmic RNA preparation

Recommendations

1. Remember that RNA is notoriously susceptible to degradation and special care is required for its isolation. RNases are difficult to inactivate as they do not require cofactors and are heat stable, spontaneously refolding following heat inactivation.

2. We recommend using the TRI REAGENT™ to isolate undegraded total RNA from splenocytes. This procedure is an improvement of the guanidine thiocyanate/phenol/ chloroform single-step extraction procedure described by Chomczynski and Sacchi (5).

3. Wear gloves at all times to avoid sample contamination with nucleases shed from the skin. Change gloves frequently.

4. Whenever possible, use sterile, disposable plasticware for handling RNA. These materials are generally considered to be RNase-free and require no treatment.

Equipment and reagents

- Tissue homogenizer (such as Brinkmann or Dounce homogenizer)
- 14 ml polypropylene tubes (Falcon, 2059) (these tubes are guaranteed RNase-free by the manufacturer and do not require pre-treatment to inactivate RNases)
- Mouse spleen(s) (see *Protocol 1*)
- TRI REAGENT™ (Sigma, T9424)
- Chloroform (Sigma, C2432)
- Isopropanol (Sigma, I9516)
- 75% (v/v) ethanol, ice-cold
- Nuclease-free water: add diethyl pyrocarbonate (DEPC; Sigma D-5758) to distilled water to a final concentration of 0.1% (v/v) (i.e. 1 ml/litre). Incubate overnight at room temperature or at 37 °C

for 2 h in a fume hood. Autoclave for 20 min at 115 °C and 15 psi to remove any traces of DEPC.
- TE buffer pH 7.4: 10 mM Tris–HCl pH 7.4 (10 ml of 1 M Tris–HCl), 1 mM EDTA pH 8.0 (2 ml of 0.5 M EDTA pH 8.0), adjust the volume to 1 litre with distilled water
- 50 × TAE (Tris–acetate EDTA) electrophoresis buffer: 2 M Tris base (242.2 g), 2 M glacial acetic acid (57.1 ml), and 50 mM EDTA (18.61 g $Na_2EDTA.2$ H_2O), adjust to 1 litre with distilled water
- 10 mg/ml ethidium bromide (Sigma, E-1510)
- Agarose (UltraPure) powder (100 g, Life Technologies, 15510-019)

Method

1. Homogenize spleen(s) in 10 ml of TRI REAGENT™ (see *Protocol 1*) in a glass homo-genizer or using a mechanical homogenizer. Allow samples to stand for 5 min at room temperature.

2. Add 2 ml of chloroform,[a] shake vigorously for 15 sec, and allow to stand for 2–15 min at room temperature.

3. Centrifuge the resulting mixture at 12 000 g for 15 min at 4 °C.

4. Transfer the aqueous phage (colourless upper phase) to a fresh tube, add 5 ml of isopropanol, and mix.

5. Allow the sample to stand for 5–10 min at 4 °C.

Protocol 2 continued

6. Centrifuge at 12 000 g for 10 min at 4 °C. The RNA precipitate will form a pellet on the side and bottom of the tube.

7. Remove the supernatant and wash the RNA pellet by adding 10 ml (minimum) of 75% (v/v) ethanol. Centrifuge at 7500 g for 5 min at 4 °C.

8. Carefully and completely remove the ethanol and briefly dry the RNA pellet for 5–10 min by air drying.[b]

9. Resuspend the RNA pellet with 1 ml of nuclease-free water or TE buffer[c] and store at −20 °C.[d]

10. Quantitate the RNA by UV absorption spectrometry. Add 50 μl of total RNA to 450 μl of nuclease-free water in a 0.6 ml quartz cuvette. Cover with Parafilm and mix. Read the sample at 260 nm[e] and 280 nm. Multiply the 260 nm reading × 10 × 40 to determine the yield of total RNA in μg/ml.[f]

11. Calculate the purity of the RNA preparation by dividing the 260 reading by the 280 reading.[g]

12. Determine the integrity of the isolated RNA[h] by denaturing agarose gel electrophoresis,[i] followed by staining in ethidium bromide,[j] and visualization on a UV light box.[k]

13. Proceed to the mRNA purification step (see *Protocol 3*).

[a] The chloroform used for phase separation should not contain isoamyl alcohol or other additives.

[b] Do not allow the pellet to dry out completely, as this will make the pellet very difficult to resuspend. Do not dry the RNA pellet by centrifugation under vacuum (Speed-Vac).

[c] In some instances, heating at 60 °C for 10–15 min may be required to resuspend the pellet; however, heating should be kept to the minimum required.

[d] For long-term storage, store the RNA as an ethanol precipitate: add sodium acetate pH 5.0, to a final concentration of 0.25 M, then add 2.5 vol. of ethanol, and store at −70 °C.

[e] An absorbance of one at a wavelength of 260 nm, with a 1 cm light path, corresponds to a concentration of 40 μg/ml of single-stranded RNA.

[f] Typically, a mouse spleen yields around 8.0 mg of RNA per gram of tissue.

[g] Pure RNA will exhibit A_{260}/A_{280} ratios of 2.0. More realistically you should expect to obtain RNA having A_{260}/A_{280} ratios ranging between 1.7 and 2.0. If the RNA exhibits a ratio lower than 1.7, further purify the RNA by repeating steps 8–11.

[h] The 28S and 18S eukaryotic ribosomal RNAs should exhibit a near 2:1 ratio of ethidium bromide staining, indicating that no significant degradation of RNA has occurred. In RNA samples that have been degraded, this ratio will be reversed since the 28S ribosomal RNA is characteristically degraded to an 18S-like species.

[i] We recommend using a 1% (w/v) agarose gel in 1 × TAE buffer containing 0.1% (w/v) SDS. The running buffer consists of 1 × TAE buffer containing 0.1% (w/v) SDS. The RNA samples are heated at 65 °C for 2 min and quickly chilled on ice before loading onto the agarose gel.

[j] CAUTION: ethidium bromide is a powerful mutagen. Gloves should be worn when handling solutions of this dye and stained gels. Decontamination of ethidium bromide solutions is described in Sambrook *et al.* (6) page 6.16–6.17.

[k] CAUTION: avoid skin and eye exposure to UV light. Protect your eyes with goggles.

4 Poly(A$^+$) mRNA isolation

The quality of mRNA is important in generating recombinant antibodies. It is particularly important to ensure minimal ribosomal RNA contamination when cDNA is random primed for library construction to reduce the proportion of clones which represent non-mRNA species. The spleen mRNA of immunized mice is presumably derived mainly from plasma cells, as the level of Ig mRNA in these cells is up to 1000-fold greater than in resting B cells. For spleens, extraction of total RNA followed by mRNA isolation on oligo(dT)–cellulose is recommended in order to effectively process the large number of cells present in a spleen. Cytoplasmic RNA Extraction Kit (see *Protocol 2*) followed by mRNA Purification Kit is highly recommended.

Protocol 3

mRNA purification

Recommendations

1. Observe special precautions to avoid degradation of mRNA by RNases (see *Protocol 2*, *Recommendations*).

Equipment and reagents

- RNase-free microcentrifuge tubes
- QuickPrep® mRNA Purification Kit (Pharmacia Biotech, 27-9258-01 or 02)
- Nuclease-free water (see *Protocol 2*)
- 95% (v/v) ethanol, ice-cold or chilled at −20 °C

- 3 M sodium acetate pH 5.2: dissolve 408.3 g sodium acetate.3H$_2$O in 800 ml of distilled water, adjust to pH 5.2 with glacial acetic acid, adjust the volume to 1 litre with distilled water

Method

1. Follow the manufacturer's instructions for preparation of an oligo(dT)–cellulose spun column (Procedure A).

2. Load up to 1 mg of total RNA onto one spun column.[a]

3. Perform one round of purification following the manufacturer's instructions (Procedure B: Isolation of mRNA).

4. After elution, add 10 μl of glycogen solution (10 mg/ml, QuickPrep® mRNA Purification Kit) per 500 μl of mRNA.

5. Quantitate the mRNA by UV absorption spectrometry (see *Protocol 2*, step 10).[b,c]

6. Ethanol precipitate the mRNA.[d] Mix and chill at −20 °C for at least 20 min.

7. Spin in a microcentrifuge at full speed for 10 min at room temperature.

8. Wash the precipitate twice with ice-cold 95% (v/v) ethanol and air dry the precipitate briefly.

9. Resuspend the pellet in 20 μl of RNase-free water prior to use in the first strand cDNA synthesis reaction (see *Protocol 4*).[e]

[a] At least one round of affinity chromatography using oligo(dT)–cellulose should be carried out.

[b] Be careful not to use too much of your sample simply for estimation.

[c] Alternatively you can assume that mRNA is approx. 0.1% the weight of the spleen.

[d] With 0.1 vol. of 3 M sodium acetate pH 5.2, and 2–2.5 vol. of 95% (v/v) ethanol chilled at −20 °C.

[e] If the mRNA is not being used immediately, it should be stored frozen at −70 to −80 °C, preferably as an ethanol precipitate.

5 Reverse transcriptase reaction

Since PCR amplifies DNA, you will need to make a cDNA copy of your RNA before you can amplify your sequence of interest. The advantage of having isolated RNA rather than DNA at the outset is that genomic DNA contains introns, portions of genes that are unnecessary in the final decoding to protein, and are deleted in the corresponding mRNA. This is especially relevant with respect to antibodies and their ubiquitous rearrangement. Thus by using the cDNA in PCR amplification, you will obtain a product of the correct size and direction for further protein studies.

To produce scFv molecules, you will need to do two first strand cDNA synthesis reactions: one for the heavy chain and one for the light chain. Since the spleen mRNA comes from an hyperimmunized mouse, we are only interested in the preparation of IgG DNA and not IgM DNA.

Protocol 4

Preparation of cDNA:mRNA hybrid

Reagents

• First Strand cDNA Synthesis Kit (Pharmacia Biotech, 27-9261-01)

Method

1. Set up the following reverse transcription mix in duplicate in RNase-free microcentrifuge tubes. Label one tube as light chain and the second one as heavy chain:
   • Bulk First Strand Reaction mix (component of the Kit)   11 μl
   • DTT solution (component of the Kit)   1 μl
   • FOR[a] primer[b] (10 pmol/μl; see *Table 1*)   1 μl

2. Place the mRNA sample[c] in a 500 μl RNase-free microcentrifuge tube and add RNase-Free Water (component of the Kit), if necessary, to bring the mRNA to 20 μl final volume.

3. Heat the mRNA at 65 °C for 10 min and cool immediately on ice.[d] Start the first strand cDNA reaction within 2 min after placing on ice.

Protocol 4 continued

4. Add to the RNA the reverse transcription mix and incubate for 1 h at 37 °C.[e]

[a] FOR refers to 'toward the 5' end of the antibody gene'.

[b] For hyperimmunized spleen mRNA, we suggest using 3' specific primers. The primers anneal to the 3' end (i.e. within the constant regions of the heavy or light chain). One reaction is set up containing a mixture of the two heavy chain primers MIgG1/2 For and MIgG3 For (tube labelled heavy chain). Another reaction is set up containing a unique light chain primer MCK For (tube labelled light chain). Alternatively you can use random hexadeoxynucleotides (pd(N)$_6$) but we don't recommend the use of oligo(dT).

[c] For spleen mRNA use 200 ng to 1 µg per reaction. Don't forget to set up duplicate samples, one for the heavy chain and one for the light chain.

[d] To remove secondary structure which could hinder the reaction.

[e] The reaction can be stored at −20 °C for up to a week.

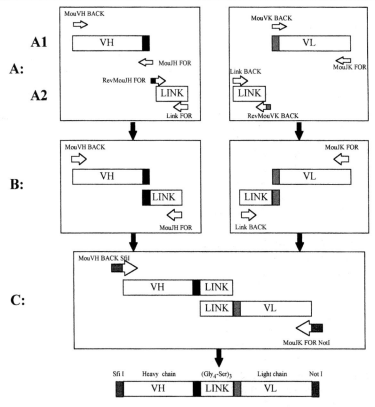

Figure 1 Construction of mouse scFv libraries using a 'jumping-PCR-assembly' method. (A1) Primary amplification PCR of Fv fragments with a mixture of V gene and J segment primers. (A2) Preparation of linkers. (B) The linker assembly. (C) The final assembly of scFv constructs by splicing overlap extension where the two genes overlap via the (Gly$_4$-Ser)$_3$ linker. The final step in (C) shows a normal PCR ensuing after the overlap extension. Arrows represent primers; boxes represent PCR-amplified DNA fragments. VH = heavy chain variable region. VL = light chain variable region. LINK = linker DNA fragment encoding the short peptide (Gly$_4$-Ser)$_3$. Marked zones within boxes represent regions of overlapping complementarity.

6 Primary PCR of antibody genes

Full-length rearranged VH and VK genes were PCR amplified separately using an equimolar mixture of the appropriate family-based back and forward primers (see *Table 1* and *Figure 1 A1*). The set of mouse primers we are using is that described by Amersdorfer *et al.* (ref. 7, with permission) which has been recently published and therefore should be optimum, covering all mouse VH and VK sequences. VH and VK genes-specific primers were designed to anneal to the first 23 nucleotides comprising framework 1 and JH and JK gene segment-specific primers were designed to anneal to the final 24 nucleotides.

Table 1 Oligonucleotide primers used for construction of mouse scFv antibody libraries[a]

A. First strand cDNA synthesis

Mouse heavy chain constant region primers

MuIgG1/2 For	5′ CTG GAC AGG GAT CCA GAG TTC CA 3′
MuIgG3 For	59 CTG GAC AGG GCT CCA TAG TTC CA 3′

Mouse kappa constant region primer

MuCK For	59 CTC ATT CCT GTT GAA GCT CTT GAC 3′

B. Primary PCRs

Mouse VH Back primers

MuVH1 Back	5′ GAG GTG CAG CTT CAG GAG TCA GG 3′
MuVH2 Back	5′ GAT GTG CAG CTT CAG GAG TCR GG 3′
MuVH3 Back	5′ CAG GTG CAG CTG AAG SAG TCA GG 3′
MuVH4/6 Back	5′ GAG GTY CAG CTG CAR CAR TCT GG 3′
MuVH5/9 Back	5′ CAG GTY CAR CTG CAG CAG YCT GG 3′
MuVH7 Back	5′ GAR GTG AAG CTG GTG GAR TCT GG 3′
MuVH8 Back	5′ GAG GTT CAG CTT CAG CAG TCT GG 3′
MuVH10 Back	5′ GAA GTG CAG CTG KTG GAG WCT GG 3′
MuVH11 Back	5′ CAG ATC CAG TTG CTG CAG TCT GG 3′

Mouse JH Forward primers

MuJH1 For	5′ TGA GGA GAC GGT GAC CGT GGT CCC 3′
MuJH2 For	5′ TGA GGA GAC TGT GAG AGT GGT GCC 3′
MuJH3 For	5′ TGC AGA GAC AGT GAC CAG AGT CCC 3′
MuJH4 For	5′ TGA GGA GAC GGT GAC TGA GGT TCC 3′

Mouse VK Back primers

MuVK1 Back	5′ GAC ATT GTG ATG WCA CAG TCT CC 3′
MuVK2 Back	5′ GAT GTT KTG ATG ACC CAA ACT CC 3′
MuVK3 Back	5′ GAT ATT GTG ATR ACB CAG GCW GC 3′
MuVK4 Back	5′ GAC ATT GTG CTG ACM CAR TCT CC 3′
MuVK5 Back	5′ SAA AWT GTK CTC ACC CAG TCT CC 3′
MuVK6 Back	5′ GAY ATY VWG ATG ACM CAG WCT CC 3′
MuVK7 Back	5′ CAA ATT GTT CTC ACC CAG TCT CC 3′
MuVK8 Back	5′ TCA TTA TTG CAG GTG CTT GTG GG 3′

Mouse JK Forward primers

MuJK1 For	5′ TTT GAT TTC CAG CTT GGT GCC TCC 3′
MuJK2 For	5′ TTT TAT TTC CAG CTT GGT CCC CCC 3′
MuJK3 For	5′ TTT TAT TTC CAG TCT GGT CCC ATC 3′
MuJK4 For	5′ TTT TAT TTC CAA CTT TGT CCC CGA 3′
MuJK5 For	5′ TTT CAG CTC CAG CTT GGT CCC AGC 3′

Table 1 *Continued*

C. Linker assembly

Reverse mouse JH Forward for scFv linker

RevMouJH1 For	5′ GGA CCA CGG TCA CCG TCT CCT CAG GTG G 3′
RevMouJH2 For	5′ GCA CCA CTC TCA CAG TCT CCT CAG GTG G 3′
RevMouJH3 For	5′ GGA CTC TGG TCA CTG TCT CTG CAG GTG G 3′
RevMouJH4 For	5′ GAA CCT CAG TCA CCG TCT CCT CAG GTG G 3′

Reverse mouse VK Forward for scFv linker

RevMouVK1 Back	5′ GGA GAC TGT GWC ATC ACA ATG TCA GAT CCG CC 3′
RevMouVK2 Back	5′ GGA GTT TGG GTC ATC AMA ACA TCA GAT CCG CC 3′
RevMouVK3 Back	5′ GCW GCC TGV GTY ATC ACA ATA TCA GAT CCG CC 3′
RevMouVK4 Back	5′ GGA GAY TGK GTC AGC ACA ATG TCA GAT CCG CC 3′
RevMouVK5 Back	5′ GGA GAC TGG GTG AGM ACA WTT TSA GAT CCG CC 3′
RevMouVK6 Back	5′ GGA GWC TGK GTC ATC WBR ATR TCA GAT CCG CC 3′
RevMouVK7 Back	5′ GGA GAC TGG GTG AGA ACA ATT TGA GAT CCG CC 3′
RevMouVK8 Back	5′ CCC ACA AGC ACC TGC AAT AAT GAA GAT CCG CC 3′

(Gly$_4$-Ser)$_3$ linker Back for jumping-PCR-assembly

LinkerBack	5′ GGT GGC GGT GGC TCG GGC GGT 3′

(Gly$_4$-Ser)$_3$ linker Forward for jumping-PCR-assembly

LinkerFor	5′ AGA TCC GCC GCC ACC CGA CCC 3′

D. Final assembly with primers containing restriction sites[b]

Mouse VH Back *Sfi*I primers

MouVH1Back*Sfi*I	5′ACT GCg gcc cag ccg gcc ATG GCC GAG GTG CAG CTT CAG GAG TCA GG 3′
MouVH2Back*Sfi*I	5′ACT GCg gcc cag ccg gcc ATG GCC GAT GTG CAG CTT CAG GAG TCR GG 3′
MouVH3Back*Sfi*I	5′ACT GCg gcc cag ccg gcc ATG GCC CAG GTG CAG CTG AAG SAG TCA GG 3′
MouVH4/6Back*Sfi*I	5′ACT GCg gcc cag ccg gcc ATG GCC GAG GTY CAG CTG CAR CAR TCT GG 3′
MouVH5/9Back*Sfi*I	5′ACT GCg gcc cag ccg gcc ATG GCC CAG GTY CAR CTG CAG CAG YCT GG 3′
MouVH7Back*Sfi*I	5′ACT GCg gcc cag ccg gcc ATG GCC GAR GTG AAG CTG GTG GAR TCT GG 3′
MouVH8Back*Sfi*I	5′ACT GCg gcc cag ccg gcc ATG GCC GAG GTT CAG CTT CAG CAG TCT GG 3′
MouVH10Back*Sfi*I	5′ACT GCg gcc cag ccg gcc ATG GCC GAA GTG CAG CTG KTG GAG WCT GG 3′
MouVH11Back*Sfi*I	5′ACT GCg gcc cag ccg gcc ATG GCC CAG ATC CAG TTG CTG CAG TCT GG 3′

Mouse JK Forward *Not*I primers

MouJK1For*Not*I	5′ GAG TCA TTC TCG ACT Tgc ggc cgc TTT GAT TTC CAG CTT GGT GCC TCC 3′
MouJK2For*Not*I	5′ GAG TCA TTC TCG ACT Tgc ggc cgc TTT TAT TTC CAG CTT GGT CCC CCC 3′
MouJK3For*Not*I	5′ GAG TCA TTC TCG ACT Tgc ggc cgc TTT TAT TTC CAG TCT GGT CCC ATC 3′
MouJK4For*Not*I	5′ GAG TCA TTC TCG ACT Tgc ggc cgc TTT TAT TTC CAA CTT TGT CCC CGA 3′
MouJK5For*Not*I	5′ GAG TCA TTC TCG ACT Tgc ggc cgc TTT CAG CTC CAG CTT GGT CCC AGC 3′

E. Reamplification of assembled scFv

Universal *Sfi*I Back primer

5′ TTACTCAAGATAAATGGATGGGTATATTGCTCACATTGGTGCCGAAACCTATTC ACT GCg gcc cag ccg gcc ATG GCC 3′ — 78 mer

Universal *Not*I For primer

5′ CAGGTATTTTAGATATGAATACTGTAGTGTTACAAGAAGGTGAACACTTATATT GAG TCA TTC TCG ACT Tgc ggc cgc 3′ — 78 mer

[a] Sections A and B are taken from ref. 7, with permission.

[b] The underlined sequences in Section D represent the mouse VH Back and mouse JK Forward primers, respectively. The lower case letters are the cloning sites (*Sfi*I for 5′ and *Not*I for 3′ termini, respectively).

34

Protocol 5

PCR amplification of mouse variable regions

Equipment and reagents

- GeneAmp® PCR 0.5 ml reaction tubes (PE Applied Biosystems, N801-0180)
- GeneAmp® dNTPs: separate 10 mM stock solutions of dATP, dCTP, dGTP, and dTTP in distilled water, titrated to pH 7.0 with NaOH (PE Applied Biosystems, N808-0007)
- Deep Vent$_R$® DNA polymerase: 200 units, 2 U/µl (New England Biolabs, 258S)
- 10 × ThermoPol reaction buffer: 100 mM KCl, 100 mM $(NH_4)_2SO_4$, 200 mM Tris–HCl pH 8.8, 20 mM $MgSO_4$, 1% Triton X-100 (supplied with Deep Vent$_R$® DNA polymerase)
- 50 × TAE (see *Protocol 2*)
- NuSieve® GTG® agarose powder (25 g, FMC BioProducts, 50081): low melting temperature ($< 65\,°C$) agarose resolving PCR products ranging from 10–1000 bp
- Wizard™ PCR Preps DNA Purification System (Promega, A7170)
- Ethidium bromide (see *Protocol 2*)
- 1 kb PLUS DNA Ladder™ (Life Technologies, 10787-018)
- 0.25 M HCl solution: mix in the following order, 979.45 ml distilled water with 20.55 ml concentrated HCl (11.6 M)

Method

1. Set up the following 100 µl reactions for each PCR and negative control, i.e. one reaction for each of the nine family-specific V_H BACK primers (V_H1 to V_H11 Back primers, see *Table 1*) and eight family-specific V_K BACK primers (V_K1 to V_K8 Back, see *Table 1*), together with controls (no DNA) for each; a total of 34 reactions.

   - H_2O 74 µl
   - 10 × ThermoPol reaction buffer (20 mM $MgSO_4$)   10 µl
   - 10 mM[a] (250 µM each dNTP final concentration)   10 µl
   - Individual BACK[b] primer (10 pmol/µl)   2 µl
   - Equimolar mixture FOR[c] primer[d] (10 pmol/µl)   2 µl
   - Final volume   98 µl

2. Add 1 µl cDNA:mRNA hybrid[e] (from *Protocol 4*), and overlay with 100 µl of mineral oil.[f]

3. Place the tubes in a DNA thermal cycler pre-set at 94 °C.

4. Add 1 µl Deep Vent$_R$® DNA polymerase[g] to each reaction beneath the mineral oil, using separate pipette tips for each addition.

5. Return the tubes to the thermal cycler pre-set at 94 °C and immediately cycle at 94 °C for 1 min, 60 °C for 1 min, and 72 °C for 1 min over 30 cycles. Follow the last cycle with a final extension step at 72 °C for 10 min before cooling at 4 °C (storage). Use a ramp time of 2.5 min between the annealing (60 °C) and extension (72 °C) steps and a 30 sec ramp time between all other steps of the cycle.

Protocol 5 continued

6. Separate each PCR product[h] by electrophoresis on a 3.0% (w/v) NuSieve GTG agarose gel[i] ($\sim$ 5 mm thick), containing 0.5 µl/ml ethidium bromide,[j] prepared in 1 $\times$ TAE buffer[k] alongside 1 µl of 1 kb PLUS DNA ladder.

7. Visualize the bands on a UV light box,[l] cover the screen with Saran Wrap to prevent nucleic acid cross-contamination.

8. Carefully excise the V_H and V_K bands and transfer each gel band slice to a separate sterile 1.5 ml microcentrifuge tube.[m]

9. Use the Wizard™ PCR Preps DNA Purification System[n] to purify each PCR amplified DNA product. Recover the DNA in 50 µl H_2O per original PCR.

[a] 10 mM dNTP mix is an equimolar mixture of dATP, dCTP, dGTP, and dTTP with a total concentration of 10 mM nucleotide (i.e. 2.5 mM each dNTP).

[b] BACK refers to 'toward the 3' end of the antibody gene'.

[c] See *Protocol 4*, footnote *a*.

[d] FOR primers shown in *Table 1*. Equimolar mixture of JH Forward (JH1 to JH4 For) for the heavy chains or JK Forward (JK1 to JK5 For) the light chains.

[e] This is the minimal amount needed, you may find it necessary to use more.

[f] The use of mineral oil can be avoided when using a DNA thermal cycler equipped with a heated lid.

[g] Contains a strong 3' $\rightarrow$ 5' proof-reading exonuclease activity. *Pfu*, another proof-reading DNA polymerase exhibiting the lowest error rate of any thermostable DNA polymerase can also be used with an optimized Mg^{2+} concentration and reaction buffer supplied by the manufacturer (Stratagene).

[h] Make sure not to include any mineral oil when you transfer the PCR reactions. To remove mineral oil from a reaction before loading the sample on a gel, pipette the sample onto a piece of Parafilm. The mineral oil will be attracted to the film and the aqueous sample will form a drop in the centre. The sample can then be easily pipetted away from the oil, using a fresh tip.

[i] To avoid nucleic acid cross-contamination the agarose gel box, gel tray, combs, and spacers should be depurinated with 0.25 M HCl (overnight) and rinsed with water prior to use. Prepare the gel immediately before use. We recommend using NuSieve® GTG® agarose (FMC BioProducts).

[j] See *Protocol 2*, footnote *j*.

[k] The best buffer to use when DNA is to be recovered from agarose gel is TAE.

[l] Minimize the exposure of the DNA to the UV light as best you can. Too long of an exposure can render the DNA unusable. The addition of 1 mM guanosine or cytidine to the gel and electrophoresis buffer is effective in protecting DNA against UV-induced damage. Also, see *Protocol 2*, footnote *k*.

[m] Using a fresh sterile razor blade for each chain to prevent contamination. The V_H and V_K bands are approximately 340 and 325 bases long, respectively.

[n] Used according to the manufacturer's instructions.

7 Preparation of linker

The linker gene coding for the peptide $(Gly_4$-$Ser)_3$ (1) was amplified from a modified version of the construct scFv D1.3 (see *Protocol 6*, footnote *c*) using primers designed to create complementarity between the amplified linker sequence and the relevant heavy or light chain primers (i.e. MouJH For and MouVK Back, respectively). To make the scFv linker DNA, 32 separate PCR reactions should be performed using each of the four reverse JH primers in combination with each of the eight reverse VK oligonucleotides. By using 'universal' primers which are located at the 5' and 3' ends (Link Back and Link For primers, respectively) of the $(Gly_4$-$Ser)_3$ linker, this number can be reduced to 12, using each of the four reverse JH primers (BACK primers in *Protocol 6*) in combination with the Link For primer (four reactions) in one instance and each of the eight reverse VK oligonucleotides (FOR primers in *Protocol 6*) in combination with the Link Back primer (eight reactions) in a second instance (see *Figure 1 A2* and *Table 1*).

Protocol 6

Preparation of scFv linker DNA

Equipment and reagents

- Water-bath at 65 °C
- GeneAmp® PCR 0.5 ml reaction tubes (see *Protocol 5*)
- GeneAmp® dNTPs (see *Protocol 5*)
- Deep Vent$_R$® DNA polymerase (see *Protocol 5*)
- NuSieve® GTG® agarose (see *Protocol 5*)
- 5 × TBE (Tris, borate, EDTA) electrophoresis buffer: 445 mM Tris base (54.0 g), 445 mM boric acid (27.5 g), and 10 mM EDTA (3.72 g $Na_2EDTA.2H_2O$), adjust to 1 litre with distilled water

- 10 × ThermoPol reaction buffer (see *Protocol 5*)
- TE buffer (see *Protocol 2*)
- 100 ml Tris–HCl saturated phenol (Life Technologies, 15513-039)
- 70% (v/v) ethanol, ice-cold
- 4 M LiCl
- 20 mg/ml glycogen (Life Technologies, 10814-010)
- Ethidium bromide (see *Protocol 2*)
- 1 kb PLUS DNA ladder (see *Protocol 5*)

Method

1. Set up 50 μl linker reactions mix:[a]
   - H_2O 33.5 μl
   - 10 × ThermoPol reaction buffer (20 mM $MgSO_4$)   5.0 μl
   - 10 mM dNTP mix[b] (250 μM each dNTP final concentration)   5.0 μl
   - BACK primers (10 pmol/μl) (see text above)   1.0 μl
   - FOR primers (10 pmol/μl) (see text above)   1.0 μl
   - pSW1-scFvD1.3[c] (10 ng/μl)   1.0 μl
   - Deep Vent DNA polymerase (2 U/μl)   0.5 μl
   - Final volume   50.0 μl

Protocol 6 continued

2. Overlay with 100 μl of mineral oil[d] and place the tubes in a DNA thermal cycler pre-set at 94 °C. Amplify the linker DNA by PCR for 25 cycles of 94 °C for 1 min, 65 °C for 1 min, and 72 °C for 2 min.

3. Incubate at 60 °C for a further 5 min in the heating block.

4 Separate each PCR product[e] by electrophoresis on a 4% (w/v) NuSieve GTG agarose gel, containing 0.5 μl/ml ethidium bromide,[f] prepared in 1 × TBE buffer alongside 1 μl of 1 kb PLUS DNA ladder.

5. Visualize the bands on a UV light box,[g] cover the screen with Saran Wrap to prevent nucleic acid cross-contamination.

6. Cut out a slice of gel containing the fragment of interest (68 bp) in the smallest possible volume. Introduce the slice in a tarred microcentrifuge tube.[h] Remove any buffer that may have come in with the slice. Estimate the volume of the slice from its weight and add 1 vol. of TE buffer. Place the tube in a 65 °C water-bath for 5–10 min.

7. Make sure that the agarose is fully melted by pipetting the solution up and down a few times. Immediately add 1 vol. of Tris-buffered phenol (no chloroform) at room temperature and mix thoroughly by inversion.

8. Spin for 3 min at 10 000 to 12 000 r.p.m. Transfer the aqueous phase to a fresh tube containing 1 vol. of Tris-buffered phenol.[i] Mix.

9. Spin sample as above. Transfer the aqueous phase to a new tube containing 0.1 vol. of 4 M LiCl. Mix by inversion. A white precipitate appears immediately. Place the tube on ice for 2 min. Spin as above for 3 min.

10. Transfer supernatant to a new tube, leaving the transparent pellet behind.[j] Add 2.5 vol. of cold ethanol. Mix by inversion and leave for 5–30 min at −70 °C. Spin as above for 10 min. Wash the pellet with 1 ml of cold 70% (v/v) ethanol. Air dry the DNA pellet and resuspend it into 10–20 μl water or TE buffer.

[a] It is often convenient to set up ten reactions at one time for each pair of primers.

[b] See *Protocol 5*, footnote *a*.

[c] Where the linker DNA sequence was modified for better codon usage in *E. coli*. The linker was originally prepared by complementarity oligonucleotide synthesis to generate the sequence:

*Bst*EII cut   Link Back primer   *Sac*I cut
5′*gtcaccgtctcctca*GGTGGCGGTGGCTCGGGCGGTGGTGGGTCGGGTGGCGGCGGATCT*gacatcgagct*
*g*cagaggagtCCACCGCCACCGAGCCCGCCACCA<u>CCCAGCCCACCGCCGCCTAGA</u>ctgtag*c* 5′
                                                                Link For primer

This construct was inserted as a *Bst*EII–*Sac*I DNA fragment in pSW1-VHD1.3-VKD1.3-TAG1 (8) to create pSW1-scFvD1.3TAG1 expression vector which was used in this protocol. However, the linker oligonucleotide sequence can be cloned into any convenient vector for subsequent PCR amplification.

[d] See *Protocol 5*, footnote *f*.

[e] See *Protocol 5*, footnote *h*.

[f] See *Protocol 2*, footnote *j*.

[g] See *Protocol 5*, footnote *l*.

[h] The protocol used to purify the linker DNA is that described by Favre (9). In our hands, this method works at least as well but often better than commercially available kits based on phenol-free methods.

[i] Do not be too concerned about picking up small amounts of powdered agarose that form at the interface because it will be subsequently eliminated.

[j] It is a wise precaution to add a carrier molecule at this stage, e.g. 1 μl of 20 mg/ml glycogen.

8 Assembly of VH and VK gene fragments with linker DNA

Procedures for the assembly of H and L chain variable regions (Fv) via a linker fragment has originally been described by Clackson *et al.* (10). This procedure consists of submitting the H and L chain Fv and linker fragments to several cycles of annealing–denaturations followed by PCR amplification of the mixture in the presence of added oligo primers complementary to the opposite ends of the Fv fragments. We and others, however, have found this method to be technically difficult. Therefore as an alternative to this method, we designed a modified version of the 'jumping-PCR-assembly method' described by Ørum *et al.* (11). First, the H and L chain Fv gene fragments were amplified (see Section 6 and *Figure 1 A1*) and the linker DNA prepared (see Section 7 and *Figure 1 A2*). Secondly, the linker fragment was joined to the H and L chain Fv fragments in two separate PCR reactions by submitting the fragments to a PCR protocol in the presence of primers complementary to the opposite ends of the two fragments (see this section and *Figure 1B*). A similar strategy was used in the final assembly, this time using the linker-assembled H and L chain Fv fragments (see Section 9 and *Figure 1C*). In essence this method represents the processes that occur in a single tube in the PCR and splicing by overlap extension method, but we have found that separation of the assembly into individual steps allows greater control without significantly increasing the time or complexity of the procedure. This stepwise approach also allows rapid detection and correction of experimental errors during the procedure by gel analysis of aliquots at each stage.

Protocol 7

Linker assembly

Reagents

- GeneAmp® PCR 0.5 ml reaction tubes (see *Protocol 5*)
- AmpliTaq® DNA polymerase: 250 units, 5 U/μl (PE Applied Biosystems, N801-0060)
- 1.5 ml GeneAmp®10 × PCR buffer: 500 mM KCl, 100 mM Tris–HCl pH 8.3, 15 mM MgCl$_2$, and 0.01% (w/v) gelatin (PE Applied Biosystems, N808-0006)
- GeneAmp® dNTPs (see *Protocol 5*)
- NuSieve® GTG® agarose (see *Protocol 5*)
- 50 × TAE buffer (see *Protocol 2*)
- Wizard™ PCR Preps DNA Purification System (see *Protocol 5*)
- 1 kb PLUS DNA ladder (see *Protocol 5*)

Method

1. Estimate the quantities of VH and VL DNA prepared by the primary PCR reactions (*Protocol 5*) and the quantity of linker DNA prepared in *Protocol 6*, using a dot spot technique.[a] Adjust the concentrations of VH and VL PCR products to 5 ng/μl and that of linker DNA to 2 ng/μl.

2. Set up two sets of jumping PCR reactions, the first for the VH products involving nine reactions (one for each of the nine primary PCR reactions from *Protocol 5*) and the second for the VK products involving eight reactions (one for each of the eight primary PCR reactions from *Protocol 5*). For each of the 17 jumping PCR reactions, set up the following 100 μl mixture:

   * VH or VL DNA (5 ng/μl) (from *Protocol 5*)   3.0 μl (see step 1)
   * Linker DNA (2 ng/μl) (from *Protocol 6*)   3.0 μl (see step 1)
   * GeneAmp® 10 × PCR buffer (15 mM MgCl$_2$)   10.0 μl
   * 10 mM dNTP mix[b]   8.0 μl
   * BACK primer (25 pmol/μl)   1.0 μl
   * FOR primer (25 pmol/μl)   1.0 μl
   * Sterile H$_2$O   73.5 μl
   * Final volume   95.5 μl

3. Overlay with 100 μl of mineral oil[c] and place the tubes in a DNA thermal cycler pre-set at 94 °C.

4. Add 0.5 μl of AmpliTaq® DNA polymerase[d] to each reaction beneath the mineral oil, using separate pipette tips for each addition.

5. Return the tubes to the thermal cycler pre-set at 94 °C[e] and immediately cycle at 94 °C for 1 min, 55 °C for 1 min, and 72 °C for 2 min over 25 cycles. Follow the last cycle with a final extension step at 72 °C for 10 min before cooling to 4 °C (storage).

6. Separate each PCR product[f] by electrophoresis on a 3.0% (w/v) NuSieve GTG agarose gel[g] (~ 5 mm thick), containing 0.5 μl/ml ethidium bromide,[h] prepared in 1 × TAE buffer[i] alongside 1 μl of 1 kb PLUS DNA ladder.

7. Visualize the bands on a UV light box,[j] cover the screen with Saran Wrap to prevent nucleic acid cross-contamination.

8. Carefully excise the V$_H$–linker and linker–V$_K$ bands and transfer each gel band slice to a separate sterile 1.5 ml microcentrifuge tube.[k]

9. Use the Wizard PCR Preps DNA Purification System[l] to purify each PCR amplified DNA product. Recover the DNA in 50 μl H$_2$O per original PCR.

[a] The DNA is present in quantities too small to measure spectrophotometrically, so we estimate the DNA concentration by comparing the ethidium bromide-mediated fluorescence of the unknown sample with that of samples of known DNA concentrations under short-range (300 nm) UV light.

[b] See *Protocol 5*, footnote *a*.

[c] See *Protocol 5*, footnote *f*.

[d] For jumping PCR, we recommend using AmpliTaq® DNA polymerase which in our hands works better than Deep Vent$_R$®DNA polymerase.

[e] As an alternative to manual Hot Start, AmpliTaq Gold™ DNA polymerase (250 units, PE Applied Biosystems, N808-0241) can be used without having to re-optimize the PCR conditions.

[f] See *Protocol 5*, footnote *h*.

[g] See *Protocol 5*, footnote *i*.

[h] See *Protocol 2*, footnote *j*.

[i] See *Protocol 5*, footnote *k*.

[j] See *Protocol 5*, footnote *l*.

[k] Using a fresh sterile razor blade for each PCR product to prevent contamination. The V_H–linker and linker-V_K bands are approx. 385 (340 + 45) and 370 (45 + 325) bases long, respectively.

[l] See *Protocol 5*, footnote *n*.

9 Assembly of single chain Fv antibody fragments

In the final assembly reaction, the VH–link and link–VK DNA are joined into a single chain Fv antibody fragment and restriction sites are added. These restriction sites are used to clone the scFv into a phagemid vector as a *Sfi*I–*Not*I DNA fragment. The primers contain either *Sfi*I (Mouse VH Back *Sfi*I primers) or *Not*I (Mouse JK Forward *Not*I primers) restriction sites. They anneal to the 5′ end of the heavy chain (*Sfi*I restriction site) or the 3′ end of the light chain (*Not*I restriction site). Because of the complementarity of the linker DNA between the two PCR fragments, a fill-in reaction is primed in the presence of AmpliTaq® DNA polymerase (see *Figure 1C*). For this reaction to proceed efficiently, approximately 10 ng of VH–link and 10 ng of link–VK DNA should be added to the reaction.

Protocol 8

Final assembly

Equipment and reagents

- GeneAmp® PCR 0.5 ml reaction tubes (see *Protocol 5*)
- GeneAmp® dNTPs (see *Protocol 5*)
- GeneAmp®10 × PCR buffer (see *Protocol 7*)
- AmpliTaq® DNA polymerase (see *Protocol 7*)
- NuSieve® GTG® agarose (see *Protocol 5*)
- 50 × TAE buffer (see *Protocol 2*)
- Ethidium bromide (see *Protocol 2*)
- Wizard™ PCR Preps DNA purification System (see *Protocol 5*)
- 1 kb PLUS DNA ladder (see *Protocol 5*)

Method

1. Estimate the quantities of VH–linker and linker–VL DNA prepared by the 'jumping' PCR reactions (*Protocol 7*) using a dot spot technique.[a] Adjust the concentrations of VH–linker and linker–VL PCR products to 5 ng/μl.

Protocol 8 continued

2. Set up 100 μl reactions for assembling VH–linker with linker–VK:

    - VH–linker DNA (5 ng/μl) (from *Protocol 7*)   2.0 μl (see step 1)
    - Linker–VK DNA (5 ng/μl) (from *Protocol 7*)   2.0 μl (see step 1)
    - GeneAmp®10 × PCR buffer (15 mM MgCl$_2$)   10.0 μl
    - 10 mM dNTP mix[b]   8.0 μl
    - Mix Mouse VH Back *Sfi*I primers (25 pmol/μl)   1.0 μl
    - Mix Mouse JK Forward *Not*I primers (25 pmol/μl)   1.0 μl
    - Sterile H$_2$O   71.5 μl
    - Final volume   95.5 μl

3. Overlay with 100 μl of mineral oil[c] and place the tubes in a DNA thermal cycler pre-set at 94 °C (Hot Start).

4. Add 0.5 μl of AmpliTaq® DNA polymerase[d] to each reaction beneath the mineral oil, using separate pipette tips for each addition.

5. Return the tubes to the thermal cycler pre-set at 94 °C[e] and immediately cycle at 94 °C for 1.5 min, 69 °C for 1 min, and 72 °C for 2 min over 25 cycles. Follow the last cycle with a final extension step at 72 °C for 10 min before cooling to 4 °C (storage).

6. Separate each PCR product[f] by electrophoresis on a 2.5% (w/v) NuSieve GTG agarose gel[g] (~ 5 mm thick), containing 0.5 μl/ml ethidium bromide,[h] prepared in 1 × TAE buffer[i] alongside 1 μl of 1 kb PLUS DNA ladder.

7. Visualize the bands on a UV light box,[j] cover the screen with Saran Wrap to prevent nucleic acid cross-contamination.

8. Carefully excise the assembled scFv product[k] and transfer the gel band slice to a sterile 1.5 ml microcentrifuge tube.

9. Use the Wizard PCR Preps DNA Purification System[l] to purify the assembled DNA product. Recover the DNA in 50 μl H$_2$O per original PCR.

[a] See *Protocol 7*, footnote *a*.

[b] See *Protocol 5*, footnote *a*.

[c] See *Protocol 5*, footnote *f*.

[d] For final assembly PCR, we recommend using AmpliTaq® DNA polymerase which in our hands works better than Deep Vent$_R$®DNA polymerase.

[e] See *Protocol 7*, footnote *e*.

[f] See *Protocol 5*, footnote *h*.

[g] See *Protocol 5*, footnote *i*.

[h] See *Protocol 2*, footnote *j*.

[i] See *Protocol 5*, footnote *k*.

[j] See *Protocol 5*, footnote *l*.

[k] A predominant band of ~ 750 base pair (bp) in size corresponding to the full-length scFv DNA fragment should be present. Some VH–linker and linker–VL monomers may also be visible.

[l] See *Protocol 5*, footnote *n*.

10 Reamplification of assembled scFv DNA

A portion of the assembled scFv product (from *Protocol 8*) is used to reamplify the fully assembled product. This step is carried out for two reasons, first to obtain more material for subsequent procedures; secondly and mainly to extend the assembled scFv fragments with 59 and 70 nucleotides beyond the recognition sequences of the *Sfi*I and *Not*I restriction enzymes (respectively) in order to improve the cutting efficiencies of these enzymes. The primers used (see *Table 1E*) were designed with a 100% match to the constant 5′ portion of the primers used in the final assembly reaction (see *Table 1D*) so that a single pair of primers can be used and no biased amplification is introduced. This approach allows greater monitoring of the restriction enzymatic digestion by agarose gel analysis of aliquots at each stage (i.e. the difference of molecular weight between restriction enzymes digested and undigested scFv DNA fragments is clearly visible on the gel).

Protocol 9

Reamplification of scFv fragments

Equipment and reagents

- GeneAmp® PCR 0.5 ml reaction tubes (see *Protocol 5*)
- GeneAmp® dNTPs (see *Protocol 5*)
- Deep Vent$_R$® DNA polymerase (see *Protocol 5*)
- 10 × concentrated ThermoPol reaction buffer (see *Protocol 5*)
- TE buffer (see *Protocol 2*)
- 100 ml phenol:chloroform:isoamyl alcohol (25:24:1, by vol.) (Life Technologies, 15593-031)
- Agarose (UltraPure) (see *Protocol 2*)
- 50 × TAE buffer (see *Protocol 2*)
- 1 kb PLUS DNA ladder (see *Protocol 5*)

Method

1. Estimate the quantities of assembled scFv purified from *Protocol 8* using a dot spot technique.[a]

2. Add the following components to a 0.5 ml microcentrifuge tube:[b]

   - Assembled scFv DNA (from *Protocol 8*; 1–50 ng)   X μl
   - 10 × ThermoPol Reaction buffer (20 mM MgSO$_4$)   10.0 μl
   - 10 mM dNTP mix[c]   8.0 μl
   - Universal *Sfi*I Back primer (see *Table 1E*; 25 pmol/μl)   1.0 μl
   - Universal *Not*I For primer (see *Table 1E*; 25 pmol/μl)   1.0 μl
   - Sterile H$_2$O   to 99.0 μl

3. Overlay with 100 μl of mineral oil[d] and place the tubes in a DNA thermal cycler preset at 94 °C (Hot Start).

Protocol 9 continued

4. Add 1 μl of Deep Vent$_R$® DNA polymerase to each reaction beneath the mineral oil, using separate pipette tips for each addition.

5. Return the tubes to the thermal cycler pre-set at 94 °C[e] and immediately cycle at 94 °C for 1.5 min, 69 °C for 1 min, and 72 °C for 2 min over 20 cycles. Follow the last cycle with a final extension step at 72 °C for 10 min before cooling to 4 °C (storage).

6. Analyse 5 μl aliquots of the reaction by electrophoresis on a 1.2% (w/v) agarose/1 × TAE gel[f] next to 1 μl of 1 kb PLUS DNA ladder. The PCR reactions exhibiting a strong band at approx. 880 bp[g] are pooled.[h]

7. Add an equal volume of phenol:chloroform:isoamyl alcohol[i] (25:24:1). Mix the contents of the tube until an emulsion forms. Centrifuge the mixture at 12 000 g for 5 min in a microcentrifuge at room temperature. Transfer the upper aqueous phase in a fresh tube. Add an equal volume of TE pH 7.8 to the organic phase and interface. Mix well. Centrifuge as above. Combine the second aqueous phase with the first.

8. Recover the DNA contained in the aqueous extract by precipitation with ethanol.[j] Wash the pellet twice with 70% (v/v) ethanol. Air dry the DNA pellet and resuspend with 100 μl of water ready for digestion (proceed to *Protocol 10*).

[a] See *Protocol 7*, footnote *a*.

[b] Set up ten PCR reactions together with controls (no DNA).

[c] See *Protocol 5*, footnote *a*.

[d] See *Protocol 5*, footnote *f*.

[e] See *Protocol 7*, footnote *e*.

[f] See *Protocol 5*, footnote *i*.

[g] 880 bp represents the size of the extended scFv, i.e. 750 bp (assembled scFv) plus 59 bp (nucleotides beyond the *Sfi*I site) plus 70 (nucleotides beyond the *Not*I site).

[h] For each positive PCR reaction, carefully transfer the aqueous (lower) phase leaving the mineral oil behind.

[i] Equilibrated against TE buffer.

[j] Precipitate the DNA by adding 0.1 vol. of 3 M sodium acetate pH 5.2 and 2 vol. of ethanol, as described in Sambrook *et al.* (6) on page E.10.

11 Restriction enzyme digestion of assembled scFv

The assembled scFv (from *Protocol 9*) is digested with *Sfi*I (located at the 5′ end of the heavy chain) and *Not*I (located 3′ of the light chain) restriction enzymes to allow cloning into the phagemid vector pHEN-1 cut with the same two enzymes (see *Protocol 12*). For mouse antibody scFv, *Sfi*I and *Not*I sites are used because there are extremely rare cutters.

Protocol 10

Restriction digestion of scFv DNA fragments

Equipment and reagents

- Restriction endonuclease SfiI, 20 U/µl (New England BioLabs, 123L)
- Restriction endonuclease NotI, 10 U/µl (New England BioLabs, 189L)
- 1 × NEBuffer 2: 50 mM NaCl, 10 mM Tris–HCl, 10 mM MgCl$_2$, and 1 mM DTT (pH 7.9 at 25 °C)
- 1 × NEBuffer 3: 100 mM NaCl, 50 mM Tris–HCl, 10 mM MgCl$_2$, and 1 mM DTT (pH 7.9 at 25 °C)
- Phenol:chloroform:isoamyl alcohol (25:24:1, by vol.) (see Protocol 9)
- NuSieve® GTG® agarose (see Protocol 5)
- 50 × TAE buffer (see Protocol 2)
- Wizard™ PCR Preps DNA Purification System (see Protocol 5)
- Ethidium bromide (see Protocol 2)

Method

1. Estimate the quantities of assembled scFv purified from Protocol 9 using a dot spot technique.[a]

2. Add the following components in the order stated to a 1.5 ml microcentrifuge tube:

   - Water   up to 50.0 µl
   - 10 × NEBuffer 2   5.0 µl
   - 100 × BSA acetylated (10 mg/ml)   0.5 µl
   - Purified scFv DNA product (1–2 µg)   X µl
   - SfiI (20 U/µl)   2.0 µl (40 U)

3. Mix gently by pipetting the reaction mixture up and down. Follow with a quick ('touch') spin-down in a microcentrifuge.[b]

4. Overlay the mix with mineral oil to prevent evaporation. Incubate at 50 °C for at least 12 h or overnight.[c]

5. Extract with phenol:chloroform:isoamyl alcohol, precipitate with ethanol, wash, and dry.[d] Resuspend in 42.5 µl of water.

6. Set up second digest:

   - SfiI cut DNA   42.5 µl
   - 10 × NEBuffer 3   5.0 µl
   - 100 × BSA acetylated (10 mg/ml)   0.5 µl
   - NotI (10 U/µl)   4.0 µl (40 U)

7. Mix as described in step 3 and incubate at 37 °C for 4 h.

8. Analyse for size[e] a 5 µl aliquot of the reaction by electrophoresis on a 2.5% (w/v) NuSieve GTG agarose/1 × TAE gel alongside undigested assembled scFv.

Protocol 10 continued

9. Gel purify the *Sfi*I–*Not*I cut scFv DNA fragment electrophoresis on a 2.5% (w/v) NuSieve GTG agarose/1 × TAE using the Wizard PCR Preps DNA Purification System.[f]

10. Determine the concentration of the recovered DNA using a dot spot technique.[a]

11. Ethanol precipitate the DNA and resuspend with water at 100 ng/μl.

[a] See *Protocol 7*, footnote *a*.

[b] Do not vortex.

[c] Small aliquots can be removed during the course of the reaction and analysed on a minigel to monitor the progress of the digestion. There is approx. a 60 bp difference between uncut and *Sfi*I digested assembled scFv DNA. See Section 10.

[d] As described in *Protocol 9*, steps 7 and 8.

[e] There is approx. a 130 bp difference between uncut and double cut DNA. See Section 10.

[f] See *Protocol 5*, footnote *n*.

12 Purification of pHEN-1 vector by equilibrium centrifugation in CsCl ethidium bromide gradients

pHEN-1 (12) is a phagemid vector containing both *E. coli* and phage origins of replication in addition to an ampicillin resistance gene. The phagemid vector pHEN-1 allows the cloning of scFv antibody genes as *Sfi*I/*Not*I fragments in fusion with gene III protein (pIII) of filamentous bacteriophage fd. The transcription of antibody–pIII fusions is driven from the inducible *lacZ* promoter and the fusion protein targeted to the periplasm by means of the pelB leader (13). Between the polylinker and the coding sequence of gene III is an in-frame fusion of a c-myc epitope tag (14) for the detection of recombinant scFv antibodies. Phagemid vectors require 'rescue' with an helper phage to provide proteins for replication and packaging (15) but are preferred for library construction due to the greater transformation efficiencies of these pUC-based vectors (10^8 to 10^9 clones per μg of DNA) compared with phage fd vectors (10^6 to 10^7 clones per μg of DNA).

For library construction it is essential to use vector DNA of the highest purity, this is why we highly recommend to prepare the pHEN-1 phagemid vector by caesium chloride isopycnic centrifugation. This procedure is performed in the presence of the intercalating dye, ethidium bromide, which is bound in smaller amounts by the covalently closed circular plasmid DNA than by the chromosomal DNA. Since the bound dye decreases the buoyant density of the DNA, the plasmid molecules band at a denser position in the gradient ('lower band' in step 15) and the RNA is pelleted. This procedure provides a very effective method of removing all traces of chromosomal DNA and RNA from the phagemid preparation.

Protocol 11

DNA prep using CsCl banding

Equipment and reagents

- 50 ml blue screw-capped polypropylene tubes (Falcon, 2070)
- 14 ml polypropylene tubes (see *Protocol 2*)
- VTi 65.1 vertical tube rotor (Beckman, 362759)
- Quick-Seal Ultra Clear tubes,13.5 ml filling capacity ($\times$ 50, Beckman, 344322)
- STE solution: 0.1 M NaCl, 10 mM Tris–HCl pH 8.0, 1 mM EDTA (dissolve 5.84 g NaCl, 1.21 g Tris base, and 0.37 g $Na_2EDTA.2H_2O$ in 800 ml distilled water, adjust to pH 8.0 with HCl, adjust the volume to 1 litre with distilled water)
- TE buffer (see *Protocol 2*)
- Solution I: 50 mM glucose, 25 mM Tris–HCl pH 8.0, 10 mM EDTA (dissolve 3.03 g Tris base, 3.72 g $Na_2EDTA.2H_2O$ in 800 ml, add 25 ml of a 2 M D(+) glucose (anhydrous) solution, adjust to pH 8.0 with HCl, adjust the volume to 1 litre with distilled water)

- Solution II: 0.2 M NaOH, 1% (w/v) SDS (dissolve 8.0 g NaOH pellets in 950 ml distilled water and add 50 ml of a 20% (w/v) SDS solution)
- Solution III: 3 M potassium acetate pH 5.5 (dissolve 29.4 g potassium acetate in 50 ml of distilled water, adjust to pH 5.5 with glacial acetic acid ($\sim$ 11.5 ml), adjust the volume to 100 ml with distilled water)
- 2 $\times$ TY medium: dissolve 16 g Bacto tryptone, 10 g Bacto yeast extract, and 5 g NaCl in 800 ml of distilled water, adjust to pH 7.0 with 1 M NaOH, adjust the volume to 1 litre with distilled water; sterilize by autoclaving
- *E. coli* TG1 strain (Pharmacia Biotech, 27-9401-01) carrying the phagemid vector, pHEN-1[a]
- Ethidium bromide (see *Protocol 2*)
- Caesium chloride (Sigma, C-3032)
- Lysozyme (Sigma, L-6876)

Method

1. Inoculate 30 ml of 2 $\times$ TY medium containing 50 μg/ml ampicillin (2TY/Amp) with a single colony of *E. coli* TG1 strain carrying the phagemid vector pHEN-1 from a freshly streaked plate. Grow this starter culture to an OD_{600} of approx. 0.6 or overnight at 37 °C.

2. Use 25 ml of this late-log culture to inoculate 500 ml of 2TY/Amp medium in a 2 litre flask. Incubate this secondary culture with vigorous shaking[b] until the OD_{600} reaches approx. 0.4 and add 2.5 ml of a solution of chloramphenicol.[c] Incubate the culture for a further 12–16 h at 37 °C with vigorous shaking.[b]

3. Pellet bacteria by centrifuging at 4000 r.p.m. for 20 min at 4 °C. Decant supernatant carefully and stand the open centrifuge bottle in an inverted position to allow all of the supernatant to drain away.

4. Resuspend the bacterial pellet in 100 ml of ice-cold STE solution[d] and collect the bacterial cells by centrifugation as in step 3.

5. Resuspend the washed pellet in 9 ml of solution I. Add 1 ml of freshly prepared solution of lysozyme (10 mg in 1 ml of solution I). Transfer to two 50 ml 'Falcon' tubes. Stand at room temperature for 5 min.

Protocol 11 continued

6. To each tube add 10 ml of fresh solution II. Mix thoroughly by gently inverting the tubes several times. Incubate at room temperature for 10 min.

7. To each tube add 15 ml of ice-cold solution III. Mix well by shaking several times and incubate on ice for 10 min.[e]

8. Centrifuge at 4000 r.p.m. for 20 min at 4 °C. Allow the rotor to stop without braking.

9. Carefully decant supernatant into a fresh 50 ml 'Falcon' tube, leaving behind the pellet and floating precipitate.[f] Add 0.6 vol. of isopropanol, mix well, and incubate for 10 min at room temperature.

10. Spin at 5000 r.p.m. for 20 min at room temperature, decant the supernatant carefully, and invert the open tube to allow the last drops of supernatant to drain away.[g]

11. Resuspend the DNA in 10 ml TE buffer.

12. Weigh out 10 g of CsCl into a tube and add to the DNA solution.[h] Warm to 30 °C to dissolve salt. Mix gently.

13. Add 0.8 ml ethidium bromide solution[i] and mix immediately. Wrap the tube in silver foil.[j] Give a quick spin at 8000 r.p.m. for 5 min and immediately transfer the clear, red solution under the scum to a Beckman Quick-Seal tube using a 10 ml disposable syringe fitted with an 18 gauge hypodermic needle. Tare tubes as precisely as possible. Fill the remaining of the tube with mineral oil and seal tubes. Avoid bubbles at the top.

14. Centrifuge at 45 000 r.p.m. for 16 h or longer at 20 °C in a Beckman vertical high speed VTi65.1 rotor.

15. Carefully take the tube out of the rotor. Two bands of DNA, located in the centre of the gradient, should be visible in ordinary light. Collect the lower band (closed circular plasmid DNA) into a 14 ml polypropylene tube using a 5 ml syringe and 18 gauge hypodermic needle.[k]

16. Add an equal volume of 1-butanol saturated with water to the solution of DNA and mix the two phases by vortexing.

17. Centrifuge the mixture at 1500 r.p.m. for 3 min at room temperature.

18. Transfer the lower, aqueous phase to a clean 14 ml tube. Repeat the extraction (step 16 and 17) four times until all the pink colour disappears from both aqueous and organic phases.

19. After final extraction, add 3 vol. of water and precipitate the DNA by adding 2 vol. of ethanol.[l] Leave on ice for 15 min and centrifuge at 10 000 g for 15 min at 4 °C.

20. Dissolve the precipitated DNA in 0.5 ml of TE buffer.[m] Measure the OD_{260} and calculate the concentration of DNA. Store the DNA in aliquots at −20 °C.

[a] pHEN-1 vector (12) was reconstructed from its commercial derivative pCANTAB 5E phagemid (Pharmacia Biotech, 27-9401-01) by substituting the gene III signal and E tag sequences by the pelB signal (13) and c-myc epitope tag (14) sequences, respectively, using basic molecular biology techniques. Alternatively, the pHEN-1 vector can be obtained directly from Dr Greg Winter at the MRC Centre for Protein Engineering, Hills Road, Cambridge CB2 2QH, UK.

Protocol 11 continued

[b] 300 cycles/min on a rotary shaker.

[c] At 34 mg/ml in ethanol (final concentration at 170 μg/ml). Chloramphenicol inhibits bacterial replication. This reduces the bulk and viscosity of the bacterial lysate and greatly simplifies purification of the plasmid.

[d] Extra wash step before cell lysis in order to remove all traces of phage supernatant.

[e] A flocculent white precipitate should form.

[f] This can be done easily by filtering through an absorbent gauze or four layers of cheesecloth.

[g] Add a few ml of 70% (v/v) ethanol to wash out the isopropanol. Swirl the ethanol around the edges of the 'Falcon' tube, drain off the ethanol by leaving the tube inverted on a pad of paper towels for a few minutes at room temperature to allow the final traces of ethanol to evaporate.

[h] For every millilitre of the DNA solution, add exactly 1 g of solid CsCl.

[i] At 10 mg/ml in water. Make sure that the CsCl is dissolved before adding the ethidium bromide.

[j] In high concentration of ethidium bromide, the DNA can get nicked so it must be protected from the light.

[k] As described in Sambrook *et al.* (6) on page 1.43, step 6.

[l] The final volume equals six times the original volume of the undiluted aqueous phase.

[m] If the final preparation of DNA contains significant quantities of ethidium bromide, as judged from its colour, extract it once with phenol and once with phenol:chloroform, and then precipitate the DNA with ethanol.

13 Restriction digestion of the phage display vector, pHEN-1

The pHEN-1 vector (see Section 12) contains the restriction sites *Sfi*I (located 3' of the pelB leader and *Not*I (located 5' of the c-myc epitope tag) for directional cloning of antibody scFv genes cut with the same two enzymes (see *Protocol 10*). The pHEN-1 vector is first digested with *Sfi*I restriction enzyme followed by digestion with *Not*I. Finally the double digested *Sfi*I/*Not*I pHEN-1 vector is cut with *Pst*I (located in the polylinker, between the *Sfi*I and *Not*I sites) to reduce the vector background (i.e. religation of the vector on itself).

Protocol 12

Cutting and purification of cut pHEN-1[a]

Reagents

- Restriction endonuclease *Sfi*I (see *Protocol 10*)
- Restriction endonuclease *Not*I (see *Protocol 10*)
- Restriction endonuclease *Pst*I, 20 U/μl (New England BioLabs, 140S)
- 1 × NEBuffer 2 (see *Protocol 10*)
- 1 × NEBuffer 3 (see *Protocol 10*)
- Agarose (UltraPure) (see *Protocol 2*)
- 50 × TAE buffer (see *Protocol 2*)
- 1 kb PLUS DNA Ladder™ (see *Protocol 5*)
- CHROMA SPIN + TE-1000 columns (Clontech, K1324)

Protocol 12 continued

Method

1. Digest pHEN-1 with *Sfi*I by adding the following components, in the order stated, to a 1.5 ml microcentrifuge tube:
   - Distilled water   up to 100.0 μl
   - 10 × NEBuffer 2   10.0 μl
   - 100 × BSA acetylated (10 mg/ml)   1.0 μl
   - CsCl purified supercoiled DNA (see *Protocol 11*)   X μl (5 μg)
   - *Sfi*I (20 U/μl)   5.0 μl (100 U)

2. Mix gently by pipetting the reaction mixture up and down. Follow with a quick ('touch') spin-down in a microcentrifuge.[b]

3. Overlay the mix with mineral oil to prevent evaporation. Incubate at 50 °C for 4 h.

4. Run 2 μl (100 ng) on a 0.8% (w/v) agarose gel next to 1 μl of 1 kb PLUS DNA ladder.[c]

5. Extract with phenol:chloroform, precipitate with ethanol, wash, and dry.[d] Resuspend in 81 μl of water.

6. Set up second digest:[e]
   - *Sfi*I cut pHEN-1 vector   81.0 μl
   - 10 × NEBuffer 3   10.0 μl
   - 100 × BSA acetylated (10 mg/ml)   1.0 μl
   - *Not*I (10 U/μl)   8.0 μl (80 U)

7. Mix as described in step 3 and incubate at 37 °C for 4 h.

8. Extract with phenol:chloroform, precipitate with ethanol, wash, and dry.[d] Resuspend in 85 μl of water.

9. Set up third digest:[e]
   - Double *Sfi*I–*Not*I cut pHEN-1 vector   85.0 μl
   - 10 × NEBuffer 3   10.0 μl
   - 100 × BSA acetylated (10 mg/ml)   1.0 μl
   - *Pst*I (20 U/μl)   4.0 μl (80 U)

10. Mix as described in step 3 and incubate at 37 °C for 4 h.

11. Purify the triple digested vector using a CHROMA SPIN 1000 gel filtration spin column.[f]

12. Estimate the concentration of the digested vector by agarose gel electrophoresis along with molecular weight standards of a known concentration.

13. Adjust the concentration of digested pHEN-1 vector to 200 ng/μl with water.

[a] To reduce the background, vector pHEN-1 can be treated with calf intestine alkaline phosphatase (CIAP) rather than cut with the *Pst*I restriction enzyme, but we found that CIAP treatment reduces ligation efficiencies and therefore the size of the library.

[b] Do not vortex.

[c] Digested phagemid DNA vector should run as a 4.5 kb linear double-stranded fragment. If there is any residual uncut closed circular replicative form (which runs at the position of a ~ 2.5 kb supercoiled DNA), digest further with *Sfi*I.

[d] As described in *Protocol 9*, steps 7 and 8. Alternatively, use a clean-up system to purify the DNA from restriction enzymes and salts, we recommend using the Wizard® DNA Clean-Up System from Promega (A7280).

[e] Perform a control digest in parallel using undigested pHEN-1 DNA vector starting material (supercoiled form) to confirm that the enzyme is active (see step 5).

[f] According to the manufacturer's instructions.

14 Ligation of pHEN-1 and insert antibody scFv

The restriction digested *SfiI/NotI* antibody scFv genes (see *Protocol 10*) are directly ligated into pHEN-1 vector digested with the same two restriction enzymes (see *Protocol 12*). Set up a ligation of the cut vector in the absence of antibody scFv insert to determine the vector background (i.e. religation of the vector on itself). Test various vector:insert DNA ratios in order to find the optimum ratio. *SfiI-NotI* cut pHEN-1 vector is 4.48 kb and is prepared at 200 ng/μl (see *Protocol 12*). To calculate the appropriate amount of scFv 750 bp DNA fragment (insert) to include in the ligation reaction use the following equation:

$$\frac{200 \text{ (ng vector)} \times 0.75 \text{ (kb size insert)}}{4.48 \text{ (kb size of vector)}} \times \text{insert:vector molar ratio} = \text{ng of insert.}$$

We recommend starting with insert ratios in the range of 3:1 to 1:3. In our hands a 3:1 molar ratio of insert:vector works well. Using the above equation this is achieved by adding 100 ng of insert to 200 ng of vector.

Protocol 13

Ligation of scFv gene into pHEN-1

Reagents

- T4 DNA ligase(1 U/μl) and 5 × reaction buffer (100 units, Life Technologies, 15224-017)

- Phenol:chloroform:isoamyl alcohol (25:24:1, by vol.) (see *Protocol 9*)

- 100 ml chloroform:isoamyl alcohol (49:1) (5 Prime → 3 Prime, Inc.®, I-525538)

- 3 M sodium acetate pH 5.2 (see *Protocol 3*)

- 100% (v/v) and 70% (v/v) ethanol

Method

1. Using a 3:1 molar ratio of restriction digested pHEN-1 vector and scFv insert,[a] set up the following reaction in a 1.5 ml microcentrifuge tube:

    - Sterile distilled water   13 μl
    - Restriction digested antibody scFv gene fragment (100 ng/μl)   1 μl
    - 5 × T4 reaction buffer   4 μl
    - Restriction digested pHEN-1 (200 ng/μl)   1 μl

Protocol 13 continued

- T4 DNA ligase   1 μl
- Final volume   20 μl

2. Set up the following vector background control[b] ligation reaction with pHEN-1 and no insert in a 1.5 ml microcentrifuge tube:

   - Sterile distilled water   14 μl
   - 5 × T4 reaction buffer   4 μl
   - Restriction digested pHEN-1 (200 ng/μl)   1 μl
   - T4 DNA ligase   1 μl
   - Final volume   20 μl

3. Mix the reaction by pipetting and incubate the reactions overnight (16 h) at 4 °C.

4. Add 280 μl of TE buffer to each ligation tube. Extract the ligated scFv/pHEN-1 DNA with 300 μl of phenol:chloroform:isoamyl alcohol (25:24:1). Vortex for 1 min and centrifuge in a microcentrifuge at 12 000 g for 2 min.

5. Transfer the upper aqueous phase to a fresh tube and add 1 vol. of chloroform: isoamyl alcohol (49:1). Vortex for 1 min and centrifuge in a microcentrifuge at 12 000 g for 2 min.

6. Transfer the upper aqueous phase to a fresh tube. Add 0.1 vol. of 3 M sodium acetate pH 5.2 and 2.5 vol. of ice-cold 100% (v/v) ethanol. Mix and place at −70 °C for 30 min. Centrifuge at 12 000 g for 15 min at 4 °C.

7. Carefully pour off the supernatant, wash the pellet with 1 ml of 70% (v/v) ethanol,[c] and air dry. Resuspend the ligated DNA in 10 μl of water ready for electroporation. Store at −20 °C until further use.

[a] Molar ratio may require optimization (see text).

[b] This ligation will allow determination of the number of background colonies resulting from undigested or single cut vector alone.

[c] Failure to adequately wash will decrease transformation efficiency with electroporation.

15 Preparation of electroporation competent *E. coli* TG1 strain cells

For library construction it is absolutely necessary to use electrocompetent cells since electroporation provides a method of transforming *E. coli* to efficiencies greater than available with the best chemical methods. Using the protocol described below for preparing electrocompetent *E. coli* TG1 strain we routinely obtain 10^9 to 10^{10} transformants/μg of DNA. Alternatively *E. coli* TG1 strain electroporation competent cells with a transformation efficiency greater than 1×10^{10} transformants/μg can by purchased from Stratagene® (200123).

Protocol 14

Preparation of competent cells

Equipment and reagents

- Minimal medium agar plates (see Chapter 4, maintenance of bacterial stocks)
- Glycerol stock of *E. coli* TG1 strain cells (see *Protocol 11*)
- Glycerol (Fisher Scientific, G/0650/17)
- 2 × TY medium (see *Protocol 11*)

Method[a]

1. Using sterile techniques, streak *E. coli* TG1 strain cells from a glycerol stock onto a minimal medium plate. Grow at 37 °C for 24–36 h.[b]

2. Inoculate a single colony from the overnight *E. coli* TG1 strain, streaked on minimal medium plate, into 10 ml of 2 × YT medium and grow overnight at 37 °C in a shaking incubator.

3. Inoculate 1 litre of pre-warmed 2 × YT medium with the 10 ml overnight culture of *E. coli* TG1 strain cells and incubate at 37 °C with shaking at 250 r.p.m. for 2–3 h until an OD_{600} reading of 0.5–0.7 is reached.[c] For 1 litre culture, use two 5 litre flasks for proper aeration during growth.

4. Chill the culture on ice[d] for 1 h and then centrifuge at 4000 g for 20 min at 4 °C.

5. Pour the supernatant off[e] and resuspend the cells in 1 litre of ice-cold sterilized Milli-Q™ water taking care not to lyse them. Centrifuge as in step 4.

6. Pour the supernatant off and resuspend the cells in 500 ml of ice-cold sterilized Milli-Q™ water. Centrifuge as in step 4.

7. Resuspend the cells in 20 ml of ice-cold 10% (v/v) glycerol. Centrifuge as in step 4.

8. Finally, gently pour the supernatant off and resuspend the cells in a final volume of 2–3 ml of ice-cold 10% (v/v) glycerol.[f]

9. Dispense the electrocompetent cells in 50–100 μl aliquots in 1.5 ml polypropylene tubes and proceed as quickly as possible to the transformation protocol[g] (see *Protocol 15*).

10. Determine the efficiency of transformation of each competent cell preparation by transforming the cells with 1 ng of uncut (supercoiled) vector DNA[h] (pUC19 for example). This is carried out in parallel with transformation of competent cells with ligated DNA (see *Protocol 13*) for the library construction (proceed to *Protocol 15*).

[a] An aseptic technique should be used for all steps of the procedure.

[b] *E. coli* grow much more slowly on minimal medium than on rich LB medium. Therefore we recommend streaking the plate two days before starting the preparation of TG1 strain competent cells.

[c] It is critical that the optical density is not more than 0.7. The best results are obtained with cells that are harvested at early to mid-log phase.

Protocol 14 continued

[d] Keep the cells in an ice water-bath throughout their preparation.

[e] Remove as much of the supernatant as possible. It is better to sacrifice the yield by pouring off a few cells than to leave any supernatant behind.

[f] The cell concentration should be about 1–3×10^{10} cells/ml.

[g] For maximum transformation efficiency cells must be used for transformation within 2–3 h. Alternatively, aliquots may be frozen on dry ice and stored at $-70\,°C$.

[h] Cells should transform with an efficiency of at least 2×10^9 c.f.u./μg to be used for library construction.

16 Electroporation

To make the library, ligation mixtures (see *Protocol 13*) are electroporated in competent *E. coli* TG1 strain (see *Protocol 14*). Aliquots from each electroporation are titrated to estimate the size of the library.

Protocol 15

Electrotransformation and plating

Equipment and reagents

- Bio-Rad *E. coli* pulser electroporator (Bio-Rad, 165-2104)
- Electroporation cuvettes, 0.1 cm electrode gap (Bio-Rad, 165-2089)
- Bioassay dish $243 \times 243 \times 18$ mm (Nunc™, 240835)
- 3.0 cm blade sterile cell scraper (Falcon, 3087)
- 14 ml polypropylene tubes (see *Protocol 11*)
- Electrocompetent *E. coli* TG1 strain cells (see *Protocol 14*)
- Ligation mix (see *Protocol 13*)

- SOC medium (Life Technologies™, 15544-042)
- 20% (w/v) D(+) glucose (anhydrous); sterilize by filtration through a 0.2 μm filter (do not autoclave)
- 2 × TY agar: 2 × TY medium (see *Protocol 11*) plus 1.5% (w/v) Bacto agar; autoclave to sterilize
- 2 × TY-AMP-GLU: 2 × TY medium (see *Protocol 11*) containing 100 μg/ml ampicillin and 2% (w/v) glucose

Method

1. Add 2 μl of ligation mix to 50–100 μl of freshly prepared electrocompetent *E. coli* TG1 strain cells in ice-cold 1.5 ml microcentrifuge tubes. Mix well and let sit on ice for 1 min.

2. Set the Gene Pulser apparatus at 25 μF and 1.8 kV. Set the Pulse Controller to 200 ohms.

3. Transfer the mixture of cells and DNA to a cold 0.1 cm electroporation cuvette, and

Protocol 15 continued

shake the suspension to the bottom. Dry the cuvette with tissue and place it in a chilled electroporation chamber.

4. Pulse once at the above settings.[a]

5. Remove the cuvette from the chamber and immediately add 1 ml of SOC medium to the cuvette and quickly resuspend the cells with a Pasteur pipette.

6. Transfer the cell suspension to 17 × 100 mm polypropylene tubes and incubate at 37°C with shaking (225 r.p.m.) for 1 h to allow expression of the antibiotic resistance gene.

7. Plate every 1 ml transformation onto 2 × TY-AMP-GLU agar in 243 × 243 mm dishes.[b] Incubate for 20–24 h at 30 °C.[c] Also plate untransformed, competent TG1 cells as a negative control.

8. Determine the size of the library by plating tenfold dilutions of transformed TG1 cells from 10^{-3} to 10^{-10} onto 2 × TY-AMP-GLU agar plates.[d] Assess background vector religation by ligating and transforming triple digested pHEN-1 vector in the absence of antibody scFv insert.

9. Scrape the colonies off the plates with a sterile cell scraper by flooding the plates with 5 ml of 2 × TY-AMP-GLU medium. Transfer the cells to a sterile polypropylene tube and disperse the clumps by vortexing. Add sterile 50% (v/v) glycerol to 15% (v/v) final, mix, freeze down 1 ml aliquots on dry ice, and store at −70 °C.[e]

[a] This should produce a pulse with a time constant of 4.5–5 msec.

[b] Ideally, you want the transformed TG1 cells to be subconfluent rather than forming a uniform lawn in order to avoid competitive selection between the clones.

[c] E. coli TG1 strain cells grow slowly at 30 °C and require a relatively long period of incubation before growth is apparent.

[d] Titrate aliquots from each electroporation to estimate library size.

[e] At this point it might be advantageous to use these fresh cells for phagemid rescue (see Chapter 4, Section 1, production and PEG precipitation of phage) so that you have your library in two forms; frozen cells and infectious phage. However, the scraped cells must be diluted to an OD_{600} of 0.5 per ml and the volumes and amount of helper phage scaled up proportionally. Use sufficient numbers of bacteria to ensure that the representation is not compromised. Thus for a library of 10^7 clones, use at least 10^8 viable (i.e. colony forming) bacteria.

17 Analysis of recombinant clones from the library

In order to estimate the diversity of the original (i.e. before selection) library, colonies from the ligations are first screened for inserts by PCR followed by digestion with the frequent-cutting enzyme BstNI. Analysis of 48 clones should indicate that at least 90% of the clones have inserts. The BstNI pattern should be very complex indicating that the library is likely to be very diverse. Note though that this is only a quick test and that to really assess the clone diversity, DNA sequencing is a must.

Protocol 16

PCR screening and fingerprinting

Equipment and reagents

- Nunc inoculating needles (Life Technologies, 254399)
- AmpliTaq® DNA polymerase (see *Protocol 7*)
- GeneAmp® 10 × PCR buffer (see *Protocol 7*)
- GeneAmp® dNTPs (see *Protocol 5*)
- *Bst*NI restriction enzyme (New England Biolabs, 168)

- 1 × NEBuffer 2 (see *Protocol 10*)
- Agarose (UltraPure) (see *Protocol 2*)
- NuSeive® 3:1 agarose (25 g, FMC BioProducts, 50091)
- 5 × TBE (see *Protocol 6*)
- 50 × TAE (see *Protocol 2*)

Method

1. Prepare a bulk solution of the PCR reaction mix (sufficient for 48 clones) by combining the following ingredients:

   - H_2O 1191 μl
   - 10 × PCR buffer   150 μl
   - 10 mM dNTP mix[a] (250 μM each dNTP final concentration)   120 μl
   - RSP[b] (50 pmol/μl)   15 μl
   - Gene 3–28[c] (50 pmol/μl)   15 μl
   - AmpliTaq DNA polymerase (5 U/μl)   9 μl

2. Dispense 30 μl aliquots into a 96-well microplate.

3. Using an inoculating needle, gently 'stab' an individual colony from transformation plates taking care not to take too much colony. Twist the needle about three times in the PCR mix in the appropriate well of the microplate.

4. Overlay the PCR mixes in the microplate with one drop of mineral oil per well. Place the microplate into the PCR machine and hold at 94 °C for 10 min then cycle at 94 °C for 1 min, 55 °C for 1 min, and 72 °C for 2 min over 30 cycles. Use a ramp time of 1.5 min between the annealing (55 °C) and extension (72 °C) temperatures.

5. Run a 10 μl aliquot from each PCR mix on a 1% (w/v) agarose/1 × TAE buffer gel containing 0.5 μg/ml ethidium bromide, and determine the size of the PCR product bands on the gel.[d]

6. Prepare a restriction enzyme mix for *Bst*NI fingerprinting by combining the following ingredients:

   - H_2O 870 μl
   - NEBuffer 2   100 μl
   - 100 × BSA acetylated   10 μl
   - *Bst*NI (10 U/μl)   20 μl

7. Add 20 μl of the above mix to each well containing a PCR reaction (20 μl left after PCR screening for inserts). Incubate at 60 °C for 4 h on a thermal cycler.

8. Load 20 μl of each sample on a 4% (w/v) NuSieve 3:1 agarose /1 × TBE buffer gel containing 0.5 μg/ml ethidium bromide. Analyse the *Bst*NI restriction digest pattern obtained between the different clones.

[a] See *Protocol 5*, footnote *a*.

[b] RSP anneals 5′ of the antibody scFv insert, just before the *Hin*dIII site in pHEN-1 vector and as the following sequence: 5′ CAGGAAACAGCTATGAC 3′.

[c] Gene 3–28 anneals 3′ of the antibody scFv insert, at the beginning of the gene coding for the g3p coat protein, and has the following sequence: 5′ GTATGAGGTTTTGCTAAACAAC 3′.

[d] A PCR band of around 950 bp will be seen using the RSP and gene 3–28 primers on the pHEN-1 vector containing an insert (antibody scFv). A negative result (i.e. no insert) will produce a band of 212 bp.

References

1. Huston, J. S., Levinson, D., Mudgett, H. M., Tai, M. S., Novotny, J., Margolies, M. N., *et al.* (1988). *Proc. Natl. Acad. Sci. USA*, **85**, 5879.

2. Chester, K. A., Begent, R. H., Robson, L., Keep, P., Pedlley, R. B., Boden, J. A., *et al.* (1994). *Lancet*, **343**, 455.

3. McCafferty, J., Griffiths, A. D., Winter, G., and Chiswell, D. J. (1990). *Nature*, **348**, 552.

4. Schier, R., McCall, A., Adams, G. P., Marshall, K. W., Meritt, H., Yim, M., *et al.* (1996). *J. Mol. Biol.*, **263**, 551.

5. Chomczynski, P. and Sacchi, N. (1987). *Anal. Biochem.*, **162**, 156.

6. Sambrook, I., Fritsch, E. F., and Maniatis, T. (ed.) (1989). *Molecular cloning: a laboratory manual* (2nd edn). Cold Spring Harbor Laboratory Press, NY.

7. Amersdorfer, P., Wong, C., Chen, S., Smith, T., Deshpande, S., Sheridan, R., *et al.* (1997). *Infect. Immnun.*, **65**, 3743.

8. Ward, S. E., Gussow, D., Griffiths, A. D., Jones, P. T., and Winter, G. (1989). *Nature*, **341**, 544.

9. Favre, D. (1992). *Biotechniques*, **13**, 22.

10. Clackson, T., Hoogenboom, H. R., Griffiths, A. D., and Winter, G. (1990). *Nature*, **352**, 624.

11. Ørum, H., Andersen, P. S., Øster, A., Johansen, L. K., Riise, E., Bjørnvad, M., *et al.* (1993). *Nucleic Acids Res.*, **21**, 4491.

12. Hoogenboom, H. R., Griffiths, A. D., Johnson, K. S., Chiswell, D. J., Hudson, P., and Winter, G. P. (1991). *Nucleic Acids Res.*, **19**, 4133.

13. Lei, S. P., Lin, H. C., Wang, S. S., Callaway, J., and Wilcox, G. (1987). *J. Bacteriol.*, **169**, 4379.

14. Munro, S. and Pelham, H. R. (1986). *Cell*, **46**, 291.

15. Viera, J. and Messing, J. (1987). In *Methods in enzymology* (ed. R. Wu and L. Grossman), Vol. 153, p. 3. Academic Press, London.

Appendix

1 The design of the humanized antibody

Humanization by CDR grafting (also called reshaping) is now a well-established procedure for reducing the immunogenicity of monoclonal antibodies (mAbs) from xenogeneic sources (commonly rodent) and for improving their activation of the human immune system; in fact there are many humanized mAbs in clinical trials and a few have been given approval to be used as drugs. Although the mechanics of producing the engineered mAb using the techniques of molecular biology are relatively straightforward, simple grafting of the donor rodent complementarity determining regions (CDRs) into human acceptor frameworks does not always reconstitute the binding affinity and specificity of the original mAb. The design of the engineered mAb is now the critical step in reproducing the function of the original molecule. This design includes various choices: the extents of the CDRs, the human frameworks to use, and the substitution of residues from the rodent mAb into the human framework regions (backmutations). The positions of these backmutations have been identified principally by sequence/structure analysis or by analysis of an homology model of the variable regions 3D structure. Recently, phage libraries have been used to vary the amino acids at chosen positions (1, 2). Similarly, many approaches have been used to choose the most appropriate human frameworks in which to graft the rodent CDRs. Early experiments used a limited subset of well-characterized human mAbs (often where the structure was available), irrespective of the sequence identity to the rodent mAb (the so-called 'fixed frameworks' approach). Some groups use variable regions with high amino acid sequence identity to the rodent variable regions ('homology matching' or 'best-fit'); others use consensus or germline sequences, while still others select fragments of the framework sequences within each light or heavy chain variable region from several different human mAbs. There are also approaches to humanization developed which replace the surface rodent residues with the most common residues found in human mAbs ('resurfacing' or 'veneering') and those which use differing definitions of the extents of the CDRs.

Not surprisingly, some rodent mAbs have proved difficult to humanize using standard protocols (3). The design and engineering of humanized mAbs are still very much areas of research, as much for the light they shed on protein structure and function as well as for the potential therapeutic and diagnostic benefits. For this reason, an iterative protocol called the 'Humanization design cycle' (see *Protocol 1*) is described in the accompanying sections. These sections additionally supply data and raise issues of concern to the prospective antibody designers to help them through the bewildering variety of choices. The data is also available on the WWW under either of the two following 'Humanization bY Design' urls:

http://mathbio.nimr.mrc.ac.uk/jsaldan or

http://www.cryst.bbk.ac.uk/~ubcg07s

with hypertext links to the various resources for database search, sequence analysis, and structural modelling. Search forms for Sequence and Text input are also available under these urls to aid navigation through the list of previously humanized antibodies taken from the literature.

Protocol 1

The humanization design cycle

Equipment

- Any computer of reasonable power (for instance Pentium processor, 32 Mb RAM) which is connected to the WWW through a standard browser (e.g. Netscape) is needed to accomplish the design protocol

Method

1. Examine the source donor amino acid sequences for each chain (see Section 1.1.1).

2. Produce a structural model of the Fv regions (see Section 1.1.2).

3. Consider the human acceptor framework sequences for each chain (see Section 1.1.3).

4. Identify putative backmutations in the chosen frameworks (see Section 1.1.4).

5. Reconsider framework choice and design humanized antibody sequence.

6. Construct humanized and chimeric antibody sequences.

7. Test constructs.

8. Success? If 'No' then return to step 3.

1.1 Issues to consider

1.1.1 The source donor amino acid sequences

In the source donor sequence for each chain (heavy and light) it is important to consider:

(a) The complementarity determining regions. These contain the residues most likely to bind antigen and thus they must be retained in the humanized antibody (unless the precise residues involved in binding are known which is unlikely). CDRs are defined by sequence according to the Kabat scheme or by structure according to the Chothia scheme (see *Table 2*). The advantage of using the latter definitions is that the CDRs are shorter, therefore the humanized antibody will have fewer xenogeneic fragments. However, the experience of other groups (4) has shown that using the Kabat definition generally requires fewer iterations in the humanization design cycle. A third scheme for defining CDR extents (see *Table 2*) called the 'contact definition' from Andrew Martin is based on an analysis of antibody/antigen interactions. It is likely to be useful in humanization but is as yet untested.

(b) Canonical residues. These are key residues which determine to a large extent the structural conformation of the CDR loop. They were originally defined by Chothia, but have now been revised by Andrew Martin. His WWW form at url: `http://www.biochem.ucl.ac.uk/~martin/abs/chothia.html` allows input of the variable region sequences and automatic identification of the canonical structure class and important residues. Canonical residues are almost always retained in the humanized antibody if they are different to those in the human frameworks. Numerous examples of the importance of canonical residues can be found in the literature which can be searched for in the 'Humanization bY Design' web sites.

(c) Interchain packing residues. Defined by Novotny and Haber, and Chothia at the interface between the variable heavy and variable light chain domains (see *Table 3*). Generally, unusual packing residues should be retained in the humanized antibody if they differ from those in the human frameworks. Their importance is illustrated in the humanization of mAb 1B4 (5).

(d) Unusual framework residues. These can be located using Andrew Martin's WWW form at url: `http://www.biochem.ucl.ac.uk/~martin/abs/seqtest.html`. For mouse antibodies, it is better to determine the subgroup using the form on the Kabat Database WWW site at url: `http://immuno.bme.nwu.edu/subgroup.html` and identify the residue positions which differ from the consensus. These donor-specific differences may point to somatic mutations which enhance activity. Unusual residues close to the binding site may possibly contact the antigen. However, if they are not important for binding, then it is desirable to remove them because they may create immunogenic neoepitopes in the humanized antibody. Note that sometimes unusual residues in the donor sequence are actually common residues in the human acceptor sequence. Unusual residues in the acceptor frameworks are not desirable because of the possibility of immunogenicity.

(e) Potential *N*- or *O*-glycosylation sites. It is well known that potential *N*-glycosylation sites are specific to a consensus pattern N-X-S/T. These sites can be located in the donor sequence using the Prosite WWW form at url: `http://www.expasy.ch/sprot/scnpsite.html`. Removal of potential *N*-glycosylation sites has not destroyed the binding avidity of a humanized antibody thus far, and in one case actually increased it (6). *O*-glycosylation sites are usually found in helical segments which means that they are not common in the beta-sheet structure of antibodies. They have no consensus pattern, but differences in *O*-linked glycosylation between chimeric and humanized forms of BrE-3 was hypothesized to cause the increased binding of the humanized mAb (7).

Table 2 CDR loop definitions (numbering according to Kabat *et al.*, ref. 15)

Loop	Kabat	Chothia	Contact
CDR-L1	L24–L34	L26–L32	L30–L36
CDR-L2	L50–L56	L50–L52	L46–L55
CDR-L3	L89–L97	L91–L96	L89–L96
CDR-H1	H31–H35	H26–H32	H30–H35
CDR-H2	H50–H65	H52A–H55	H47–H58
CDR-H3	H95–H102	H95–H101	H93–H101

Table 3 Residues at the VL/VH domain interface[a]

	Kabat number[b]	Mouse[c]	Human[c]
VK	34	H678 N420 A408 Y147 E114	A531 N147 D66
	36	Y1653 F198 L96	Y748 F80
	38	Q1865 H47	Q799 H22
	44(+)	P1767 V132 I40	P839 L5
	46	L1381 R374 P97	L760 V37
	87	Y1457 F448	Y795 F41
	89	Q1170 L206 F144	Q687 M107
	91	W376 S374 G356 Y295 H182	Y404 R115 S105 A84
	96(+)	L537 Y380 W285	L134 Y215 F78 W73 I71
	98(+)	F1724	F654
VH	35	H1001 N636 S402	E184 S527 H340 G167 A143
	37	V2336 I200	V1037 I477 L27
	39	Q2518 K67	Q1539 R16
	45(+)	L2636 P16	L1531 P24
	47	W2518 L64 Y50	W1534 Y21
	91	Y2149 F479	Y1429 F116
	93	A2202 T222 V102	A1346 T90 V71
	95	Y399 G375 S340 D340 R226	D268 G266 R109 E100
	100k(+)	F1285 M450	F540 M109 L33
	103(+)	W1469	W323

[a] The positions of interdomain residues were as defined by Chothia *et al.* (16).

[b] Numbering is according to Kabat *et al.* (15). Residues underlined are in the framework, other residues are in the CDRs. (+) residues are the six that form the core of the VL/VH interface according to Chothia *et al.* (16).

[c] The number following the one-letter amino acid code is the frequency taken from Kabat (November 1997 dataset; see url in the text). The number of sequences examined is identical for each domain and equal to 1997 and 852 for mouse and human sequences, respectively; except for residue 103 in the heavy chain where numbers are 1651 and 369 for mouse and human sequences, respectively.

1.1.2 Modelling the Fv regions

A carefully built model of the Fv regions of the source antibody and in some cases of the designed humanized antibody, is a common denominator in many of the successful humanization experiments. It is possible to build a model completely automatically using commercial packages or academic servers such as Swiss-Model at url: `http://www.expasy.ch/swissmod/SWISS-MODEL.html`. However, the danger of allowing a computer to make all the decisions is highlighted in the humanization of antibody AT13/5 (8) where the interaction between residues at positions 29 and 78 (Kabat numbering) in the heavy chain was not modelled correctly. The experience of an expert in protein sequence analysis and modelling is always worthwhile.

1.1.3 Choice of human acceptor frameworks

Despite this being the most critical part of the humanization design cycle, there are no hard-and-fast rules for choosing the acceptor frameworks into which to graft the donor CDRs. This is because the benefit of the various choices (in terms of immunogenicity in the patient) has not clearly been proved in the clinic. Therefore, there are only a set of approaches which have arisen from the collective experience of previous humanization experiments.

(a) Fixed frameworks or homology matched. Some groups prefer to use fixed frameworks (usually antibody NEW for the heavy chain and REI for the light chain since their structures are solved) for all their humanized antibodies. Other groups try to use the frameworks that are most similar to their donor sequence (homology matching or best-fit). A subtle comparison of the two methods in terms of the ease of producing a functional humanized antibody appears to favour homology matching (9). Note that it is entirely possible to 'mix and match' framework regions (i.e. choose frameworks 1–4 from different human antibodies according to sequence similarity) although this approach hasn't been tested exhaustively.

(b) Light/heavy chain frameworks from the same/different clone. In general, light and heavy chains from the same antibody are more likely to form a functional binding site than light and heavy chains from different antibodies. However, since the interface between variable domains is so well conserved (see *Table 3*), this is not usually a problem. A comparison of the two approaches once again appears to favour the latter (10).

(c) Human subgroup consensus or individual framework sequences. Being limited to frameworks from particular human antibodies runs the risk of the somatic mutations creating immunogenic epitopes, even though the frameworks are human. An alternative approach is to use the frameworks from human consensus sequences where idiosyncratic somatic mutations will have been 'evened out'. The two approaches have been compared; in one case showing no difference in binding avidities (11) and in the other case showing better binding with individual frameworks (12). Consensus sequences for human and mouse subgroups can be found under urls:

http://mathbio.nimr.mrc.ac.uk/jsaldan/cons/cons.html; or
http://www.cryst.bbk.ac.uk/~ubcg07s/cons/cons.html.

(d) Frameworks from germline sequences. Consensus sequences are artificial and although having no idiosyncratic residues, may create unnatural sequence motifs which are immunogenic. An alternative is to use human germline sequences which have been compiled in the database VBase at url: http://www.mrc-cpe.cam.ac.uk/imt-doc/public/INTRO.html.

Having decided on an approach to take in order to choose the human frameworks; which particular human antibody, consensus, or germline sequence should be used? This is simple in the fixed framework approach since the choice is always NEW for the heavy chain and REI for the light (although since modifications to both frameworks are apparent from the literature (13) even there a choice needs to be made!). The choice is usually made by performing a search for the most homologous sequence over the appropriate database. There are many database search programs available, the two most common being *BLAST* and *FASTA* which can be found at the following two urls respectively: http://www.ncbi.nlm.nih.gov/BLAST/ and http://www2.ebi.ac.uk/fasta3/. *BLAST* is often preferable since it allows searching over the Kabat database. In order to search the consensus or germline sequences, they need to be downloaded onto the local computer. Choice of the particular human frameworks for the light and heavy chain variable regions should be made by trying to match the length of the CDRs, the canonical residues and the interface packing residues (as described in Section 1.1.1), as well as trying to find the highest percentage identity between the donor and acceptor sequences. Unusual residues in the acceptor frameworks are not usually tolerated since they may lead to immunogenic epitopes in the resultant humanized antibody. Try to find human frameworks which are similar (in terms of percentage identity) to the source sequences and also require the least number of backmutations (see next section).

1.1.4 Identify putative backmutations

This is the most difficult and unpredictable procedure in the humanization of monoclonal antibodies. It is also the area which throws much light on protein structure and function. A solid body of data for helping to identify strategic alterations (using the sequence and text search forms) is available on the Humanization bY Design web sites under mirrored urls:
http://mathbio.nimr.mrc.ac.uk/jsaldan/selhum.html;
http://www.cryst.bbk.ac.uk/~ubcg07s/selhum.html.

(a) Analysis of donor and acceptor sequences. Putative point backmutations from the human framework residue back to the original source residue will already have been identified from the analysis of the donor and acceptor sequences for canonical residues, interchain packing residues, unusual residues, and glycosylation sites. Experience has shown that it is especially important to retain the source's canonical and interchain packing residues (though unfortunately not in all cases).

(b) Vernier zone and CDR-H3. Consider residues in the Vernier zone (see *Table 4*) (14) which form a 'platform' on which the CDRs rest and therefore potentially affect their conformation. Due to its extreme variability, special attention should be paid to CDR-H3, analysing the structural model for residues which may potentially affect its conformation.

(c) Proximity to binding site. Using the structural model, analyse residues within 5–6 Å of any CDR residues. These residues are likely to bind to antigen, especially if they are unusual as classified by the analysis of the source sequence.

(d) Glycosylation sites. Consider removal of any potential glycosylation sites, especially if they are on the surface of the structural model, since experience has shown that this can enhance the avidity of the humanized antibody.

Table 4 Residues in the VERNIER zone

Heavy chain	Light chain
2	2
27–30	4
47–49	35–36
67	46–49
69	64
71	66
73	68–69
78	71
93–94	98
103	–

Having decided on the residues to backmutate, return to the question of human acceptor frameworks. It is not unlikely that an overlooked human framework may actually contain the backmutations which are to be retained. If this is the case, then there is no need to introduce residues from the source sequence, thus making the humanized antibody more 'human'.

References

1. Rosok, M. J., Yelton, D. E., Harris, L. J., Bajorath, J., Hellstrom, K.-E., Hellstrom, I., *et al.* (1996). *J. Biol. Chem.*, **271**, 22611.
2. Baca, M., Presta, L. G., O'Connor, S. J., and Wells, J. A. (1997). *J. Biol. Chem.*, **272**, 10678.
3. Pichla, S. L., Murali, R., and Burnett, R. M. (1997). *J. Struct. Biol.*, **119**, 6.
4. Presta, L. G., Lahr, S. J., Shields, R. L., Porter, J. P., Gorman, C. M., Fendly, B. M., *et al.* (1993). *J. Immunol.*, **151**, 2623.
5. Singer, I. I., Kawka, D. W., DeMartino, J. A., Daugherty, B. L., Elliston, K. O., Alves, K., *et al.* (1993). *J. Immunol.*, **150**, 2844.
6. Co, M. S., Scheinberg, D. A., Avdalovic, N. M., McGraw, K., Vasquez, M., Caron, P. C., *et al.* (1993). *Mol. Immunol.*, **30**, 1361.

7. Couto, J. R., Blank, E. W., Peterson, J. A., Kiwan, R., Padlan, E. A., and Ceriani, R. L. (1994). In *Antigen and antibody molecular engineering in breast cancer diagnosis and treatment* (ed. R. L. Ceriani), p. 55. Plenum Press, New York.

8. Ellis, J. H., Barber, K. A., Tutt, A., Hale, C., Lewis, A. P., Glennie, M. J., *et al.* (1995). *J. Immunol.*, **155**, 925.

9. Graziano, R., Tempest, P., White, P., Keler, T., Deo, Y., Ghebremariam, H., *et al.* (1995). *J. Immunol.*, **155**, 4996.

10. Roguska, M. A., Pedersen, J. T., Henry, A. H., Searle, S. M. J., Roja, D. M., Avery, B., *et al.* (1996). *Protein Eng.*, **9**, 895.

11. Kolbinger, F., Saldanha, J., Hardman, N., and Bendig, M. (1993). *Protein Eng.*, **6**, 971.

12. Sato, K., Tsuchiya, M., Saldanha, J., Koishihara, Y., Ohsugi, Y., Kishimoto, T., *et al.* (1994). *Mol. Immunol.*, **31**, 371.

13. Riechmann, L., Clark, M., Waldmann, H., and Winter, G. (1988). *Nature*, **332**, 323.

14. Foote, J. and Winter, G. (1992). *J. Mol. Biol.*, **224**, 487.

15. Kabat, E. A., Wu, T. T., Perry, H. M., Gottesman, K. S., and Foeller, C. (1991). In *Sequences of proteins of immunological interest* (5th edn). NIH publication No. 91-3242. US Department of Health and Human Services, Public Health Service, National Institutes of Health, Bethesda, MD.

16. Chothia, C., Novotny, J., Bruccoleri, R., and Karplus, M. (1985). *J. Mol. Biol.*, **186**, 651.

Chapter 3

Selection of antibodies from phage libraries of immunoglobulin genes

Jane K. Osbourn

Cambridge Antibody Technology, The Science Park, Melbourn, Cambridgeshire, SG8 6JJ U.K.

1 Introduction

This chapter explains the various approaches it is possible to take in the selection of phage antibodies with specific binding characteristics from large phage display libraries. The particular selection approach taken will be determined by the form and availability of the starting antigen and the desired properties of the selected antibody. One of the main benefits of the phage display approach to antibody generation is the ability to tailor the selection regime to generate antibodies with particular characteristics such as: high affinity, neutralization potency, ability to recognize specific epitopes, or to recognize specific cell types. A variety of screening regimes can also be employed to identify antibodies that can be utilized for particular applications such as: Western blot reagents, immunocytochemistry reagents, or species cross-reactivity. The phage system provides a route to the generation of antibodies to antigens that are normally inaccessible using conventional immunization techniques, for example mAbs to toxic moieties, carbohydrate-containing molecules, and anti-self antigens can easily be selected using the appropriate conditions. Methods for the generation of phage antibody libraries from immunized rodents are described elsewhere (Chapter 2) and these techniques are equally applicable to the generation of human phage antibody libraries if appropriate primers are used (1). The selection methods described here can be employed for immunoglobulin genes displayed in scFv, Fab, or diabody formats. The methods described provide a necessarily brief outline of the possible approaches and it should be stressed that there are no hard and fast rules for selection protocols; investigators should use as much imagination as possible in designing selection procedures which will best suit the final application of the antibody. Possible selection methods such as *in vivo* selection (2), selective infective phage selections (3, 4), and various deselection methods (e.g. 5) have not been discussed in detail here and reference should be made to the appropriate literature for further ideas on these approaches.

2 Preparation and storage of phage library stocks

2.1 Phagemid libraries

Early phage display antibody libraries were generated using phage vectors which carried all the genetic information required for the phage life cycle (6, 7) but now phagemids have become a more popular type of vector for display. Phagemids are small plasmid vectors that contain both a plasmid origin of replication and a filamentous phage origin of replication along with the phage gene III with appropriate cloning sites allowing fusion of the scFv (8–10). Phagemids have higher transformation efficiencies than those achieved by phage vectors enabling very large antibody repertoires to be generated. In many phagemids expression of the antibody–gene III protein fusion is driven by the *lacZ* promoter. This necessitates removal of glucose, the catabolite repressor of the *lacZ* promoter, in order to generate sufficient fusion product to produce phage particles expressing at least one copy of the antibody gene. The use of phagemid vectors also requires the addition of helper phage (such as M13KO7) to supply the various structural proteins required to package the phagemid DNA into phage particles. Since the helper phage genome encodes wild-type gene III protein rescued phage will consist of infectious phage particles that will display a mixture of wild-type gene III protein and the gene III–antibody fusion protein.

Protocol 1

Production and polyethylene glycol (PEG) precipitation of phage

Equipment and reagents

- Shaking incubator at 30 °C
- Shaking incubator at 37 °C
- M13KO7 helper phage (Pharmacia)
- 2TY media: 16 g tryptone, 10 g yeast extract, 5 g NaCl per litre; autoclave
- 2TYG: 2TY supplemented with 2% (w/v) filter sterilized glucose
- Ampicillin: 100 mg/ml in filter sterilized water
- Kanamycin: 50 mg/ml in filter sterilized water
- PEG/NaCl: 20% polyethylene glycol 8000, 2.5 M NaCl; autoclave
- TE: 10 mM Tris pH 8.0, 1 mM EDTA
- Glycerol stock of phage library (see Chapter 2 for library generation protocols)

Method

1. Add 100 μl of glycerol library stock to each of two 1 litre flasks containing 500 ml of 2TYG with 100 μg/ml ampicillin. Grow at 37 °C with rapid shaking (300 r.p.m.) for 1–2 h or until A_{600} is between 0.5–1.0.

2. Add M13KO7 to a final concentration of 5×10^8 plaque forming units (p.f.u.)/ml. This should give a multiplicity of infection (m.o.i.) of approximately 10.

3. Incubate cells at 37 °C for 30 min without shaking and then for 30 min with moderate shaking (200 r.p.m.) to allow infection.

4. Transfer cultures to 250 ml centrifuge pots and centrifuge to pellet the cells (e.g. Sorvall RC-5B rotor, 5000 r.p.m., 10 min). Decant off the supernatant, resuspend the bacterial pellet in 250 ml of 2TY (no glucose) with 100 µg/ml ampicillin and 50 µg/ml kanamycin, and transfer to fresh flasks. Grow overnight with rapid shaking (300 r.p.m.) at 30 °C.

5. Since the phage titre after overnight growth is usually around 1×10^{12}/ml it is generally sufficient to precipitate phage from only 100 ml of the overnight culture. Pellet the cells from 100 ml culture (50 ml from each flask) (e.g. Sorvall RC-5B rotor, 8000 r.p.m., 15 min).

6. Transfer the supernatant to a sterile flask and add 30 ml of PEG/NaCl per 100 ml of supernatant. Swirl to mix and leave on ice for 1 h. Transfer to ten 14 ml centrifuge tubes and pellet the phage particles (e.g. Sorvall SM 24 rotor, 8000 r.p.m., 15 min, 4 °C).

7. Decant as much supernatant as possible, briefly re-spin the tubes, and remove residual supernatant with a pipette. Resuspend each phage pellet in 1 ml of TE and pool the phage to give a 10 ml phage stock.

8. Centrifuge the phage to remove any remaining bacterial debris (e.g. Sorvall SM 24 rotor, 8000 r.p.m., 5 min) and transfer the supernatant to a fresh tube. This is your purified and concentrated phage stock which can be used in a first round of selection. It is stable for approx. one week if kept at 4 °C. The phage titre should be approx. 1×10^{13}/ml.

2.2 Phage libraries

If the library you are working with is a phage rather than phagemid library there is no need to add helper phage to rescue the library. A glycerol stock of the phage library can simply be added to 2TYG containing the appropriate antibiotic and grown overnight at 30 °C as in *Protocol 1*. *Protocol 1* can then be followed exactly from step 5.

3 Maintenance of bacterial stocks and titration of phage preparations

Filamentous bacteriophage is the most common type of phage used for antibody display. Filamentous phage (such as fd or M13) are non-lytic and infect strains of *E. coli* containing the F conjugative plasmid. Phage particles attach to the tip of the F pilus, which is encoded by the genes of the F plasmid, and the phage DNA genome is translocated into the bacterial cytoplasm. The *E. coli* strain TG1 is routinely used as a host for filamentous phage. TG1 is also a supE bearing strain which reads the amber stop codon (TAG) as glutamic acid. This generates antibody gene III protein fusion products in phagemid vectors which carry amber codons between the antibody and gene III protein coding regions (e.g. pCantab 6 described in ref. 11). For the F conjugative plasmid to be maintained by the bacteria it is important to grow the bacterial stock on minimal plates supplemented with thiamine.

The phage library preparation protocols generally give a phage titre in the region of 1×10^{13} phage/ml for the PEG precipitation and up to 1×10^{14} phage/ml after caesium banding. It is important to verify these titres to ensure the phage preparation procedure has been successful, and this can easily be achieved by preparing six or seven hundred-fold serial dilutions of the phage stocks which can be used to infect E. coli, and the number of resulting colonies counted.

Protocol 2

Maintenance of bacterial stocks and titration of phage

Reagents

- Glycerol stock of E. coli TG1 stored at $-70\,°C$
- Minimal medium plates prepared as follows. Prepare stock solutions and sterilize by filtration through a 0.22 μm filter: 1 M $MgCl_2.6H_2O$, 1 M $CaCl_2.2H_2O$, 1 M thiamine hydrochloride. Make up 6 g Na_2HPO_4 (dibasic), 3 g $KH_2.PO_4$

(monobasic), and 1 g NH_4Cl with 500 ml of distilled water. In a separate bottle place 20 g Bacto agar in 500 ml distilled water. Sterilize both bottles by autoclaving, when cool combine the contents, and add 1 ml of 1 M $MgCl_2.6H_2O$ and 1 ml of 1 M $CaCl_2.2H_2O$. Pour plates and allow them to set.

Method

1. Immediately before using the minimal plates spread 50 μl of 1 M thiamine hydrochloride onto the plate with a spreader and allow to dry.

2. Streak the plate with a loop inoculated with TG1.

3. Grow at 37 °C for 36–48 h. Once grown seal the plate with Parafilm and store at 4 °C for up to four weeks.

4. To titre phage stocks take a single colony of TG1 from the freshly prepared minimal plate, inoculate 50 ml of 2TYG (Protocol 1), and grow at 37 °C with rapid shaking (300 r.p.m.) for approx. 4 h, or until the bacteria reach mid-log phase (A_{600} is between 0.5–1.0).

5. Make serial dilutions of phage stocks in TE and add around 100 μl to 5 ml of TG1 in a 50 ml Falcon tube.

6. Incubate the cells at 37 °C for 30 min without shaking and then at 30 °C with moderate shaking (200 r.p.m.) to allow infection.

7. Plate out a fixed volume of cells on 2TYG plates containing 100 μg/ml ampicillin (or appropriate antibiotic) (2TYAG plates) and incubate at 30 °C overnight.

8. Count the resulting bacterial colonies and from the known volumes and phage dilutions used calculate the phage titre.

4 Selection of phage libraries on purified, immobilized antigen

4.1 Immobilization of antigen on immunotubes

Once the phage library stocks have been prepared and titred the next step is to select the repertoire against an appropriate antigen. If purified antigen is available it can be immobilized on a solid support, such as an immunosorb tube, and the phage antibody library allowed to bind to the antigen in a process called 'panning'. Non-binding phage can be washed away and the remaining phage containing the antigen-specific population eluted from the tube and recovered ready for a second round of amplification and selection. It is possible to immobilize the antigen on various types of solid supports. If antigen is in good supply the method of preference is an immunotube, as described, but it is also possible to coat antigen directly onto the well of a microtitre plate to allow coating in a smaller volume. Antigen can also be immobilized on resin and packed into a column that the phage library can then be passed over to achieve selection. When carrying out selections on peptides or haptens successful selection is hard to achieve by direct coating of such molecules onto immunosorb tubes or other solid matrices. It is highly recommended to couple peptides and haptens to carrier proteins such as bovine serum albumin (BSA) before immobilization. Coupling can be carried out using maleimide-activated BSA which can be purchased as part of a conjugation kit from Pierce and used following the manufacturer's instructions.

Protocol 3

Selection of phage antibody libraries by panning on immunotubes

Equipment

- Maxisorb immunotubes (Nunc; 75 × 12 mm)

Method

1. Coat antigen onto an immunotube by the addition of 1 ml of an appropriate concentration of antigen in PBS. Proteins are routinely coated at 10 μg/ml, but for scarce or expensive antigens this concentration can be lowered as far as 0.2 μg/ml. For some proteins coating may be better in 50 mM $NaHCO_3$ pH 9.6, than in PBS. Coating is usually overnight at 4 °C, but coating temperature is antigen-dependent and room temperature or 30 °C will often give good coating depending on the stability of the antigen at these temperatures.

2. Rinse the tube three times with PBS and block the tube by filling it to the brim with PBS containing 3% (w/v) skimmed milk powder and incubating for 1–2 h at 37 °C. If the antigen is unstable at 37 °C block at room temperature for 2 h.

Protocol 3 continued

3. Rinse the tube three times with PBS.

4. Add 1×10^{12} phage in 1 ml of PBS containing 3% (w/v) skimmed milk and incubate for 1–2 h at room temperature.

5. Rinse the tube by completely filling it ten times in PBS containing 0.1% Tween 20, then ten times in PBS using a wash bottle.

6. Elute the phage which have bound to the tube by adding 1 ml of freshly made 100 mM triethylamine and incubating at room temperature for 10 min.

7. Transfer the eluted phage to a 1.5 ml Eppendorf tube and neutralize immediately after elution with 0.5 ml of 1 M Tris–HCl pH 7.4. Infect half the eluted phage into 5 ml of mid-log *E. coli* TG1 cells at 37 °C for 30 min without shaking and then at 30 °C with moderate shaking (200 r.p.m.) to allow infection. Store the remaining eluted phage at 4 °C as a back-up.

8. Make four tenfold serial dilutions of the infected cells into 2TY broth and plate 100 μl of each dilution on to 2TYAG plates. This allows determination of the titre of the recovered phage.

9. Pellet the remaining cells at 3500 r.p.m. in a Sorvall RT7 benchtop centrifuge (PN11053 bucket with a 00438 insert), resuspend the cell pellet in approx. 0.5 ml of 2TY, and spread on one 243 × 243 mm 2TYGA plate. Incubate the plates at 30 °C overnight.

4.2 Elution conditions

The protocol described above includes a high pH phage elution step, but a number of other elution conditions can be equally effective, for example acidic elution using 100 mM glycine at pH 3. It is also possible use excess antigen as an elution reagent to try and enrich the eluted population for antigen-specific phage, as oppose to any background 'sticky' phage that may be bound to the immunosorb tube or other support matrix. When using antigen to elute the phage, antigen should be added in an approximately tenfold excess over the amount used to coat the support. It is also advisable to leave the phage eluting for longer than the standard acid or alkali elution regimes, since phage with longer off rates will need time to dissociate from the immobilized antigen and be captured by the solution phase antigen.

4.3 Storage and rescue of the phagemid population after selection

Once eluted phage have been recovered and transfected to *E. coli* it is necessary to rescue and amplify the selected phage population to provide an input population for a further round of selection. It is also necessary to generate glycerol stocks of the infected bacteria from each round of selection.

Protocol 4

Storage and rescue of phagemid antibodies after selection

Reagents

- M13KO7 helper phage (Pharmacia)

Method

1. Scrape the bacterial colonies from the large (243 mm × 243 mm) selection plate generated in *Protocol 3* by adding 6 ml of 2TY broth to the surface of the plate and scraping with a disposable spreader. Collect the colony suspension using a pipette into a 50 ml Falcon tube.

2. Add 3 ml of sterile 50% (v/v) glycerol and mix by placing the tube on an end-over-end rotator for 10 min at room temperature. Freeze 3 × 1 ml aliquots of the glycerol stock at −70 °C. These can be stored indefinitely and provide a back-up stock for each selection round.

3. Inoculate 25 ml of 2TYGA with 50–100 μl of the above plate scrape (or glycerol stock) and grow at 37 °C with moderate shaking until the bacteria reach mid-log phase (A_{600} = 0.5–1.0).

4. Add M13KO7 helper phage to a final concentration of 5×10^8 p.f.u./ml and infect the cells at 37 °C for 30 min stationary, followed by 30 min moderate shaking (200 r.p.m.).

5. Transfer the bacteria to a 50 ml Falcon tube and centrifuge at 3500 r.p.m. in a Sorvall RT7 benchtop centrifuge for 10 min, then resuspend the bacterial pellet in 25 ml pre-warmed 2TY containing 50 μg/ml kanamycin and 100 μg/ml ampicillin. Transfer to a 250 ml flask and grow for 2 h (or overnight) with rapid shaking (300 r.p.m.) at 30 °C to produce phage particles.

6. Transfer 1 ml of culture to a 1.5 ml Eppendorf tube and centrifuge at 13 000 r.p.m. in a standard microcentrifuge for 2 min. Transfer the phage-containing supernatant to a fresh tube. The phage titre should be around 1×10^{11}/ml after 2 h growth and 1×10^{12}/ml after overnight growth. Phage prepared in this manner should be stored at 4 °C and used immediately or at most within 24 h.

7. Perform the next round of selection as described in *Protocol 3* using this phage population as the input.

4.4 Choice of number of selection rounds

The number of rounds of selection carried out on the antigen may vary depending on the desired properties of the antibodies required from the selection. If a diverse population of antibodies is required it may be advisable to carry out only two rounds of selection. After two rounds of selection the total percentage of phage antibodies positive for the antigen may be lower than after a third or fourth round, but the number of different positive antibodies may be higher. After a third or fourth round of selection it is common for one or two clonal phage antibodies to dominate the selected population. These dominant clones may not necessarily be the highest affinity antibodies; they may be antibodies which are more readily expressed and correctly folded within the bacterial periplasmic space, or are the least toxic to the host bacterial strain.

5 Selection of phage libraries on biotinylated antigen

5.1 Biotinylation of antigen

The use of antigen in solution phase during selections can allow a greater degree of control over the selection process. The most common method of carrying out phage selections in solution phase is to use biotinylated antigen, to which the phage antibody binds, and then the antigen–phage complex can be recovered using streptavidin-coated magnet beads. Beads can be pulled out of solution using a magnet, can be thoroughly washed, and then used either directly to infect *E. coli* or phage can be eluted from the beads and the eluted phage then used to infect bacteria.

In some circumstances it is possible to purchase antigen which has already been biotinylated, but this is obviously not possible in many cases. Biotinylation reactions can be carried out using NSH esters of biotin (available from Pierce) which biotinylate primary amine groups, or alternatively biotin-LC hydrazide (Pierce) can be used to couple biotin to carbohydrates via the activator EDC (Pierce). For biotinylation protocols see the manufacturer's instructions. Biotinylation need not be limited to protein antigens. A number of examples exist in the literature where haptens, peptides, or other small molecules have been biotinylated and used to select phage antibodies (12). *Protocol 5* describes the biotinylation of a peptide by coupling of an NHS ester to lysine residues, but this is readily adaptable to proteins or haptens. The biotinylation reagent used in *Protocol 5* has a cleavable disulfide bond which allows elution of the phage particles from any streptavidin matrix by the addition of dithiothreitol (DTT).

Protocol 5

Biotinylation of a peptide

Reagents
• NHS-SS biotin (Pirece)

Method

1. Dissolve 1 mg of NHS-SS-biotin in 1 ml of sterile distilled water, resulting in a stock solution of 1.65 nM/µl. The NHS reagent should dissolve readily in water, and if it does not dissolve within 10 min it may have absorbed water due to inappropriate storage and will not couple effectively. Always try and use fresh reagent.

2. Add an equimolar amount of peptide to the NHS-SS biotin to give a reaction volume of around 200 µl.

3. Add 50 µl of 1 M NaHCO$_3$ pH 8.5 and make up the total reaction volume to 1 ml with sterile distilled water.

4. Incubate on ice for 2 h. The sample can be used without removal of unincorporated biotin if coupling is performed at low density, since a vast excess of streptavidin is used when capturing the phage. If desired, however, unincorporated biotin can be removed from the sample using, for example, column chromatography.

5.2 Selections using biotinylated antigen

The choice of selection using biotinylated antigen in solution is often taken when antibodies of high affinity are desired as output from the selection regime. Selection in solution allows the possibility of precisely controlling the concentration of antigen in solution and the amount of available antigen can be lowered after successive selection rounds, so biasing the selection for antibodies with higher affinity (which by definition are those that bind antigen at the lowest antigen concentration). The principle of the method is described in detail by Hawkins and colleagues (13). One possible approach is to carry out a first round of selection on antigen directly immobilized on a solid matrix to capture all antigen-specific phage and then to use the population from this selection as input for a series of more stringent rounds of selection using biotinylated antigen in solution phase.

Protocol 6

Selection of phage antibody libraries by selection on biotinylated antigen

Equipment and reagents

- Dynabeads (M-280, streptavidin-coated, Dynal)
- Magnetic rack (Dynal MPC-E, Dynal)

Method

1. Make up an aliquot of the phage library (approx. 1 × 10^{12} phage) to 500 µl in PBS containing 3% skimmed milk and block for 30 min at room temperature.

2. Add biotinylated antigen and incubate at room temperature for 1 h with end-over-end rotation. The amount of biotinylated antigen you wish to add may vary. Antigen should be added to a concentration approximately equivalent to the desired K_d of the resulting selected antibodies and in molar excess over phage (1 × 10^{12} phage/ml is 10 nM).

3. Dispense 100 μl of streptavidin-coated magnetic beads into a 1.5 ml Eppendorf tube and collect the beads at the bottom of the tube using the magnetic rack. Aspirate the supernatant and resuspend the beads in 1 ml PBS containing 3% skimmed milk. Block for at least 1 h with end-over-end rotation.

4. Capture the blocked streptavidin beads using the magnetic rack, aspirate off the block solution, and resuspend the beads in the phage–antigen solution. Incubate at room temperature for 15 min with rotation.

5. Capture the beads using the magnetic rack as before, aspirate the supernatant, and resuspend the beads in 1 ml PBS containing 0.1% Tween 20.

6. Recover the beads on the magnet again and repeat the wash steps. Wash the beads four times in 1 ml PBS/0.1% Tween, followed by two 1 ml PBS washes.

7. After the last PBS wash step resuspend an aliquot of the beads in 100 μl of 50 mM DTT, mix, and incubate for 5 min at room temperature to elute the phage. Store the remaining beads in PBS at 4 °C in case you need to return to them.

8. Add 5 ml log phase *E. coli* cells to the eluted phage and carry out transfections and titrations as described in *Protocol 3*, steps 7 and 8.

6 Cell surface selections

In many cases it may not be possible to obtain purified target antigen, and the antigen may only be available presented on a cell surface. Alternatively it may also be the goal of the selection to generate antibodies that are specific for a particular cell type without prior knowledge of a specific antigen. In such cases it is necessary to carry out cell surface based selections. There is a wide range of selection techniques available for such situations, and the choice of technique will depend on the cell type being worked with. Direct panning on cell surfaces carrying the antigen may be carried out on adherent cells grown as monolayers, or on intact cells in suspension (14–16). This may fortuitously select for antigen-specific phage (if the antigen is known), however it may be possible to bias the selection towards a particular antigen by using depletion/subtraction methods, cell sorting using flow cytometry (17) or magnetic bead-based systems (18), competitive elution with an antigen-specific ligand, or selection by alternating between different cell types which all carry the antigen of interest. Often the concentration of the antigen of interest in these situations is very low and it may be necessary to try and boost the available density of membrane-associated antigens by making cell membrane preparations and coating these onto micro-titre wells at high concentrations, resulting in an improved antigen density compared to the standard cell surface selection. The basic selection regime described below should be considered a starting framework on which additional selection pressures can be included if required. Selections need not be limited to homogeneous cell populations if a method sorting or capturing the cell type of interest from the background population is available. The investigator may also wish to consider selecting on immobilized tissues sections or even on cells *in vivo* (2).

6.1 Adherent cell selections

Selections on adherent cell lines can be carried out in an almost identical manner to selections on immobilized antigen. Cells are grown on chamber slides, and may or may not be fixed depending on the desired specificity of the final antibody. The advantage of fixation is that the cells can be subjected to more rigorous washing procedures, whereas unfixed cells can be washed away unless the washing procedure is very gentle, or they have been grown on pre-treated (e.g. gelatin-coated) slides. Unfixed cell, however, have the possible advantages that the cells are not permeabilized, hence antibodies against internal proteins should not be selected, and that they retain their membrane proteins in a native configuration. The choice of fixation or not is largely governed by the robustness of the cell line and it must also be borne in mind that it may be necessary to screen the selected antibodies by cell ELISA which requires vigorous washing procedures. Elution conditions for the adherent cell selection may also be varied. The protocol outlined below involves elution with triethylamine, as for purified antigen, but it is also possible to either elute the phage with PBS containing 0.1% Triton X-100 which lyses the cells, or by directly adding bacteria to the cell chambers and allowing a direct infection event to occur.

Protocol 7

Selection of phage antibody libraries by panning adherent cells

Equipment
- 16-well chamber slides (Nunc)

Method

1. Grow the cell line under normal culture conditions on chamber slides to approx. 80% confluence (approx. 1×10^5 to 1×10^6 cells per well). Cells can either be left unfixed or fixed with a variety of fixation procedures, for example treatment with 0.1% glutaraldehyde for 15 min at room temperature. If unfixed cells are used keep the cells at 4 °C at all times, since membrane proteins may be internalized at higher temperatures.

2. Block the cells in 100 μl PBS containing 3% skimmed milk for 1–2 h at 4 °C for unfixed cells, or room temperature for fixed cells.

3. Gently wash the cells with three changes of PBS.

4. Add 1×10^{12} phage in 100 μl of PBS containing 3% (w/v) skimmed milk and incubate for 1–2 h at 4 °C or room temperature.

5. Gently wash the cells as in step 3.

Protocol 7 continued

6. Elute the phage which have bound to the cells by adding 100 μl of freshly made 100 mM triethylamine and incubating at room temperature for 10 min.

7. Transfer the eluted phage to a 1.5 ml Eppendorf tube and neutralize immediately after elution with 50 μl of 1 M Tris–HCl pH 7.4.

8. Infect and titre the phage exactly as described in *Protocol 3*.

6.2 Cells in suspension

It is also possible to carry out selections on non-adherent cell lines or cell preparations in suspension. The basic protocol described above (*Protocol 7*) can be applied using pre-blocked cells mixed in suspension with the phage library. Occasional mixing by gentle pipetting is required and it may be necessary to increase the selection volume from 100 μl up to 500 μl depending on the cell concentration (between 1×10^5 and 1×10^6 cells per selection is appropriate). Washes are carried out by pelleting the cells by low speed centrifugation and resuspending in PBS. The final elution step can be dispensed with and the cells used directly to infect the bacteria.

6.3 Screening the output of cell surface selections

Once the output has been recovered from a number of rounds of cell surface selection it is necessary to screen for those antibodies which are specific for the cell line in question. Perhaps the easiest way of achieving this is using a cell ELISA technique that is a modification of *Protocol 12*, using immobilized cells rather than immobilized purified antigen. It is also possible to screen clones either as phage particles or scFv using flow cytometry or immunocytochemistry, although these can be lower throughput methods. Ideally a control cell line, such as a non-transfected CHO cell line or other related cell line that does not express the antigen of interest, should be used in the screening process.

7 Proximity selections

Section 6 described basic protocols for the selection of antibodies to cell surface antigens in a non-guided manner. A cell surface is, however, antigenically complex which makes selection *in situ* difficult to control given that several selections on many different accessible antigens may be occurring simultaneously. All phage antibodies that bind to the cell type in question are selected for and the antibodies must be screened to identify those that are cell type, or antigen, specific. Given these difficulties a method has recently been developed to improve the efficiency and specificity of the cell surface selection procedure by targeting the selection of phage antibodies which bind in close proximity to a known target molecule on the cell surface. This greatly narrows the scope of the selection and increases the probability of selecting antibodies of interest.

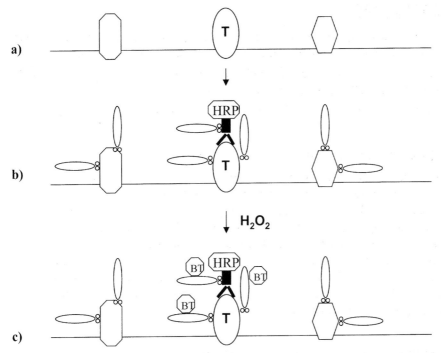

Figure 1 (a) T is a target antigen on the surface of a cell. (b) Phage antibodies bind to the cell surface along with an HRP-conjugated ligand (either a natural ligand or antibody) to T. (c) In the presence of hydrogen peroxide biotin tyramine molecules (BT) become covalently linked to phage antibodies binding around the HRP–ligand complex. These antibodies are likely to be antibodies that recognize the target antigen (T) or the ligand–target-antigen complex.

The method is referred to as proximity (ProxiMol) selection and is outlined in schematic form in *Figure 1*.

The method relies on the use of catalysed enzyme reporter deposition (CARD) which has been previously been used as a means of signal amplification in immunocytochemistry, ELISA, and blotting formats (19–22). CARD is based on the use of horseradish peroxidase (HRP)-conjugated reagents, such as antibodies, in conjunction with a biotin tyramine substrate; when HRP is added, the enzyme catalyses formation of biotin tyramine free radicals, which then react with proteins in the immediate vicinity of the enzyme, hence biotinylating them. By then adding streptavidin–HRP complex, the number of molecules of HRP at a specific site is increased resulting in signal enhancement when the enzyme is detected colorimetrically. When CARD is used as an amplification system in immunocytochemistry no detectable loss of image resolution is apparent, demonstrating that the deposition occurs in close association with the catalytic enzyme. The proximity selection approach is a modification of the CARD methodology that uses the ability of HRP, in conjunction with biotin tyramine, to biotinylate phage binding around the site of enzyme activity at a specific site on a cell surface. Biotinylated phage can be recovered on streptavidin-coated magnetic beads. Proximity selections can be carried out on any cell type, but also on antigen in many other forms, such as purified antigen, cell extracts, or membrane preparations. The technique relies on the availability of a ligand to the target antigen or cell type that can be HRP conjugated and used to guide the selection process. This ligand could be a naturally occurring ligand such as a growth factor, chemokine, existing antibody, or any other type of molecule known to bind the cell type of interest. The method directs biotinylation of phage particles that bind up to an approximate radius of 25 nm from the original binding site and results in very low numbers of phage particles being recovered (in some cases only a few hundred). The low number of recovered phage particles is due to the highly specific nature of the selection procedure and often means that it is only necessary to carry out a single round of selection.

7.1 Proximity selection using an existing antibody

In some selection cases an antibody may already exist which binds the antigen or cell type of interest, but may not recognize the desired epitope, or may be a mouse monoclonal, whereas a human antibody is required. In such cases it is possible to use the existing antibody as a guide molecule in a proximity (ProxiMol) selection to generate a population of antibodies which bind very close to, but not at the same epitope as, the original antibody. Such an approach has been used to generate human anti-carcinoembryonic antigen (CEA) antibodies using a commercially available mouse mAb as the guide molecule (23). The guide antibody must be conjugated either directly, or indirectly using a secondary anti-species conjugate, to HRP for the selection to work. Direct conjugation of HRP to antibodies can be carried out using a Pierce HRP conjugation kit following the manufacturer's instructions. Use of a secondary anti-species HRP conjugate appears to be equally successful and avoids the need for direct conjugation, although this approach may slightly broaden the area over which phage binding to the cell surface are captured because the HRP may be further away from the cell membrane. In some cases this may be desirable. The following protocol for proximity selection using a guide antibody takes the secondary anti-species HRP conjugate approach and has been written for selection on adherent cell types, but can easily be adapted for use with a direct conjugate and/or cells in suspension. The same issues regarding cell fixation and phage elution conditions apply as discussed under the general cell selection section.

Protocol 8

Proximity selection using a guide antibody on an adherent cell line

Equipment and reagents

- Biotin tyramine (NEN, available as part of the Renaissance TSA kit)
- Streptavidin-coated magnetic beads (Dynal) with magnetic rack

Method

1. Grow the cell line under normal culture conditions on chamber slides to approx. 80% confluence (approx. 1×10^5 to 1×10^6 cells per well). Cells can either be left unfixed or fixed with a variety of fixation procedures, for example treatment with 0.1% glutaraldehyde for 15 min at room temperature. If unfixed cells are used, carry out all the following steps at 4 °C.

2. Block the cells in 100 μl PBS% containing 3% skimmed milk for 1–2 h at room temperature.

3. Gently wash the cells with three changes of PBS.

4. Add the primary guide antibody at an appropriate concentration in PBS containing 3% skimmed milk. Incubate for 1 h at room temperature.

5. Wash the cells as in step 3.

6. Add 1×10^{12} phage in 100 μl of PBS containing 3% (w/v) skimmed milk and incubate for 1–2 h at room temperature.

7. Gently wash the cells as in step 3.

8. Add the secondary anti-species-HRP conjugate at an appropriate concentration (normally 1:1000 to 1:5000) in PBS containing 3% skimmed milk and incubate for 1 h at room temperature.

9. Wash as in step 3, then add 0.4 μl of biotin tyramine (approx. 1 mg/ml) in 100 μl of 50 mM Tris–HCl pH 7.4 containing 0.03% H_2O_2 and incubate at room temperature for 10 min.

10. Wash the cells as before and elute all phage bound to the cells by adding 100 μl of freshly made 100 mM triethylamine and incubating at room temperature for 10 min.

11. Transfer the eluted phage to a 1.5 ml Eppendorf tube and neutralize immediately after elution with 50 μl of 1 M Tris–HCl pH 7.4.

12. Add 20 μl of streptavidin-coated magnetic beads (which have been pre-blocked in PBS containing 3% skimmed milk) to the eluted phage and rotate for 15 min at room temperature.

13. Pellet the beads using the magnetic rack and wash three times in 1 ml of PBS containing 0.1% Tween, followed by three washes in 1 ml PBS.

14. Resuspend the beads in 100 μl of PBS and use 50 μl of this to infect and titre the phage exactly as described in *Protocol 3*. There is no need to elute the phage off the beads for successful infection to occur.

7.2 Proximity selection using natural ligands

In many situations it is not possible to obtain an existing antibody specific to the antigen in question, but another type of known binding molecule, such as a natural ligand may be available. A number of natural ligands have been used as guide molecules for proximity selections which have generated panels of anti-bodies which include antibodies to the ligand's known receptors. Examples include the use of biotinylated sialyl Lewis X to generate E and P-selectin anti-bodies (23), and the chemokine macrophage inflammatory protein-1α (MIP-1α) to generate antibodies to the CCR5 receptor (24). In theory there is no limit to the type of molecule that could be used as a selection guide. The only limitation is the fact that the ligand must be either HRP-conjugated directly or tagged in some way to allow indirect conjugation to HRP, e.g. by biotinylation followed by the addition of streptavidin–HRP. It must be ensured that the biotinylation, or other modification process, does not affect the ability of the ligand to recognize its cognate receptor. The following protocol describes the use of a biotinylated ligand to guide selection of antibodies to a cell surface receptor on cells in sus-pension. Again it is possible to modify this method for use with adherent cells, or other methods of antigen presentation, and for use with directly conjugated ligands.

Protocol 9

Proximity selection using a natural ligand on cells in suspension

Reagents

- Streptavidin–HRP conjugate (Amersham)

Method

1. Prepare approx. 1×10^5 to 1×10^6 cells in solution in an approx. volume of 100 μl PBS containing 3% skimmed milk powder and incubate for 1 h at room temperature.

2. Add biotinylated ligand at an appropriate concentration (e.g. for selections on CD4$^+$ cells using biotinylated MIP-1α the ligand was added to a final concentration of 375 nM) and incubate for 1 h at room temperature.

3. Gently wash the cells by low speed centrifugation with three changes of PBS.

4. Add streptavidin–HRP conjugate at a dilution of 1:1000 in PBS containing 3% skimmed milk powder and incubate at room temperature for 1 h.

5. Wash the cells as in step 3.

6. Add 1×10^{12} phage in 100 μl of PBS containing 3% (w/v) skimmed milk and incubate for 1–2 h at room temperature.

7. Gently wash the cells as in step 3.

9. Add 0.4 μl of biotin tyramine (approx. 1 mg/ml) in 100 μl of 50 mM Tris–HCl pH 7.4 containing 0.03% H_2O_2 and incubate at room temperature for 10 min.

10. Wash the cells as before and resuspend the cells in 100 μl Tris–EDTA (TE) containing 0.5% Triton X-100. This is an alternative to TEA elution which lyses the cells to release the phage.

11. Add 20 μl of streptavidin-coated magnetic beads (which have been pre-blocked in PBS containing 3% skimmed milk) to the cell lysates and rotate for 15 min at room temperature.

12. Pellet the beads using the magnetic rack and wash three times in 1 ml of PBS containing 0.1% Tween, followed by three washes in 1 ml PBS.

13. Resuspend the beads in 100 μl of PBS and use 50 μl of this to infect and titre the phage exactly as described in *Protocol 3*.

7.3 Step back selections

The selections described above using either existing antibodies or natural ligands as guide molecules result in panels of antibodies that bind close to the site of antibody or ligand binding, but which do not overlap with it. This is because the selection method relies on the presence of the guide molecule on the cell surface to direct the selection. It is possible, however, to generate antibodies which block ligand or antibody binding to the cell by carrying out a second round of proximity selection using the output from the first round of selection to guide a further round. In theory either biotinylated scFv or phage particles can be used as guide molecules following similar protocols to those described. Antibodies which block MIP-1α binding to CD4$^+$ cells have been generated using this approach (24).

8 Screening of selected phage

8.1 Basic screening assays

The outcome of any selection procedure is a population of antibodies with a variety of different properties in terms of, for example, their antigen specificity, affinity, or expression levels. It may be necessary to screen large numbers of variants in order to identify those antibodies with the most optimal characteristics for the application required. It is, therefore, important initially to develop screening systems which are high-throughput and which require unpurified phage particles or unpurified antibody. It is also important to bear in mind the eventual use of the antibody and to try and link the screening as closely as possible to this use. If the desired antibody is required for use as a Western blot reagent then it may be inappropriate to carry out an ELISA screening regime in which native or partially denatured, rather than denatured protein is used as the basis of the assay.

8.1.1 Phage ELISA

A useful sensitive primary screening process for antibodies on the basis of their binding to immobilized antigen is the phage ELISA. Individual colonies generated by the selection process can be picked into 96-well format and phagemid rescue carried out. The resultant phage particles can then be used directly in an

ELISA assay to determine those which recognize the antigen of interest and do not recognize an unrelated antigen. Once potential positive clones have been identified using this method it is advisable to re-screen the clones on the test antigen plus a panel of unrelated antigens to establish specificity of the binding reaction. Clones are usually specific for the antigen of interest, but some may cross-react with unrelated antigens; these are poly-reactive clones and should be discarded.

Protocol 10

Rescue of phagemid clones in 96-well format for use in ELISA assay

Equipment

• Sterile polycarbonate 96-well microtitre plates (Corning)

Method

1. Pick individual colonies with a sterile toothpick or pipette tip. Transfer the toothpick into 100 μl of 2TYGA in a 96-well microtitre plate and grow with shaking (100 r.p.m.) overnight at 30 °C. This is the master plate.

2. Prepare a replica 96-well microtitre plate with 100 μl of 2TYGA in each well. Inoculate the replica plate from the master plate. Add 50 μl of 50% glycerol to each well of the master plate and store at −70 °C. Grow the replica plate with shaking (100 r.p.m.) at 37 °C for approx. 5 h, or until the broth becomes turbid.

3. Add 10 μl of M13K07 helper phage in 2TYGA (at 5×10^{10} p.f.u./ml, which should give a m.o.i. of approx. 10) to each well of the replica plate. Incubate the plate at 37 °C for 30 min without shaking, and then for 30 min with shaking (100 r.p.m.) at 37 °C to allow superinfection of the helper phage.

4. Centrifuge the plate at 2000 r.p.m. for 10 min in a Sorvall RT7 benchtop centrifuge (plate spinner PN11093). Pipette off the supernatant and resuspend the bacterial pellets in 100 μl of 2TY containing 50 μg/ml kanamycin and 100 μg/ml ampicillin. Grow overnight with shaking (100 r.p.m.) at 30 °C to produce phage particles for ELISA.

Protocol 11

Phage ELISA

Equipment and reagents

• Sterile 96-well flexible microtitre plates (Falcon)

• Anti-fd–HRP conjugate (Pharmacia)

• 3,3′,5,5′, tetramethylbenzidine (TMB) (Sigma) buffer

Method

1. On the same day as performing the rescue (*Protocol 10*) coat one flexible 96-well plate with 50 μl of antigen (at between 1 μg/ml and 10 μg/ml depending on antigen availability) per well in PBS. Coat a second plate with 50 μl per well of BSA, or other unrelated antigen, at a similar concentration to that of the test antigen. Leave to coat at 4 °C overnight.

2. Flick out the coating material from the two plates and rinse each three times with PBS. Block with 200 μl PBS containing 3% skimmed milk for 2 h at room temperature (or 37 °C if the antigen is stable at this temperature).

3. Centrifuge the overnight culture plate from *Protocol 10* at 2000 r.p.m. for 10 min in a Sorvall RT7 benchtop centrifuge. Transfer 100 μl of each culture supernatant to the appropriate well of a microtitre plate containing 20 μl of 6 × PBS containing 18% (w/v) skimmed milk powder. Mix by pipetting and leave phage to block for 1 h at room temperature.

4. Wash the coated blocked plates three times in PBS.

5. Transfer 50 μl of the pre-blocked phage to each of the two plates and allow to bind for 1 h at room temperature.

6. Wash plates for 2 min per wash with three changes of PBS containing 0.1% Tween 20, followed by three washes with PBS.

7. To each well add 50 μl of a 1:5000 dilution of anti-fd–HRP conjugated mAb. Dilution should be carried out in PBS containing 3% (w/v) skimmed milk powder.

8. Wash as in step 6.

9. Add 50 μl of development solution and leave for approx. 10 min, or until colour starts to develop. Stop the development reaction by the addition of 25 μl of 1 M H_2SO_4. Read the ELISA plates at 405 nm.

8.1.2 Soluble (scFv) ELISA

The phage ELISA screening protocol described is useful for identifying any antibody that recognizes the antigen of interest. Phage particles may be expressing more than one copy of a particular antibody and this introduces an avidity effect which increases the possibility of identifying lower affinity antibodies using a phage ELISA screen. Soluble expression of antibody fragments has the advantage of minimizing the avidity effects since only one antibody binding site (in the case of a scFv) is normally present, although it is worth noting that some scFv's show a variable tendency to dimerize. ScFv binding is detected in ELISA by secondary antibodies that are specific to the peptide tag present at the amino-terminal of the scFv molecule. A variety of tags are used, one of the most common being the myc tag which is recognized by the mAb 9E10. The binding of the secondary antibody to the scFv can be detected by the use of an enzyme-conjugated anti-species mAb.

Protocol 12

Soluble scFv ELISA assay

Equipment and reagents

- Sterile polycarbonate 96-well microtitre plates (Corning)
- Sterile flexible 96-well microtitre plates (Falcon)
- Anti-myc tag antibody 9E10 (Cambridge Bioscience)
- Goat anti-mouse IgG alkaline phosphatase conjugate (Pierce)
- p-Nitrophenyl phosphate (PNPP) tablets (Sigma)
- PNPP buffer: 0.1 M glycine, 1 mM $MgCl_2$, 1 mM $ZnCl_2$ pH 10.4
- Saline solution: 0.9% (w/v) NaCl

Method

1. Pick individual colonies with a sterile toothpick or pipette tip. Transfer the toothpick into 100 µl of 2TYGA in a 96-well microtitre plate and grow with shaking (100 r.p.m.) overnight at 30 °C. This is the master plate.

2. Prepare a replica 96-well microtitre plate with 100 µl of 2TYGA in each well. Inoculate the replica plate from the master plate. Add 50 µl of 50% glycerol to each well of the master plate and store at −70 °C. Grow the replica plate with shaking (100 r.p.m.) at 37 °C for approx. 5 h, or until the broth becomes turbid.

3. Centrifuge the plate at 2000 r.p.m. for 10 min in a Sorvall RT7 benchtop centrifuge. Pipette off the supernatant and resuspend the bacterial pellets in 100 µl of 2TY containing 1 mM IPTG and 100 µg/ml ampicillin (NB: no glucose should be present). Grow overnight with shaking (100 r.p.m.) at 30 °C to induce antibody expression.

4. On the same day as inducing scFv production coat one flexible 96-well plate with 50 µl of antigen (at between 1 µg/ml and 10 µg/ml depending on antigen availability) per well in PBS. Coat a second plate with 50 µl per well of BSA, or other unrelated antigen, at a similar concentration to that of the test antigen. Leave to coat at 4 °C overnight.

5. Flick out the coating material from the two plates and rinse each three times with PBS. Block with 200 µl PBS containing 3% skimmed milk for 2 h at room temperature (or 37 °C if the antigen is stable at this temperature).

6. Centrifuge the overnight culture plate at 2000 r.p.m. for 10 min in a Sorvall RT7 benchtop centrifuge. Transfer 100 µl of each culture supernatant to the appropriate well of a microtitre plate containing 20 µl of 6 × PBS containing 18% (w/v) skimmed milk powder. Mix by pipetting and leave scFv to block for 1 h at room temperature.

7. Wash the coated blocked plates three times in PBS.

8. Transfer 50 µl of the pre-blocked scFv to each of the two plates and allow to bind for 1 h at room temperature.

9. Wash plates for 2 min per wash with three changes of PBS containing 0.1% Tween 20, followed by three washes with PBS.

10. To each well add 50 μl of a 1:200 dilution of the anti-myc mAb (9E10). Dilution should be carried out in PBS containing 3% (w/v) skimmed milk powder.

11. Wash as in step 9.

12. Add 50 μl of alkaline phosphatase-conjugated anti-mouse mAb at a dilution of 1:5000 in PBS containing 3% skimmed milk powder.

13. Wash as in step 9, then wash once with saline solution.

14. Dissolve one 20 mg PNPP tablet in 20 ml of PNPP buffer and add 100 μl of this to each well. Leave at room temperature until sufficient colour develops and read at 405 nm.

8.2 Affinity screening

Once phage or scFv which specifically bind the antigen of interest have been identified it may be necessary to carry out further screening tests to assess which antibodies from the positive population have the highest affinities. There are a variety of ways of achieving this including surface plasmon resonance (BIAcore) screening, fluorescence quench measurement, and competition ELISAs. Descriptions of these protocols are beyond the scope of this chapter.

9 Soluble scFv production and purification

Antibodies cloned in scFv format can be used as immunological reagents for a variety of applications including ELISA, Western blotting, and immunocyto-chemistry. The rescued phage themselves can be used as detection reagents, or either crude or purified scFv can be generated. ScFv prepared by periplasmic extraction as described in *Protocol 13* can be used as a crude extract with no further purification for some applications, and is generally best used within 12 hours. The production of more stable, purified scFv requires a purification scheme based on the use of nickel–chelate chromatography for scFv cloned in polyhistidine-containing vectors (such as pCantab 6). This generates scFv that can be stored at 4 °C for one to two weeks, or for longer with the addition of glycerol and storage at −70 °C.

Protocol 13
Preparation of scFv periplasmic extracts

1. Inoculate 10 ml of 2TYGA in a 50 ml Falcon tube with an individual colony of the clone of interest and grow overnight at 30 °C with shaking (200 r.p.m.).

2. Add 0.5 ml of the overnight culture to 50 ml of 2TYGA pre-warmed to 30 °C in a 250 ml flask. Grow with rapid shaking (300 r.p.m.) until A_{600} is between 0.2–0.3 (normally between 1–2 h).

Protocol 13 continued

3. Split the culture between two 50 ml Falcon tubes and centrifuge at 4000 r.p.m. for 15 min at 4 °C in a pre-chilled Sorvall SS 34 rotor.

4. Pour off the supernatant and resuspend the cell pellet in 50 ml of 2TY containing 100 μg/ml ampicillin and 1 mM isopropyl β-δ-thiogalactopyranoside (IPTG) pre-warmed to 30 °C.

5. Return the cells to a 250 ml flask and grow for 4 h at 30 °C with shaking.

6. Pellet the cells as before and resuspend the cell pellet in 1 ml of 50 mM Tris–HCl pH 8.0 containing 20% (w/v) sucrose and 1 mM EDTA. Transfer the suspension into a 1.5 ml Eppendorf tube and leave on ice for 15 min.

7. Centrifuge at 13 000 r.p.m. in a standard microcentrifuge for 10 min at 4 °C to pellet the cell debris.

8. Carefully take off the supernatant and transfer to a fresh tube. This is your crude soluble scFv preparation.

Protocol 14

Purification of scFv by nickel–chelate chromatography

Equipment and reagents

- Mini-His columns (Qiagen)
- Wash buffer: 50 mM sodium phosphate, 300 mM NaCl, 10 mM imidazole pH 8.0
- Elution buffer: 50 mM sodium phosphate, 300 mM NaCl, 250 mM imidazole pH 8.0

Method

1. Pre-equilibrate a mini-His column with 600 μl of wash buffer.

2. Cut off the cap and centrifuge at 2000 r.p.m. for 2 min in a standard microcentrifuge.

3. Add $MgCl_2$ to the crude periplasmic extract (prepared as in *Protocol 13*) to a final concentration of 1 mM and apply 600 μl to the equilibrated column.

4. Centrifuge as in step 2.

5. Wash the column twice with 600 μl wash buffer. Centrifuge as in step 2 after each application.

6. Elute the scFv from the column with two successive applications of 600 μl elution buffer. Centrifuge as in step 2 after each application.

7. For long-term storage of the purified scFv preparation it is advisable to buffer exchange the sample to remove the imidazole which can lead to sample instability. This can be achieved either by dialysis (against PBS) or by use of a Napp 5 Sephadex G25 column (Pharmacia Biotech) following the manufacturer's instructions.

References

1. Vaughan, T. J., Williams, A. J., Pritchard, K., Osbourn, J. K., Pope, A. R., Earsnshaw, J. C., *et al.* (1996). *Nature Biotechnol.*, **14**, 309.
2. Pasqualini, R. and Ruoslahti, E. (1996). *Nature*, **380**, 364.
3. Spada, S., Krebber, C., and Pluckthun, A. (1997). *Biol. Chem.*, **378**, 445.
4. Malmborg, A. C., Soderlind, E., Frost, L., and Borreback, C. A. (1997). *J. Mol. Biol.*, **273**, 544.
5. Parsons, H. L., Earnshaw, J. C., Wilton, J., Johnson, K. S., Schueler, P. A., Mahoney, W., *et al.* (1996). *Protein Eng.*, **9**, 1043.
6. Clackson, T., Hoogenboom, H. R., Griffiths, A. D., and Winter, G. (1991). *Nature*, **352**, 624.
7. Griffiths, A. D., Malmqvist, M., Marks, J. D., Bye, J. M., Embleton, M. J., McCafferty, J., *et al.* (1993). *EMBO J.*, **12**, 725.
8. Garrard, L. J., Yang, M., O'Connell, M. P., Kelley, R. F., and Henner, D. J. (1991). *BioTechnology*, **9**, 1373.
9. Hoogenboom, H. R., Griffiths, A. D., Johnson, K. S., Chiswell, D. J., Hudson, P., and Winter, G. (1991). *Nucleic Acids Res.*, **19**, 4133.
10. Bass, S., Greene, R., and Wells, J. A. (1990). *Proteins*, **8**, 309.
11. McCafferty, J., FitzGerald, K. J., Earshaw, J., Chiswell, D. J., Link, J., Smith, R., *et al.* (1994). *Appl. Biochem. Biotechnol.*, **47**, 157.
12. Pope, A., Pritchard, K., Williams, A., Roberts, A., Hackett, J. R., Mandecki, W., *et al.* (1996). *Immunotechnology*, **2**, 209.
13. Hawkins, R. E., Russell, S. J., and Winter, G. (1992). *J. Mol. Biol.*, **226**, 889.
14. Cai, X. and Garen, A. (1995). *Proc. Natl. Acad. Sci. USA*, **92**, 6537.
15. Marks, J. D., Ouwehand, W. H., Bye, J. M., Finnern, R., Gorick, B. D., Voak, D., *et al.* (1995). *BioTechnology*, **11**, 1145.
16. Palmer, D. B., George, A. J., and Ritter, M. A. (1997). *Immunotechnology*, **91**, 473.
17. De Kruif, J., Terstappen, L., Boel, E., and Logtenberg, T. (1995). *Proc. Natl. Acad. Sci. USA*, **92**, 3938.
18. Siegel, D. L., Chang, T. Y., Russell, S. L., and Bunya, V. Y. (1997). *J. Immunol. Methods*, **206**, 73.
19. Bobrow, M. N., Harris, T. D., Shaughnessy, K. J., and Litt, G. J. (1989). *J. Immunol. Methods*, **125**, 279.
20. Bobrow, M. N., Litt, G. J., Shaughnessy, K. J., Mayer, P. C., and Colon, J. (1992). *J. Immunol. Methods*, **150**, 145.
21. Merz, H., Malisius, R., Mannweiler, S., Zhou, R., Hartmann, W., Orscheschek, K., *et al.* (1995). *Lab Invest.*, **73**, 149.
22. Adams, J. C. (1992). *J. Histochem. Cytochem.*, **40**, 1457.
23. Osbourn, J. K., Derbyshire, E. J., Vaughan, T. J., Field, A. W., and Johnson, K. S. (1998). *Immunotechnology*, **3**, 293.
24. Osbourn, J. K., Earnshaw, J. C., Johnson, K. S., Parmentier, M., Timmermans, V., and McCafferty, J. (1998). *Nature Biotechnol.*, **16**, 778.

Chapter 4

ARM complexes for *in vitro* display and evolution of antibody combining sites

Michael J. Taussig, Maria A. T. Groves, Margit Menges, Hong Liu, and Mingyue He
Laboratory of Molecular Recognition, The Babraham Institute, Babraham, Cambridge CB2 4AT, U.K.

1 Introduction

Antibody V_H and V_L region genes are readily obtained by PCR and can be recombined at random to produce large libraries of single chain fragments (1). Such libraries may be obtained from normal or immune B lymphocytes, or constructed artificially from cloned gene fragments with synthetic H-CDR3 regions generated *in vitro* (2). Libraries can also be generated by mutagenesis of cloned DNA fragments encoding specific V_H/V_L combinations and screened for mutants having improved properties of affinity or specificity.

It is clear that for efficient antibody display/selection it is necessary to have a means of producing, and selecting from, very large libraries, e.g. single chain antibody libraries are potentially of a size of $> 10^{10}$ members. However, the size of such libraries can exceed by several orders of magnitude the ability of cell-based technologies to display all the members. Thus, the generation of phage display libraries (3) requires bacterial transformation with DNA, but the limitation of DNA uptake efficiency by bacteria means that a typical number of transformants which can be obtained is only 10^7 to 10^9 per transformation. There are also additional factors which reduce library diversity in all cell-based methods, e.g. certain antibody fragments may not be secreted, may be proteolysed, or form inclusion bodies, leading to the absence of such binding sites from the final library.

In order to avoid these limitations, alternative display technologies have been sought, in particular using *in vitro* systems. One such approach, known as *ribosome* or *polysome display*, consists of the display of single chain antibodies as nascent proteins on the surface of ribosomes, such that stable complexes with the encoding mRNA are formed; complexes are selected with antigen, and the genetic information is obtained by reverse transcription of the isolated or trapped mRNA (*Figure 1*) (4, 5). Since ribosome display libraries are generated wholly by

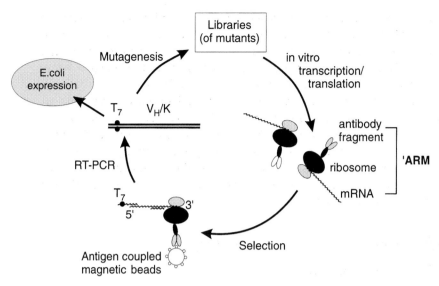

Figure 1 Ribosome display cycle for antibody selection (ARM display). ARM = antibody–ribosome–mRNA complex.

in vitro techniques (PCR), their size is limited mainly by the number of ribosomes which can be brought into the reaction mixture and the amount of DNA which can be handled conveniently per reaction.

Initially, display of a peptide library on prokaryotic polysomes in the *E. coli* S30 system was described (6, 7). To produce a population of stalled polysomes, agents such as rifampicin or chloramphenicol, which block prokaryotic translation, were used. (The term 'stalled' implies that the process of translation has been interrupted such that the ribosome, mRNA, and nascent protein remain associated.) The main example was of screening a large peptide library with $\sim 10^{12}$ members and selection of epitopes by a specific antibody. Peptide libraries have also been expressed and selected in the wheat germ system (8).

More recently, prokaryotic ribosome display was modified to select single chain antibody fragments (5, 9), with the introduction of additional features to make the method more suited to display of whole proteins in the prokaryotic system. One innovation was the stalling of the ribosome through the absence of a stop codon; stalling occurs because of the requirement for the stop codon to be recognized by release factors, which terminate translation by causing release of the nascent polypeptide chain (10). A number of additions were also made to improve the yield of mRNA after the polysome display cycle, including stem-loop structures at the 5' and 3' ends of the mRNA, vanadyl ribonucleoside complexes as nuclease inhibitors, protein disulfide isomerase (PDI) for folding of disulfide-bridged proteins, and an antisense nucleotide to inhibit ssrA RNA, which in the prokaryotic system causes release and degradation of proteins synthesized without a stop codon.

We have developed an alternative method of ribosome display based on eukaryotic *in vitro* expression and demonstrated its application to selection of antibodies (4). We have termed the method 'ARM display', since the selection particles consist of **a**ntibody–**r**ibosome–**m**RNA (ARM) complexes. The ARM method derives from two experimental results, namely the functional production of single chain antibodies *in vitro* in rabbit reticulocyte lysates (11) and the stabilizing effect of stop codon deletion, such that individual nascent proteins remain associated with their corresponding mRNA as stable ternary polypeptide–ribosome–mRNA complexes (12–14). We have demonstrated the specificity of ARM complexes and efficient selection of complexes carrying specific combining sites from artificial library mixtures (see Section 3). ARM display is also a suitable system for evolution of antibodies *in vitro* through mutagenesis and selection and for antibody engineering.

2 Ribosome display methodology

2.1 Outline of procedure

Figure 1 shows the ARM cycle. ARM complexes are generated by an *in vitro* coupled transcription/translation system to provide, in the absence of a 3′ stop codon, the linkage between individual antibodies and their encoding mRNA. Selection by antigen simultaneously captures the specific combining site and its relevant genetic information as mRNA. The latter can then be converted by RT-PCR to DNA for further manipulation. The antibody is in the form of the single chain fragment termed V_H/K (below), but the method is in principle equally applicable to other single chain forms.

The stages in the procedure are:

(a) Generation of ARM complexes by *in vitro* coupled transcription/translation of a DNA library.

(b) Removal of input DNA by DNase I digestion.

(c) Selection of ARM complexes by antigen immobilized on Dynabeads.

(d) Washing to remove non-selected complexes.

(e) RT-PCR amplification of selected mRNA.

(f) Gel analysis of recovered DNA.

One cycle is completed in about seven hours, including gel analysis.

2.2 Primer design and single chain antibody (V_H/K) construction for ARM display

Single chain Fv (sFv) fragments, in which the V_H and V_L domains are linked by a flexible linker peptide, have been widely used in phage display and bacterial expression (3, 15). Another type of single chain fragment, and the one that we have used principally, is V_H/K, in which the V_H domain is linked to the complete light chain, i.e. V_H-linker-V_L-C_L (16). The linker sequence is derived from the 'elbow' region between the V_H and C_H1 domains of the heavy chain. The use of

Table 1 Primers used to generate single chain V$_H$/K antibody construct

---

A. Primers T7Ab/back and Hlinker/for are used to generate the heavy chain V$_H$ domain

T7Ab/back

5′-GCAGC*TAATACGACTCACTATAGGAACAGACCACC****ATG***(C/G)AGGT(G/C)CA(G/C)CTCGAG(C/G)AGTCTGG-3′

The T7 promoter sequence is italicized.

ATG (bold) encodes the initiation methionine.

The underlined sequence is an *Xho*I site for cloning.

Hlinker/for

Human antibodies

5′-GGAGACTG(C/G)GTCATC(A/G)C(C/A)AT(T/G)TCGGAGACGAGGGGGAAAAG-3′

Mouse antibodies

5′-GGAGACTG(C/G)GTCATC(A/G)C(C/A)AT(T/G)TCGAGCTCGGCCAGTGGATAGACAGATGG-3′

B. Primers V$_L$/back and C$_K$/for are used to generate the complete κ light chain

VL/back

5′-GA(C/A)AT(T/G)G(T/C)GATGAC(C/G)CAGTCTCC-3′

CK/for

Human antibodies

5′-GCTCTAGAACACTCTCCCCTGTTGAAGCT-3′

Mouse antibodies

5′-GCTCTAGAACACTCATTCCTGTTGGAGCT-3′

---

the V$_H$/K fragment has a number of advantages, including stability of expression in *E. coli*, and the presence of the C$_L$ domain to act as a spacer for ribosome display and as a tag in detection and purification systems, such as ELISA, Western blotting, and affinity columns. A C-terminal spacer is a necessary element in constructs for ribosome display, since part of the nascent protein (about 26 amino acids) remains within the ribosome at the end of translation (17).

The V$_H$/K DNA construct is generated either by RT-PCR (mRNA as template) or PCR (DNA as template). For *in vitro* transcription/translation, a T7 promoter and protein initiation sequence are added using an upstream primer which also includes degenerate sequences complementary to antibody 5′ sequences (*Table 1*), the latter based on Marks *et al.* (1). To generate stable ARM complexes, the stop codon is removed from a downstream primer (CK/for) complementary to 3′ sequences of κ light chain (*Table 1*).

Construction of V$_H$/K from hybridomas and lymphocyte mRNA is carried out by single-step RT-PCR on total mRNA isolated from mouse spleen cells. V$_H$ and κ genes are produced separately, the V$_H$ domain DNA being obtained using the primer pair T7Ab/back and Hlinker/for and the complete κ chain DNA using primers V$_L$/back and CK/for (*Table 1, Protocol 1A*). Full-length V$_H$/K constructs are then produced by random pairing of V$_H$ and κ DNA using PCR assembly, in which joining of V$_H$ to κ is obtained through an overlap of 23 base pairs between the linker and V$_L$ (*Table 1, Protocol 1B*). The diversity of V$_H$/K libraries can be evaluated by cloning and sequencing or by PCR screening (1).

Protocol 1

Construction of V$_H$/K fragments from hybridomas and lymphocyte mRNA

Equipment and reagents

- mRNA purification kit (Pharmacia Biotech, Cat. No. 27-9255-01)
- dNTP stock solutions: 10 mM or 2.5 mM of each dNTP (Pharmacia, Cat. No. 27-2050/60/70/80/-01)
- Titan™ one tube single-step RT-PCR system (Boehringer Mannheim, Cat. No. 1888 382)

- Qiagen gel extraction kit (Cat. No. 0297.1)
- *Taq* polymerase (Boehringer Mannheim, Cat. No. 1146165)
- Agarose (Sigma, Cat. No. A-9539)
- 5 × gel loading buffer: 40% (w/v) sucrose, 0.25% bromophenol blue
- Sterilized (DEPC-treated) distilled water

A. Preparation of V$_H$ and κ chain DNA by single-step RT-PCR

1. Isolate mRNA from lymphocytes or hybridoma cells using Pharmacia mRNA purification kit (instructions provided with kit).

2. Prepare solutions for single-step RT-PCR as follows.

   (a) Solution 1:
   - 100 mM DTT (in kit)   5 μl
   - 10 mM dNTPs   2 μl
   - Upstream primer (16 μM)
     (i.e. T7Ab/back for V$_H$, or V$_L$/back for κ, *Table 1*)   3 μl
   - Downstream primer (16 μM)
     (i.e. Hlinker/for for V$_H$ or Cκ/for for κ, *Table 1*)   3 μl
   - Distilled H$_2$O to   50 μl

   (b) Solution 2:
   - 5 × RT-PCR buffer (with Titan kit)   20 μl
   - Distilled H$_2$O   28 μl

   Both solutions can be stored at −20 °C ready for use.

3. Set up RT-PCR reactions for V$_H$ and κ.[a]
   - Solution 1   25 μl
   - Solution 2   25 μl
   - 1-50 ng mRNA   1–2 μl
   - Enzyme mix from Titan kit   0.5 μl[b]
   - Cover with layer of mineral oil   10–20 μl[c]

4. Carry out thermal cycling using the following program.

   (a) One cycle of 48 °C for 45 min (reverse transcription), 94 °C for 2 min (AMV RT inactivation and mRNA/cDNA/primer denaturation).

   (b) Then 30–40 cycles of: 94 °C for 30 sec (denaturation), 54 °C for 1 min (annealing), 68 °C for 2 min (extension).

   (c) Finally, one cycle of 68 °C for 7 min (extension), then hold at 4 °C.

5. Analyse products by agarose gel electrophoresis. Prepare a 1% agarose gel containing 0.5 μg/ml ethidium bromide. Remove a 5–10 μl aliquot from each RT-PCR sample and add 1.5 μl of 5 $\times$ loading buffer. Mix well, load gel, and electrophorese samples. Visualize DNA bands under UV transillumination. Isolate V_H and κ chain DNA from the gel using Qiagen extraction kit.

B. Generation of V_H/K construct by PCR assembly of V_H and κ light chain

1. Combine equal amounts of V_H and κ chain DNA, in total 10–50 ng DNA.

2. Set up the first PCR (V_H/K template construction) as follows:
   - V_H DNA   x μl
   - κ DNA   y μl
   - 10 $\times$ PCR buffer (supplied with *Taq*)   2.5 μl
   - 2.5 mM dNTPs   1 μl
   - *Taq* polymerase   1 U
   - Distilled H_2O to   25 μl

3. Carry out seven to ten cycles of thermal cycling: 94 °C for 1 min, 63 °C for 4 min. Finally, hold at 4 °C.

4. Set up the second PCR (to amplify the V_H/K template made in step 3) as follows:
   - The above PCR product   2 μl
   - 10 $\times$ PCR buffer   5 μl
   - 2.5 mM dNTPs   4 μl
   - Primer T7Ab/back (16 μM)   1.5 μl
   - Primer Cκ/for (16 μM)   1.5 μl
   - *Taq* polymerase   1–2.5 U
   - Distilled H_2O to   50 μl

5. Carry out 30 cycles of thermal cycling: 94 °C for 30 sec, 54 °C for 1 min, 72 °C for 1 min. Finally, one cycle of 68 °C for 7 min (extension), then hold at 4 °C.

[a] The mixture volume can be scaled up to 100 μl or reduced to 10 μl.

[b] Alternatively, 0.5 μl (4–5 U) AMV reverse transcriptase from Promega (Cat. No. M5101) and 0.5 μl (2 U) of *Taq* DNA polymerase from Boehringer Mannheim (Expand™ high fidelity PCR system, Cat. No. 1732641) can be mixed and added to the reaction mixture.

[c] Thermal cyclers with features such as an overlying heat lid to prevent condensation require no mineral oil over the samples. Consult manufacturer's instructions.

2.3 Generation of ARM complexes by coupled transcription/translation *in vitro*

A number of *in vitro* transcription/translation systems are commercially available, including rabbit reticulocyte, wheat germ, and *E. coli* S30 extracts (18). There is also a variety of rabbit reticulocyte systems, including nuclease-treated, non-nuclease-treated, DTT-deficient, and coupled transcription/translation. We use the 'TNT T7 Quick' coupled transcription/translation system (Promega); this avoids separate isolation of mRNA, which is costly in time and materials. We have shown that plasmid, PCR products, or mRNA (either purified from cells or produced by a separate *in vitro* transcription system) can be used directly in this system to generate ARM complexes. PCR DNA without the 3' stop codon is used to produce ARM complexes. Where it is problematic to engineer stop codon deletion, as in cDNA or mRNA libraries, an alternative method would be the use of suppressor tRNA, charged with an amino acid, which recognizes and reads through the stop codon, thereby preventing the action of release factors (19).

Protocol 2 describes the procedure for generating ARM complexes. For display of a single antibody species, 1–100 ng of purified or non-purified PCR fragment is used. The amount of DNA used may depend on antibody affinity: antibodies with higher affinity are more sensitively recovered and thus require less input DNA for display.

In order to eliminate possible input DNA contamination of the selected ARM complexes, DNase I digestion is performed in one of two ways:

(a) DNase can be added to the transcription/translation mixture after translation (*Protocol 2*, step 3).

(b) DNase can be added to the Dynabeads (the carriers of the immobilized antigen) after ARM selection (*Protocol 4*, step 3).

Protocol 2

Generation of ARM complexes by coupled transcription/translation *in vitro*

Equipment and reagents
All solutions used must be sterilized.

- TNT® T7 Quick coupled transcription/ translation system (Promega, Cat. No. L1170 or L1171)
- PCR V$_H$/K DNA (see *Protocol 1*)
- 25 mM Mg acetate
- 10 × DNase I buffer: 400 mM Tris–HCl pH 7.5, 60 mM MgCl$_2$, 100 mM NaCl

- RNase-free DNase I (Boehringer Mannheim, Cat. No. 778 785 or Promega, Cat. No. M6101)
- Siliconized RNase-free microcentrifuge tubes (size 500 × 0.5 μl, Ambion, Cat. No. 12350)
- Sterilized (DEPC-treated) distilled H$_2$O

Method

1. Set up *in vitro* coupled transcription/translation mixture in a siliconized tube.[a]

• TNT system	20 μl
• PCR DNA	1–100 ng
• 1 mM methionine (included in TNT kit)	0.5 μl
• Mg acetate	0.5 μl
• Distilled H$_2$O to	25 μl

2. Incubate the mixture at 30 °C for 60 min.

3. Carry out DNase I digestion either by the following procedure, in the TNT transcription/translation mixture, or alternatively as in *Protocol 4*, step 3 after antigen selection. To the 25 μl of TNT transcription/translation mixture, add:

   - DNase I   2–5 U[b]
   - 10 × DNase I digestion buffer   3 μl
   - Distilled H$_2$O to   30 μl

4. Incubate at 30 °C for 15 min, then dilute with 30 μl cold PBS. The mixture is then ready for *antigen selection* (see *Protocol 4*).

[a] The reaction volume can be scaled up and down. As little as 5 μl of the TNT system has been successfully used for *in vitro* transcription/translation.

[b] The amount of DNase I used depends on the amount of input DNA. 2–5 U of DNase I is adequate for up to 100 ng PCR DNA. The reaction volume of the digestion can also be varied according to the TNT volume.

2.4 Antigen selection of ARM complexes

Specific ARM complexes are selected by binding to antigen-coupled magnetic beads (Dynabeads). Antigen can be coupled either directly onto tosyl activated Dynabeads or indirectly onto streptavidin-linked Dynabeads through protein biotinylation (see *Protocol 3*). The coating of antigen to the Dynabeads is critical for successful ARM display, as inadequately coated Dynabeads tend to trap ARMs, causing a high background during RT-PCR recovery.

Selection is carried out by adding the antigen-coated Dynabeads directly to the TNT translation mixture (*Protocol 4*). Non-binding complexes are removed by washing the beads. A magnetic particle concentrator (Dynal MPC) is used to wash and recover the beads carrying bound ARM complexes.

Protocol 3

Preparation of antigen- or streptavidin-coupled magnetic beads

Equipment and reagents

Sterilize all buffer solutions by autoclaving.

- Magnetic particle concentrator (Dynal MPC)
- Buffer A : 0.1 M Na phosphate buffer pH 7.4
- Buffer D: buffer A with 0.1% BSA (Sigma, Cat. No. A-4503)
- Buffer E: 0.2 M Tris–HCl pH 8.5 with 0.1% BSA
- PBS (phosphate-buffered saline) pH 7.4
- Antigen solution (0.5–1 mg/ml) in PBS

- Tosyl activated Dynabeads M-280 (Dynal UK): 6.7 × 10^8/ml or 10 mg/ml (Product No.142.03)
- Dynabeads M-280 Streptavidin (Dynal UK): 6.5 × 10^8/ml or 10 mg/ml (Product No.112.05/06)
- EZ-link™ sulfo-NHS-LC-LC-biotin (Pierce, Cat. No. 21338), 1 mg/ml in water
- Sterilized (DEPC-treated) distilled H$_2$O

Protocol 3 continued

A. Coupling of antigen to tosyl activated Dynabeads M-280 (see also manufacturer's instructions)

1. A convenient amount to prepare is 0.5–1 mg (50–100 μl) beads. Wash beads with 0.5–1 ml buffer A.

2. Mix with antigen in proportions of 20 μg antigen:1 mg beads and vortex for 1 min. Incubate for 16–24 h at 37 °C with slow rotation.

3. Remove the supernatant and wash the beads twice with 0.5 ml buffer D for 5 min at 4 °C, then treat with 0.5 ml buffer E for 4 h at 37 °C to block free tosyl groups.

4. Finally wash once with 0.5 ml buffer D for 5 min at 4 °C and resuspend in the original volume in buffer D containing 0.02% Na azide.

5. Coated beads can be stored for several months at 4 °C.

B. Coupling of biotinylated proteins to Dynabeads Streptavidin

1. Biotinylation of protein. Mix antigen solution in PBS (pH 7–8.5) with sulfo-NHS-biotin solution in proportions of 25 μg antigen:1 μg sulfo-NHS-biotin. Incubate the mixture at room temperature for 30 min.

2. Remove unreacted biotin by dialysis against 2×500 ml PBS overnight at 4 °C. Store at -20 °C.

3. Coupling. Wash Dynabeads M-280 Streptavidin three times with buffer A. Add biotinylated proteins in PBS and incubate the mixture at room temperature for 30 min, using a ratio of biotinylated protein:beads of 10 μg:1 mg.

4. Remove the supernatant and wash the beads three times with 50 μl PBS. Resuspend beads in the original volume in buffer D containing 0.02% Na azide.

5. Use SDS–PAGE analysis of the biotinylated protein before and after coating to confirm adsorption.

Protocol 4

Antigen selection of ARM complexes

Equipment and reagents
Sterilize all buffers by autoclaving.

- Antigen-coupled Dynabeads (see *Protocol 3*)
- PTM washing buffer: PBS containing 1% Tween 20 and 5 mM Mg acetate
- $10 \times$ DNase I buffer: 400 mM Tris–HCl pH 7.5, 60 mM $MgCl_2$, 100 mM NaCl
- RNase-free DNase I (Boehringer Mannheim, Cat. No. 776 785 or Promega, Cat. No. M6101)
- Sterilized (DEPC-treated) distilled H_2O

Protocol 4 continued

Method

1. Add 2 μl antigen-coupled beads to 10–20 μl of the TNT translation mixture obtained from *Protocol 2*, step 3. Incubate at room temperature for 60 min with gentle vibration.

2. Wash beads three times with 50 μl PTM buffer and twice with 50 μl H_2O.

3. Optional. For DNase digestion after antigen selection (an alternative to *Protocol 2*, step 3) add 2–5 U of DNase I in 50 μl of 1 × DNase I digestion buffer to the antigen-selected, washed beads. Incubate at 37 °C for 10 min. Remove DNase by washing the beads three times with 50 μl PTM and once with 50 μl sterilized distilled H_2O.

4. Resuspend beads in 10 μl sterilized distilled H_2O ready for RT-PCR (*Protocol 5*). The beads can be stored either at 4 °C or −20 °C.

2.5 Recovery and amplification of DNA from antigen-selected ARM complexes

To recover and amplify the genetic material after selection, a single-step RT-PCR is performed on ribosome-bound mRNA (*Figure 1*). Since the ribosome occupies the 3′ end of the mRNA at the termination of translation, a downstream primer D2, designed to hybridize at least 60 nt upstream of the 3′ end of the mRNA, is used in combination with the upstream primer T7Ab/back in the RT-PCR reaction (*Figure 2* and *Table 2*). Since the use of D2 produces a shortened DNA

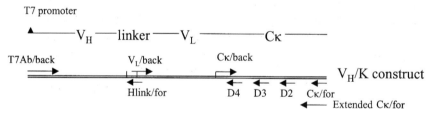

Figure 2 Single chain antibody construct V_H/K for ribosome display showing sites of primer hybridization.

Table 2 Downstream primers for DNA recovery by RT-PCR

Human antibodies

D2: 5′-GCTCAGCGTCAGGGTGCTGCT-3′
D3: 5′-GGGATAGAAGTTATTCAGC-3′
D4: 5′-GAAGACAGATGGTGCAGC-3′

Mouse antibodies

D2: 5′-CGTGAGGGTGCTGCTCAT-3′
D3: 5′-GGGGTAGAAGTTGTTCAAGAAG-3′
D4: 5′-CTGGATGGTGGGAAGATGG-3′

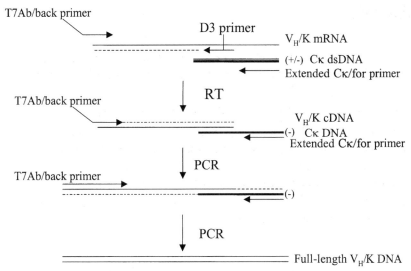

T7Ab/back primer

D3 primer

V_H/K mRNA

(+/-) Cκ dsDNA
Extended Cκ/for primer

RT

T7Ab/back primer

V_H/K cDNA

(-) Cκ DNA
Extended Cκ/for primer

PCR

T7Ab/back primer

(-)

PCR

Full-length V_H/K DNA

Figure 3 Full-length recovery of V_H/K DNA by single tube RT-PCR.

fragment in the first cycle, another primer D3, which hybridizes upstream of D2, is used for recovery in the second cycle; similarly in the third cycle a primer D4 hybridizing upstream of D3 is used (*Figure 2* and *Table 2*). The recovered DNA thus becomes progressively shorter in each cycle, but full-length V_H/K can be regenerated in any cycle by recombinational PCR. The shortening only affects the constant domain of the light chain. *Protocol 5A* describes the RT-PCR procedure to recover DNA from ARM complexes.

Recently, we have developed a procedure to recover full-length cDNA by single-step RT-PCR (*Figure 3*). Here, primer D3 is used to make first strand cDNA. A DNA fragment encoding the complete Cκ region, which overlaps the first strand cDNA, is included for full-length DNA extension by PCR. The primer T7Ab/back and an extended Cκ/for primer (*Table 3*) are used to amplify the full-

Table 3 Primers for generating Cκ DNA fragment for full-length DNA recovery by PCR

Human antibodies

Cκ/back: 5'-ACTGTGGCTGCACCATCTG-3'

Cκ/for: 5'-GC<u>TCTAGA</u>ACACTCTCCCCTGTTGAAGCT-3'

Extended Cκ/for: 5'-CG<u>GAATTC</u><u>TCTAGA</u>*GTGATGGTGATGGTGATG*GTAGACTTTGTGTTTCTCGTAGT
CTGCTTTGCTCAGCGTCAGGGTGCTGCT-3'

Mouse antibodies

Cκ/back: 5'-AAACGGGCTGATGCTGCA-3'

Cκ/for: 5'-GC<u>TCTAGA</u>ACACTCATTCCTGTTGGAGCT-3'

Extended Cκ/for: 5'-GC<u>TCTAGA</u>GGCCTCACAAGGTATAGCTGTTATGTCGTTCATACTCGTCCT
TGGTCAACGTGAGGGTGCTGCTCAT-3'

Underlined sequences are restriction enzyme sites for *Xba*I (TCTAGA) and *Eco*RI (GAATTC) for cloning.
Italicized sequence is a (His)$_6$ tag.

length DNA by PCR. In order to avoid possible interference by remaining D3 primer in the PCR process, the annealing temperature in PCR is increased to 68 °C at which D3 does not hybridize (*Protocol 5B*).

Protocol 5

Recovery and amplification of genetic information from antigen-selected ARM complexes

Equipment and reagents

- Titan™ one tube RT-PCR system (Boehringer Mannheim, Cat. No. 1888 382)
- 10 mM dNTP stock solutions
- Primers T7Ab/back, D2, D3, D4, Cκ/for, Cκ/back, extended Cκ/for (*Tables 2, 3*)
- Sterilized (DEPC-treated) distilled H_2O

A. Using individual upstream primers D2, D3, D4

1. Prepare RT-PCR solutions.

    (a) Solution 1:

• 100 mM DTT (in Titan kit)	5 μl
• 10 mM dNTPs	2 μl
• T7Ab/back (16 μM)	3 μl
• D2 primer (16 μM)	3 μl
• Distilled H_2O	37 μl

    This solution is for the first ARM cycle; in the second cycle, replace primer D2 by D3, and in the third by D4, etc.

    (b) Solution 2:

• 5 × RT-PCR buffer (with Titan kit)	20 μl
• Distilled H_2O	28 μl

2. Set up RT-PCR mixture as follows:

• Solution 1	25 μl
• Solution 2	25 μl
• Enzyme mix	0.5 μl (also see *Protocol 1*, footnote *b*)

3. For each individual RT-PCR, dispense 10–20 μl of the RT-PCR mixture into each reaction tube. Add 2 μl of the beads and mix well. Thermally cycle the sample using the program in *Protocol 1A*, step 4.

4. Analyse the sample by agarose gel electrophoresis using 5–10 μl of the sample, as in *Protocol 1A*, step 5.

B. Recovery of full-length DNA by RT-PCR

1. Generate Cκ DNA by PCR on κ chain DNA (e.g. the V_H/κ construct made in *Protocol 1*) using primers Cκ/back and Cκ/for (*Table 3*).

2. Prepare solutions for RT-PCR.

    (a) Solution 1:

• 100 mM DTT	5 μl
• 10 mM dNTPs	2 μl
• T7Ab/back (16 μM)	3 μl
• D3 primer (16 μM)	3 μl
• Cκ DNA	∼ 10–100 pg
• Extended Cκ/for primer (16 μM)	3 μl
• Distilled H$_2$O to	50 μl

    (b) Solution 2:

• 5 × RT-PCR buffer (with Titan kit)	20 μl
• Distilled H$_2$O	28 μl

3. Set up RT-PCR mixture as follows:

    • Solution 1   25 μl

    • Solution 2   25 μl

    • Enzyme mix   0.5 μl (also see *Protocol 1*, footnote *b*)

4. For each individual RT-PCR, dispense 10–20 μl of the RT-PCR mixture into a re-action tube. Add 2 μl of the beads and mix well. Thermally cycle the sample using the following program.

    (a) One cycle of 48 °C for 45 min (reverse transcription), 94 °C for 2 min (AMV RT inactivation and RNA/cDNA/primer denaturation).

    (b) 40 cycles of 94 °C for 30 sec (denaturation), 68 °C for 2 min (annealing and extension).

    (c) Finally, one cycle of 68 °C for 7 min (extension). Then 4 °C (hold).

5. Analyse the sample by agarose gel electrophoresis using 5–10 μl of the sample, as in *Protocol 1A*, step 5.

2.6 Further ARM cycles and cloning

The DNA recovered in one ARM cycle can be used as the starting point for a second cycle or can be cloned in *E. coli*. To carry out a second cycle, the PCR products from the first cycle (*Protocol 5A or 5B*), with or without gel purification, are added to the TNT system (*Protocol 2*). Alternatively, recovered faint bands can be extracted and reamplified by PCR before carrying out further cycles.

A cloning technology we have used is the Stratagene TA vector system (Cat. No. K20030-01/40/J10). The ARM method can be used for screening and identification of functional clones prior to *E. coli* expression analysis. Plasmid DNA is prepared and amplified by PCR for ARM display. In order to quantify recovery and rule out differences caused by the variable efficiency of RT-PCR, an internal control in the form of an mRNA carrying the same flanking sequences but of shorter length than the targeting fragment, is used as a standard in RT-PCR (see *Protocol 6*). The internal standard can be prepared by PCR deletion of ~300 bp from the V_H/K construct, so that standard and targeting DNA are readily distinguished on gel electrophoresis.

2.7 Analysis of clones encoding antibodies by ARM display

It is also possible to derive information on binding site characteristics of individual clones by ARM display. An ARM complex is a bifunctional entity in which the antibody fragment binds to specific ligand and the attached mRNA is used as a sensitive detection signal for amplification by RT-PCR. ARM binding activity can be estimated by scanning the intensity of the RT-PCR-generated DNA on an agarose gel (also using the internal standard above as reference). The use of free antigens as inhibitors (competitive ARM display) reveals data on specificity and affinity. Thus, the anti-progesterone antibody DB3 expressed as an ARM complex has shown similar properties to its bacterially expressed V_H/K fragment analysed by ELISA.

Protocol 6

Screening of clones by ARM display using an internal standard in RT-PCR

Equipment and reagents

- Titan™ one tube RT-PCR system (Boehringer Mannheim, Cat. No. 1888 382)
- 10 mM dNTP stock solutions
- Primers T7Ab/back, D3 (*Table 2*)
- mRNA internal standard produced by *in vitro* transcription of shortened V_H/K DNA
- Sterilized (DEPC-treated) distilled H_2O

Method

1. Carry out ARM display on individual clones and isolate complexes on beads (*Protocols 2–4*).

2. Analyse ARM mRNA by RT-PCR. Prepare solutions as follows.

    (a) Solution 1:

• 100 mM DTT	5 μl
• 10 mM dNTPs	2 μl
• T7Ab/back primer (16 μM)	3 μl
• D3 primer (16 μM)	3 μl
• mRNA internal standard	0.1–1 ng
• Distilled H_2O to	50 μl

    (b) Solution 2:

• 5 × RT-PCR buffer (included in Titan kit)	20 μl
• Distilled H_2O	28 μl

3. Set up RT-PCR mixture as follows:

   - Solution 1   25 μl
   - Solution 2   25 μl
   - Enzyme mix   0.5 μl (also see *Protocol 1*, footnote *b*)

4. For each individual RT-PCR, dispense 10–20 μl of the RT-PCR mixture into each reaction tube. Add 2 μl of the beads and mix well. Thermally cycle the sample using the program in *Protocol 1A*, step 4.

5. Analyse the sample by agarose gel electrophoresis using 5–10 μl of the sample, as in *Protocol 1A*, step 5.

3 Examples of ARM display

3.1 ARM specificity

We have used extensively the anti-progesterone antibody DB3 as a model for ribosome display. The crystallographic structure of the DB3 Fab' fragment has been determined and the fine-specificity of the antibody has been studied in structural detail by X-ray crystallography (20, 21). DB3, and the corresponding V_H/K fragment, have an affinity of $\sim 10^9$ l/M for free progesterone, but bind testosterone with a 7000-fold lower affinity (16). They also distinguish progesterone derivatives in which hemisuccinyl (HMS) or carboxymethyl (CMO) groups are attached at different positions (C11, C6, C3, C21) on the steroid nucleus, the order of reactivity of such steroids being C11-HMS > C3-CMO > C6-HMS > C21-HMS (16).

In addition to DB3 itself, we have also used two mutant forms of DB3 in order to investigate antibody specificity and selection from ARM libraries. The mutant designated $DB3^R$, in which residue H100 is mutated from tryptophan to arginine, binds progesterone with an affinity similar to that of native DB3. *Figure 4A* demonstrates antigen-specific ARM selection, in which the V_H/K fragment of $DB3^R$ was translated *in vitro* and ARMs were exposed to beads coupled to progesterone-11α-BSA, testosterone-3-BSA, or BSA alone. RT-PCR of the washed beads led to recovery of a single DNA fragment of the expected size (~ 1 kb) from progesterone-BSA coupled beads (track 2), but not from testosterone-BSA or BSA beads (tracks 4 and 6). The identity of the band was confirmed by direct DNA sequencing. There was no recovery when PCR alone was carried out on the beads after selection (tracks 3, 5, 7), or when the procedure was performed with non-translated $DB3^R$ mRNA (track 1), showing the absence of DNA or mRNA carryover but the requirement for protein translation. The $DB3^R$ ARM was also captured on beads coated with rat anti-mouse κ, confirming the expression of the Cκ domain 'tag' (*Figure 4B*, track 2).

DB3 ARM binding was sensitively inhibited by free progesterone-11α-hemisuccinate (IC_{50} 3×10^{-9} M). When the series of progesterone derivatives was used, the order of reactivity (above) was the same as that with the native anti-

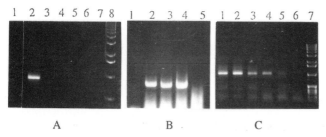

Figure 4 (A) Specific selection of DB3^R ARMs by progesterone-11α-BSA coupled beads. Track 1: RT-PCR of non-translated DB3^R mRNA selected by progesterone-11α-BSA beads. Tracks 2, 4, 6: RT-PCR of DB3^R ARMs selected by progesterone-11α-BSA, testosterone-3-BSA, or BSA coupled beads respectively. Tracks 3, 5, 7: PCR of DB3^R ARMs selected by progesterone-11α-BSA beads, testosterone-3-BSA, or BSA coupled beads. Track 8: 1 kb DNA marker. (B) Non-binding of a DB3^{H35} ARM library to progesterone-11α-BSA coupled beads. Track 1: RT-PCR of solution control. Tracks 2, 3: RT-PCR of DB3^R ARMs selected by rat anti-κ coupled beads or progesterone-11α-BSA beads respectively. Tracks 4, 5: RT-PCR of DB3^{H35} ARMs selected by anti-κ beads or progesterone-11α-BSA beads respectively. (C) Selection of DB3^R from ARM libraries containing different ratios of DB3^R and DB3^{H35} mutants. Selection was with progesterone-11α-BSA coupled beads. Track 1, ratio of DB3^R:DB3^{H35} of 1:10; track 2, 1:10^2; track 3, 1:10^3; track 4, 1:10^4; track 5, 1:10^5. Track 6, DB3^{H35} mutant library alone; track 7, 1 kb DNA marker.

body, confirming that specificity and affinity closely resembled those of the soluble V_H/K. Since antibody fine-specificity is dependent on correct folding, the result indicates that nascent DB3 V_H/K in the form of an ARM complex folds appropriately in the TNT *in vitro* system.

3.2 Selection of DB3^R V_H/K from libraries

In contrast with DB3^R, mutants at position H35 bind progesterone weakly or not at all. In these mutants, the H35 codon AAC was mutated to [C/G][T/A/G]A, encoding a set of six amino acid substitutions (Leu, Gln, Arg, Val, Glu, Gly). When the DB3^{H35} mutant library alone was displayed as ARM complexes, no DNA band was recovered after selection with progesterone-11α-BSA beads (*Figure 4B*, track 5); translation of DB3^{H35} was confirmed by the band obtained with beads coated with anti-κ antibody (track 4). When DNA mixtures containing DB3^R and DB3^{H35} mutants in ratios ranging from 1:10 to 1:10^5 were displayed as ARM libraries, a band of V_H/K size was in all cases recovered after a single cycle (*Figure 4C*, tracks 1–5). To identify the selected band, RT-PCR fragments were cloned before and after selection and identified through restriction enzyme mapping (a unique *Hinc*II enzyme site was removed from DB3^{H35} but left in DB3^R) (He *et al.*, in preparation). DB3^R constituted 90% or more of clones selected from the 1:10 to 1:10^3 library mixtures, 70% of clones from a 1:10^4 library, and 40% of those from a 1:10^5 library. The enrichment factor in one cycle was thus ~ 10^4 fold, in agreement with the data from direct sequencing of PCR mixtures (4).

Consecutive cycles of ARM display enabled selection from libraries in which the binding fragment was present at a lower frequency. Thus a 1:10^6 DB3^R:

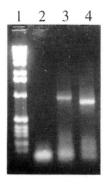

Figure 5 Enrichment of DB3^R from a 1:10^6 (DB3^R:DB3^{H35}) library by repeated ARM display cycles. Selection was with progesterone-11α-BSA coupled beads. Track 1: 1 kb DNA marker. Tracks 2, 3, 4: RT-PCR after first, second, and third cycles respectively.

DB3^{H35} library did not produce a detectable RT-PCR band after one cycle (*Figure 5*, track 2), but two further cycles of ARM generation and selection led to recovery of a V$_H$/K band, with increased intensity at each repetition (tracks 3, 4) (4). Selection of DB3^R was confirmed by sequencing.

4 Troubleshooting

4.1 Background

Non-specific DNA recovery can occur for a number of reasons, including carry-over of input DNA despite DNase digestion, non-specific sticking of ARMs through the washing procedure, or inadequate blocking of the Dynabeads. PCR (instead of RT-PCR) on the selected Dynabeads can be used to detect con-tamination by input DNA. Further DNase I digestion on the Dynabeads can then be carried out to remove DNA completely. Increasing the number of washes, particularly with distilled H$_2$O, can reduce background due to sticking of complexes. Dynabeads which are uncoated or inadequately coupled tend to bind ARMs non-specifically.

4.2 No DNA recovery

Sometimes no DNA is recovered by RT-PCR after the ARM cycle. As in *Figure 5*, track 2, this may merely indicate that the species being selected is rare and that further cycles are required. However, other factors can account for lack of recovery, including the efficiency of DNA transcription/translation or RT-PCR, RNA degradation, or functionality of the nascent protein. Variability of the transcription/translation reaction can be monitored by use of the positive control provided in the kits. The use of internal control mRNA can indicate the efficiency of the RT-PCR (*Protocol 6*). It is possible to include RNase inhibitors into the washing buffer to prevent degradation.

5 Summary

ARM display is the application of eukaryotic ribosome display to selection and evolution of antibody combining sites. It potentially displays very large libraries and enables rapid isolation of antibodies *in vitro*. Cycles of ARM display can be run consecutively to select rare antibody specificities. For antibody engineering, mutations can be introduced sequentially until the desired properties are achieved; the operation of the cycle itself also introduces a low level of random mutation through the errors of PCR, which can lead to 'protein evolution' through selection of improved properties of affinity and specificity (9). In principle, ribosome display can also be applied to other proteins or peptides, with potential applications in functional genomics, e.g. in isolation of genes from cDNA or mRNA libraries, identification of new functional genes, and discovery of protein–protein interactions. *Protocol* 7 summarizes the entire procedure.

Protocol 7

Summary of procedure for the ARM display cycle

Equipment and reagents

- See *Protocols 1–6*

Method

1. Set up *in vitro* coupled transcription/translation in a siliconized PCR tube. A standard mixture would be:

    - TNT Quick system   20 µl
    - PCR DNA   1–100 ng
    - 1 mM methionine   0.5 µl
    - 25 mM Mg acetate   0.5 µl
    - Distilled H_2O to total   25 µl

    Incubate at 30 °C for 60 min.

2. (Option: if this step is chosen, step 5 can be omitted.) Add 2–5 U DNase I together with 10 × DNase I buffer and incubate at 30 °C for a further 15 min.

3. Dilute the mixture with an equal volume of cold PBS and add 1–2 µl antigen-linked Dynabeads. Incubate at room temperature for 60 min with gentle vibration.

4. Wash the Dynabeads three times with 50 µl PBS, 1% Tween 20, 5 mM Mg acetate, followed by two washes with 50 µl sterilized H_2O.

5. (Option: if this step is chosen, step 2 can be omitted.) Add 2–5 U DNase I together with DNase I digestion buffer to the Dynabeads and incubate at 37 °C for 10 min, followed by washing as described in step 4.

6. Resuspend the beads in 10 µl H_2O for RT-PCR.

7. Set up 10 µl RT-PCR mixture for each sample and carry out thermal cycling.

8. Analyse products by agarose gel electrophoresis.

References

1. Marks, J. D., Griffiths, A. D., Malmqvist, M., Clackson, T. P., Bye, J. M., and Winter, G. (1992). *Bio/Technology*, **10**, 779.
2. Hawkins, R. E., Russell, S. J., and Winter, G. (1992). *J. Mol. Biol.*, **226**, 889.
3. Winter, G., Griffiths, A. D., Hawkins, R. E., and Hoogenboom, H. R. (1994). *Annu. Rev. Immunol.*, **12**, 433.
4. He, M. and Taussig, M. J. (1997). *Nucleic Acids Res.*, **24**, 5132.
5. Hanes, J. and Plückthun, A. (1997). *Proc. Natl. Acad. Sci. USA*, **94**, 4937.
6. Mattheakis, L. C., Bhatt, R. R., and Dower, W. J. (1994). *Proc. Natl. Acad. Sci. USA*, **91**, 9022.
7. Mattheakis, L. C., Dias, J. M., and Dower, W. J., (1996). In (Abelson J. ed) *Methods in enzymology* Vol. 267, p. 195. (Academic Press).
8. Gersuk, G. M., Corey, M. J., Corey, E., Stray, J. E., Kawasaki, G. H., and Vessella, R. L. (1997). *Biochem. Biophys. Res. Commun.*, **232**, 578.
9. Hanes, J., Jermutus, L., Weber-Bornhauser, S., Bosshard, H. R., and Plückthun, A. (1998). *Proc. Natl. Acad. Sci. USA*, **95**, 14130.
10. Stansfield, I., Jones, K. M., and Tuite, M. F. (1995). *Trends Biochem. Sci.*, **20**, 489.
11. Nicholls, P. J., Johnson, V. G., Andrew, S. M., Hoogenboom, H. R., Raus, J. C. M., and Youle, R. J. (1993). *J. Biol. Chem.*, **268**, 5302.
12. High, S., Gorlich, D., Wiedmann, M., Rapoport, T. A., and Dobberstein, B. (1991). *J. Cell. Biol.*, **113**,
13. Fedorov, A. N. and Baldwin, T. O. (1995). *Proc. Natl. Acad. Sci USA*, **92**, 1227.
14. Jungnickel, B. and Rapaport, D. (1995). *Cell*, **82**, 261.
15. Huston, J. S., Mudgett-Hunter, M., Tai, M. S., McCartney, J., Warren, F., Haber, E., *et al.* (1991). In *Methods in enzymology* (ed. J. J. Langone), Vol. 203, p. 46. (Academic Press).
16. He, M., Kang, A. S., Hamon, M., Humphreys, A. S., Gani, M., and Taussig, M. J. (1995). *Immunology*, **84**, 662.
17. Makeyev, E. V., Kolb, V. A., and Spirin, A. S. (1996). *FEBS Lett.*, **378**, 166.
18. Tymms, M. J. (ed.) (1995). *Methods in molecular biology*, Vol. 37.
19. Ellman, J., Mendel, D., Anthony-Cahill, S., Noren, C. J., and Schultz, P. G. (1991). In *Methods in enzymology* (ed. J. J. Langone), Vol. 202, p. 301. (Academic Press).
20. Arevalo, J. H., Stura, E. A., Taussig, M. J., and Wilson, I. A. (1993). *J. Mol. Biol.*, **231**, 103.
21. Arevalo, J. H., Hassig, C. A., Stura, E. A., Sims, M. J., Taussig, M. J., and Wilson, I. A. (1994). *J. Mol. Biol.*, **241**, 663.

Chapter 5

Human monoclonal antibodies to blood group antigens

Belinda M. Kumpel

International Blood Group Reference Laboratory, Southmead Road,
Bristol BS10 5ND, U.K.

1 Introduction

Many human alloantigens on red blood cells are not recognized by the mouse immune system, and therefore murine monoclonal antibodies to many desired blood group specificities cannot be produced using conventional murine hybridoma technology. Among these is the Rh (rhesus) D antigen, which is the most immunogenic of the blood group alloantigens and clinically is the most important after the ABO blood groups. Most human monoclonal blood group antibodies have been made to the Rh D polypeptide. Monoclonal anti-D has been produced to replace or supplement plasma-derived anti-D for blood grouping (1) and for therapeutic Rh D prophylaxis for prevention of haemolytic disease of the newborn (2). Also, there are more individuals with high titre anti-D than with antibodies to other blood groups, who can donate source material. For this reason, the methods described in this chapter are those that have been used for the production of monoclonal anti-D.

Human monoclonal antibodies have been produced by three procedures, each with minor variations between laboratories.

(a) Human B cells have been immortalized with Epstein–Barr virus (EBV) and cloned. This was the method initially used about twenty years ago. Some cell lines are very stable, with continuous production of antibody (2). However, there is great variation between cell lines, and many either cease to grow or to secrete antibody, and for this reason alternative strategies have been developed.

(b) EBV-transformed B cells have been fused with other cell lines, generally a mouse myeloma, to form heterohybridomas (3). These cells grow rapidly but care must be taken to ensure stability of antibody secretion, by repetitive cloning.

(c) Less often, human B cells have been directly fused with other cell lines.

With heterohybridomas, it should be borne in mind that the glycosylation of the human antibody will be dictated predominantly by the myeloma fusion partner, and thus will be characteristic of mouse rather than human antibodies if a murine myeloma is used.

Although both IgG and IgM monoclonal anti-D have been produced by these procedures, EBV transformation alone favours IgG.

The following sequence of procedures is usually undertaken:

- Selection of donor
- Immortalization of B cells with EBV
- Selection or enrichment of specific antibody-producing cells (optional)
- Fusion (optional)
- Cloning
- Screening culture supernatants and characterization of monoclonal antibodies

2 General equipment and reagents required

The following equipment, materials, and reagents can be obtained from many suppliers.

(a) Equipment
- Refrigerator
- Class II laminar flow cabinet
- Bench centrifuge, rotor radius 15–20 cm
- 37 °C incubator with a humidified atmosphere of 5% CO_2 in air
- Liquid nitrogen storage tanks for cryopreservation of cells
- Freezer −20 °C
- Inverted microscope with phase-contrast optics
- Microscope with phase-contrast optics
- Neubauer counting chamber (haemocytometer)

(b) Sterile disposables
- Tissue culture flasks, 50 ml and 250 ml
- Tissue culture plates, 24-well, 48-well, and 96-well
- Pipettes, 1 ml and 10 ml
- Centrifuge tubes, 10 ml and 25 ml (Universals)
- Glass bottles, 30 ml
- Plastic cryopreservation tubes, 1–2 ml
- Filters, 0.22 μm
- Syringes, 10 ml
- Needles, 21 gauge
- Instruments, forceps, scalpels
- Petri dishes

(c) Reagents

- RPMI-1640
- Fetal calf serum (FCS)
- L-Glutamine, 200 mM (100 ×)
- Penicillin, 10 000 U/ml (100 ×)
- Streptomycin, 10 000 µg/ml (100 ×)
- RPMI/FCS: complete culture medium containing RPMI-1640 supplemented with 10% FCS, 2 mM L-glutamine, 100 U/ml penicillin, and 100 µg/ml streptomycin
- Phosphate-buffered saline (PBS)

3 Selection of donor

Human immune cells must be obtained from an immunized donor. Peripheral blood is usually the only source of material. For Rh D, immunization of a Rh D negative individual will have occurred by infusion of incompatible (Rh D positive) blood by either:

(a) Transfusion (this may occur if a large emergency transfusion is required and sufficient Rh D negative blood is unavailable).

(b) Deliberate immunization of Rh D negative males with Rh D positive red cells (to produce anti-D for Rh prophylaxis).

(c) Pregnancy, when a mother may be immunized by fetal blood, usually at delivery.

The immune status of the donor is very important in determining a successful outcome. B lymphocytes activated by antigen are only in the peripheral circulation for a relatively short time after immunization, so that it is desirable—indeed almost mandatory—that the donor is re-immunized before donating blood. Ideally, this is achieved deliberately. In practice, a feto-maternal haemorrhage is usually negligible and is difficult to quantify. It has been found that Rh D-specific heterohybridomas were only generated using blood taken from donors about two to four weeks after re-immunization with Rh D positive red cells (4). Thus it may be futile to attempt to produce human monoclonal antibodies using blood that is not obtained within one to six weeks after boosting, even though the donor has serum antibodies.

B cells obtained from lymphoid organs such as the spleen and tonsil are more suitable but rarely available.

3.1 Preparation of lymphocytes

Carry out all procedures under strict aseptic conditions. Collect blood into anti-coagulant and process immediately or store at ambient temperature (not refrigerated) for up to two days.

Protocol 1

Preparation of peripheral blood mononuclear cells

Reagents

- Ficoll-Triosil (Histopaque, Lymphoprep) density 1.077
- Heparin (sodium salt) 1000 U/ml

- RPMI-1640 containing 10 U/ml heparin (Hep-RPMI)

Method

1. Layer blood carefully over Ficoll-Triosil in centrifuge tubes, using a ratio of two volumes of blood to one volume density gradient medium.

2. Centrifuge the tubes at 1100 g for 20 min at room temperature.

3. Collect a sample of plasma and store frozen to act as a control for the desired antibody.

4. Using a pipette, collect the layer of white lymphocytes from the interface between the plasma and Ficoll, and place in a fresh centrifuge tube. Some medium and plasma will be collected as well, but ensure that the tube is less than half filled with cell suspension.

5. Fill the tubes with Hep-RPMI and centrifuge at 800 g for 5 min to pellet the cells.

6. Discard the supernatant, resuspend the cell pellet by gently tapping the tube, and fill with Hep-RPMI. Centrifuge at 500 g for 5 min.

7. Repeat this step.

8. Resuspend in 10 ml RPMI/FCS and count the cells. The cell preparation—peripheral blood mononuclear cells (PBMC)—will be a mixture of B lymphocytes, T lymphocytes, NK cells, and monocytes, and some platelets.

9. If using spleen, place the tissue in a sterile Petri dish and gently disaggregate with forceps and a scalpel. Layer the cell suspension over Ficoll and proceed as above.

4 EBV transformation of B cells

Epstein–Barr virus is a human herpes virus and infects human B cells by binding to the C3d complement receptor (CR2, CD21). This receptor is not expressed on other leucocytes. B cells are initially activated by the virus, a minority then become transformed, and subsequently a fraction of these develop immortality, when the EBV genome has integrated into the host cell DNA. Thus EBV-transformed B cell lines (B-lymphoblastoid cell lines, B-LCL) will develop from less than 1% of the original B cells. Greater efficiency of immortalization can be achieved with the use of EBV of high titre.

Steps must be taken to remove or counter cytotoxic T cells which may kill the virally-infected B cells. About 90% of adults have been infected with EBV and will have developed T cell immunity. This T cell abrogation can be achieved by

T cell depletion or, more simply, by addition of PHA (phytohaemagglutinin, which activates T helper cells) or cyclosporin A (which depletes cytotoxic T cells). Greater success was obtained with PHA than with cyclosporin A in the production of monoclonal antibodies (5).

4.1 Preparation of EBV

The B95-8 marmoset B cell line is used as the source of EBV, as it secretes infectious virus into culture medium at high levels, unlike human B-LCL. The cell line can be obtained from national cell culture collections and should be treated as hazardous material. Greater amounts of virus are secreted if the cells are put under environmental stress.

Protocol 2

Preparation of EBV

Reagents

- Human umbilical cord blood, taken after delivery of a healthy term baby
- B95-8 cell line

A. Culture of EBV

1. The B95-8 cells are cultured in RPMI/FCS until there are about 10×10^6 cells.

2. Stress the cells to produce high titre virus. Dispense at 2×10^5 cells/ml in RPMI-1640 containing 2% FCS. Incubate at 33 °C in a humidified atmosphere of 5% CO_2 in air for two to three weeks, without feeding, until the culture medium is yellow.

3. Harvest the supernatant medium; pour into centrifuge tubes and centrifuge at 800 g for 5 min.

4. Filter the supernatant through 0.2 μm sterile filters.

5. Dispense aliquots of 1–2 ml into cryotubes.

6. Place in racks in liquid nitrogen storage tank.

B. Determination of EBV titre

1. Collect cord blood aseptically into anticoagulant.

2. Prepare PBMC (*Protocol 1*).

3. Divide leucocytes into five aliquots of 5×10^6 cells each, in centrifuge tubes.

4. Thaw 6 ml of the B95-8 supernatant prepared above and serially titrate tenfold (1 ml into 9 ml RPMI/FCS) to give dilutions of 1/10, 1/100, and 1/1000.

5. To each aliquot of cord PBMC, add either 5 ml of RPMI/FCS, 5 ml of undiluted B95-8 supernatant, 5 ml of 1/10 dilution, 5 ml of 1/100 dilution, or 5 ml of 1/1000 dilution, and mix.

6. Incubate cells and virus at 37 °C for 1 h, resuspending the cells gently after 30 min. (Infection should occur in 20 min.)

Protocol 2 continued

7. Centrifuge the tubes, discard the supernatant, and resuspend each aliquot in 10 ml RPMI/FCS.

8. Dispense 2 ml aliquots into wells of 24-well plate.

9. Culture for three weeks at 37 °C in humidified atmosphere of 5% CO_2, feeding twice weekly by removing 1 ml medium from each well without disturbing the cells and replacing with 1 ml fresh RPMI/FCS.

10. Assess visually for the presence of proliferating B-LCL cells, growing in cell clusters.

11. The titre of the EBV is the highest dilution at which cord B cells are transformed.

12. Discard all the stocks of frozen B95-8 supernatant if the titre of EBV is less than 1/100 (i.e. if cells only grow if infected with neat or 1/10 supernatant) and repeat this protocol.

4.2 EBV transformation of B cells and growth of B-LCL

Only use EBV with a high titre. After EBV transformation, cells may be grown in 0.5 μg/ml cyclosporin A in place of PHA.

Protocol 3

EBV transformation of human B cells

Reagents

- Human PBMC (*Protocol 1*)
- EBV (*Protocol 2*)
- Phytohaemagglutinin (PHA) (sterile kidney bean extract)

- RPMI/FCS containing 1% (v/v) PHA (approx. 50–100 μg/ml)

A. EBV transformation

1. Loosen a pellet of 10×10^6 PBMC in a centrifuge tube by gentle agitation, add 1 ml EBV, and mix.

2. Incubate 1 h at 37 °C (the EBV should infect the B cells in 20 min).

B. Culture of B-LCL

1. Resuspend the cells at 0.5×10^6 cells/ml in RPMI/FCS containing PHA, mix.

2. Dispense either 2 ml aliquots into wells of a 24-well plate, or 200 μl into wells of a 96-well plate.

3. Feed twice weekly; remove half the medium and replace with an equal volume of RPMI/FCS. Cells proliferating in the first week are mainly T cells, which then die and clusters of B-LCL are evident after two weeks.

4. After three weeks, screen the culture supernatants for antibody activity (*Protocol 8*) and expand positive cultures by transferring cells into larger wells or small flasks.

5. The use of penicillin and streptomycin is recommended in the culture media to maintain sterility. If gross bacterial, yeast, or fungal contamination occurs, destroy the cultures. If slight contamination occurs and it is desired to keep the cell lines, treatment with gentamycin (50 μg/ml) or amphotericin B (2.5 μg/ml) can be attempted with caution.

5 Selection of antigen-specific B cells by rosetting

B cells and B-LCL express surface immunoglobulin and this can be exploited to select cells secreting antibody with the desired specificity. For blood group antibodies, sterile red cells of the appropriate blood group are conveniently used. If the antigen is resistant to proteolytic enzymes (papain, bromelain, trypsin) these can be used to reduce the surface charge of the cells which enhances contact with B cells.

Protocol 4

Selection of B cells by rosetting

Reagents

- Red cell lysing buffer: 0.01 M Tris (Tris[hydroxymethyl]aminomethane), 0.14 M ammonium chloride (dissolve 0.121 g Tris and 0.749 g NH_4Cl in 100 ml distilled water, adjust pH to 7.4, filter sterilize)

- 0.2% (v/v) papain, filter sterilized
- Ficoll-Triosil 1.077

Method

1. Incubate equal volumes of papain and packed red cells at 37°C for 15 min in a centrifuge tube.

2. Wash the red cells three times with sterile PBS.

3. Resuspend to a 5% suspension.

4. Wash B-LCL in RPMI/FCS and discard the wash medium.

5. Add 0.8 ml of 5% red cells to a pellet of approx. 10×10^6 B-LCL.

6. Incubate at 4°C for 30 min.

7. Centrifuge the cells at 70 g for 3 min.

8. Incubate at 4°C for 30 min.

9. Add 1 ml RPMI/FCS, gently resuspend the cells, and layer over 2 ml Ficoll-Triosil in a fresh centrifuge tube.

10. Centrifuge at 1200 g for 10 min.

11. Aspirate medium, including non-rosetted B cells at the interface, and discard.

Protocol 4 continued

12. Add 1 ml pre-warmed lysing buffer to the red cell pellet, mix, and incubate at 37 °C for 2–5 min until the red cells have haemolysed.

13. Wash the cells by filling the tube with RPMI/FCS and centrifuging at 500 g for 5 min.

14. Resuspend the rosetted B-LCL cells at 1×10^5 cells/ml in RPMI/FCS and dispense into wells of a 24-, 48-, or 96-well plate.

15. Culture the cells and feed twice weekly, expanding when cell growth permits.

This rosetting procedure may be repeated at approximately three to four weekly intervals, to increase the proportion of antigen-specific B-LCL and to increase the affinity of the antibodies selected (5). Rosetting has also been employed with PBMC to select out anti-D producing B cells with desired D variant specificity, prior to EBV transformation (6).

6 Fusion of B-LCL with murine myeloma cells (P3X63Ag8.653)

If it is found that the EBV-transformed B-LCL fail to grow rapidly or that they have lost antibody secretion, fusion of human B-LCL with mouse myeloma may be attempted to stabilize the cell lines. However, stable heterohybrids are more likely to be produced if B-LCL are fused at about a month after EBV transformation, rather than older cell lines. Human B-LCL or (less commonly) B cells have been fused with myeloma or (less often) heterohybridoma cell lines. Here the fusion of B-LCL with P3X63Ag8.653 is described (3, 7).

Human B cells (sensitive to ouabain) and murine myeloma cells (sensitive to HAT) are mixed and fused using polyethylene glycol, and heterohybridomas selected by growth in medium containing oubain and HAT. The cells are grown in conditioned medium or over feeder layers of mouse peritoneal macrophages.

Protocol 5

Preparation of feeder cells

Prepare feeder cells one to three days before fusion.

Reagents
- Mice
- 70% ethanol

Method

1. Kill mice by cervical dislocation.

2. Soak the abdomen in 70% EtOH to sterilize. Dissect the skin but leave the peritoneal membrane intact.

Protocol 5 continued

3. Inject 5 ml Hep-RPMI into the peritoneal cavity, using a needle and syringe, withdraw 5 ml, and dispense into a centrifuge tube.

4. Centrifuge at 200 g for 5 min, resuspend the peritoneal macrophages in 10 ml RPMI/FCS.

5. Dispense 0.1 ml of the cells into wells of a 96-well plate.

6. Culture for one to three days.

Protocol 6

Preparation of cells and reagents for fusion

Equipment and reagents

- Autoclave
- Feeder cells (*Protocol 5*) *or* conditioned medium (Hybridoma cloning factor or Hybridoma enhancing supplement) from murine macrophage cell line
- PEG 4000 (polyethylene glycol, M_r approx. 4000)
- DMSO (dimethyl sulfoxide)

- HAT media supplement (50 ×): 5 mM hypoxanthine, 0.02 mM aminopterin, 0.8 mM thymidine
- 1 mM ouabain (100 ×)
- B-LCL
- Myeloma cells (P3X63Ag8.653)

Method

1. Split and feed both the B-LCL and the myeloma cells one day prior to fusion to ensure they are in log phase of growth.

2. On the day of fusion, replace the medium over the feeder cells with RPMI-1640 containing 20% FCS plus HAT and 10 μM ouabain, *or* place 0.1 ml of conditioned medium in 96-well plates.

3. Prepare the fusion mix:
    (a) Weigh 0.5 g PEG 4000 and place in a glass Universal and autoclave.
    (b) Add 0.05 ml DMSO to 0.5 ml PBS, warm to 37 °C.
    (c) Add the DMSO to the hot PEG 4000, mix, and maintain at 37 °C.

4. Warm 20 ml PBS to 37 °C.

5. Wash the B-LCL twice in PBS and count.

6. Wash the myeloma cells twice in PBS and count.

7. Mix equal numbers of B-LCL and myeloma cells (between 1×10^6 and 15×10^6) in a plastic Universal, fill with PBS, centrifuge at 250 g for 5 min, and decant the PBS.

8. Gently resuspend the cell pellet and warm the Universal to 37 °C.

Protocol 7

Fusion of B-LCL and myeloma cells

Equipment and reagents

- Small water-bath at 37 °C to fit inside laminar flow cabinet
- Mixed B-LCL and myeloma cells (warm)
- Fusion mix (warm)
- RPMI (warm)

- RPMI with 20% FCS plus HAT and 10 μM ouabain
- RPMI with 20% FCS plus HT and 10 μM ouabain (HT is as HAT but without aminopterin)

Method

1. Add the warm fusion mix dropwise slowly—over 60 sec—to the warm cells, with constant agitation.

2. Agitate for a further 60 sec at 37 °C.

3. Slowly add the warm RPMI, taking 30 sec for 1 ml, then 30 sec for 3 ml, and finally 60 sec for 16 ml.

4. Centrifuge at 250 g for 5 min.

5. Resuspend the cells at 1×10^5 cells/ml in RPMI with 20% FCS plus HAT and 10 μm ouabain and dispense 0.1 ml to the wells of the 96-well plate which either contain murine macrophages *or* conditioned medium.

6. Set up control wells consisting of the two parental (unfused) cell lines, in the same medium, to check that they are killed by HAT or ouabain.

7. Culture at 37 °C in a humidified atmosphere of 5% CO_2 in air.

8. Feed every two to three days by replacing 50 μl medium with fresh medium (RPMI with 20% FCS and HAT and ouabain) until day 14.

9. On day 14, change the medium to RPMI with 20% FCS plus HT plus ouabain and continue to feed with this medium.

10. On day 21, change the medium to RPMI with 20% FCS until the cell lines are established. When cells occupy about one-quarter of the surface area of the well, transfer to larger wells and then flasks. Feed every two to three days with RPMI/FCS (i.e. 10% FCS).

7 Screening techniques

Use the method most applicable to the antibody under study. For instance, anti-D can be detected by agglutination of native (for IgM) or papainized (for IgG) D positive red cells. The culture supernatant may be tested undiluted or titrated; the titre will give an indication of antibody concentration. Quantitation of total immunoglobulin concentration is not informative until the cells are cloned.

False positive reactions may be obtained if screening EBV-transformed cells during the first three weeks after infection and culture initiation. Antibody may be secreted by B cells which were activated by EBV but which are not subsequently immortalized, so that the antibody will be secreted only temporarily. Also, PHA directly agglutinates red cells, and the concentration of PHA will only reduce gradually each time the cells are fed.

Protocol 8

Microtitre assay

Equipment and reagents

- 96-well V-well plates
- Frame with a 60° slope to hold plates
- Single- and multichannel pipettes
- Pipette tips
- PBS with 1% FCS (PBS/FCS)

Method

1. Dispense 50 μl of PBS/FCS into columns 2–12 of the plate, using a multichannel pipette, i.e. leaving the first column of eight wells empty.

2. Using a single-channel pipette, add 50 μl of culture supernatant to each of the first and second well of a row. Eight supernatants can be tested on one plate.

3. Using a multichannel pipette, serially titrate 50 μl from column 2 to column 12.

4. Add 50 μl of 0.5% red cells suspended in PBS/FCS. The cells should be pre-treated as appropriate for the antibody, e.g. papain treated (*Protocol 4*, steps 1–3) for IgG anti-D.

5. Incubate at 37 °C for 60 min.

6. Centrifuge the plates at 70 g for 3 min.

7. Place the plates on the slope for 5–10 min to read the result; unagglutinated cells stream down, whereas agglutinated cells remain in a clump.

8 Cloning

Cloning by limiting dilution in 96-well plates is the most efficient method. A feeder cell or conditioned medium (*Protocol 5*) must be used.

Select cultures to be cloned preferably on the basis of high antibody titre. A titre of 1/1000 will contain approximately 1000 times more specific antibody-producing cells than a titre of 1/1. Titres of B-LCL may be enhanced by rosetting (*Protocol 4*) over a period of weeks or months. Titres of heterohybridomas may be highest three weeks after fusion.

Protocol 9

Cloning

Reagents

- Feeder cells of murine macrophages (*Protocol 5*) *or* conditioned medium
- RPMI with 20% FCS

Protocol 9 continued

Method

1. Ensure B-LCL or heterohybridomas are in log phase of growth by splitting and feeding one to two days before cloning.

2. Harvest the cells, count, and resuspend at 100, 50, 10, and 5 cells/ml, in 20 ml volumes of each. Medium is either RPMI with 20% FCS if plating out over murine macrophages, or RPMI with 20% FCS and 20% conditioned medium.

3. Remove medium from murine macrophages.

4. Dispense 0.2 ml of the diluted cell suspension to the wells, using one plate for each dilution. The cell concentration will be 10, 5, 1, and 0.5 cells/well.

5. Place in the incubator. Do not feed the B-LCL, but feed the heterohybridomas twice weekly, with the same medium as used originally.

6. After two to three weeks, colonies of cloned cells can be observed initially microscopically, then by eye. If 25% or fewer wells in a plate have growing colonies, these can be considered to be probably monoclonal.

7. Screen the supernatants for antibody activity (*Protocol 8*).

8. Transfer cells positive for antibody into 48- or 24-well plates, when either the medium turns yellow/orange (B-LCL) or the heterohybridoma cells occupy about one-quarter of the surface area of the well.

9. Expand the cell lines, using RPMI/FCS, and monitor antibody secretion by haemagglutination and ELISA.

10. Re-clone if the antibody titre falls during expansion and culture of the cell lines.

11. Re-clone if many clones are negative for antibody.

9 Cryopreservation of cells

Samples of cells should be frozen in liquid nitrogen at all phases during the culture process, to provide a reserve in case of contamination or loss of antibody. Cells must be frozen slowly but thawed rapidly.

Protocol 10

Cryopreservation of cells

Equipment and reagents

- Liquid nitrogen storage tank
- Plastic cryopreservation ampoules
- RPMI/FCS with 20% DMSO

A. Freezing cells

1. Harvest approx. 10×10^6 cells from culture, and resuspend the cell pellet in 0.5 ml RPMI/FCS.

Protocol 10 continued

2. Add 0.5 ml RPMI/FCS containing 20% DMSO so that the final concentration of DMSO is 10%.

3. Mix and transfer to a labelled cryopreservation tube.

4. Place the tube either in a freezer with controlled temperature reduction, *or* in the neck of a liquid nitrogen Dewar, *or* at $-80\,^{\circ}$C for 2 h.

5. Transfer to rack in liquid nitrogen storage vessel.

B. Thawing cells

1. Place ampoule in 37 °C water-bath until just thawed. Do not totally immerse.

2. Swab the neck of the ampoule with 70% EtOH and immediately transfer cells to 1 ml RPMI/FCS in a centrifuge tube, mix, then fill with RPMI/FCS.

3. Centrifuge, discard medium, and resuspend cells in RPMI/FCS. Count the cells. Culture at approx. 0.5×10^6 cells/ml (B-LCL) or 0.5×10^5 cells/ml (heterohybridomas).

References

1. Rouger, P., Noizat-Pirenne, F., and Le Pennec, P. Y. (1997). *Transfus. Clin. Biol.*, **4**, 345.

2. Kumpel, B. M., Goodrick, M. J., Pamphilon, D. H., Fraser, I. D., Poole, G. D., Morse, C., *et al.* (1995). *Blood*, **86**, 1701.

3. Thompson, K. M., Hough, D. W., Maddison, P. J., Melamed, M. D., and Hughes-Jones, N. (1986). *J. Immunol. Methods*, **94**, 7.

4. Melamed, M. D., Thompson, K. M., Gibson, T., and Hughes-Jones, N. C. (1987). *J. Immunol. Methods*, **104**, 245.

5. Kumpel, B. M., Poole, G. D., and Bradley, B. A. (1989). *Br. J. Haematol.*, **71**, 125.

6. Leader, K. A., Kumpel, B. M., Poole, G. D., Kirkwood, J. T., Merry, A. H., and Bradley, B. A. (1990). *Vox Sang.*, **58**, 106.

7. Hancock, R. J. T., Martin, A., Laundy, G. J., Smythe, J., Roberts, I., Cooke, H., *et al.* (1988). *Hum. Immunol.*, **22**, 135.

Chapter 6

Laboratory based methods for small scale production of monoclonal antibodies

Bryan Griffiths

Sc & P, 5 Bourne Gardens, Porton, Salisbury, Wiltshire SP4 0NU, U.K.

1 Introduction

Although hybridoma cells are spherical and well adapted to suspension culture they do grow well in stationary culture resting on the substrate, sometimes with very light attachment. This means that a wide selection of culture vessels and systems are available for their culture ranging from a simple tissue culture plate or flask to highly sophisticated bioreactors with full instrumentation to control the physiological environment (1–3). Commercially hybridoma cell lines are grown at scales up to 2000 litres and beyond in culture units scaled-up from laboratory size vessels. Laboratory, pilot, and production scale is ill-defined so in this chapter laboratory scale will be taken as 10 litre volume cultures, and below.

The format of this chapter will be to review basic parameters for culturing hybridoma cells, stationary culture methods, which are analogous to those for anchorage-dependent cell lines, suspension culture methods using stirred or other mixing modes, and finally perfusion methods. Perfusion culture is based on immobilizing the cells so that a continuous supply of medium is fed through the culture giving a continuous harvest of antibody over several weeks, even months. This method is favoured by many commercial companies and 'turn-key' equipment is available. The choice of culture method depends upon required yield of antibody, facilities and technical expertise available, and of course budget. These factors will be considered for the following range of methods described.

2 General principles

2.1 Culture parameters

Hybridoma cells should not be considered any differently from other cell lines when it comes to culture, and the principles described in many cell culture books can be implemented (e.g. refs 4–7). The important issue is that the culture system supports production of monoclonal antibody, and this may require slightly different culture conditions to those optimized for growth. An example of this is oxygen levels where it has been found that a reduced pO_2 level (less than 5% rather than 20%) increases mAb production (8–10). The kinetics of antibody production varies between different hybridomas in that it may be:

(a) Growth associated, i.e. specific rate of secretion is highest during late lag and logarithmic growth phases.

(b) Secreted in the stationary phase, or even only from dying cells.

(c) Constant production throughout the culture cycle and independent of growth rate (10).

This factor needs to be determined for each hybridoma line if production is to be maximized, although the commonest pattern is during the stationary and dying phase (i.e. b above) (typically 80% of hybridoma cell lines). Antibody is susceptible to cleavage by cellular proteases thus culture times should not be extended more than is absolutely essential to maximize yields. Perfusion systems are an advantage in that they reduce product retention time compared to batch cultures.

Stability of antibody secretion from a hybridoma is always an issue, both from the actual yield and changes in specificity of the antibody. Changes can occur over extended culture periods and it is recommended that at regular intervals (two to three months) a new culture is initiated from a frozen cell bank. In a production situation, especially where considerable seed expansion takes place to the final production culture, each batch should originate from a frozen ampoule in the seed bank. This basic culture rule should be observed as it also minimizes the risk of microbial contamination overtaking the culture—particularly of mycoplasma and viruses. Cell Culture Collections (such as ECACC, ATCC, Riken) report a very high percentage of hybridoma cell lines received as being mycoplasma contaminated.

A problem with *in vitro* cultured hybridomas is low antibody yield, typically in the range 20–100 mg/litre compared to the ascites *in vivo* route of 5–10 g/litre. Regulation in the use of experimental animals in many countries now means that the ascitic route is not an option, but to put matters in perspective 1 g of mAb obtainable from 100 mice can be produced in 10–20 litres of culture. This difference can be dramatically reduced if perfusion systems with high cell densities (over 5×10^7/ml) are used. When mAbs are needed for research or diagnostic purposes then batches up to only 200 mg are needed and this can be achieved in 2–5 litres of culture. However for *in vivo* diagnostics and therapeutics where batch sizes of 10–100 g are required then the scale-up to fermenter/airlift reactor or perfusion systems is a necessity.

There are some general guidelines, many of which are basic and obvious, which if adopted will increase the success rate for culturing hybridoma cells and producing mAbs:

(a) Always inoculate with healthy actively growing (not stationary phase) cells.

(b) Inoculate cells into medium which is at the required temperature (37 °C) and pH (6.9–7.3). Do not allow the pH to drift outside these limits especially when initiating a culture.

(c) Maintain adequate inoculum levels—even if the culture starts growing long lag periods always result in a lower final cell yield.

(d) Choose the medium after comparative tests and screen serum before using a particular batch, for both growth and mAb production.

(e) Check regularly for contamination, particularly mycoplasma. Initiating new cultures regularly from a frozen cell bank will reduce this risk.

(f) Take care when growing many hybridoma lines that no cross-contamination occurs. Do not handle cells at the same time in cabinets, or mix different lines in the same multiwell plates etc., and always label the culture.

(g) Handle cells in class 2 Hepa air filtered cabinets, not on the open bench, and follow good aseptic technique when handling media etc.

(h) Most probable growth-limiting factors are oxygen, glutamine, and glucose respectively, and should the culture be under-performing look to supply extra concentrations of these nutrients.

(i) Do not physically stress cells by using too high a rate of centrifugation and stirring speeds. These should be optimized for each cell line and particular set of conditions

2.2 Medium and serum

A range of conventional media can be used but the commonest choices are Dulbecco's MEM (DMEM), Iscove's modified DMEM (IMDM), RPMI-1640, or a 1:1 mixture of DMEM and Ham's F12 medium.

Serum is the main problem in the growth medium. Apart from the obvious disadvantages of using fetal calf serum (cost, contamination, culture foaming) it does contain immunoglobulins and creates difficulties in downstream processing. It is very difficult to adapt cells to serum-free media and even with the addition of growth factors and other vitamin and mineral supplements growth yields are usually significantly less than with serum present. A compromise used by many is to adapt the cells to reduced levels of serum (1–3%). The degree of serum- or protein-free levels in the medium largely depends upon the intended use of the mAb, i.e. whether it is intended for research, and whether it is hoped that it maybe commercialized in a diagnostic kit. It is important to reduce the serum levels, and protein levels, as far as practically possible by trying adaptation techniques, such as the one described in *Protocol 1*, or the many published procedures (11–13). It could be more rewarding at the time of clone selection to put the cells in serum-free media and select for ability to grow in these conditions as well as on antibody production levels.

The adaptation process involves serial transfer through a series of progressively lower serum concentrations and:

(a) addition of commercial complex serum substitutes, or

(b) addition of defined proteins and metabolites, or

Table 1 Serum substitutes and supplements for serum-free media

Complex serum replacements
HL1 (Sigma)
CPSRl-5 (Sigma)
Nutridoma (Boehringer)
NuSerum (Collaborative Research)
Ultroser HY (Life Technology)
Bovine serum albumin (BSA), hydrosylated proteins, and yeast extracts (Quest Intl)

Defined supplements	
Transferrin (25 µg/ml), **I**nsulin (10 µg/ml), **S**elenium (5×10^{-8} M), **E**thanolamine (20 µM), **L**actoferrin (35 µg/ml), 2-**M**ercaptoethanol (10 µM)—used in mixtures as ITS, ITES, ILES, TES, MILES, and MITES (46)	
Hormones:	insulin, steroids, peptide growth factors (FGF, EGF, PDGF), hydrocortisone
Binding proteins:	albumin, fetuin, peptide hormones
Lipids:	cholesterol, oleic acid, linoleic acid, high and low density lipoproteins
Trace elements:	selenium, magnesium, iron, zinc
Amino acids:	Eagle's non-essential, $\times 2$ concentration of all amino acids

(c) addition of only small molecular weight molecules to achieve a protein-free medium (*Table 1*).

High (5×10^5/ml) cell densities should be maintained at each step until the cells have adapted to each condition, and different combination of additives should be tried in replicate cultures at each stage. The whole process could take two to four months or even longer, and different cell lines will adapt at different rates—in fact with some cell lines one can leave out some of the steps given in *Protocol 1*. There is also the option of weaning cells into protein-free, low protein, and serum-free commercial media produced by companies such as Life Technologies, Boehringer, BioWhittaker, Sigma, Hyclone, PAA Laboratories Cell Culture Technologies GmbH etc.

Protocol 1

Adaptation of cells to low serum or serum-free medium[a]

1. Reduce the serum level to 7.5% and passage the cells until the same kinetics are achieved as at 10%.

2. Reduce the serum level to 5% and repeat the process—there should be very little difference in cell performance at this level.

3. Reduce the serum level to 2% and plate cells at 5×10^5 with different additives to the medium (see *Table 1*). Select the best conditions and grow enough cells to be able to freeze store a few ampoules.

4. Reduce the serum to 1% and passage a few times. If necessary try additional supplements if the cells do not respond well. If satisfactory growth is achieved again freeze store a sample.

5. Reduce the serum further—to 0.5% and then to 0.2% using a high seed density (5×10^5/ml). This may require centrifuging cultures to build-up a high enough cell density rather than reducing the culture volume progressively.

6. When the cells have adapted to 0.2% serum, reduce the seeding density to normal levels. Once the number of cell doublings equal the control conditions (more important than actual cell density) again freeze an aliquot.

7. Proceed to serum-free conditions inoculating at a high level, and maintaining them at this level (by centrifugation and concentration if necessary) until they start dividing. The use of conditioned medium is usually beneficial at this stage.

8. As the cells adapt and grow reduce the level of conditioned medium, possibly with higher concentrations of the additives or extra additives. This step will probably take two months before true serum-free culture is achieved.

[a] Fuller information is given in ref. 12 on which the above procedure is based.

3 Stationary cultures

3.1 Tissue culture plates and flasks

Growth of hybridoma cells in small scale tissue culture plates and flasks will be a familiar technique as this is used during the initiation of new hybridoma cell lines following fusion. The scale-up from cloning will go through a series of steps from a well of a microtest plate, through multidishes, to small scale flasks (25 cm^2 followed by 75 cm^2). A minimum cell density of 10^4/ml is required to initiate a culture, although densities of 2–5×10^4/ml are recommended, and during the initial cloning and build-up stages feeder cell layers of macrophages (14) or irradiated thymocytes are used to maintain the critical cell density. Also conditioned medium (medium from confluent cultures diluted with fresh medium 1:1), or media enriched with additional nutritional supplements (non-essential amino acids, growth factors, vitamins, etc.) are recommended when low cell densities are used. Once the hybridoma is growing strongly in culture one can expect an inoculum of 5×10^4/ml to reach almost 10^6/ml within two to four days. The cells will stick to the surface and the strength of the attachment will increase with time but the cells can always be detached with a light tap of the culture vessel. If attachment is a problem one can always use the 'bacteriological' rather than the 'tissue culture' grade plastic vessels.

As the first step in scale-up is (or should be) to produce enough cells to lay down a cell bank in liquid nitrogen *Protocol 2* describes the procedure for growing cells in stationary monolayer cultures with the purpose of creating a cell bank.

Protocol 2

Growth of hybridoma cells in stationary monolayer cultures

1. Add 1 ml DMEM (or similar medium) to each well of a multidish (4 × 1 ml wells with a 1.9 cm^2 surface area) and add minimum of 2 (preferably 5) ×10^4 cells. Cover and place in a 37 °C CO_2 incubator for four days.

2. Harvest cells and medium from wells, giving a slight tap to dislodge cells and get them in homogeneous suspension. Pool cells, carry out a haemocytometer count, and add to a 25 cm^2 flask containing 7 ml pre-warmed medium at 2–5 × 10^4 cells/ml. Gas with 5% CO_2 and incubate for a further four days.

3. Harvest the culture; expected yield 5–7 × 10^6 cells. Add to a 75/80 cm^2 flask 30 ml medium and 1–2 ×10^6 cells. There should be enough cells to set up at least three flasks. Gas with 5% CO_2 and incubate for a further four days.

4. Harvest the culture, pool cells, carry out a haemocytometer count. The object is to store five to ten ampoules containing 5 × 10^6 cells per ampoule, leaving enough cells over to continue a new culture in a 75 or 175 cm^2 flask.

5. Cryopreservation of cells is well described (15) and covered in Chapter 2.

The range of plastic tissue culture vessels is huge, and available from companies such as Nunc A/S, Bibby Sterilin, Costar, Corning, and Becton Dickinson, and distributors such as Life Technologies. Examples of the range are:

- Microwells—96 × 0.2 ml (0.3 cm^2)
- Multidishes—4 × 1 ml (1.9 cm^2), 12 × 2 ml (3.5 cm^2), 24 × 1 ml (1.9 cm^2), 48 × 0.5 ml (1.1 cm^2)
- Flasks—25 cm^2 (7 ml), 75/80 cm^2 (30 ml), 175 cm^2 (70 ml)
- Triple flasks—500 cm^2 (200 ml)

This is the simplest and most straightforward method of culturing cells but is confined to very small scale operation and should be considered as part of the initiation process aimed at preparing cells for larger scale processes.

3.2 Specialized (scale-up) culture systems

An extension of the flask culture principle for giving large unit surface areas are the Nunc A/S Cell Factories (16). These are polystyrene plates of 632 cm^2 which are available in single, double (1264 cm^2), 10 tray (6320 cm^2), and 40 tray (25280 cm^2) units. Each tray level requires 200 ml medium and can be filled, harvested, etc. in one operation per factory unit. For large scale commercial processes the system can be semi-automated with the Manipulator and Robot System.

Protocol 3

Growth of hybridoma cells in Nunc Cell Factories

1. Prepare Cell Factory in a Class II cabinet by removing packaging and fixing the provided air filter and tube connector to the two open adaptor caps.

2. Prepare cell suspension in aspirator bottle (2 litres for a 10 tray unit with 5–10 × 10⁷ cells homogeneously mixed) and connect to the tube connector in the Cell Factory.

3. Turn Cell Factory on its side with the supply tube at the bottom, and with the aspirator above the Cell Factory open the clamp and allow cell suspension to flow in gently under gravity (alternatively a minimal amount of air pressure may be applied using a pump).

4. Allow levels within the unit to equalize and then turn the Cell Factory through 90° with the filling connector at the top. The medium will equalize between the chambers. Now turn the unit to the horizontal position with the connectors at the top. Disconnect from aspirator and place in incubator for three to four days at 37 °C (if a CO_2 incubator is not used then gently gas the unit with 5% CO_2 and seal).

5. Inspect the unit daily for contamination, maintenance of pH, and visual signs of cell growth.

6. To harvest attach the connector to a sterile aspirator bottle and turn the unit onto its long side with the connector at the bottom. Open the clamp and allow the cell suspension to flow out under gravity.

7. If the operation is carried out carefully and aseptically the unit can be refilled immediately with a new suspension of the *same* cell line and the process repeated.

The unit can be considered as a giant flask and is an easy and convenient way of producing 2 litres of antibody supernatant. The 40 tray unit is not recommended as it is large to handle and requires a hot room or special incubator. It is more convenient to use four to eight 175 cm² flasks for production than the one and two tray units which are designed for development work and controls in a large process This is a good method for laboratories that are not equipped, or experienced with, more complex and advanced high productivity units and requires no investment in capital equipment, although individual units are expensive.

4 Stirred cultures

4.1 Spinner flasks

The spinner flask is a glass vessel with a suspended stirrer shaft and magnetic bar and is designed to be used for replicate cultures on a multibased magnetic stirrer (*Figure 1*). Sizes range from 10 ml to 20 litres, but for ease of handling and

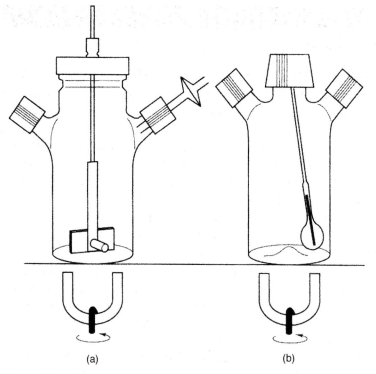

(a)   (b)

Figure 1 Spinner flask cultures. (a) Conventional magnet bar stirrer. (b) Radial or pendulum stirrer.

safety it is recommended that 10 litres is the upper limit. There are a wide range of spinners available commercially, varying in their geometry, design, and stirring principles (4). The commonest type has a magnetic bar, side-arms to allow gassing, and access to allow addition or removal of cells, medium components, etc. (NB: spinner flasks with large sail-like spinner bars are intended for microcarrier culture but can be used for hybridomas which are fragile or only grow poorly in stirred suspension as these stirrers allow mixing at speeds of only 30–90 r.p.m.) Spinner flasks with slightly convex bottoms are recommended as it increases the mixing efficiency.

The conventional culture media can be used (DMEM, IMDM, etc.) but additional components can be added to increase the success of the culture if considered necessary. Pluronic F-68 (polyglycol) (BASF, Wyandot) can be added at 0.1% to protect the cells against mechanical damage, especially at reduced serum concentrations. Carboxymethyl cellulose (CMC) (15–20 cP) can also be used at 0.1% for this purpose. Antifoam (6 p.p.m.) (Dow Chemical Co.) is recommended if foaming occurs which usually happens if the serum concentration is above 5%. The use of Hepes buffer can also be considered to stabilize pH during set up and if gassing facilities are not available.

Protocol 4

Spinner flask culture

1. Add 200 ml medium to a 1 litre spinner flask, gas with 5% CO_2, and warm to 37 °C.

2. Add $2-4 \times 10^7$ cells ($1-2 \times 10^5$/ml) harvested from a growing culture of cells (i.e. not a stationary or dying culture). NB: this is a higher inoculum than used for stationary cultures.

3. Place spinner vessel on magnetic stirrer at 37 °C and set at 100–200 r.p.m. (this is variable depending upon the vessel size, geometry, fluid volume, and individual cell. Set a speed which visually shows complete homogeneous mixing of the cells throughout the medium—150 r.p.m. is usually a safe choice).

4. Monitor the growth daily by removing a small sample through the side-arm in a Class II cabinet and carrying out a cell count and visual inspection for cell morphology and lack of microbial contamination.

5. Monitor the pH (colour change), especially in closed (non-gassed) systems, and should the culture become acidic (under pH 6.8) add sodium bicarbonate (5.5% stock solution) or re-gas with 5% CO_2. After three days a better option would be to allow the cells to settle out, remove 40–70% of the medium, add fresh pre-warmed medium to the culture, and continue stirring.

6. After four, possibly five days the cells should reach their maximum density ($7-10 \times 10^5$/ml) and can be harvested. The actual timing will depend upon the kinetics of the cell (whether mAb is growth or dying phase associated—see Section 1.1).

7. Clumping or attachment to the culture vessel should not be a problem with hybridomas but should it occur the vessels can be siliconized with Dow Corning 1107 or Repelcote (dimethydichlorosilane) (Hopkins and Williams), or media with reduced Ca^{2+} and Mg^{2+} can be used.

Spinner vessels available are:

- Conventional vessels with spin bar (Bellco, Wheaton).
- Radial (pendulum) stirring system (*Figure 1*) for improved mixing at low speeds (under 100 r.p.m.) (Techne, Integra Biosciences CellSpin).
- Bellco dual overhead drive system (radial stirring and permits perfusion).
- Techne Br-06 floating impeller system allows working volumes of 500 ml to 3 litres to be used and increased during culture.

Spinner culture is the simplest form of suspension culture available, relatively cheap to set up, and with reasonable scale-up potential (larger units and increasing the number of replicates).

4.2 SuperSpinner

The SuperSpinner (17) was developed to increase the productivity within a spinner flask without going to the complexity of the hollow fibre and membrane systems described below. It consists of a 1 litre Duran flask (Schott) equipped with a tumbling membrane stirrer moving a polypropylene hollow fibre through the medium to improve the oxygen supply. The small device is placed in a CO_2 incubator and a small membrane pump is connected via a sterilizing filter with the membrane (stirrer). The culture is placed on a magnetic stirrer and has a working volume from 300 ml to 1000 ml. Three different membrane lengths (1, 1.5, and 2 m) are available depending upon the degree of oxygenation the cell line needs. Cell concentrations in excess of 2×10^6/ml are achieved with mAb production of 160 mg/litre (specific productivity c. 50 μg/10^6 cells). Repeated batch culture (80% media replacement every two to three days) over 32 days produced 970 mg mAb (17). This relatively inexpensive modification of a spinner culture is available from B. Braun, Biotech International.

4.3 Stirred bioreactors

Laboratory scale bioreactors (fermenters) are available from 1 litre to 15 litres (for commercial production this range increases to 10 000 litres). The purpose of using a bioreactor rather than a spinner flask is that it is far more controllable (pH, oxygen, redox, mixing), and also as a developmental culture which can be sequentially scaled-up without altering the culture parameters. The disadvantage is the greatly increased cost. There is a great choice of systems from the various fermenter suppliers (e.g. Braun, Bioengineering, Applikon) but the basic differences to a spinner system are:

- stainless steel, not glass, vessel with a curved bottom, no baffles, and a smooth interior finish
- top driven marine impeller
- height to diameter ratio between 1:1.5 and 1:2
- water jacket temperature control
- sophisticated environmental control systems, giving better oxygenation, pH, and mixing control and continuous records of culture parameters
- possibility of using fed-batch and perfusion modes of culture
- options such as draft tubes, spin filters, oxygenation devices
- usually plumbed in to water and steam supply for *in situ* cleaning and sterilization (necessary above 10 litres volume)

The basic principles for growing cells in these bioreactors is similar to that given in *Protocol 4* for spinner flasks and also well described in the literature (18, 19) and by the manufacturers, and as they are on the borderline of being laboratory methods the reader is referred to these for experimental details. A modified stirrer system, the Celligen, is available from New Brunswick with adaptations specifically for hybridoma culture. The Celligen was developed to grow shear-sensitive cell lines in a highly oxygenated culture using a gas exchanger impeller where the culture is lifted into the tube of the impeller shaft and expelled through discharge ports in a continuous recirculation loop (20). An aeration screen keeps any foam which is generated separate from the cells, thus preventing cell entrapment and damage. The Celligen Plus was then developed in which a fibrous basket is loaded with polyester discs for cell immobilization and which can be perfused. The small laboratory scale unit is 1.4 litres working volume with a total vessel volume of 2.2 litres—other sizes available are 3.5 and 7.5 litres working volumes.

Fermenter cultures are undoubtedly more productive, and give more reproducible results than the spinner systems which lack control units for environmental factors. However they are expensive to purchase, need special facilities, and are for the more expert cell culture practitioner. Nevertheless the smaller laboratory model packages are a worthwhile investment where gram quantities of mAb are needed, especially where the possibility of a current or future market application is envisaged.

5 Dynamic (non-stirred) culture systems

5.1 Roller bottle culture

Plastic disposable roller bottles are available in sizes from 670 cm^2 to 1750 cm^2 (depending upon manufacturer; Bibby, Costar, Sarstadt, Becton Dickinson). They can be considered as a modification of a standard flask in ease and simplicity of use but have the advantage of a higher unit surface area (as the whole internal surface is used) and better aeration as the medium is continuously moving. A specialized roller culture unit is needed giving speeds between 5–60 r.p.h. and small ones are available which will operate inside an incubator and take four of the smaller flasks (Bellco, Integra Biosciences, Wheaton, New Brunswick). Greater numbers need a unit placed in a hot room. Although roller culture is mainly for anchorage-dependent cells and is widely used in large scale production of several pharmaceuticals, it is equally a simple (*Protocol 5*) first scale-up step for hybridomas from flasks.

Protocol 5

Roller bottle culture

1. Add to a 700 cm^2 plastic roller bottle 150 ml of pre-warmed medium (DMEM, IMDM, etc.) and gas with 5% CO_2 in air. A general guideline is a minimum of 1 ml medium/5 cm^2 culture surface area.

2. Add 5–10 × 10^4 cells/ml medium from a logarithmically growing culture.

3. Screw on bottle cap and place on roller culture apparatus at 37 °C.

4. Rotate bottle at 5–10 r.p.h.

5. Grow for three to four days visually checking daily for growth, pH, and contamination. If necessary add additional fresh medium (30% of original volume) or carry out a partial (30–70%) medium change.

6. Harvest cell by siphoning out (or pouring) the culture fluid into a centrifuge pot and spin out the cells at 500 g.

5.2 Airlift fermenters

The airlift fermenter relies upon the bubble column principle to both agitate and aerate the culture (21–23). Air bubbles are introduced into the bottom of the culture vessel and rise through a central inner draft tube (*Figure 2*). Aerated medium has a lower density than non-aerated so the medium rises up the draft tube. As the cells utilize oxygen the medium increases in density and falls down the outer ring of the culture. This provides a very low energy input gentle mixing system with no shear forces, and is thus suitable for fragile cells which may be damaged by stirring. In addition the constant stream of air bubbles gives efficient oxygenation of the culture which is often a limiting factor in other systems.

An airlift fermenter requires a high aspect ratio to be efficient (at least 8:1, optimally 12:1), and for animal cells an airflow rate of about 300 ml/min. Temperature is controlled by a circulating water jacket and pH can be controlled by automatic addition of CO_2 in the sparged gas and by addition of sodium hydroxide. Oxygen can also be controlled by altering the gas ratios in the sparged gas. Foaming is always a problem (if the medium contains serum or other proteins) and antifoam and Pluronic F-68 need to be added. The smaller laboratory units can be operated with less sophisticated control systems. Units are available from 2 litres upwards (to 2000 litres) but it is advisable to use a 5 litre, or better still, 10 litre size as it is a system which tends to operate better at higher volumes. To control air supply in the small units a pump gives more control at low flow rates than air-mixing valves. The unit is operated as a batch culture with conventional seeding densities (1×10^5/ml) with a five to six day culture time.

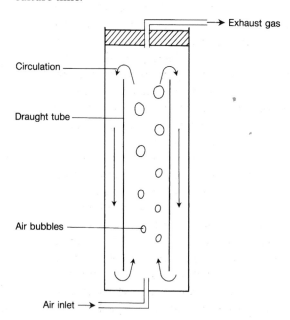

Figure 2 Airlift bioreactor for the cultivation of suspended mammalian cells.

In conclusion this is a system for fragile cells, or where high oxygenation is needed, and which with no moving parts, seals, etc. gives good reliability and freedom from contamination. However at the laboratory level there is not much choice in supply of airlift apparatus as it is considered a high volume system. Fermenter manufacturers such as Bioengineering AG do supply small units (10 litres) but it is also relatively simple to get them made up by a glass-blower and to purchase the control systems from any fermenter manufacturer. There are a number of excellent reviews on airlift fermenters (21–23).

6 Perfusion (high cell density) systems

Cells are packed to densities of over 3×10^9/cc in tissue and it has always been an ambition to culture cells at such densities. Traditionally densities of around 2×10^6/ml was all that could be achieved due to oxygen and nutrient limitation, low pH, and toxic metabolite accumulation. The circulatory system of the living organism is needed to achieve high unit density. In culture the perfusion system was developed to mimic this. There is a long history of perfusion devices in cell culture (see review 24), but the starting point for the current successful perfusion technology was the work of Knazek with ultrafiltration capillary hollow fibres (25). The first commercial units (Amicon Vitafiber) were available in 25, 250, and 2500 cm^2 sizes with extracapillary (cell compartment) of 2.5, 25, and 250 ml respectively. The fibres had a spongy wall (50–70 μm thick) and a central lumen of 200 μm diameter. The external surface was very porous but the lumen lining was a thin ultrafiltration skin with various molecular weight cut-offs (1000–10 000). A unit consisted of thousands of fibres 'potted' at either end in a cylindrical housing capable of supporting 10^7 cells/cm^2 or 10^8/ml. Subsequently over the years many commercial versions have appeared, and some disappeared, and many ingenious modifications made to overcome the drawback of small scale and lack of homogeneity. The scale-up problem was that pressure and concentration gradients built up along the length of the cartridge due to transmembrane flux being redirected back into the lumen for the latter half of the flowpath, resulting in necrotic cells and limiting the operation time of the culture. A move to filtration rather than ultrafiltration grade fibres allowed a greater flux rate but did not allow the product to concentrate in the cell compartment. Adaptations to overcome these problems were to bundle air only fibres amongst the medium supply ones, alternate medium flow, and the use of flat bed rather than cylindrical vessels (reviewed in refs 1 and 24).

Perfusion systems have many advantages over standard batch culture. The cells being at such high density have a greatly reduced need for serum thus it is much easier to use a serum-, even protein-free medium. In fact if the culture is initiated with low serum concentration to get the cells growing but left out of the perfusion medium it is virtually washed out by the time the cells have grown to production density and mAb is starting to be harvested. Also cells often have a higher specific production rate at high density (26) and this, coupled with the elevated cell number produces a much higher product concentration so facilitating downstream processing. Some systems (e.g. Ultrafiltration grade fibres and membranes) allow extra concentration of the product in the cell compartment, however this does limit the scale-up size and efficiency of long-term operation. As perfused cells are immobilized they are well protected from shear forces or other mechanical stresses. The fact that cells are concentrated 50-fold higher than conventional cultures means that one can have in the laboratory a system equivalent to a pilot plant batch culture operation. This is a reason for them being popular with start-up biotechnology companies, especially as most have excellent environmental control and give a constant readout and record of many culture parameters which is needed to manufacture a licensed product.

6.1 Hollow fibre (Acusyst) culture

The Acusyst (Endotronics, now Cellex Biosciences Inc.) system made the breakthrough by using pressure differentials to simulate *in vivo* arterial flow (27). The unit has a dual medium circuit, one passing through the lumen, the other through the extracapillary (cell) space. By cyclically alternating the pressure between the two circuits, media is made to pass either into, or out of, the lumen. This allows a flushing of media through the cell compartment and overcomes the gradient problem, and also gives the possibility of concentrating the product for harvesting. There are a range of Acusyst units, the Acusyst-R and -Jr being the laboratory models using respectively 2–3 and 3–8 litres medium per day, and capable of running continuously for 50–150 days (four to five months being a typical run length). There is an initial growth phase of 10–25 days depending upon inoculum size, medium composition, and cell line, and when a density of $1–2 \times 10^{10}$ cells is reached (1.1 m^2 cartridge, e.g. Acusyst-Jr) production starts and daily harvesting of mAb begins. Cells cannot be sampled during the culture so reliance is placed upon a number of process measurements made on a regular basis—on-line pH and oxygen; off-line glucose, lactate, and if necessary ammonia and LDH (for loss of viability) at least weekly. The cultures are purchased as a turn-key unit with all accessories and full operating instructions so it is inappropriate here to produce a protocol. An objective description of the operation of the Acusyst-Jr is published by Hanak and Davis (28) and some of their data is presented in *Table 2* to show the efficiency of operation. Other hollow fibre systems are available from Cell-Pharm (3, 19, 35 ft^2 units), Amicon (Mini Flo-Path), Asahi Medical Co. (Cultureflo), Kinetek, Cellco.

6.2 Tecnomouse

The Tecnomouse (29) is a hollow fibre bioreactor containing up to five flat culture cassettes containing hollow fibres surrounded by a silicone membrane that gives uniform oxygenation and nutrient supply of the culture and ensures homogeneity within the culture. The system comprises a control unit (media supply), a gas and medium supply unit, and the five culture cassettes.

The culture is initiated with cells from 3×225 cm^2 flasks (5×10^7) and grows to 5×10^8 cells. The continuous medium supply is controllable and program-

Table 2 Performance data of the Acusyst-Jr bioreactor (from Hanak and Davis (28) and Cellex Biosciences Inc.)

Parameter	Cell a	Cell b	Cell c	Cellex
Medium	IMDM 1% FCS	IMDM S. Free	IMDM S. Free	
mAb conc. T-flask (μg/ml)	8	4	80	
mAb conc. Acusyst (μg/ml)	160	400	1330	100–300
mAb mg/day	38	96	287	135–500
mAb g/month	1	2.5	7.5	4–20
Medium feed rate (ml/h)	50	100	80	125–300
Total cells $\times 10^{10}$	1–2	1–2	1–2	1
Bioreactor surface area (m^2)	1.1	1.1	1.1	1.1

mable with typical flow rates (perfusion or recirculation) of 30 ml/h increasing to 70 ml/h in 5 ml steps. After six days harvesting was started and repeated every two to three days for 30–70 days. Each cassette gives 7 ml culture with 2.5–5.5 mg/ml mAb (equivalent to an average of 10 mg/day). Thus about 10 litres will yield 400 mg mAb/month and 1.5 g/month using the five cassettes. (Data based on manufacturer's information.)

6.3 Membrane culture systems (miniPERM)

There have been a number of membrane-based culture units developed (24), many based on dialysis tubing (1), and even available as large fermenters (Bioengineering AG Membrane Laboratory Fermenter with a Cuphron dialysis membrane of 10 000 dalton molecular weight cut-off forming an inner chamber). One of the more successful and currently available systems is described here—the miniPerm Bioreactor (Heraeus Instruments).

The miniPERM Bioreactor (30, 31) consists of two components, the production module (40 ml) containing the cells and the nutrient module (600 ml of medium). The modules are separated by a semi-permeable dialysis membrane (MWCO 12.5 kDa) which retains the cells and mAb in the production module but allows metabolic waste products to diffuse out to the nutrient module. There is a permeable silicone rubber membrane for oxygenation and gas exchange in the production module. The whole unit rotates (up to 40 r.p.m.) within a CO_2 incubator. It can be purchased as a complete disposable ready to use unit or the nutrient module (polycarbonate) can be autoclaved and reused at least ten times.

Protocol 6

MiniPERM Bioreactor[a]

1. Add 35 ml of cell suspension at $1\text{–}2 \times 10^6$ cells/ml in a syringe to the production module through the Luer lock.

Protocol 6 continued

2. Add 300–400 ml culture medium to the nutrient module.

3. Place the bioreactor on the bottle turning device in the gassed incubator and set the speed to 10 r.p.m.

4. Sample daily by removing aliquots of culture with a syringe via the Luer lock and carry out cell counts (and optionally glucose measurements).

5. At a cell density of 6×10^6/ml replace the medium in the nutrient module every four days, at higher densities every day. Serum-free medium can be used at this stage.

6. Typically a cell density of $15–30 \times 10^6$ cells/ml is maintained after day six with a daily yield of 0.8 mg mAb/ml (range for various hybridomas 0.5–12 mg/day).

7. Keep the culture going until viability is lost (after 40 days).

[a] Data from Falkenberg *et al.* (30).

6.4 Packed bed systems

The use of beds packed with glass spheres (3–5 mm) through which medium is continuously perfused by peristaltic pump or airlift has been used for anchorage-dependent cells for many years and for a range of products (32). The system was then extended to anchorage-independent cells, such as hybridomas (33) using 1 mm borosilicate glass beads in a tubular reactor (2.5 cm diameter × 13 cm) fed from a 1 litre holding vessel sparged with CO_2 and oxygen. The bed was inoculated with 30–60 ml at 3×10^6 cells/ml and then perfused at 35–50 ml/h. The cells grew to 28×10^6/ml of void volume in 12 days. Glass bead bed

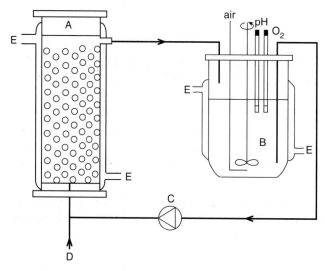

Figure 3 A glass bead bioreactor. A, glass bead bed; B, reservoir; C, pump; D, inoculation and harvest line; E, temperature-controlled jackets.

reactors are very easy to construct, the glass beads are re-utilizable and a cheap substrate, and the whole culture simple in construction and ease of use (*Figure 3*). Its simplicity make it reliable and it can be recommended for this reason. For further details and description of the apparatus see refs 4 and 34.

The next developmental step was to use microporous borosilicate glass beads (Siran, Schott Glasswerke) (35, 36). The same apparatus as described for solid glass beads can be used or commercially purchased (Meredos, Germany) in bed sizes from 0.1–1 litre. The Meredos fixed bed stands inside the conditioning vessel which has a tenfold higher medium volume than the bed. An example of its operation is published (29). Siran porous beads can be used in sizes from 3–5 mm, they have a pore size of 60–300 μm, and a 60% open internal volume in which cells are entrapped and protected from shear and mechanical stresses. 5 mm beads are preferred as they give an open bed structure (1 cm^2 channels) which minimizes blockages due to biomass build-up, uneven distribution of the inoculum, and media channelling within the bed (37, 38).

Typical results in the author's laboratory (37–39) with *Protocol 7* are 2.75×10^{10} viable cells per litre giving an average yield of 166 mg/litre/day (compared to stirred reactor and airlift cultures of the same hybridoma of 25.5 and 18.5 mg/litre respectively). This method is a low investment introduction to high productivity production of mAb which is simple to use and reliable with low maintenance, at least for the first 50 days of culture.

Protocol 7

Fixed porous bead (Siran) bed reactor

Protocol based on a 1 litre fixed bed (water jacket temperature control) of 5 mm Siran beads perfused from a 15 litres (11 litres working volume) Applikon stirred tank with pH and oxygen control as a medium reservoir.

Method

1. Preliminary preparation. Beads: boil in distilled water for 2 h, rinse, and oven dry (4 h at 100 °C). Rinse again and heat dry, then heat sterilize (180 °C for 2 h) in a sealed beaker, add aseptically to autoclaved culture vessel. Reservoir vessel: calibrate probes, assemble, and autoclave.

2. Suspend inoculum (5×10^9 cells per litre bed volume) in 800 ml pre-warmed medium (it is recommended that the medium is circulated for several hours through the bed to both equilibrate the system, check the probes, and condition the beads before inoculation).

3. Add cell suspension to the inoculation vessel, connect to the bottom connector of the glass bead vessel, and slowly add the inoculum by air pressure (by hand pump) being careful not to introduce air bubbles.

4. Drain and refill the bed twice.

5. Start perfusing the medium at 40 ml/min or a linear flow velocity of 2 cm/min.

Protocol 8 continued

6. Monitor the reservoir daily for glucose concentration and when the level falls below 2 mg/ml change the medium for a fresh supply (the Harvest). Alternatively the system can be run in a continuous mode where the medium feed rate is adjusted to maintain a glucose concentration of 2 mg/ml.

7. Run the culture until the cells lose viability or the bed becomes blocked with cell debris, usually after 50–60 days. This can be delayed further if occasionally the perfusion rate is significantly increased for a short time, or the culture is set up for alternate up- and down-flow.

8. At the end of the culture should a cell count be needed remove a measured aliquot of beads (e.g. 20 cc), add 100 ml of 0.1% citric acid and 0.2% Triton X-100, place on an orbital shaker (150 rev/min) at 37 °C for 2 h. Collect supernatant, measure the volume, wash beads with PBS, add crystal violet to the supernatant (0.1% final concentration), and carry out a nuclei count in a haemocytometer.

6.5 Fluidized bed bioreactors

Porous microcarrier technology, currently the most successful scale-up method for high density perfused cultures, was pioneered by the Verax Corporation (now Cellex Biosciences Inc.). Turn-key units were available from 16 ml to 24 litres fluidized beds (40). The smallest system in the range, Verax System One, is a benchtop continuous perfusion fluidized bioreactor suitable for process assessment and development, and also for laboratory scale production of mAbs. Cells are immobilized in porous collagen microspheres, weighted to give a specific gravity of 1.6, which allowed high recycle flow rates (typically 75 cm/min) to give efficient fluidization. The microspheres have a sponge-like structure with a pore size of 20–40 μm and internal pore volume of 85% allowing immobilization of cells to densities of 1–4×10^8/ml. They are fluidized in the form of a slurry.

The Verax system comprises a bioreactor (fluidization tube), a control system (for pH, oxygen, medium flow rates), gas and heat exchanger, and medium supply and harvest vessels. The system is run continuously for long periods (typically over 100 days). In the authors laboratory it produced 15×10^{10} cells/litre and 540 mg mAb/litre/day (compared to 166 in the fixed bed described above, 25.5 in a stirred reactor, and 18.5 mg in an airlift fermenter). Protocols for its operation come with the equipment and versions have been published (41). In summary it is probably the most productive system available giving the cells a very high specific production rate but does require some skill to operate to its maximum potential.

An alternative commercial system is the Cytopilot (Pharmacia) which is a fluidized system using polyethylene carriers (Cytoline) and supports 12×10^7 cells/ml carrier (42). It is available as the Cytopilot-Mini (400 ml bed) for laboratory scale operation as well as sizes up to 25 litres. The unit has a magnetic stirrer that drives the medium up through a distribution plate into the upper

Table 3 Microporous microcarriers

Microcarrier	Manufacturer	Material	Diameter (microns)	Culture mode
Cellsnow	Kirin Ltd.	Cellulose	800–1000	Stirred
Cytocell	Pharmacia	Cellulose	180–210	Stirred
Cultispher	Hyclone	Gelatin	170–270	Fluidized
Cytoline 1	Pharmacia	Polyethylene	1200–1500	Fluidized CHO cells
Cytoline 2	Pharmacia	Polyethylene	1200–1500	Fluidized hybridomas
ImmobaSil	Ashby Scientific	Silicone rubber	1000	Stirred
Siran	Schott Glaswerke	Glass	400–5000	Fixed bed

chamber in which the microcarriers are lifted by hydrodynamic pressure. The degree of fluidization is controlled by the stirrer speed and a clear boundary layer is kept at the top of the culture so that clear medium can flow through the internal central circulation tube (loop) back to the stirrer. The culture is oxygenated by means of a mini-sparger delivering microbubbles and a medium feed rate of up to 25 bed matrix volumes per day giving high productivity. The unit can also be used as a packed bed bioreactor if the circulation system is reversed.

There is a range of porous microcarriers available (*Table 3*), which allow one to design their own fluidized system, or alternatively some of the carriers are designed for stirred cultures.

7 Harvesting and concentration

Purification of mAbs is described in Chapter 7. However after the culture has been terminated and before purification can be carried out the supernatant has to be rendered cell-free, and in many instances the mAb concentrated.

7.1 Harvesting and clarification

The culture fluid contains cells and also cell debris, especially in cultures with protracted stationary phases for maximization of mAb production from dying cells. This cellular material needs removing not only because the proteins will interfere with subsequent purification but also cell proteolytic enzymes will be released which may destroy the product during purification. Thus the first down-stream processing step, supernatant clarification, must be carried out immediately to minimize product degradation. With cultures of free-suspension cells the choice of clarification is usually between centrifugation and filtration.

At the laboratory scale centrifugation is a suitable method. When harvesting cells g forces of about 500 are normally used but at this stage unless the cells are needed for further culture (not recommended as it will be an ageing population) much higher centrifugal force is needed to ensure all cells are sedimented and also as much cell debris as possible. A minimum centrifugal force of 2000 g should be used for 20 min, but speeds giving up to 6000 g will increase the efficiency of the process.

Filtration systems used for sterilization (membrane filters) can be used to separate cells from the supernatant but only for relatively small volumes as they do rapidly become clogged, and for this sort of volume centrifugation is usually easier but not as efficient at removing small cell debris. Tangential flow filtration is used to reduce problems of filter blockage by continuous recirculation of the suspension across the membrane, and is recommended for larger volumes (several litres). Low volume systems are available from recognized filtration suppliers (e.g. Millipore, Sarstadt, Pall). Filtration systems have been reviewed (43).

An advantage of high density perfusion systems described above is that as the cells are immobilized the supernatant is relatively cell-free and thus more easily filtered. Also the mAb is at a far higher density reducing the degree of concentration that is needed. However perfusion systems can produce high daily volumes of supernatant requiring concentration to achieve more manageable volumes for purification.

7.2 Concentration

The classical concentration method for culture supernatants is precipitation, ammonium sulfate being the commonest used material. Other agents used are sodium sulfate and PEG (44). A method using ammonium sulfate is give in *Protocol 8* (11, 45).

Protocol 8

Ammonium sulfate precipitation

This concentration method is used to precipitate gammaglobulins and remove albumins which is present if serum was dead to the medium.

Method

1. Prepare a saturated stock solution of $(NH_4)_2SO_4$ by adding 100 g to 100 ml distilled water, stir for 12–24 h, and filter (Whatman No.1 paper).

2. Add $(NH_4)_2SO_4$ at 4 °C slowly to the culture supernatant with stirring until a 35–40% ammonium sulfate concentration is reached.

3. Stir for 20 min then centrifuge (minimum 2000 g, preferably 3000 g or more) for 20 min.

4. Wash and dissolve precipitate in a small volume of PBS.

5. Repeat steps 2–4.

6. Dissolve precipitate in PBS and dialyse against 5 mM phosphate buffer if required immediately, or 0.1 M Tris–HCl buffer pH 7.4 for short-term storage.

7. If sodium sulfate is used the above protocol can be followed but using 18%, followed by 14% saturated Na_2SO_4 (45).

This method does have many disadvantages and is more commonly used for concentrating ascitic fluid rather than culture supernatants where the initial low concentration of mAb results in losses and poor yields (10). Ultrafiltration methods are preferred (hollow fibres, tangential flow devices) as being far more efficient, especially with serum/protein-free media. 50- to 100-fold concentration with almost full recovery of mAb is possible with ultrafiltration devices using membranes of less than 100 kDa molecular weight cut-offs. In addition these systems are a good starting step for subsequent purification procedures replacing gel filtration at the beginning of the purification schedule. Suitable devices are widely available from many manufacturers and come with full operational instructions.

8 Summary

Culture procedures have been described from the simple flask and spinner cultures through increasing degrees of sophistication to the highly productive, but complex and expensive, perfusion bioreactors. The choice of culture system depends upon how much mAb is needed, the quality needed (i.e. for what purpose is it intended), and the expertise and cell culture facilities available. In order to choose on the basis of mAb yield the data in *Table 4* summarizes each system described in this chapter with values for cell density and mAb production. Although somewhat hypothetical the values are based on an average

Table 4 Comparison of culture systems for hybridoma cells

Working volume litre	Culture volume /size	Culture unit	Total cells	Antibody yield mg[a]	Normalized yield mg/litre[a]
Monolayer culture units					
0.07	175 cm^2	Flask	5.0×10^7	2	30/4d
0.2	500 cm^2	Triple flask	1.5×10^8	6	30/4d
0.1	1750 cm^2	Roller bottle	2.2×10^8	9	30/4d
2.0	6320 cm^2	Cell factory 10	1.5×10^9	60	30/5d
Stirred cultures					
0.2	1 litre	Spinner	1.8×10^8	5	30/5d
1.0	5 litres	Spinner	9.0×10^8	25	30/5d
1.0	1 litre	Superspinner	2.0×10^9	160	50/5d
10.0	14 litres	Fermenter	1.0×10^{10}	1600	50/5d
Perfusion cultures					
2.5/d	1.1 m^2	Acusyst-Jr	2.0×10^{10}	100 mg/d	40/d
0.35/d	35 ml	Technomouse	5.0×10^8	16 mg/d	40/d
0.60/d	40 ml	MiniPERM	3.0×10^7	12 mg/d	20/d
3.0/d	1 litre	Siran fixed bed	3.0×10^{10}	55 mg/d	166/d
0.15/d	16 ml	Verax System 1	6.0×10^9	75 mg/d	540/d

[a] Average theoretical yields based on published data but with different hybridoma cell lines.

from published data. Obviously direct comparisons cannot be made accurately as different hybridomas have been used which all have a wide range of secretion rates, but it does show the potential for each system.

Some perfusion systems are aimed at the small organizations needing to produce marketable mAbs in small GMP standard production facilities. The degree of environmental control and monitoring they give is needed in order to have a reproducible and reliable process, and to provide the data for product licensing records. It is a bonus that there are less sophisticated (and less expensive) perfusion cultures available for standard laboratory use thus the cell culturist has a genuine choice from all the culture strategies which have been described and which cover a wide range of culture types.

References

1. Griffiths, J. B. (1988). In *Animal cell biotechnology* (ed. R. E. Spier and J. B. Griffiths), Vol. 3, p. 179. Academic Press, London.
2. Prokop, A. and Rosenberg, M. Z. (1989). *Adv. Bioch. Eng./Biotech.*, **39**, 29.
3. Griffiths, J. B. (1990). In *Methods in molecular biology* (ed. J. W. Pollard and J. M. Walker), Vol. 5, p. 49. Humana Press Inc., Clifton, NJ.
4. Griffiths, J. B. (1986). In *Animal cell culture: a practical approach* (ed. R. I. Freshney), p. 48. IRL Press, Oxford.
5. Griffiths, J. B. (1993). In *Methods of immunological analysis* (ed. R. F. Masseyeff, W. H. Albert, and N. A. Staines), Vol. 2, p. 349. VCH, Weinheim.
6. Freshney, R. I. (1994). *Culture of animal cells*. Wiley-Liss, New York.
7. Doyle, A. and Griffiths, J. B. (1997). *Mammalian cell culture—essential techniques*. John Wiley & Sons, Chichester.
8. Phillips, H. A., Scharer, J. M., Bols, N. C., and Moo-Young, M. (1987). *Biotechnol. Lett.*, **9**, 745.
9. Miller, W. M., Wilke, C. R., and Blanch, H. W. (1987). *J. Cell. Physiol.*, **132**, 524.
10. Merten, O.-W. (1990). In *Animal cell biotechnology* (ed. R. E. Spier and J. B. Griffiths), Vol. 4, p. 257. Academic Press, London.
11. Newell, D. G., McBride, B. W., and Clark, S. A. (1988). *Making monoclonals*, p. 47. PHLS, Oxford.
12. Kan, M. and Sato, J. D. (1993). In *Cell and tissue culture: laboratory procedures* (ed. A. Doyle, J. B. Griffiths, and D. G. Newell), p. 2C: 2.1. John Wiley & Sons, Chichester.
13. Kan, M. and Sato, J. D. (1998). In *Laboratory procedures in biotechnology* (ed. A. Doyle and J. B. Griffiths), p. 92. John Wiley & Sons, Chichester.
14. McBride, B. (1993). In *Cell and tissue culture: laboratory procedures* (ed. A. Doyle, J. B. Griffiths, and D. G. Newell), p. 2D: 2.1. John Wiley & Sons, Chichester.
15. Doyle, A. and Morris, C. B. (1993). In *Cell and tissue culture: laboratory procedures* (ed. A. Doyle, J. B. Griffiths, and D. G. Newell), p. 4C: 1.1. John Wiley & Sons, Chichester.
16. Mazur-Melnyk, M. (1998). In *Laboratory procedures in biotechnology* (ed. A. Doyle and J. B. Griffiths), p. 254. John Wiley & Sons, Chichester.
17. Heidemann, R., Riese, U., Lutkeymeyer, D., Buntemeyer, H., and Lehmann, J. (1994). *Cytotechnology*, **14**, 1.
18. Lang, A. B. and Schurch, U. (1993). In *Cell and tissue culture: laboratory procedures* (ed. A. Doyle, J. B. Griffiths, and D. G. Newell), p. 28B: 4.1. John Wiley & Sons, Chichester.
19. Handa-Corrigan, A. (1991). In *Mammalian cell biotechnology: a practical approach* (ed. M. Butler), p. 139. IRL Press, Oxford.
20. Shevitz, J., Reuveny, S., LaPorte, T. L., and Cho, G. H. (1989). In *Advances in biotechnological processes*, Vol. 11, p. 81. Alan R. Liss. Inc., New York.

21. Merchuk, J. C. and Siegal, M. H. (1988). *J. Chem. Technol. Biotechnol.*, **41**, 105.

22. Rhodes, M., Gardner, S., and Broad, D. (1991). In *Animal cell bioreactors* (ed. C. Ho and D. Wang), p. 253. Butterworth-Heinemann, Boston.

23. Katinger, H. W. D., Scheirer, W., and Kromer, E. (1987). *Ger. Chem. Eng.*, **2**, 31.

24. Griffiths, J. B. (1990). In *Animal cell biotechnology* (ed. R. E. Spier and J. B. Griffiths), Vol. 4, p. 149. Academic Press, London.

25. Knazek, P. M., Gallino, P. O., Kohler, P. O., and Dendrick, R. L. (1972). *Science*, **178**, 65.

26. Griffiths, J. B., Looby, D., and Racher, A. J. (1992). *Cytotechnology*, **9**, 3.

27. Tyo, M. A., Bulbulian, B. J., Menken, B. Z., and Murphy, T. J. (1988). In *Animal cell biotechnology* (ed. R. E. Spier and J. B. Griffiths), Vol. 3, p. 357. Academic Press, London.

28. Hanak, J. A. J. and Davis, J. M. (1993). In *Cell and tissue culture: laboratory procedures* (ed. A. Doyle, J. B. Griffiths, and D. G. Newell), p. 28D: 3.1. John Wiley & Sons, Chichester.

29. Singh, R. P., Fassnacht, D., Perani, A., Simpson, N. H., Goldenzon, C., Portner, R., *et al.* (1998). In *New developments and new applications in animal cell technology* (ed. O.-W. Merten, P. Perrin, and J. B. Griffiths), p. 235. Kluwer Academic Publishers, Dordrecht.

30. Falkenberg, F. W., Weichart, H., Krane, M., Bartels, I., Palme, M., Nagels, H.-O., *et al.* (1995). *J. Immunol. Methods*, **179**, 13.

31. Nolli, M. L., Rossi, R., Soffientini, A., Zanette, D., and Quarta, C. (1998). In *New developments and new applications in animal cell technology* (ed. O.-W. Merten, P. Perrin, and J. B. Griffiths), p. 409. Kluwer Academic Publishers, Dordrecht.

32. Murdin, A. D., Thorpe, J. S., Groves, D. J., and Spier, R. E. (1989). *Enzyme Microb. Technol.*, **11**, 341.

33. Ramirez, O. T. and Mutharasan, R. (1993). *Biotechnol. Bioeng.*, **33**, 1072.

34. Looby, D., Racher, A. J., Griffiths, J. B., and Dowsett, A. B. (1989). In *Physiology of immobilized cells* (ed. J. A. M. deBont, J. Visser, B. Mattiasson, and J. Tramper), p. 255. Elsevier Science Publishers, Amsterdam.

35. Racher, A. J., Looby, D., and Griffiths, J. B. (1990). *J. Biotechnol.*, **15**, 129.

36. Park, S. and Stephanoulos, G. (1993). *Biotechnol. Bioeng.*, **41**, 25.

37. Griffiths, B. and Looby, D. (1990). In *Animal cell bioreactors* (ed. C. Ho and D. Wang), p. 167. Butterworth-Heinemann, Boston.

38. Racher, A. J. and Griffiths, J. B. (1993). *Cytotechnology*, **13**, 125.

39. Looby, D. and Griffiths, J. B. (1998). In *Laboratory procedures in biotechnology* (ed. A. Doyle and J. B. Griffiths), p. 268. John Wiley & Sons, Chichester.

40. Runstadler, P. W. Jr., Tung, A. S., Hayman, E. G., Ray, N. G., Sample, J. G., and DeLucia, D. E. (1989). In *Large scale mammalian cell culture technology* (ed. A. S. Lubiniecki), p. 363. Marcel Dekker, New York.

41. Looby, D. (1993). In *Cell and tissue culture: laboratory procedures* (ed. A. Doyle, J. B. Griffiths, and D. G. Newell), p. 28D: 1.1. John Wiley & Sons, Chichester.

42. Valle, M. A., Kaufman, J., Bentley, W. E., and Shiloach, J. (1998). In *New developments and new applications in animal cell technology* (ed. O.-W. Merten, P. Perrin, and J. B. Griffiths), p. 381. Kluwer Academic Publishers, Dordrecht.

43. Ball, G. D. (1985). In *Animal cell biotechnology* (ed. R. E. Spier and J. B. Griffiths), Vol. 2, p. 87. Academic Press, London.

44. van der Marel, P. (1985). In *Animal cell biotechnology* (ed. R. E. Spier and J. B. Griffiths), Vol. 2, p. 185. Academic Press, London.

45. Lovborg, U. (1985). *Monoclonal antibodies, production and maintenance.* William Heinemann Medical Books, London.

46. Murakami, H. (1989). In *Advances in biotechnological processes*, Vol. 11, p. 107. Alan R. Liss, New York.

Chapter 7

Isolation and purification of monoclonal antibodies from tissue culture supernatant

Geoff Hale

Sir William Dunn School of Pathology, University of Oxford, South Parks Road, Oxford OX1 3RE, U.K.

1 Objectives of antibody purification

Why do you need to purify an antibody? There are many possible reasons and the answer will be a major factor in determining the best strategy. Depending on the intended application, different contaminants are important and different degrees of purity are necessary (*Table 1*). Manufacture of antibodies for human therapy imposes many stringent requirements which are beyond the scope of this chapter. Anyone intending to develop a process for ultimate therapeutic application should seek detailed advice at an early stage.

A great deal of good research has been carried out using monoclonal antibodies in crude culture supernatant. They generally remain stable for many years, but it is necessary to know the antibody concentration. Usually this would be measured by ELISA (*Protocol 12*), which is fine for most purposes, but needs careful calibration if absolute accuracy is required. Purification becomes more important when antibodies are labelled (e.g. with biotin, fluorochromes, radioisotopes, enzymes), of if they are used as affinity capture reagents in

Table 1 Antibody applications and required purity

Application	Key requirements	Purity
Indirect immunostaining	Adequate (saturating) concentration	n/a
Direct immunostaining	Avoid non-specific labelling	> 50%
Affinity purifications	Maximize capacity	> 80%
Coating reagent for ELISA	Maximize capacity	> 80%
Cell activation/culture etc.	Free from cytokines, endotoxin, preservative, etc.	> 80%
In vivo experiments	Free from above and other immunogenic proteins	> 95%
Quantitative activity assays	Accurate antibody concentration	> 95%
General research reagent	All of the above	> 95%
Human diagnostic reagent	All of the above and reproducible quality	> 95%
Human therapeutic reagent	Free from a wide range of contaminants	> 99%

chromatography, ELISA, or similar applications. Then the antibody should at least be the major protein, but removal of trace contaminants is usually not crucial.

For cell culture experiments or injection into animals there are important bioactive contaminants to be avoided. Bacterial endotoxin causes a plethora of undesirable effects and can be lethal. It is better to avoid it than to attempt to remove it later. This is possible by using disposable plasticware throughout cell culture and purification and using aseptic technique, with appropriate sanitation of chromatography columns. Other bioactive substances such as cytokines may be present in serum, or secreted by hybridoma cells. Traditionally, isotype-matched control antibodies were used to detect the non-specific biological effects of such contaminants, but it is wise to reduce their levels by appropriate purification methods. Antibodies for sale either as research or *in vitro* diagnostic agents should be rigorously purified and quality controlled since they might be required for many different applications.

In principle, the whole spectrum of protein purification technologies can be applied to antibodies. The best established fall into the following categories:

- Salt fractionation
- Euglobulin/phase separations etc.
- Affinity chromatography
- Ion exchange chromatography
- Size exclusion chromatography
- Metal–chelate chromatography
- Hydrophobic interaction chromatography

A superb description of the theoretical and practical aspects of antibody purification is given by Gagnon (1). I strongly recommend this book to anyone wanting to understand the principles underlying separation techniques or seeking to optimize a purification process for a large scale application.

2 Essential information about the antibody

Before starting to develop a purification scheme you need some basic information about the antibody. What is the species, class, and subclass? Are there any unusual genetic modifications? What is the concentration of antibody in the culture supernatant? If it is genetically engineered, the basic structural properties are predictable. Antibodies from hybrid myelomas need to be characterized. There are several ways to identify class and subclass; one of the most simple and convenient is by Ouchterlony double diffusion using appropriate subclass-specific antisera (e.g. The Binding Site Ltd.). An alternative, more sensitive method is haemagglutination using red cells coupled with subclass-specific antibodies (Serotec Ltd.). Both tests are available as kits; just follow the instructions. Assay by ELISA is also possible, and this allows a more quantitative measure of antibody concentration. Armed with these results, it will be easier to choose the most appropriate purification method.

3 Problems with purifying antibodies from culture supernatant

3.1 Low concentration of antibody in culture supernatant compared with serum

Most purification techniques were originally developed using serum or ascitic fluid as the starting material. Monoclonal antibodies should now be made from cell culture supernatant and this poses a new constraint because of the relatively low concentrations (1–100 μg/ml) and therefore larger volumes which need to be processed. Good choice of cell line and culture medium makes the purification easier. Selection of a high-secretor should be obvious, but it is equally important that secretion should be stable, which may require re-cloning of the line (mouse hybridomas) or maintenance of the selection pressure (transfectants). Hybridomas generally produce about 50 μg/ml of antibody in spent culture, but genetically engineered transfectants may be much less productive, often producing less than 1 μg/ml. This can rise to 100 μg/ml with suitable selection and expression systems (2) but it is hard or impossible to obtain high secretion of some engineered antibodies. This might be due to imperfect association of heavy and light chains which are not selected *in vivo*.

If a low-secreting cell line is the best available and the antibody is important, the only option is to grow a large quantity of cells and to concentrate the culture supernatant before further processing. These objectives can both be accomplished in a hollow fibre fermenter. Alternatively, supernatant from conventional cultures can be concentrated in a variety of devices. Hollow fibre kidney dialysis cartridges are particularly convenient, being reliable, free from endotoxin, easy to use, and cheap.

Protocol 1°

Concentration of culture supernatant using a hollow fibre device

Equipment and reagents

- Cellulose di-acetate haemodialyser of choice (e.g. Altra-Flux-14, 10 kDa cut-off; Althin Medical Ltd., Cat. No. 211-712G)
- Silicone tubing for pump and fluid connections; male and female Luer tubing connectors, stoppers, and caps (e.g. CP Instrument Co.)

- Peristaltic pumps of choice (e.g. Masterflex with size 17 pump head)
- 0.2 μm air vent filter
- Phosphate-buffered saline (PBS)
- Culture supernatant, preferably contained in a single flexible bag (e.g. Stedim)

Method

1. Use sterile items and aseptic technique if you require sterile products. Otherwise, add sodium azide (0.1% final) to the culture supernatant.

Protocol 1 continued

2. Assemble the system as in *Figure 1*. Rinse with 5 litres of PBS at 250 ml/min to remove preservatives from the hollow fibre device.

3. Reassemble the tubing as in *Figure 2*. Adjust the permeate rate to approx. 40 ml/min.

4. Replace the PBS bag with the product bag. Concentrate until the volume of product is about 100 ml less than required. Check progress by measuring the antibody concentration. Check that there is no antibody in the permeate.

5. Reverse the permeate pump and backflush with 50 ml of permeate at 5 ml/min to desorb antibody from the fibres.

6. Adjust the product pump to 50 ml/min. Close the product bag clamp and open the air filter clamp. Pump air to flush out the product. Avoid frothing.

7. Repeat steps 5 and 6.

8. Measure the volume and concentration of product. Recovery should be better than 80%.

Notes:

(a) Volumes, dimensions, and flow rates may be adjusted pro rata to suit the scale of your process.

3.2 Potential contamination by bovine IgG

Despite the availability of serum-free media, the majority of cell lines are cultured in medium with fetal bovine serum (FBS). On average, FBS contains

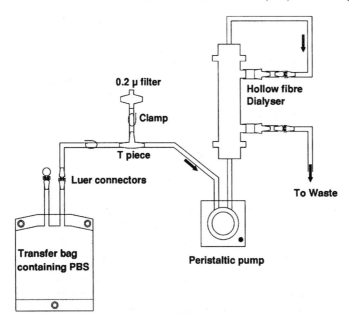

Figure 1 Hollow fibre dialyser assembly ready for flushing with PBS.

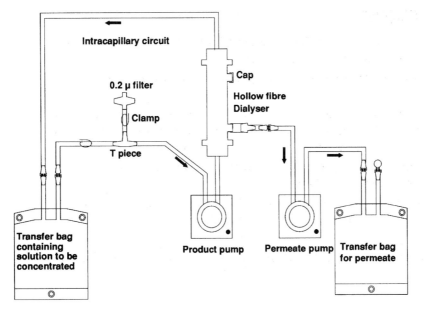

Figure 2 Hollow fibre dialyser assembly ready for concentration of antibody.

about 100 μg/ml of IgG so a culture supernatant with 10% FBS may contain more bovine IgG than the monoclonal antibody itself. If you need to purify such an antibody, reduce the bovine IgG by one of the following methods:

(a) Use serum-free medium (might reduce yield).

(b) Deplete IgG by passing serum through a Protein A or Protein G affinity column (3).

(c) Purchase serum which is already depleted of IgG (4) (e.g. Life Technologies Ltd.).

In our opinion, the last option is the simplest and best; however you must evaluate whether the depleted serum meets your requirements for growth promotion, country of origin, freedom from adventitious agents, etc.

4 Equipment for antibody purification

From the practical point of view, separation techniques fall into two categories:

(a) Precipitation methods require a centrifuge, preferably as large as possible. Its size is the main limitation on the amount of supernatant which can be processed in one batch. Continuous flow centrifuges can handle much larger quantities, but are not commonly available in a research laboratory.

(b) Chromatographic methods require pumps and ancillary equipment, which might include valves, monitors, and a fraction collector. The system can be as simple or as complex as you please. Simple separations can be carried out on a small scale with no equipment other than the chromatography column.

Sample and buffer flow can be driven by gravity and fractions collected manually. For larger scale experiments, it is more convenient to pump the samples and buffers.

4.1 Equipment for chromatography

4.1.1 Pump

Peristaltic pumps are ideal because they do not dilute or contaminate the sample and are self-priming. Sizes are available to suit all applications. The only drawback is that they cannot develop the pressures required for high-resolution HPLC. This is not a limitation for preparative chromatography since high pressure columns are prohibitively expensive in any case.

4.1.2 Fraction collector

A fraction collector allows you to leave the system running unattended, but make sure it is set up correctly with sufficient tubes! The Frac-100 (Amersham Pharmacia) is simple and reliable for general applications, but other models are probably equally good.

4.1.3 Monitors

A UV monitor is not essential, you can measure absorbance of fractions in a conventional spectrophotometer. However, an in-line monitor linked to a chart recorder and/or integrator is valuable if you are carrying out a large number of different separations. It speeds up method scouting and allows the system to be used in an analytical mode to measure antibody purity. At a preparative scale, the concentration of eluted protein often exceeds the linear range of the in-line monitor so integration should be checked by off-line measurements on a diluted sample. Conductivity and pH monitors are not usually included in a simple chromatography system but can be useful during method development and to check the reproducibility of buffers and samples. Modern UV and conductivity monitors are very reliable and require little attention. In-line pH probes can be more troublesome because they need regular calibration and the glass electrode is damaged by NaOH, which is commonly used as a cleaning and sanitizing agent.

4.1.4 Integrated chromatography systems

A complete chromatography system includes a computer controller, gradient mixer, pumps, valves, monitors, and peak integrator and fraction collector. Complex methods can be programmed to run automatically and chromatograms can be stored and analysed. A widely used system is the FPLC (Pharmacia), still a versatile workhorse, ideal for laboratory purification of antibodies. It is now superseded by the 'A'kta family of systems which are designed both for method scouting and production at the 1–20 g scale. These systems are versatile and include a sample loading device, the 'superloop', which is very convenient for sample volumes in the range 1–150 ml. Pharmacia systems are comparatively expensive. Less versatile, but quite practical systems from Bio-Rad would suit the more limited budget.

Protocol 2

General principles for liquid chromatography

Equipment and reagents

- Chromatographic equipment of choice, including controller, pumps, gradient mixer, valves, monitors, fraction collector, and integrator as required

- 0.5 M NaOH
- 20% ethanol

Safety notes

1. Maximum column pressures decrease with increasing diameter. Pumps can exceed these pressures, possibly leading to the tube shattering. Do not exceed the recommended limits.

2. Wear eye protection to reduce the risk from liquids which might leak from connectors under pressure. Take especial care with sodium hydroxide and other hazardous chemicals.

Method

1. Prepare buffers by weighing out the appropriate acid and base according to the proportions given in standard tables (5). Errors can be avoided by using the pH meter to check the buffer rather than to titrate it with acid or alkali.

2. Pass buffers and samples through a 0.45 μm or 0.2 μm filter to reduce the risk of columns fouling.

3. Ensure that buffers are equilibrated at the required temperature before starting.

4. Degassing buffers is usually not necessary if they are freshly made. If you encounter problems with air bubbles in the UV detector, put a pressure restrictor downstream.

5. Estimate the required column size from the manufacturer's binding capacity (often about 10–20 mg protein/ml gel) allowing that actual capacity may be 20–80% of this value. To determine the actual capacity, collect fractions of the flow-through during sample loading and measure antibody concentration. Stop when this exceeds 10% of the starting value. If the antibody is sufficiently concentrated (> 0.5 mg/ml) this can easily be seen from the UV trace.

6. Most media are packed into columns with a bed height of 1–10 cm. Size fractionation media require a bed height of 50–60 cm.

7. Set the flow rate according to the manufacturer's recommendations. Typically this is about 60 cm/h (e.g. for Sepharose Fast Flow gels from Pharmacia). Use linear flow rates (cm/h) to ensure consistent results independent of column dimensions. If the media support high flow rates, this may be useful for equilibration and washing; however higher capacity and sharper peaks are often obtained with slower rates for application and elution of the sample.

8. Use 0.5 M NaOH for cleaning and sanitizing the fluid path and columns, except where it is incompatible with the matrix (e.g. porous glass) or ligand (e.g. Protein G, antibodies). Treatment for 30–60 min will remove most strongly adsorbed molecules, destroy microbes and viruses, and remove endotoxin (6). Removal of the NaOH is easy to monitor and disposal is not a problem.

9. Use 20% ethanol for storing columns and fluid path in a sanitary condition unless otherwise recommended by the manufacturer.

5 Precipitation methods

Fractionation of antibody by precipitation has been used for 150 years and is well established in commercial processes for manufacture of polyclonal antibodies. A variety of different methods can be used, exploiting different physicochemical properties of antibodies which distinguish them from typical contaminants (*Table 2*). Control of pH is important because proteins are least soluble at their isoelectric point. Compared with other serum proteins, most antibodies are comparatively basic (high pI), but of course there is considerable variation between different monoclonals.

Salt precipitation with ammonium sulfate is the most commonly used of these methods. Different ions were ranked according to their lyotropic/chaotropic ('salting-out'/'salting-in') effects by Hoffmeister in 1888 (*Table 3*). Ammonium sulfate ranks among the most effective precipitating salt and is convenient to use because of its high solubility which is only minimally dependent on temperature (7).

Antibodies are more hydrophobic than other serum proteins and are pre-

Table 2 Types of precipitation methods

	Application	Precipitates	Mechanism[a]
Inorganic salts	All Ig	Antibody	Sharing of protein hydration shells
Polyethylene glycol	IgM, some IgG	Antibody	Sharing of protein hydration shells
Electrolyte depletion	IgM, some IgG	Antibody	Protein–protein electrostatic interactions
Organic acid	All Ig	Contaminants	Decreased solubility of acidic proteins
Ethacridine	IgG	Contaminants	Formation of insoluble ionic complexes

[a] Precise mechanisms are often poorly understood. See ref. 1 for further discussion.

Table 3 The Hoffmeister series of lyotropic and chaotropic ions[a]

	Increasing lyotropic effect ('salting-out') $\Rightarrow \Rightarrow \Rightarrow$							
SCN^-	ClO_4^-	NO_3^-	Br^-	Cl^-	COO^-	SO_4^{2-}	PO_4^{3-}	
Ba^{2+}	Ca^{2+}	Mg^{2+}	Li^+	Cs^+	Na^+	K^+	Rb^+	NH_4^+
	$\Leftarrow \Leftarrow \Leftarrow$ Increasing chaotropic effect ('salting-in')							

[a] Modified from ref. 1.

cipitated at lower salt concentrations. Monoclonals differ one from another, so for optimum results you should carry out trials to find the optimum concentrations. There is always a compromise between yield and purity; e.g. hydrophobic contaminants can be removed by a preliminary treatment with a lower salt concentration than optimal to precipitate the antibody, but this will still result in some losses. The main difficulty with culture supernatants (compared with serum or ascitic fluid) is the low concentration of antibody. Because salt precipitation is an equilibrium process, it is less efficient at low protein concentrations. Also the precipitate is fine and hard to sediment. Therefore it may be preferable to carry out the process in two stages—first using a single relatively high salt concentration to precipitate most of the antibody without concern for purity, and then to redissolve the precipitate in a smaller volume and carry out a series of cuts at increasing salt concentration to optimize purity and yield.

Protocol 3

Precipitation with ammonium sulfate

Equipment and reagents

- Centrifuge with suitable buckets and bottles (preferably at least 250 ml each); use bottles with conical bottoms if available (e.g. Corning, Cat. No. 25350-250)

- Large magnetic stirrer; use the most powerful available

- Cold room (4 °C); alternatively use a large fridge with power supply for stirrer

- Saturated ammonium sulfate: mix 1 kg of $(NH_4)_2SO_4$ (AR or better) with 1 litre of water (or pro rata). Heat to 50 °C and mix to obtain maximal dissolution. (It can be

heated in an autoclave with liquid cycle, but thorough mixing whilst still warm is important.) Fresh crystals should form on storage at 4 °C. Adjust to pH 6.5–7.5. (Note: a 100% saturated solution is approx. 4.05 M.)

- Phosphate-buffered saline (PBS)

- Culture supernatant: you can use unconcentrated supernatants, but it is more efficient to concentrate the supernatant at least tenfold in advance (*Protocol 1*)

Safety notes

1. Balance centrifuge bottles with an equivalent ammonium sulfate solution, *not* water. Density differences cause imbalance under running conditions.

2. Ammonium sulfate corrodes centrifuge rotors and buckets. Rinse them thoroughly after use.

Method

1. Keep all materials at 4 °C. Work in a cold room or chilled cabinet or keep the solutions in a bowl of ice. Take samples at each step to check antibody recovery.

2. Measure the volume of culture supernatant. Call this '1 volume'. Transfer it to a flask or bottle large enough to contain 2 volumes.

3. Measure 1 volume of saturated $(NH_4)_2SO_4$ and add it slowly (over 1–5 min) with continuous stirring. Avoid frothing.

4. Continue to stir overnight (not critical, but allows the precipitate to coalesce more easily).

5. Centrifuge at maximum speed (3000–10 000 g) for 1 h at 4 °C. Resuspend the precipitate in PBS to give a final of 0.1 volume.

6. Add 0.043 volumes of saturated $(NH_4)_2SO_4$ as in step 3 to give a final concentration of 30%. Stir for 1 h. Centrifuge as before and discard the precipitate (which should be small).

7. Add a further 0.057 volumes of saturated $(NH_4)_2SO_4$ as in step 3 to give a final concentration of 50%. Stir for 1 h. Centrifuge as before and resuspend the precipitate in PBS to a final of 0.1 volume.

8. Dialyse the product extensively against PBS. Removal of ammonium ions is critical for some procedures, e.g. biotin or FITC labelling. A simple test kit is available to estimate ammonium content (BDH/Merck, Cat. No. 31526 2R).

Notes:

(a) Although we carry out this process at 4 °C, others find equally good results at room temperature (1, 6). Required concentrations of $(NH_4)_2SO_4$ may be a little different.

(b) The amounts of ammonium sulfate were optimized for a rat IgM antibody. Different concentrations may be optimal for other antibodies. Check the antibody content of each fraction before discarding anything.

(c) For processing large volumes it is more economical to add solid $(NH_4)_2SO_4$ in the first step. Use 300 g/litre as a starting point. This procedure introduces a greater risk of antibody denaturation. Add it very slowly with steady stirring to avoid a local high concentration.

6 Affinity chromatography

A straw poll of commercial suppliers indicates that affinity chromatography on Protein A or Protein G is the method of choice for isolating monoclonal antibodies sold for research. I expect the same is true for academic laboratories. It is obvious why these methods are so popular. Antibody can be captured from dilute culture supernatant, usually without need to adjust pH or salt concentration. Yield and purity are high and the methods are applicable to the majority of IgG antibodies in common use. Generic binding and elution buffers are usually satisfactory so there is little need for process optimization. This is particularly convenient for academic researchers and commercial suppliers who are producing a wide range of reagents. Nevertheless, the apparent simplicity of affinity

Protocol 4 continued

7. Wash the column with 5 CV binding buffer. Repeat steps 1-6 to process further batches of the same antibody.

8. If required, clean the column with 1 CV 0.5 M NaOH. The contact time must be no more than 15 min. Neutralize by washing with binding buffer. This step may not be suitable for all Protein A media (particularly those based on porous glass)!

[a] Modified from the methods given in refs 1 and 12.

Notes:

(a) Choice of binding and elution buffers is very wide and those suggested are not necessarily ideal for each situation. 1.0 M sodium sulfate could be substituted for 3 M NaCl in buffer B2.

(b) A lower elution pH results in sharper peaks, but increases the risk of denaturation. Adjust the elution buffer to the highest pH which still gives a satisfactory peak.

(c) Some investigators find improved capacity for low affinity antibodies at 4 °C (11).

(d) A high salt concentration in the binding buffer may help to dissociate charged molecules (e.g. DNA) which can adsorb to antibodies.

(e) Although some IgM antibodies bind to Protein A, they frequently appear to be severely damaged by the low pH elution.

Protocol 5

Affinity chromatography on Protein G

Equipment and reagents

- Protein G media of choice, packed into a suitable column of choice
- Binding buffer: phosphate-buffered saline (PBS)
- Elution buffer: 0.1 M sodium citrate pH 2.8, or 0.1 M glycine–HCl pH 2.5
- Neutralization buffer: 1.0 M Tris–HCl pH 8.0
- Cleaning buffer: 1.0 M acetic acid
- Culture supernatant (which may be concentrated or not): pass through 0.45 μm or 0.2 μm filter to remove any particulate material

Method

1. Prepare the materials, equilibrate the column, bind, and elute the antibody as in *Protocol 4*.

2. Addition of neutralization buffer to the collection tubes in advance is strongly recommended. Resolution of bovine IgG is unlikely to be possible.

3. Between batches of different antibodies, clean the column with 1 CV 1.0 M acetic acid. Neutralize by washing with binding buffer.

Notes:

(a) The same comments apply as in *Protocol 4*.

(b) Since the required elution pH is lower than for Protein A, there is a greater risk of antibody denaturation. An alternative to neutralizing the fractions is to divert the eluted peak directly through a desalting column (e.g. Sephadex G25) pre-equilibrated with the desired final buffer (e.g. PBS). If you have the necessary valves and chromatography controller, this is a very good option.

6.3 Immunoaffinity purification

For antibodies which do not bind efficiently to either Protein A or G, alternative affinity methods have been developed. The most versatile is anti-immuno-globulin, which has especially been used for purification of rat IgGs (13). A wide range of polyclonal and monoclonal anti-Igs are available, specific for Ig species, isotype, or domain. In principle these give the advantages of affinity chroma-tography (high purity, high yield, capture from physiological solutions) and also allow the separation of the desired antibody from contaminating Ig (e.g. bovine) as well as the purification of antibody fragments (Fab, scFv, etc.). The drawback is the relative expense of the reagents which limits the scale of separation.

A new reagent is Protein L from *Peptococcus magnus* which binds to Ig kappa chains (14, 15). In principle this provide an alternative to Protein A and G as well as being appropriate for Fab and scFv fragments. As well as being specific for kappa light chains, Protein L is reported to bind human Ig better than rat or mouse. As yet there is no widespread experience—however, it appears to be very useful for IgM antibodies and other antibodies and their fragments which do not bind Protein A or G.

7 Ion exchange chromatography

Ion exchange chromatography separates molecules according to charge differ-ences. Charge-carrying proteins bind to a matrix of opposite charge. They can be eluted either by a change of pH (to make the charge on the protein more similar to the matrix) or by an increase in salt concentration (providing counter-ions which compete with the protein for binding sites). It is an extremely versatile technology which can be applied to all antibodies, and has been widely used since the earliest days. Ion exchange chromatography offers high capacity, high resolution, good yield, effective concentration, and is robust and cheap. In principle it is possible to separate molecules which have a single charge difference, so it can be used to separate antibodies from other contaminating Ig, to resolve complex mixtures of antibodies, and even to separate different glycoforms (1, 16). However, a consequence of this versatility is that more method development needs to be done since each monoclonal can behave differently.

Both anion and cation exchangers can be used and they offer different advantages according to the properties of the individual monoclonal. In the past, weakly ionizing groups such as diethyl-aminoethyl (DEAE, anion exchanger) and carboxymethyl (CM, cation exchanger) were usually used. Nowadays, the stronger basic or acidic groups like quaternary aminoethyl (QAE) or sulfopropyl (SP) are more common because they offer consistently high capacity over a wide range of pH.

As a general rule, antibodies have a higher pI than most other serum proteins (i.e. they have a greater positive charge at neutral pH). This means that they tend to bind more strongly than other proteins to a cation exchanger and less strongly to an anion exchanger. Cation exchange chromatography is more useful for capturing antibody from culture supernatant, since most other molecules pass straight through and so the capacity of the column to adsorb antibody remains high. In contrast, anion exchange chromatography is good for removing impurities, since most of the undesirable contaminants in culture supernatant (including DNA, endotoxin, phenol red, viruses) are negatively charged at physiological pH.

When designing buffer conditions for ion exchange chromatography you should try to ensure that there is only one ion of the opposite charge to the exchange resin. So for cation exchange, buffer ions will typically be negatively charged carboxylic acids and the counter-ion will be Na^+. If low concentrations of other cations are present (e.g. K^+, Ca^{2+}, Mg^{2+}, NH_4^+), they will compete for binding to the resin and it will take a long time to equilibrate. Binding and elution of the antibody may be unpredictable at different scales. Conversely for anion exchange chromatography; the purist would use a positively charged buffer salt, e.g. Tris and a single anion, e.g. Cl^- or PO_4^{3-}, but not both. However, there are successful examples of antibody separations carried out in a mixture of buffer anions (12). Note that the popular bacteriostatic agent sodium azide is unsuited to ion exchange chromatography; at low pH it decomposes and anion exchange resins will remove it from solution. Better to make up buffers fresh (from 10 × concentrates if required).

7.1 Cation exchange chromatography

It is not normally possible to capture antibodies direct from culture supernatant because the concentration of competing salt is too high. A preliminary buffer exchange step is necessary. A wide range of buffers might be suitable and buffer exchange is usually achieved by dialysis or diafiltration. Some antibodies precipitate under these conditions which can reduce the overall yield. Several options might be explored in such a case (1):

- In-line dilution of the culture supernatant as the sample is loaded
- Change of pH and/or salt concentration
- Addition of 2 M urea

Protocol 6

Cation exchange chromatography[a]

Equipment and reagents

- Strong cation exchange media of choice (e.g. sulfopropyl)
- Binding buffer: 50 mM sodium acetate pH 4.5
- Elution buffer: 50 mM sodium acetate, 1.0 M NaCl pH 4.5
- If required, stripping buffer: 2 M NaCl
- If required, 1.0 M sodium acetate pH 4.0

A. For antibodies which are fully soluble in the binding buffer

1. Dialyse or diafilter the culture supernatant against the binding buffer. Extensive changes are not necessary.

2. Remove any visible precipitate by centrifugation (e.g. 3000 g for 30 min). Check that it does not contain antibody.

3. Filter the supernatant through a 0.45 μm or 0.2 μm filter.

4. Equilibrate the column with 5 CV of binding buffer.

5. Apply the culture supernatant.

6. Wash with 2 CV binding buffer, or until there is no protein detectable by the UV trace.

7. Elute with a linear gradient of 10 CV from binding buffer to 50% elution buffer. Collect fractions and analyse to identify the antibody peak.

8. Strip with 100% elution buffer and/or stripping buffer.

9. Clean and store the column as in *Protocol 2*.

B. For antibodies which are poorly soluble in binding buffer

1. Equilibrate the column with 5 CV of binding buffer.

2. Immediately before loading, adjust the pH by adding 5% (v/v) of 1.0 M sodium acetate pH 4.0.

3. Load the sample by in-line dilution with four parts of binding buffer. This requires two pumps and a mixer.

4. Continue as in part A, steps 6–9.

[a] Modified from the method given in ref. 1.

Notes:

(a) Trials of different binding and elution buffers and different gradients will be necessary to obtain optimal results. Stepwise elution may be practicable in many cases.

(b) Antibody may equally well be eluted by an increase in pH. If this is the final or only purification step, it is convenient to elute the antibody with PBS (see *Protocol 7*).

Cation exchange is a useful secondary step to remove residual contaminants after affinity chromatography. In principle it is possible to load the low pH eluate directly onto the cation exchange column. At the typical pH (e.g. 3.0) for elution from the affinity matrix, the antibody will bind tightly almost irrespective of salt concentration. This binding can be too tight and we have found it to be essentially irreversible for some antibodies, probably as a result of partial denaturation on the column. It has been overcome by simply equilibrating the column in a binding buffer of less extreme pH, e.g. pH 4.0.

Protocol 7

Cation exchange chromatography following Protein A

Equipment and reagents

- Binding buffer: 0.1 M sodium citrate pH 4.0
- Elution buffer: PBS
- Strong cation exchange media of choice (e.g. sulfopropyl)

Method

1. Equilibrate the column with 5 CV of binding buffer.

2. Elute the antibody from Protein A with 0.1 M sodium citrate pH 3.0–3.5 (*Protocol 4*). Do not adjust the pH, but load it directly on the cation exchange column.

3. Wash with 2 CV binding buffer, or until there is no protein detectable by the UV trace.

4. Elute with 2 CV PBS. Collect the antibody in a single peak.

5. Clean and store the column as in *Protocol 2*.

Notes:

(a) If the buffer strength of PBS is insufficient resulting in a broad peak, use 0.1 M sodium phosphate pH 7.2 for elution.

7.2 Anion exchange chromatography

Under physiological conditions antibodies do not normally bind to anion exchangers, whereas several important contaminants bind very strongly. A good way to remove residual DNA or endotoxin from a purified antibody is to pass it through a small amount of anion exchanger. Porous filters (e.g. 'Sartobind', Sartorius) are particularly convenient, being available in disposable or reusable units fitted with Luer lock connectors. The potential flow rate is high and the antibody can just be pushed through with a syringe.

Protocol 8

Removal of contaminants by anion exchange chromatography

Equipment and reagents

- Sartobind-Q filters (Sartorius), size of choice (e.g. Q15)
- Chromatography pump or syringe
- Phosphate-buffered saline (PBS), pyrogen-free
- 0.5 M NaOH

Method

1. Rinse the filter with 20 ml of 0.5 M NaOH, leave for 1 h, then rinse with 100 ml PBS.
2. Pass the antibody solution through the filter at approx. 20 ml/min.
3. Regenerate the filter as in step 1.

Notes:

(a) This method is applicable to purified antibodies where relatively small amounts of contaminants like endotoxin or DNA need to be removed. Most buffers in the pH range 5–8 are suitable. See the manufacturer's information for other ion exchange applications of these filters.

(b) A similar method could be used with other anion exchange resins.

8 Immobilized metal affinity chromatography (IMAC)

IMAC encompasses all types of metal ion interactions with proteins, but one of the most common applications exploits the interaction of nickel ions with histidine residues. It is commonly used to capture a variety of recombinant proteins with a genetically attached histidine-rich motif, but is also applicable to the majority of IgG antibodies because there is a cluster of His residues at the junction of the CH2 and CH3 domains (1, 12, 17). Nickel is linked to the matrix by a carboxylic chelating agent and can form a coordination complex with several amino acids, but principally histidine. The interaction is relatively independent of salt concentration, but becomes weaker as the imidazole ring is protonated when the pH is reduced from 8 to 4. This allows elution by a simple change in pH. Proteins can also be eluted by imidazole (or histidine or histamine) leaving the nickel bound to the column, or by chelating agents such as EDTA, releasing nickel from the column.

The potential advantages of IMAC are that it can be used to capture antibodies from dilute supernatants and may allow a high degree of purification, similar to affinity chromatography. However, there are a number of pitfalls:

(a) Cell culture medium typically contains histidine and other small molecules which interfere with protein binding and/or cause leakage of the metal ions.

Desalting by dialysis or diafiltration is required to obtain reproducible results.

(b) Other proteins in the culture supernatant, e.g. bovine albumin, can also bind to the column, albeit more weakly than IgG. However, this means that time has to be spent in determining the optimal binding and elution conditions, much as for ion exchange chromatography.

(c) Even when the antibody is eluted by a simple pH gradient, there is the possibility of contamination with toxic nickel ions. A nickel scavenging step is recommended, either passage through an uncharged chelating column or size exclusion chromatography in the presence of a mixture of EDTA and imidazole (to inhibit metal–protein binding).

(d) The 'conserved' histidine residues which bind nickel also form an important part of the Protein A binding domain. Antibodies which don't bind protein A (e.g. human IgG3, allotype G3m) may be impossible or difficult to purify by IMAC.

Protocol 9

Immobilized metal affinity chromatography[a]

Equipment and reagents

- Immobilized iminodiacetic acid media of choice
- Buffer A: 50 mM sodium phosphate, 0.5 M NaCl pH 8.0
- Buffer B: buffer A with 50 mM EDTA
- Buffer C: 0.1 M sodium acetate, 0.5 M NaCl pH 4.5
- Buffer D: buffer C with 0.1 M nickel chloride
- Buffer E: 0.2 M histidine, 50 mM EDTA pH 7.0

Safety notes

1. Nickel is a potent sensitizer and frequent cause of allergic reactions. Avoid contact with skin or mucous membranes.

Method

1. Dialyse or diafilter the culture supernatant against buffer A.

2. Strip the iminodiacetic acid column with 2 CV buffer B. Wash with 2 CV buffer C.

3. Charge the column with nickel by loading buffer D until the bed is a uniform colour.

4. Wash off excess nickel with 5 CV buffer C. Equilibrate with 2 CV buffer A.

5. Load the sample. Wash with 2 CV buffer A.

6. Elute with a 10 CV linear gradient to buffer C. Collect fractions and pool those which contain the antibody.

7. Add 5% (v/v) buffer E to the pooled fractions to scavenge any nickel which has leached from the column.

8. Desalt the antibody by dialysis or gel filtration.

[a] Modified from the method given in ref. 1.

Notes:

(a) Some protocols advise that the supernatant can be loaded directly onto the column, but since it is likely to contain histidine, it is probably wise to carry out the initial salt exchange step.

(b) Although the method is compared with affinity chromatography for specificity and yield, bovine albumin can also bind to the column so optimization of the binding and elution buffers is desirable as with ion exchange chromatography.

(c) If required, the amount of residual nickel can be measured with a simple colorimetric assay (e.g. BDH/Merck, Cat. No. 16964 3J).

9 Size exclusion chromatography (SEC)

Size exclusion chromatography (or 'gel filtration') has a unique set of attributes which sets it apart from other methods. Its applications are limited because it has low resolution, low capacity, and is slow. However, it is the only practical separation method based on size, which is a relatively invariant property of immunoglobulins in contrast to charge, hydrophobicity, and other specific binding interactions. The resolution achieved by SEC may not be spectacular; however separations are predictable and nearly independent of buffer conditions, so the need for method development is minimal.

SEC is commonly used for purification of IgM antibodies, which are among the largest serum proteins and easily resolved from smaller molecular weight contaminants. Furthermore, this can be accomplished in physiological buffers, whereas IgM monoclonal antibodies are often poorly soluble under the pH and salt conditions required for affinity or ion exchange chromatography. The main drawback is that sample volume can only be a small percentage of the column bed volume (preferably less than 5%), so either a large column is required (expensive) or the sample has to be concentrated before it is loaded. Pouring a size exclusion column used to be technically demanding but the development of crosslinked agarose (e.g. Sepharose CL series), mixed agarose-polyacrylamide (e.g. Sephacryl), and then highly crosslinked dextrans (e.g. Superdex) has made this much easier and also permits higher pressure and flow rates at a preparative scale. Nevertheless, it is still important to pour columns with care to obtain a homogeneous bed which gives the optimum resolution. Pay close attention to the manufacturer's guidelines!

Protocol 10

Purification of IgM by size exclusion chromatography

Equipment and reagents

- Size exclusion media of choice (e.g. Sepharose 6B or Superdex 200, Amersham Pharmacia)
- Elution buffer: phosphate-buffered saline (PBS)

Method

1. Concentrate the sample to 10–50 mg/ml by a suitable preliminary step, e.g. precipitation with ammonium sulfate (*Protocol 3*).

2. Equilibrate the column with 2 CV of elution buffer.

3. Apply the sample in a final volume of not more than 5% of the bed volume.

4. Elute with 2 CV of elution buffer. Collect fractions, and identify the antibody which will normally be the first peak eluted.

Notes:

(a) Composition of the elution buffer is not critical to the separation. It should be chosen to suit the stability of the antibody and its intended use.

(b) It is important not to allow air to enter the column. This would severely disrupt the packing and reduce resolution. It is a good idea to check the column performance by running a sample of 1% acetone or 1 mM tyrosine and comparing the width and asymmetry of the eluted peak with that specified by the manufacturer.

The other main application of SEC is for resolving monomeric IgG from dimers and higher order aggregates. Some monoclonal antibodies spontaneously form stable oligomers, either at the low pH encountered in other purification steps or even as they are synthesized by the cell. The oligomers have different physiological properties, particularly with regard to effector functions such as binding to Fc receptors. Monomer and dimer IgG can be resolved by SEC and this is particularly recommended when IgG antibodies are to be used for *in vivo* studies or clinical trials (*Figure 3*). Because of the low capacity of SEC, it is normally used as the last step in a series of separations, and then it is easy to use an elution buffer suitable for the intended application of the antibody.

10 Hydrophobic interaction chromatography

Hydrophobic interaction chromatography (HIC) is a powerful method for resolving all types of biomolecules. Immunoglobulins have strongly hydrophobic domains compared with the majority of typical contaminants and so this technique is especially appropriate. My impression is that it is not routinely used by many investigators or companies who supply antibodies for the re-

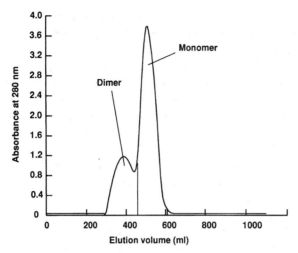

Figure 3 Separation of IgG monomer and dimer by size exclusion chromatography. Approx. 1 g of recombinant IgG1 antibody, freshly eluted from an S-Sepharose column in 70 ml of 100 mM sodium phosphate pH 7.2, was applied to a column of Superdex 200 (25 × 485 mm) and eluted with PBS.

search market. Whether this is due to a difficulty in understanding the underlying principles or an innate conservatism leading most of us to choose the more traditional techniques is not clear.

The binding of proteins to HIC media can be described in terms of two mechanisms (1). The first applies to all types of liquid chromatography and results from the fact that salt molecules are excluded from the surfaces of both proteins and chromatography media, leaving a relatively hydrated shell. As the salt concentration is increased it becomes more favourable for the protein and media to share their hydration shell, resulting in retention. The second mechanism is unique to HIC and results from the strong interactions between hydrophobic groups within the contact zone where water has been excluded. The strength of these forces is related to the hydrophobicity of the protein and the media. The consequence of this dual mechanism is that weakly hydrophobic media (e.g. butyl) require high salt to promote binding, but elution can be achieved simply by reducing the salt concentration. On the other hand, strongly hydrophobic media (e.g. phenyl) bind antibodies more readily, but elution may require the addition of organic solvents to directly compete for the hydrophobic interactions.

The ability of different salts to promote binding to HIC media is ranked according to the Hoffmeister series (*Table 3*), and this can be understood in terms of their relative abilities to penetrate the hydration shell. Salts high in the lyotropic series (NH_4^+, K^+, Na^+, PO_3^{3-}, SO_4^{2-}) are strongly excluded from protein and media surfaces and therefore promote binding, whereas chaotropic salts (Ba^{2+}, Ca^{2+}, Mg^{2+}, SCN^-) tend to penetrate the hydration shell, reduce local hydration, and diminish binding. The most commonly used salt is ammonium sulfate, but it requires care to control the pH and alternatives such as potassium phosphate, sodium glutamate, or other organic salts should be considered, especially at alkaline pH. Hydrophobic competitors have limited value because many of them tend to denature antibodies.

It is hard to predict the behaviour of an individual antibody, because it does not depend on the average hydrophobicity or charge of the protein, but rather, the antibody will be oriented in a specific fashion on the chromatography media to optimize the contacts. Regions outside this contact zone make little contribution to the net binding. This observation is obvious for affinity chromatography, but in fact it applies equally well to all forms of adsorption chromatography including ion exchange, IMAC, and HIC. The only difference in these other cases is that there may be several potential contact zones and the one which predominates will depend on the media and buffer conditions.

HIC has many important attributes. It is applicable to all isotypes and species of antibodies. By simple addition of salt, it is possible to capture antibodies from culture supernatant. Good separation from other proteins, DNA, and endotoxin can be achieved, and it is possible to resolve monoclonal antibodies from other contaminating serum immunoglobulins. Recovery and purity are potentially high. Furthermore the HIC media have high capacity, allow high flow rates, and withstand a wide range of cleaning and sanitizing agents. The main limitation is the potential for local denaturation of the antibody at the interaction zone, which is likely to be an important part of the antibody structure, commonly the antigen binding site. Fortunately, this is often reversed on elution, or can be minimized by use of less strongly hydrophobic matrices. Media stronger than phenyl are unsuitable for antibody purification. However, the problem with weaker media is that the antibodies may not be fully soluble under the high salt conditions necessary for binding. This might be overcome by in-line addition of the salt just as the antibody is being loaded onto the column, but requires a somewhat more sophisticated chromatography system. Other potential problems to be aware of when using high salt concentrations are:

• Increased viscosity which compromises mixing and flow
• Possible corrosion of pump components
• Interference in subsequent assays, particularly gel electrophoresis

One reason why many groups have not explored the use of HIC is the wide range of media and buffer conditions to be scouted in order to determine an optimal purification scheme. Nevertheless, this powerful method should not be overlooked. Some suggestions for method development are:

(a) Start with the most hydrophobic medium which does not give denaturation (e.g. phenyl). This will allow lower salt concentrations and give more reproducible results.

(b) Choose a medium which supports high flow rates.

(c) Evaluate different buffers for the most favourable selectivity, e.g. potassium phosphate pH 5.5, 7.0, or 8.5, or ammonium sulfate pH 5.5 or 7.0.

(d) For reliable results, ensure that temperature is controlled, especially on scale-up.

Protocol 11

Purification of IgG antibodies by HIC[a]

Equipment and reagents

- HIC media of choice (it is possible to obtain a kit of different media for development)

- Buffer A: 0.1 M sodium phosphate, 2 M $(NH_4)_2SO_4$ pH 7.0
- Buffer B: 0.1 M sodium phosphate pH 7.0

Method

1. Dilute the sample with 2 vol. of buffer A.

2. Equilibrate the column with 2 CV, 80% buffer A, 20% buffer B.

3. Apply the sample.

4. Wash with 2 CV, 80% buffer A, 20% buffer B.

5. Elute with a 10 CV gradient from 20–80% buffer B. Collect fractions and pool those which contain the antibody. Albumin and transferrin are eluted at the beginning of the gradient; IgG towards the end.

6. Strip with 5 CV buffer B.

[a] Modified from the method given in ref. 12.

Notes:

(a) As with other protocols, the buffer composition and elution gradient can be optimized for individual antibodies.

(b) This method could be used after a preliminary purification step, e.g. precipitation with ammonium sulfate, ion exchange, or IMAC.

(c) For phenyl columns, reduce the concentration of $(NH_4)_2SO_4$ in buffer A to 1.0 M.

11 Other chromatographic methods

Hydroxyapatite, $Ca_{10}(PO_4)_6(OH)_2$, is a unique substance which is both the re-active ligand and the matrix. It contains a combination of both positively charged (Ca^{2+}) and negatively charged (PO_4^{2-}) binding sites which interact with biomolecules in a subtle way depending on their distribution of amino and carboxyl groups. Antibodies can bind in the presence of high salt concentra-tions, which dissociates ionically bound contaminants, and are usually eluted with an increasing concentration of phosphate which competes for the strong binding Ca^{2+} sites (18). The attributes of hydroxyapatite chromatography are similar to ion exchange, HIC, and IMAC, with potential for a similar degree of purification. The original hydroxyapatite crystals were fragile but modern ceramic media are robust, easily packed, and have excellent dynamic prop-erties. They are resistant to most cleaning agents, but are severely degraded below pH 5 or by chelating agents such as EDTA.

A range of other media have been developed in recent years. Hydrophilic interaction chromatography (HILIC) is somewhat similar to HIC, but relies almost exclusively on the hydration effect rather than hydrophobic interactions (19). Antibodies bind to a hydrophilic matrix in the presence of polyethylene glycol, which like the lyotropic salts is excluded from the hydration shell of proteins and matrix. It is most applicable to IgMs but is not yet widely used. A method which has been more enthusiastically promoted is so called 'thiophilic' adsorption chromatography using ligands containing sulfone and/or thioether groups (20). These were originally thought to have a unique specificity for antibodies, but subsequent work has shown that the binding mechanism is not really different from any other form of hydrophobic interaction chromatography, so this type of matrix may more correctly be thought of as a subset of HIC.

12 Choice of method

With this wide range of techniques, it might seem difficult to make the right choice. You need to consider the ultimate application of the antibody (i.e. whether high purity or removal of specific contaminants is necessary), the scale of the purification (cost of media is unimportant at small scale, but critical for commercial manufacturing), and the number of batches required (extensive method development is of little value for a single batch). I expect that the majority of academic investigators will find that a single method like Protein A affinity chromatography meets most of their needs. However, the very diversity of antibodies is their usefulness and special approaches will be required in individual cases. When antibody production is scaled-up for a commercial application, especially as a pharmaceutical, it will pay to consider carefully the full range of technologies from first principles.

12.1 Purification of antibody fragments

There are so many derivatives of antibodies, proteolytic fragments, and genetically engineered variants that it is impossible to give specific recommendations for their purification; however the general principles will still be applicable. One comment however; it is not generally appreciated that many antibody F(ab')$_2$ fragments still bind to Protein A and Protein G and so the results of affinity chromatography can be very surprising. I would recommend size exclusion chromatography, e.g. on Superdex 200. This gives a predictable separation of intact IgG, the desired fragments, and low molecular weight peptides.

13 Storage of antibodies

Monoclonal antibodies are inherently stable proteins. Being secretory products they are stabilized by disulfide bonds and not readily susceptible to damage by oxidation. In our experience, culture supernatants containing IgG antibodies remained active for at least 15 years at 4 °C when Hepes and sodium azide were

added to stabilize the pH and prevent microbial growth. However, some purified antibodies are not so stable and are prone to precipitate, particularly IgM.
Usually they are formulated in phosphate-buffered saline pH 7.2 since this is
convenient for many applications, but this might not be the optimal buffer for
long-term stability. Useful additives include EDTA (to inhibit metal-dependent
proteases) and sodium azide (as an antimicrobial agent). IgM antibodies may be
stabilized by additional protein, e.g. serum albumin. For long-term storage we
recommend that IgG antibodies solutions are frozen in small aliquots at $-25\,°C$
or below. Accurate documentation and labelling of samples is vital. Get in the
habit of identifying the batch, concentration, solvent, who prepared it, and the
date, as well as the clone number of the antibody. Computer-printed labels are a
good way to save the chore of writing the same details on many tubes. It is a
good idea to label each tube with a unique day book and page reference where
all the relevant details are recorded. This is especially important if the antibody
is passed on to another investigator.

14 Analysis of purity and activity

It should be routine to check the concentration and purity of the purified
antibody; the range of tests will depend on its application. Usually the most
important test is a functional measure of the antigen binding activity; many
different techniques could be used including ELISA (for purified antigens), flow
cytometry (for cellular antigens), or complement-mediated cell lysis. Antibody
activity is normally assessed in terms of titre; the antibody is serially diluted and
the concentration required to give 50% of the maximum binding (or lysis) is
determined by curve fitting. The precision of these biological assays is unlikely
to be better than ± 25%. Other standard tests are shown in *Table 4*.

Table 4 Analytical tests for purified antibodies

Property	Techniques
Concentration	Absorbance at 280 nm, ELISA, HPLC
Identity	Isoelectric focusing, native gel electrophoresis
Purity	Affinity HPLC, SDS gel electrophoresis
Aggregation	Size exclusion HPLC, native gel electrophoresis
Residual Protein A, G	ELISA
Residual bovine IgG	ELISA
Endotoxin	LAL

14.1 Antibody concentration and purity

The concentration of a purified antibody is best measured by absorbance at
280 nm. For accurate results ensure that the spectrophotometer is properly
calibrated. The absorbance of a 1 mg/ml solution in a 1 cm cell is conventionally
taken to be approx. 1.4 for IgG and 1.2 for IgM. The true values will vary from

one antibody to another depending on the Tyr and Trp content, and can be calculated if the complete amino acid composition is known (21). All other indirect methods for measuring protein concentration are subject to similar variability.

Alternative methods to measure antibody concentration are required for crude cell harvest or partly purified preparations. Sandwich ELISA using a capture and detection reagent specific for the appropriate immunoglobulin species and/or class is sensitive and reasonably accurate, if somewhat laborious. A less sensitive, but more precise method is affinity HPLC using a small column of Protein A or Protein G. Even weakly binding antibodies can be captured from culture supernatant on an analytical scale and then eluted at low pH for measurement by an in-line UV monitor and integrator. Another method which is quick and simple and requires no special equipment is haemagglutination of red cells coated with a suitable antiglobulin (22).

ELISA methodology can readily be adapted to measure any particular contaminants such as bovine immunoglobulin, Protein A, or Protein G by suitable choice of reagents. The precise specificity of antibodies used in the Protein A/Protein G assays is critical to avoid interference from the monoclonal itself.

Protocol 12

Generic ELISA method for measuring immunoglobulins and other analytes

Equipment and reagents

- Flexible flat-bottomed microtitre plates (e.g. Falcon, 3912)
- 37 °C incubator (or heater block with microtitre attachment) and microtitre plate reader
- Coating buffer: phosphate-buffered saline (PBS)
- Wash buffer: PBS containing 0.05% Tween 20
- Blocking buffer: PBS containing 1% bovine serum albumin (BSA) and 0.1% sodium azide
- Dilution buffer: PBS containing 0.1% BSA
- Standard: known concentrations of analyte in dilution buffer

- Capture reagent: affinity purified polyclonal antibody directed against the analyte
- Detection reagent: same as capture reagent, but biotin labelled
- Enzyme conjugate of choice: e.g. Extravidin-peroxidase (Sigma, code E-2886)
- Substrate solution (make this fresh just before use): 10 mg as one tablet o-phenylenediamine.2HCl (e.g. Sigma, code P-8287), 5.0 ml water, 2.5 ml of 0.1 M citric acid, 2.5 ml of 0.2 M Na_2HPO_4, 4 μl of 30% (w/v) hydrogen peroxide
- Stop solution: 0.5 M sulfuric acid

Safety notes

1. The following are hazardous and must be handled with care: o-phenylenediamine (potential carcinogen), sulfuric acid, hydrogen peroxide (corrosive).

Protocol 12 continued

Method

1. Dilute the capture reagent to 5 μg/ml in coating buffer. Add 50 μl to each microtitre well. Incubate for 2 h at 37 °C. Rinse four times with wash buffer. Add 200 μl blocking buffer to each well. Incubate for 1 h at 37 °C (or store at 4 °C).

2. Rinse four times with wash buffer and tap the plates dry.

3. Add 50 μl of standards or test samples (diluted if necessary) to each well. Include some wells with dilution buffer alone as negative control. Incubate for 1 h at room temperature. Rinse as in step 2.

4. Add 50 μl of the detection reagent (optimal dilution). Incubate for 1 h at room temperature. Rinse as in step 2.

5. Add 50 μl of the enzyme conjugate (optimal dilution). Incubate for 30 min at room temperature. Rinse as in step 2.

6. Prepare the substrate. Add 50 μl per well. Incubate for 5–15 min, as required for optimal colour development. Add 50 μl stop solution. Measure the absorbance at 492 nm in a microtitre plate reader.

7. Check that positive and negative controls give acceptable results. Calculate the concentration of analyte in the test samples by interpolation from the standard curve.

Assay development:

(a) This basic assay can be used for measurement of any species or subclass of Ig by appropriate choice of capture and detection reagents. If monoclonal reagents are used, ensure they are directed against different antigenic epitopes.

(b) For measurement of Protein A or Protein G, the following are recommended:

    (i) Protein A: capture with monoclonal anti-Protein A (Sigma, code P-2921). Detect with biotin-labelled monoclonal anti-Protein A (Sigma, code B-3150).

    (ii) Protein G: capture with chicken anti-Protein G (Immunsystem AB, code 02-109). Detect with biotin-labelled chicken anti-Protein G (Immunsystem AB, code 06-109).

(c) The detection reagent and enzyme conjugate should be titrated in preliminary assays to determine the greatest dilution which still gives maximal signal.

(d) Problems with lack of sensitivity are usually due to deterioration of the enzyme conjugate. Check its activity by adding a series of dilutions direct to the substrate.

One of the most common techniques for analysing protein purity is polyacrylamide gel electrophoresis (PAGE). This subject is covered in detail in another volume in this series (23). Any convenient proprietary or home-made equipment is suitable. There are several relevant applications.

(a) SDS gel electrophoresis under reducing conditions. Antibody molecules are broken down into their constituent chains which should be visualized as two clear bands with molecular weight of approx. 50 kDa and 25 kDa (IgG) or 80 kDa and 25 kDa (IgM). Contaminating proteins, proteolytic fragments, and abnormally glycosylated forms are generally well resolved on these gels so they provide an indication both of purity and of the integrity of primary structure.

(b) Native gel electrophoresis. In the absence of detergent, antibodies migrate as a single species with a characteristic mobility proportional to both charge and molecular weight. Dimers and aggregates are easily resolved. Some antibodies are very basic and may not migrate into the gel under standard running conditions, loading near the negative electrode at slightly alkaline pH. It is more reliable to run the gel at acid pH with reversed polarity and load near the positive electrode.

(c) Isoelectric focusing. Antibodies have widely varying isoelectric points and so this is a useful confirmation of individual identity. Normally there is a spectrum of bands corresponding to the different charged species which arise from partial deamidation. These give a sensitive indication of changes which might occur during cell culture and purification.

To some extent gel electrophoresis is being superseded by capillary electrophoresis which is faster, more sensitive and precise, but requires relatively specialized and costly equipment. An alternative method for checking antibody aggregation is analytical size exclusion chromatography. This is exactly analogous to the preparative technique described earlier, but on a smaller scale it is possible to use high pressure columns and very small sample volumes which results in superior resolution. Using an HPLC system with UV monitor and integrator, it is possible to quantify the amount of aggregate, oligomers, and low molecular weight fragments.

14.2 Endotoxin contamination

Determination of possible contamination by endotoxin is too often ignored until it is realized that this might be confounding a whole series of biological experiments. In fact it is easy to measure using a commercial kit with a colorimetric readout in a standard spectrophotometer or microtitre plate reader. We recommend this for all antibodies prepared for use in cell culture or *in vivo* applications. If they contain an unacceptably high level of endotoxin then there are various methods proposed for its removal. In our experience, absorption on immobilized polymyxin (24) or extraction with TX114 detergent (25) have not been very reliable, and instead we suggest anion exchange chromatography on Sartobind filters as described above.

Protocol 13

Analysis of endotoxin by chromogenic LAL[a]

Equipment and reagents

- Disposable 96-well flat-bottomed microtitre plates (e.g. Falcon, 3072)
- Microtitre plate shaker and plate reader with 405 nm filter
- Dry block heater with microtitre plate adaptor and lid; check the temperature continuously with an accurate thermometer

- LAL test kit, QCL-1000 (Bio-Whittaker UK Ltd., Cat. No. 50-647U): this contains sterile pyrogen-free water, *E. coli* endotoxin positive control, chromogenic substrate, and limulus amoebocyte lysate; reconstitute the reagents as instructed
- 25% acetic acid

Safety notes

1. Endotoxin is toxic! Handle with care.

Method

1. Set the dry block heater to 37 ± 1 °C. Allow all reagents (except LAL) and test samples to adjust to room temperature. Draw up a plate plan as follows:

   (a) Rows A to F: standards (0.1–1.0 EU/ml), controls, and test samples.

   (b) Row G: LAL stock for pre-incubation.

   (c) Row H: substrate stock for pre-incubation.

   Culture media, water, and physiological buffers can be tested undiluted. Samples containing high endotoxin must be diluted within the working range. Test several different dilutions of unknown but suspect samples.

2. Dispense 10 µl of the standards, controls, and test samples according to the plate plan.

3. Dispense the substrate (20 µl per sample) into Row H so that it can later be quickly transferred to the sample wells with a multichannel pipetter.

4. Dispense the LAL (10 µl per sample) into Row G in the same way.

5. Equilibrate the plate in the dry block heater for 10 min.

6. Timing is critical. Use a stopwatch and add reagents in the same order, so that all wells have identical reaction time. Keep the plate in the dry block heater to maintain constant temperature.

7. At 0 min, add 10 µl LAL to each well. Mix thoroughly on the plate shaker for 15 sec.

8. At 30 min, add 20 µl substrate to each well. Mix as before.

9. At 36 min, add 20 µl of 25% acetic acid to each well. Mix as before.

10. Measure the absorbance at 405 nm against a water blank. Plot a standard curve and calculate the endotoxin level in the test samples by interpolation.

[a] Modified from the method in ref. 26.

Notes:

(a) This method is scaled down for economical use of reagents. It is less sensitive than the manufacturer's standard method but adequate for experimental work.

(b) It is important to vortex the endotoxin standard samples for 15 min when preparing the stock and making dilutions. Standards can be stored at 4 °C for up to one month, but should be vortexed for 3 min just before use because endotoxin tends to attach to the surface of containers.

(c) The LAL reagent can be stored frozen in aliquots for future use, though this is not recommended by the manufacturer.

(d) Avoid introducing contamination during the assay. Use pyrogen-free water for dilutions, and if possible work in a clean air environment. Use sterile disposable pipettes and avoid repeated sampling from stock solutions.

(e) Water is not a suitable diluent for some samples (e.g. IgM antibodies). Saline for injection may be a suitable alternative.

(f) The temperature and reaction timings are critical.

(g) Some samples contain substances which inhibit the reaction (high salt, extreme pH, organic solvents, detergents) or give a false positive (coloured substances). Validate the assay by 'spiking' a sample with 1.0 EU/ml of standard and check the recovery. If there is inhibition, the test sample can often be diluted to a point where there is no inhibition.

15 Conclusion

Production and quality control of monoclonal antibodies is only a means to an end. The experimental or clinical application is what matters. There is little point in carrying out experiments with contaminated samples of uncertain concentration or activity so effort spent on careful purification is a good investment. However, it would be futile to focus all your energy and time on these preparatory steps. Select the techniques which enable you to reach the goals most quickly; often these will be the ones where there is already local experience. Although they are so variable, antibodies are produced in abundance and are stable molecules. Purification is usually not difficult. The protocols need not be followed slavishly but can be developed to suit your own needs according to the basic chemical principles outlined.

Acknowledgements

My work is supported by the Medical Research Council, LeukoSite Inc., and the EP Abraham's Trust. I am indebted to many colleagues, particularly Herman Waldmann, Steve Cobbold, Mike Clark, Jenny Phillips, Patrick Harrison, and Pru Bird for their help and advice.

References

1. Gagnon, P. (1996). *Purification tools for monoclonal antibodies.* Validated Biosystems, Tucson, AZ.
2. Bebbington, C. R., Renner, G., Thomson, S., King, D., Abrams, D., and Yarranton, G. T. (1992). *Bio/Technology*, **10**, 169.
3. Underwood, P. A. (1986). In *Methods in enzymology* (ed. Langone, J. J. and Van Vunakis, H.), Vol. 121, p. 301. Academic Press, London.
4. Torres, A. R., Healey, M. C., Johnston, A. V., and McKnight, M. E. (1986). *Human Antibodies and Hybridomas*, **3**, 206.
5. Dawson, R. M. C., Elliott, D. C., Elliott, W. H., and Jones, K. M. (ed.) (1986). *Data for biochemical research* (3rd edn). Clarendon Press, Oxford.
6. Pharmacia LKB Biotechnology. (1989). *Sanitization of BioPilot system and columns using sodium hydroxide.* Technical Note 203. Pharmacia, Uppsala.
7. Weir, D. M. (ed) (1996). *Handbook of experimental immunology* (5th edn). Blackwell Science, Cambridge, MA.
8. Diesendorfer, J. (1981). *Biochemistry*, **20**, 2361.
9. Fuglistaller, P. (1989). *J. Immunol. Methods*, **124**, 171.
10. Hale, G., Drumm, A., Harrison, P., and Phillips, J. (1994). *J. Immunol. Methods*, **171**, 15.
11. Schuler, G. and Reinacher, M. (1991). *J. Chromatogr.*, **587**, 61.
12. Pharmacia Biotech. (1997). *Monoclonal antibody purification*, Technical handbook, Cat. No. 18-1037-46. Pharmacia, Uppsala.
13. Bazin, H., Xhurdebise, L. M., Burtonboy, G., Lebacq, A. M., De-Clercq, L., and Cormont, F. (1984). *J. Immunol. Methods*, **66**, 261.
14. Vola, R., Lombardi, A., Tarditi, L., Bjorck, L., and Mariani, M. (1995). *J. Chromatogr. B*, **668**, 209.
15. Kouki, T., Inui, T., Okabe, H., Ochi, Y., and Kajita, Y. (1997). *Immunol. Invest.*, **26**, 399.
16. Clark, M., Bindon, C., Dyer, M., Friend, P., Hale, G., Cobbold, S., *et al.* (1989). *Eur. J. Immunol.*, **19**, 381.
17. Hale, J. and Beidler, D. (1994). *Anal. Biochem.*, **222**, 29.
18. Stanker, L. H., Vanderlaan, M., and Juarez-Salinas, H. (1985). *J. Immunol. Methods*, **76**, 157.
19. Alpert, A. (1990). *J. Chromatogr.*, **499**, 177.
20. Porath, J., Maisano, F., and Belew, M. (1985). *FEBS Lett.*, **185**, 306.
21. Mach, H., Middaugh, C. R., and Lewis, R. V. (1992). *Anal. Biochem.*, **200**, 74.
22. Phillips, J. and Hale, G. (1999). In *Diagnostic and therapeutic antibodies* (ed. A. J. T. George and C. E. Urch). Humana Press, Totowa, NJ.
23. Hames, D. (1998). *Gel electrophoresis of proteins: a practical approach* (3rd edn). Oxford University Press, Oxford.
24. Savelkoul, H. F. J., Vossen, A. C. T. M., Breedland, E. G., and Tibbe, J. M. (1994). *J. Immunol. Methods*, **172**, 33.
25. Liu, S., Tobias, R., McClure, S., Styba, G., Shi, Q., and Jackowski, G. (1997). *Clin. Biochem.*, **30**, 455.
26. Phillips, J., Harrison, P., and Hale, G. (1999). In *Diagnostic and therapeutic antibodies* (ed. A. J. T. George and C. E. Urch). Humana Press, Totowa, NJ.

Chapter 8
Antibody production in plants

Pascal Drake*, Eva Stoger[†], Liz Nicholson[†],
Paul Christou[†], and Julian K. C. Ma*

*Unit of Immunology, Department of Oral Medicine and Pathology, 28th Floor,
Guy's Tower, GKT Institute of Medicine and Dentistry, Guy's Hospital,
London Bridge, London SE1 9RT, U.K.

[†]Molecular Biotechnology Unit, John Innes Centre, Norwich, U.K.

1 Introduction

Plant biotechnology is a rapidly expanding area and in the last five to ten years it has become apparent that plant systems may be particularly valuable for the expression and production of recombinant proteins such as pharmaceuticals, vaccines, and in particular, antibodies. A particular attraction of this approach is the potential for producing these kinds of reagents on an agricultural scale, thereby significantly reducing the costs of production. However there are also many other advantages related to the use of plants. In this chapter, we shall provide a brief review of the advances made in plant expression of antibodies and describe the principles of plant genetic engineering as well as the major techniques that are involved.

2 Expressing recombinant proteins in plants

There are two strategies for production of recombinant proteins in plants. Genetic transformation of the plant genome can now be achieved relatively easily using *Agrobacterium* T-DNA vectors or micro-projectile bombardment. This results in transfected plants which can express recombinant proteins either constitutively or in a tissue-specific manner. As the foreign gene is integrated into the plant genome, traditional breeding techniques can then be used to generate transgenic seed stocks for easy and stable storage or distribution. Furthermore, sexual propagation between different transgenic plants can be used to accumulate multiple foreign genes in a single plant, which is a particular benefit in the production of multimeric proteins such as immunoglobulins, and has been used to significant effect in the production of secretory IgA in plants (1). Alternatively, with micro-projectile bombardment, multiple constructs can be introduced into plant cells simultaneously to achieve co-ordinate expression.

An alternative approach involves the use of viral vectors, in which plant viruses are genetically modified to encode the foreign genes. Plants are subsequently infected with the modified virus and thus act as 'viral culture vessels'. By this means, very large quantities of recombinant proteins can be generated quite rapidly, using relatively small numbers of plants, however, this strategy at present appears to be limited to peptides or small single unit proteins. There are limitations to the use of both transgenic plants and genetically modified plant viral vectors, and it is likely that both systems will prove to be important for different protein candidates. What is clear however, is that both contribute to the overall versatility of using plants for production of all kinds of recombinant proteins.

2.1 Plant hosts

To date, most examples of plant produced antibodies have been in tobacco. This is simply because it is easy to transform and rapid to regenerate, and is one of the best studied and understood of the plant species. Historically, plants of the dicotyledonous group have been the easiest to transform, using *Agrobacterium*, consequently most of the early work has been concentrated in these plants. Other examples are *Arabidopsis* (another model plant), potato, tomato, and soy. More recently, with the development of ballistic bombardment techniques for plant transformation, many other plant species have become amenable to transformation, including important crop plants such as maize, rice, and wheat. The range of plants that could be used as vehicles for antibody production is now close to including all the commercially important crop species.

2.2 Antibodies in plants

Production of a full-length IgG antibody in transgenic plants was first described in 1989, and demonstrated that co-expression of two recombinant gene products could lead to a correctly folded and assembled multimeric molecule in plants that was functionally identical to its mammalian counterpart (2). Since then, a number of groups have also expressed other antibody molecules ranging from single chain molecules to multimeric secretory antibody in whole plants as well as plant cell culture. In some cases, the aim has been to modify plant physiology or improve plant characteristics (3), but there is also a clear potential for the exploitation of plants as bioreactors for large scale production of antibodies. Economic production of kilogram quantities would open many new areas of use for antibodies including medical and veterinary applications such as passive immunization.

2.2.1 Antibody fragments

A wide range of functional recombinant antibody fragments have been described in the literature, and most of the antibody fragments described in *E. coli* have also been produced in transgenic plants. These include a single domain antibody (dAb) in tobacco (4), single chain Fv molecules (5), Fab (6), and F(ab')$_2$ production in tobacco and *Arabidopsis*. The requirements for assembly of these molecules are quite undemanding and processing through the endomembrane system is not required, thus they can be produced in *E. coli* and many other heterologous expression systems, probably more easily than in plants. Originally, only low levels of expression ($< 0.1\%$ total soluble protein) were achieved in plants. Since then, various strategies have been devised to improve yields, such as the expression of genes for antibody fragments in fusion with the endoplasmic reticulum retention signal KDEL which can increase scFv yield by tenfold (7) or up to 4–6.8% of total soluble protein (8). Another approach which has been successful is to target scFv for expression in seeds, where the antibody fragment can accumulate to 3–4% of the total soluble seed protein (8).

2.2.2 Full-length and multimeric antibodies

An important advantage of plants as a recombinant expression system for antibody production is the ability to assemble full-length heavy chains with light chains to form full-length antibody efficiently (2). Full-length antibodies are not readily assembled in bacterial expression systems, and bivalent antibody molecules can only be produced in *E. coli* by rather complex molecular engineering. Although the constant region of the immunoglobulin has no role in antigen recognition or binding, it has important effector functions and contains several important functional regions involved in glycosylation, complement activation, phagocyte binding, the hinge region, as well as the site for association with J chain and secretory component (in α and μ chains). Importantly, it allows bivalent antigen binding with flexibility of the antibody molecule at the hinge region, which may be important if aggregation is a major protective mechanism. In addition, binding antigen bivalently significantly affects the strength of the antigen/antibody interaction, by virtue of an increase in avidity.

Antigen recognition and binding is a critical and sensitive test for correct assembly as, in the vast majority of cases, individual light or heavy chains, or misfolded Ig molecules are not functional. In mammalian plasma cells, the mechanism of this assembly is partially understood. The immunoglobulin light and heavy chains are synthesized as precursor proteins, and signal sequences direct translocation into the lumen of the endoplasmic reticulum. Within the ER, there is cleavage of the signal peptides, and stress proteins, such as BiP and GRP94, as well as enzymes such as protein disulfide isomerase (PDI) function as chaperones which bind to unassembled heavy and light chains and direct subsequent folding and assembly. In plants, passage of immunoglobulin chains through the ER is also necessary, as in the absence of a signal peptide related either to the light or heavy chain gene, assembly of antibody does not take place (2). However, both plant and non-plant signal sequences from a variety of sources, are sufficient for correct targeting (9, 10). Plant chaperones homologous to mammalian BiP, GRP94, and PDI have been described within the ER (11, 12), and expression of immunoglobulin chains in plants is indeed associated with increased BiP and PDI expression. Furthermore, BiP and PDI are associated with immunoglobulin chains *in planta* (Ma *et al.*, unpublished data). Thus, it seems likely that there are broadly similar folding and assembly mechanisms for antibodies in mammals and plants.

Several IgG mAbs have been produced in transgenic plants by academic and commercial groups, that may have therapeutic applications in humans or animals. The best example is a murine IgG1 (Guy's 13) that binds to the adhesion protein of *Streptococcus mutans* that is the primary cause of dental caries. The strategy used to produce this antibody in plants was to express each immunoglobulin chain separately in different plants and to introduce the two genes together in the progeny plant by cross-pollination of the individual heavy and light chain expressing plants. This involves two generations of plants to generate an antibody-producing plant, and by this technique, the yield of recombinant antibody is consistently high, between 1–5% of total soluble plant protein (2, 13). Other groups have expressed IgG antibodies using double trans-formation techniques (6), or cloned the light and heavy chain genes together in a single agrobacterium T-DNA vector (9, 14), and this can save time and effort. Guy's 13 IgG expressed in tobacco is relatively easy to purify in large quantities, and functionally, there is no discernible difference between the antibody expressed in plants, and that expressed in other systems (13).

The ability to accumulate genes in transgenic plants by successive crosses between individually transformed parental plants is a considerable advantage in attempting to construct multimeric protein complexes, such as secretory anti-body. Until recently, attempts to produce monoclonal SIgA have been hindered by the complexity of this molecule, which consists of two basic Ig monomeric units (heavy and light chains) that are dimerized by a joining (J) chain and then associated with a fourth polypeptide, secretory component (SC) (15). These modifications are believed to enhance the activity of SIgA in the mucosal en-vironment. Dimerization by J chain increases the avidity of binding of the anti-body and enhances the potential for bacterial aggregation, whilst the secretory component confers a degree of resistance against proteolysis, an important property for antibodies within the harsh environment of the GI tract.

In order to generate a secretory antibody version of Guy's 13 in plants, the carboxyl-terminal domains of the Guy's 13 IgG antibody heavy chain were modified by replacing the Cγ3 domain with Cα2 and Cα3 domains of an IgA antibody, that are required for binding to J chain and secretory component (13). Four transgenic plants were generated to express independently either the Guy's 13 kappa chain, the hybrid IgA-G antibody heavy chain, murine J chain, or rabbit secretory component (SC). A series of sexual crosses was performed between these plants and filial recombinants in order to generate plants in which all four protein chains were expressed simultaneously. In the final quadruple transgenic plant, three forms of antibody were detectable by Western blot analysis of samples prepared under non-reducing conditions. These bands were approx. M_r 210 kDa (the expected size for monomeric IgA-G), M_r 400 kDa (IgA-G dimerized with J chain), and M_r 470 kDa (dimeric IgA-G associated with SC). The assembly was very efficient, with greater than 50% of the SC being associated with dimeric IgA-G and the SIgA-G yield from fully expanded leaves was in excess of 5% of total soluble protein, or 200–500 μg per gram of fresh weight material (1). Functional studies confirmed that the SIgA-G molecule bound specifically to its native antigen and that the binding affinity of each antigen binding site was no different to that of the native IgG. However, the avidity and functional affinity of the entire molecule was greater, which helped to confirm that a dimeric, tetravalent antibody had been assembled. Finally, in a human trial, the plant secretory Guy's 13 antibody prevented oral colonization by *Streptococcus mutans*, thereby demonstrating for the first time, the therapeutic application in humans of a recombinant product derived from plants (16).

2.3 Modified viruses for transient expression in plants

Whereas transgenic plants have the advantage of stable integration of the foreign gene into the plant nuclear chromosome, the use of genetically modified plant viruses offers an alternative, more rapid means of generating extremely high levels of recombinant proteins. In general two approaches have been used: first, foreign gene transcription in which the foreign gene is expressed as a soluble protein and secondly, by engineering of viral coat proteins (cp) in fusion with antigenic peptides or proteins whilst allowing the continuing assembly and formation of infectious virus particles that display antigen on their surface. Several plant viruses have been used, most successfully tobacco mosaic virus (TMV) and cowpea mosaic virus (CPMV).

TMV is a well characterized RNA virus with a broad host range and is thus a good candidate vector for expression of foreign genes in plants. After infection, the amount of recoverable virus is extremely high and can reach up to 50% of the dry plant weight. Following infection of tobacco, soluble foreign gene product accumulated to levels of at least 2% of the total soluble protein (17). A significant advantage of this system is the speed with which foreign gene products can be made. DNA sequences are cloned into the viral vector, transcribed into infectious RNA, passaged through a laboratory plant which acts as a packaging host to prepare chimeric virus, which are then used for infection of plants in the field. The foreign protein accumulates to high levels and can be harvested within days, the entire process occupying only a few weeks (18). Although clearly important for many recombinant proteins, these techniques appear so far to have only limited value for production of antibody fragments.

2.4 Glycosylation of recombinant proteins in transgenic plants

Protein modification by glycosylation is found in all higher eukaryotes and plant proteins contain *N*-linked as well as *O*-linked glycans. Variations between the glycans associated with native proteins and recombinant forms may complicate immunotherapy, whatever the heterologous expression system, and this is not a problem that is specific to recombinant proteins produced in plants. However, it is important to understand the differences between plant and mammalian glycans in order to evaluate their relative importance.

It had previously been demonstrated that the *N*-linked core high mannose type glycans have identical structures in plants, mammals, and other organisms (19, 20), which are subsequently modified in a number of steps to complex glycans. Native complex glycans in plant proteins can be quite heterogeneous, but they tend to be smaller than mammalian complex glycans and differ in the terminal sugar residues. For example, a xylose residue linked $\beta(1,2)$ to the β-linked mannose residue of the glycan core, and an $\alpha(1,3)$-fucose residue in place of an $\alpha(1,6)$-fucose linked to the proximal glucosamine, are frequently found in plants, but not mammals (19). On the other hand, *N*-acetyl neuraminic acid (NANA), which is a prevalent terminal residue in mammals has not been identified in plants (nor for that matter in insect cells or yeast).

A structural comparison of the glycans associated with Guy's 13 IgG expressed either in plants or in murine hybridoma cells has recently been performed (21). The results demonstrated that the same glycosylation sites were utilized in both systems, but that compared to the murine antibody, the glycans on the plant antibody were more heterogeneous. In addition to high mannose type glycans, approximately two-thirds of the plant antibodies had $\beta(1,2)$-xylose and $\alpha(1,3)$-fucose as predicted. The differences in glycosylation patterns of plant antibodies have no effect on antigen binding or specificity. However, these types of plant glycans associated with plant glycoproteins could potentially be quite immuno-genic in humans, although it remains to be demonstrated if plant glycans pre-sented by mammalian proteins are equally immunogenic. In the recent human study of oral application of plant secretory antibody, no evidence for an im-mune response to the plant recombinant glycoprotein was detected after six applications of antibody (16). Furthermore when experimental mice were immunized subcutaneously with plant-derived murine IgG (with adjuvant), no anti-plant glycan responses were detected (nor were there any anti-IgG responses) (Chargelegue, manuscript submitted).

Nevertheless, for systemic applications, it may be necessary to remove the complex glycans, or to alter the heavy chain sequence to remove the sites for *N*-linked glycosylation. An alternative, more elegant approach is also being developed using mutant plants that lack enzymes involved in the complex glycosylation pathway (22).

3 Plant transformation

The various options for expressing antibodies in plants have been outlined and we have attempted to indicate the versatility of the plant expression system. In terms of methodology, we shall focus on expressing IgG in plants as this is not only a molecule for which plant expression is well suited, but also because it demonstrates general principles involved in expressing all antibody molecules in plants from antibody fragments to multimeric secretory antibodies. It is assumed that methods for cloning immunoglobulin genes are familiar to the reader, as this is not an aspect of this technology that is specific to plants. In the following sections we shall describe the design of typical vectors used for plant transformation, and we shall also detail the two principle methods for plant transformation—*Agrobacterium tumefaciens* and particle bombardment, using tobacco and wheat as the model plant systems respectively. Regeneration of whole plants from transformed cells will be discussed as well as the specific requirements for plant screening and plant fertilization to generate homozygous IgG producing plant stocks.

Many techniques have been employed to introduce DNA into plant cells. The two most successful involve the use of *Agrobacterium tumefaciens*, a bacterium capable of transferring and integrating a portion of its DNA into the plant genome, and micro-projectile bombardment in which DNA-coated microscopic particles are accelerated into the plant nucleus. Two strategies can be used to recover transgenic plants after the initial transformation event. The first approach is to screen all the plants that have been regenerated from a transformation experiment for the presence of the foreign gene. The second, and more widely used approach, is to select transformed cells on the basis of resistance to a selective agent which is conferred by a selectable marker gene. The gene for neomycin phosphotransferase II (*npt*-II) is the most commonly used selectable marker (23). The *npt*-II gene was initially isolated from the *E. coli* transposon Tn5 and confers resistance to aminoglycoside antibiotics such as kanamycin. The latter inhibit protein synthesis in higher plants by binding to ribosomes of the mitochondria and chloroplasts. The NPT-II enzyme phosphorylates a specific hydroxyl of the antibiotics which inhibits their ability to bind to the ribosome (24). Transgenic plants with improved traits can therefore be generated by using constructs containing a gene of interest with a selectable marker gene.

A reporter gene may also be employed. These genes encode a protein whose activity can be readily detected, providing a means of optimizing transformation parameters in order to maximize the number of transgenic plants produced. Beta-glucuronidase (*gus*, *gus* A, *uid* A) is a reporter gene initially isolated from *E. coli* which has been used extensively in plant transformation experiments (25). The GUS enzyme is a hydrolase whose activity can be readily visualized by formation of a blue precipitate in the presence of the substrate X-Gluc (5-bromo-4-chloro-3-indolyl-glucuronide) in a histochemical assay, by fluorometry in the presence of 4-methylumbelliferyl glucuronide (MUG), or spectrophotometrically using *p*-nitrophenyl glucuronide as substrate. Other reporter genes for plant transformation include the chloramphenicol acetyl gene (26), luciferase genes (27, 28), and green fluorescent protein (29).

3.1 Gene constructs

Foreign genes transferred into plants are chimeric constructs, comprising the gene coding sequence fused to plant functional regulatory signals (promoter and polyadenylation sequences). The regulatory signals which allow gene expression in plant cells are frequently derived from the 19S and 35S transcripts of the cauliflower mosaic virus (30, 31) or from the *Agrobacterium* genes nopaline synthase, octopine synthase, and mannopine synthase (32–34).

The choice between constitutive promoters (such as the CaMV 35S or maize ubiquitin-1 promoters) and tissue-specific promoters needs to be evaluated on an empirical basis. In our experience, seed-specific promoters are beneficial in some plants (e.g. peas) while constitutive promoters appear to generate better results in cereals (rice and wheat). It is important to get high-level antibody accumulation in the most suitable organs for storage and extraction, usually leaves or seeds. For example, the constitutive maize ubiquitin-1 promoter provides higher level expression in rice seeds than seed-specific promoters, such as the wheat low molecular weight glutelin promoter (35). Conversely, the seed-specific legumin promoter allows higher levels of antibody accumulation in pea seeds compared to the constitutive CaMV 35S promoter (Y. Perrin, personal communication).

Some groups believe that it is important that heterologous genes expressed in plants are codon optimized for plant systems. Different species have different codon preferences, i.e. the favoured use of certain degenerate codons corresponding to a given amino acid. Animal transgenes expressed in plants usually have a different codon bias to the host, resulting in pausing at disfavoured codons and truncation, misincorporation, or frameshifting. Such effects can be avoided by introducing translationally-neutral mutations into the coding region of the transgene by site-directed mutagenesis, bringing transgene codon usage in line with the host—this is termed codon optimization.

Finally, vector design should incorporate features to allow appropriate targeting of the antibody. It appears that immunoglobulins fold correctly in the endoplasmic reticulum (ER) but incorrectly in the cytosol. For this reason, we include a signal peptide from either the murine immunoglobulin heavy of light chain loci to allow targeting to the secretory pathway. Increased antibody stability may be achieved by retaining the newly synthesized molecules in the ER. This is carried out using a 3′ KDEL sequence in the vector. When this is translated, the tetrapeptide signal causes the proteins to be retained in the ER.

3.2 *Agrobacterium tumefaciens*-mediated transformation

A. tumefaciens is a Gram-negative soil bacteria which infects a wide variety of dicotyledons, at wound sites, causing crown gall disease. Galls, which are usually produced at ground level following infection with *A. tumefaciens* are composed of disorganized vascular and parenchymatous tissues (reviewed in ref. 36). Crown gall disease often reduces plant growth and may in some cases cause host death. Galls synthesize amino acid or sugar derivatives known as opines which are absent from untransformed cells. Opines are used by *Agrobacterium* as a source of carbon and nitrogen, and also stimulate the conjugative transfer of bacterial plasmids so producing a population of bacteria with the metabolic potential to utilize the opines (37, 38). Infection is mediated by a chemotactic response of agrobacteria to plant exudates from wounded tissue (39). Agrobacteria bind to plant cell surfaces and then synthesize cellulose fibrils which anchor the bacteria to the cells. *Agrobacterium* chromosomal genes, *chv*A and *chv*B, are required for this attachment phase to occur (40). Crown gall disease is caused by the transfer and integration into the plant genome of a portion of a large plasmid called pTi (tumour-inducing plasmid) from *A. tumefaciens* (36, 41, 42).

The T-DNA (transferred DNA) region of the Ti plasmid carries the genes which are transferred to, integrated, and expressed in recipient plant cells. T-DNA from pTi contains genes which promote the synthesis of auxins, cytokinins (oncogenes), and opines. The T-DNA genes are not involved in the transfer process and can be replaced by other foreign genes of interest without affecting transfer efficiency. Two direct repeats of 24 bp at the borders of the T-DNA, which contain cleavage and excision sites, are required for its efficient transfer (43). The other essential part of the pTi plasmid is the virulence region (*vir*). The *vir* region of Ti plasmids contains genes which are involved in the excision and transfer of the T-DNA to plant cells (44). The plant phenolic compounds acetosyringone and alpha-hydroxyacetosyringone are involved in the activation of *vir* genes.

The exact mechanism of T-DNA transfer and integration into the plant genome has not been fully elucidated. T-DNA is thought to be transferred to plant cells in single-stranded form (45, 46). Molecular analysis of the genome of transformed plants demonstrates that foreign genes are often truncated and rearranged and may be present in single or multiple copies. The numerous T-DNA and target DNA rearrangements which occur during integration suggest that this process involves recombination, replication, and repair processes (47). For plant genetic engineering, the oncogenes need to be deleted from the Ti plasmid as plants cannot be regenerated from crown galls. Two alternative strategies can be used for gene integration with the *Agrobacterium* system. In a cointegrate pTi vector, T-DNA oncogenes are replaced by homologous recombination with a DNA fragment containing the foreign gene(s) of interest (48). Prior to recombination, plasmid is transferred to the *Agrobacterium* by a triparental mating procedure. The second strategy involves a binary system (49). In the construction of a binary system, the entire T region (including border sequences) is deleted from the Ti plasmid creating a disarmed strain. The gene(s) of interest is cloned between the border sequences in a second smaller plasmid (called the binary vector) which is transferred into the disarmed *Agrobacterium* either by electroporation (50) or a freeze–thaw protocol (51). The *vir* region on the disarmed vector acts in *trans* to mediate transfer of DNA between the border sequences on the binary vector. The cointegrate system has largely been superseded by the binary strategy in recent transformation protocols, as the former has several disadvantages—notably the inability to directly transfer plasmid to agrobacteria, as the low transformation efficiencies are not conducive to cointegrate formation. In addition, aberrant recombination events may occur with the cointegrate system (see ref. 51 for further information on cointegrate and binary vectors).

3.2.1 Components of a binary vector

A typical binary vector is small (*c.* 10 kb) to facilitate genetic manipulation and transformation. The basic components of a binary vector are shown in *Figure 1*. They are:

(a) A broad host range origin of replication (functional in *Agrobacterium* and *E. coli*). Transfer functions will also be required if the plasmid is to be introduced into *Agrobacterium* by conjugation, rather than by electroporation or a freeze–thaw procedure.

189

(b) T-DNA border sequences.

(c) A multiple cloning site (MCS) between the T-DNA borders. The MCS can be constructed within the alpha-complementary region of beta-galactosidase allowing insertion of the gene of interest to be tested using blue/white screening on X-Gal indicator plates (e.g. pBin 19) (52).

(d) A plant selectable marker and reporter gene between the border sequences.

(e) A bacterial selectable marker to allow recovery of *Agrobacterium* harbouring the binary vector following transformation.

Supplementary features of some binary vectors include: overdrive and T-DNA transfer stimulator sequences (to improve T-strand synthesis), and a *cos* region from lambda bacteriophage to allow maintenance of cosmid-based plant genomic libraries in *Agrobacterium* (51 and refs contained therein).

Transgenic plants are usually produced by inoculation of axenic plant material with *Agrobacterium*. Horsch *et al.* (53) developed a leaf disc transformation system in which leaf discs are immersed in bacterial suspension, blotted dry, transferred to culture medium for approximately 48 hours, and then subcultured onto medium containing appropriate antibiotics to kill the agrobacteria. This medium will also contain plant growth regulators to induce regeneration and selective agent to discriminate between transformed and untransformed plant cells. Recently, new methods have been developed to aid *Agrobacterium* infiltration into plant tissue. The application of a vacuum has been utilized to infiltrate *Arabidopsis* with *Agrobacterium* (54) and the combination of *Agrobacterium* inoculation with a short sonication treatment has enhanced DNA transfer in a number of plant species (55).

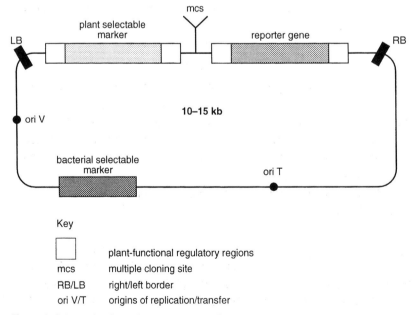

Figure 1 Schematic of a typical binary vector for *Agrobacterium*-mediated transformation.

Protocol 1

Transformation of *Agrobacterium tumefaciens* with a binary vector: freeze–thaw method[a]

Equipment and reagents

- Eppendorf tubes
- Water-bath
- Rotatory shaker
- Cuvettes and spectrophotometer
- Centrifuge tubes and centrifuge

- Liquid nitrogen
- Luria broth and semi-solidified LB medium
- 10 mM $CaCl_2$

Method

1. Inoculate 50 ml Luria broth (LB) containing appropriate antibiotics[b] with *Agrobacterium*.

2. Shake at 28 °C in the dark to an OD_{600} of 1.0.

3. Centrifuge the agrobacteria in sterile tubes at 3000 g for 15 min.

4. Remove supernatant and, while keeping the tubes on ice, resuspend the bacterial pellet in a total of 1 ml of ice-cold 10 mM $CaCl_2$ (agrobacteria should be pooled into a single tube).

5. Transfer 100 μl aliquots of the bacterial resuspension into sterile Eppendorf tubes and then place into liquid nitrogen.

6. Use the frozen agrobacteria directly for DNA transformation or store at −80 °C.

7. Pipette 10 μl of binary plasmid from a mini prep onto the surface of 100 μl competent frozen agrobacteria.

8. Incubate the DNA-bacteria mixture for 5 min at 37 °C in a water-bath.

9. Add 1 ml of LB to the bacteria and shake at 28 °C for 4 h.

10. Centrifuge the bacteria for 2 min at 12 000 g and resuspend the pellet in 100 μl of LB.

11. Spread 50 μl of the bacterial suspension onto LB medium (made semi-solid with 15 g/litre agar) containing the antibiotics required for selection of agrobacteria carrying binary and disarmed Ti helper plasmid. Seal the plates with Nescofilm and incubate for two to three days at 28 °C.

12. Re-streak resulting colonies on selective LB medium in separate Petri dishes and incubate at 28 °C for two days.[c]

[a] Adapted from ref. 51.

[b] Chromosomal and helper Ti plasmid-borne genes may confer resistance to particular antibiotics. The LB should contain antibiotics appropriate to the particular bacterial strain.

[c] Transformation can be confirmed by PCR analysis of a plasmid preparation (isolated from agrobacteria using the Qiagen miniprep protocol). *Agrobacterium* should be stored at −80 °C as a glycerol stock.

Several approaches have been employed in order to increase the virulence of *Agrobacterium*. A number of environmental stimuli have been reported to influence the induction of virulence, including acetosyringone (56), sucrose, pH, temperature (57, 58), opines (59), sugars (60, 61), and osmoprotectant compounds such as glycine betaine (62) and proline (63). *Agrobacterium tumefaciens* strain A281 carrying the plasmid pTiBo542 shows an enhanced virulence for many species (64, 65). The transfer of fragments of the *vir* region of pTiBo542 to *A. tumefaciens* has produced so-called 'supervirulent' strains (66).

3.3 Principles of particle bombardment

Particle bombardment involves the delivery of heavy metal particles into living cells. The particles carry DNA or RNA, and are accelerated towards their target by an explosive force generated by gunpowder, pressurized gas, or an electrical discharge. The DNA is thus delivered through breaches in the cell wall and membrane, generated as the particles penetrate the cell. Only DNA released from those particles that penetrate the nucleus can integrate into the host nuclear genome.

Early bombardment experiments, using a modified shotgun to accelerate tungsten particles, and onion epidermis as the target, demonstrated the viability of the biolistic concept. Transient CAT (chloramphenicol acetyltransferase) activity was detected in these cells three days after particle delivery (67). The first stable transformants resulting from particle bombardment experiments were reported in 1988 using immature soybean embryos and an electric discharge instrument (68, 69). Although the regeneration of plants was not reported, these experiments demonstrated that the technique of particle bombardment could deliver functional DNA into plant cells. Shortly thereafter, the same group reported the recovery of transgenic soybean plants by bombarding intact meristems isolated from immature soybean seed (70). The recovery of transformed tobacco (71) and maize (72, 73) callus validated the general applicability of this technology for plant transformation. By 1990, a number of plant species had been transformed using particle bombardment methodology, including cotton (74), papaya (75), and maize (76, 77). Almost all major crops have now been transformed using variations of particle bombardment.

The sections below set out important parameters for transformation by particle bombardment and discuss the different bombardment devices available. As far as apparatus and procedural matters are concerned, there are no special considerations for the delivery and expression of antibody transgenes as compared to transgenes encoding any other type of protein.

3.3.1 Parameters for particle bombardment-mediated transformation

Birch and Bower identified and discussed a number of important parameters for bombardment-mediated gene transfer. A number of these parameters were translated into principles that should be generally applicable in many diverse situations. They classified these principles in terms of apparatus design, physical, chemical, ballistic, and biological parameters, and experimental design.

These are listed below (78):

- Particle bombardment devices must be safe to operate.
- A high frequency of gene transfer is required.
- The ability to tailor bombardment conditions is important.
- Particle bombardment-mediated transformation should be of consistent efficiency.
- Reduced time required per bombardment is an advantage.
- The ability to direct particles to specific cell groups is an advantage.
- Specific desirable features for metal particles.
- The optimal size and impact velocity of particles depend on properties of the target cells.
- An effective procedure for attaching DNA to particles is probably universal to all cell types, irrespective of bombardment conditions.
- Non-optimal DNA precipitation technique is a major source of variation in gene transfer frequency.
- Particle delivery needs to be optimized for different species and explants.
- Physical damage to bombarded tissues can be reduced by instrument design.
- Vacuum treatment to maintain projectile velocity may affect plant development.
- There do not appear to be any special requirements with respect to the form of DNA to be transferred via particle bombardment (e.g. supercoiled versus linear, duplex versus single-stranded, blunt-ends versus cohesive ends).
- Transient gene expression frequency at the cellular level can be used to evaluate apparatus efficiency.
- For optimization experiments, it is best to utilize a gene construct that is expressed strongly in the target tissue.
- Target tissue characteristics substantially affect both transient expression and stable transformation frequencies.
- The osmotic concentration of the culture medium influences transgene expression frequency in, and survival of, bombarded cells.
- Optimal DNA loading rates may vary with the objective of bombardment.
- The relationship between transient and stable transformation frequencies needs to be determined experimentally for different systems.
- A high transient expression frequency is desirable but not sufficient for the production of stable transformants.
- Where particle bombardment is used to produce transgenic plants, the target tissue should be highly regenerable and readily penetrated by particles.

- Haploid tissues are potentially useful targets for introducing foreign genes into plants.

- There is interaction between the target tissue, regeneration method, and selection/screening method. These need to be considered in unison for the efficient production of transgenic plants.

- Nuclear and organelle transformation require different genetic constructs and selection conditions.

- Stable nuclear transformants vary widely in the number and arrangement of transgene copies.

- Co-transformation and co-expression can occur at a high frequency, even for genes introduced on different plasmids.

- Precautions are required to avoid bombardment artefacts in studies of gene regulatory sequences.

- Internal controls should be included for quantitative comparisons after particle bombardment.

Bower and Birch also formulated a general strategy for optimizing a stable transformation method based on the above principles. The steps they recommended are listed below:

(a) Identify penetratable target tissue capable of regeneration into complete plants.

(b) Establish bombardment conditions for efficient DNA delivery into regenerable cells based on transient expression frequency.

(c) Determine bombardment and culture conditions that maximize the recovery of stably transformed cells.

(d) Choose gene regulatory sequences appropriate for the proposed selection or screening technique.

(e) Choose an effective selectable/screenable marker gene and fine-tune the matching selection/screening conditions.

(f) Determine DNA and particle loads and constructs that result in desired integration and co-transformation patterns.

(g) Minimize tissue culture duration for target preparation, selection, and regeneration, to reduce culture-induced variation and mutation.

(h) Select techniques that are not genotype specific, for direct application to a range of cultivars.

3.3.2 Instrumentation

A variety of instruments based on diverse accelerating mechanisms are currently in use. These include the original gunpowder device (79), an apparatus based on electric discharge (80), a microtargeting apparatus (81), a pneumatic instrument (82), an instrument based on flowing helium (83, 84), and an improved version of the original gunpowder device utilizing compressed helium

(85). Hand-held devices similar in principle to the original Biolistics and Accell instruments are also in use. The most widely-used device is the Biolistics apparatus marketed by Bio-Rad, Inc., but Accell-based methodology has been particularly useful in developing variety-independent gene transfer methods for the more recalcitrant cereals and legumes. Detailed descriptions of the various acceleration devices, principles of operation, and other details may be found in the primary references.

3.3.3 Vectors for particle bombardment

One of the advantages of particle bombardment is the simplicity of vector design. Since particle bombardment is a purely physical transformation mechanism (i.e. no genes or *cis*-acting sequences are required in the vector for gene delivery), there is no intrinsic requirement for any sequences other than the transgene itself, and a selectable marker to identify and selectively propagate transgenic plant material. It is well established that co-transformation (the introduction of multiple transgenes cloned in separate vectors) results in high frequency co-transformation, and that the different transgenes tend to integrate at a single locus. There is therefore no need to use specialized cointegrate vectors (plasmids carrying multiple linked genes) to ensure that the transgene(s) and selectable markers are introduced into the plant genome. Consequently, almost any expression vector can be used to introduce the antibody transgene(s), and the selectable marker can be introduced on its own vector. This strategy avoids laborious subcloning procedures, especially if two or more transgenes are required (e.g. for the simultaneous expression of heavy and light chains).

The most important considerations for vector design therefore reflect the structure of the expression cassette. Such considerations (as discussed above) include:

- Choice of promoter: constitutive, tissue-specific, inducible, etc.
- Codon optimization.
- Sequences that allow intracellular targeting.

Since only the transgene(s) and markers are required for transformation by particle bombardment, one could ask why vectors are required at all. The vector backbone is required for vector propagation and selection in bacterial culture, and for *in vitro* manipulation. It serves no purpose in the transformation process, and there is increasing evidence that backbone sequences can have deleterious effects on transgene expression and can promote recombination events, leading to transgene rearrangement (86). Using particle bombardment, it is possible to transform plants with minimal expression cassettes excised from the vector after plasmid isolation. Recent experiments have shown that the use of such cassettes results in simpler transgene integration patterns, lower copy numbers, and less transgene silencing (X. Fu, personal communication). Co-transformation frequencies are equivalent to those obtained with whole plasmid bombardment, so the removal of backbone sequences is to be recommended (87).

Protocol 2

Preparation of plasmid DNA for particle bombardment

Equipment and reagents

- Vortex mixer
- Microcentrifuge
- Sonicator
- UV spectrophotometer
- Qiagen maxiprep kit
- 100% ethanol

- Gold particles (0.7 μm)
- 2.5 M $CaCl_2$
- TE buffer
- Spermidine
- PEG (molecular weight 8000)

Method

1. Prepare high quality plasmid DNA at a concentration of 10 mg/ml. We find the Qiagen maxiprep kit is satisfactory. Concentrate DNA by ethanol precipitation and check concentration by UV spectrophotometry.

2. Mix 2.5 mg of gold (0.7 μm particles) with 5 μg of the selectable marker plasmid and 10 μg of the plasmid carrying the antibody transgene. Add TE buffer to 100 μl. Vortex for 30 sec.

3. Add 100 μl of 100 mM spermidine to protect the DNA during the precipitation process. Vortex for 30 sec.

4. Add 100 μl of 30% PEG (MW 8000) and vortex for 30 sec.

5. Slowly add 100 μl of 2.5 M $CaCl_2$ with continuous vortexing. Vortex for a further 10 min.

6. Centrifuge at 12 000 g for 30 sec and discard supernatant.

7. Wash with 200 μl of 100% ethanol and centrifuge at 12 000 g for 30 sec.

8. Add 100 μl of 100% ethanol and sonicate.

9. Place 5–10 μl of the suspension onto the centre of a carrier disc and allow to air dry.

4 Plant transformation techniques

The inability to directly transfer a foreign gene into all the cells of a whole adult individual requires that transgenic plants be regenerated from a single, or small group, of transformed cells via an *in vitro* tissue culture process. The initial source of plant material (e.g. a leaf disc) for foreign gene transfer and subsequent plant regeneration is termed the primary explant.

4.1 *Agrobacterium*-mediated transformation of tobacco

Prior to tissue culture and transformation, plant material must be surface sterilized to remove contaminating micro-organisms.

Protocol 3

Surface sterilization of tobacco leaves

Reagents

- 2% sodium hypochlorite solution
- 70% IMS
- Sterile distilled water

Method

1. Remove leaves from tobacco plants, wash briefly under running cold tap-water.
2. Submerge leaves in 2% sodium hypochlorite for 1 min.
3. Remove leaves and submerge in 70% IMS for 1 min.
4. Wash leaves in three changes of sterile distilled water.

Plant regeneration *in vitro* can occur via two general pathways; organogenesis and somatic embryogenesis (88). Organogenesis is achieved through shoot formation directly from the primary explant, or, alternatively, shoots may be initiated from a dedifferentiated cell mass known as a callus. Whole, free-living plants are obtained from organogenesis-derived shoots following a rooting stage and *ex vitro* acclimation. Somatic embryogenesis is the process by which somatic cells develop into differentiated plants, through characteristic embryological stages mimicking those stages of development after fusion of gametes. This may occur either indirectly from callus, or directly from cells of an organized structure such as a stem segment or zygotic embryo (89). A large range of plant species have now been successfully regenerated from explants *in vitro*. In the case of species whose culture requirements are not known, optimum conditions must be determined empirically, as the biochemical and molecular events underlying regeneration are poorly understood (90). The most important factors influencing regeneration are culture medium components, explant source, and plant growth regulator combinations. Other important factors include; medium pH, temperature, and illumination (see ref. 91 for a review of the theoretical and practical considerations of plant cell culture).

Protocol 4

Preparation of culture medium for adventitious shoot regeneration on leaf discs of tobacco (Nicotiana tabacum)

Equipment and reagents

- Measuring cylinder, 500 ml glass bottle
- Murashige and Skoog basal medium (Sigma)
- Sucrose
- BAP (Sigma), NAA (Sigma)
- Distilled water
- 1 M NaOH
- Agar

Protocol 4 continued

Method

1. For 1 litre of medium, dissolve the appropriate mass of Murashige and Skoog basal medium in 800 ml of distilled water. Add 30 g of sucrose, 1 mg of BAP, 0.1 mg of NAA,[a] and stir until sucrose is fully dissolved.

2. Adjust the volume to 950 ml with distilled water.

3. Adjust to pH 5.8 with 1 M NaOH.

4. Transfer to 1 litre measuring cylinder and make volume up to 1 litre with distilled water.

5. Dispense 250 ml batches of medium to 500 ml glass bottles. Add agar to a final concentration of 0.8%. Autoclave bottles at 121 °C for 15 min.[b]

[a] For rooting medium the plant growth regulators BAP and NAA should be omitted.

[b] Semi-solid plant culture media can be melted either in a steamer or microwave and then transferred to Petri dishes for leaf disc transformation (see *Protocol* 5).

Protocol 5

Agrobacterium tumefaciens-mediated transformation of tobacco leaf discs

Equipment and reagents

- Seed tray
- 9 cm Petri dishes
- 175 ml glass jars
- Sterile inoculating loop
- Luria broth and semi-solidified LB medium

- Shoot regeneration and rooting media (see *Protocol* 4)
- Carbenicillin (Sigma), cefotaxime (Sigma)
- Selective agent for transformed plant cells
- Compost
- Plant nutrient

Method

1. Remove *Agrobacterium* (containing binary vector) from −80 °C and streak on semi-solid LB medium (with appropriate antibiotics) in a 9 cm Petri dish. Incubate at 28 °C for two days.

2. Inoculate *Agrobacterium* from Petri dish into 10 ml LB (with appropriate antibiotics) using a sterile inoculating loop.

3. Shake at 28 °C to an OD_{600} of 1.0.

4. Remove 4 ml of *Agrobacterium* suspension and add to 16 ml of sterile distilled water in a sterile 9 cm Petri dish.

5. Using a sterile scalpel cut 0.5–1.0 cm leaf discs from surface sterilized leaves; immerse for 5 min in the diluted *Agrobacterium* suspension.

Protocol 5 continued

6. Briefly dry leaf discs on sterile filter paper.

7. Place leaf discs on shoot regeneration medium (see *Protocol 4*), 10–15 discs per 20 ml of medium in each 9 cm Petri dish. Incubate for two days at 25 °C with a 16 h photoperiod.

8. Transfer leaf discs to shoot regeneration medium containing 500 mg/litre carbenicillin and appropriate concentration of selective agent for transformed plant cells.[a] Incubate for 21 days at 25 °C with a 16 h photoperiod.

9. Transfer leaf discs to shoot regeneration medium in sterile 175 ml glass jars containing 500 mg/litre cefotaxime and appropriate concentration of selective agent or transformed plant cells. Incubate at 25 °C with a 16 h photoperiod.

10. Developing shoots should be removed when they reach a convenient size (approx. 0.5 cm in length) and transferred to rooting medium (three to four shoots/40 ml medium/175 ml glass jar). Incubate for 14 days at 25 °C with a 16 h photoperiod.

11. Shoots lacking roots should be trimmed at the base and replaced in fresh rooting medium for an additional 21 days or until roots appear.

12. Rooted shoots should be transferred to compost in plant pots, watered, and supplied with nutrients. Plants should be kept in seed trays and covered with a lid for 24 h after transfer to compost, to minimize initial water loss.

[a] 200 mg/litre kanamycin is used for the selection of tobacco cells transformed with the *npt*-II selectable marker gene.

It was previously thought that *Agrobacterium* infection was limited almost exclusively to dicotyledon and gymnosperm species (92). Recent success in *Agrobacterium*-mediated transformation of monocotyledon species such as *Oryza sativa* (93), *Zea Mays* (94), and *Triticum aestivum* (95) suggests that this may not be the case, although host range restrictions remain a limitation of this transformation system.

4.2 Transformation of wheat by micro-projectile bombardment

Protocol 6

Transformation of wheat embryos

Equipment and reagents

- Bio-Rad Helium 2000 gun
- 20% sodium hypochlorite
- Sterile distilled water
- Murashige and Skoog medium
- 2,4-D
- 0.2 M mannitol
- 0.2 M sorbitol

Method

1. Dehusk wheat seeds and place in 20% sodium hypochlorite for 10 min.

2. Rinse at least three times with distilled water.

3. Excise embryos and place, scutellum-uppermost on plates containing MS medium supplemented with 2 mg/litre 2,4-D.

4. Leave plates in darkness at 24 °C.

5. At least 4 h prior to bombardment, transfer embryos to osmoticum medium (MS medium supplemented with 2 mg/litre 2,4-D, 0.2 M mannitol, and 0.2 M sorbitol).

6. The protocol for bombardment differs according to the device used. Follow manufacturer's guidelines for loading the carrier disc into the gun barrel. The manufacturer will supply instructions for the optimal use of each device. The protocol below refers to the Bio-Rad Helium 2000 gun, which is used in our laboratory.

7. Place plant material in the targeting chamber.

8. Bombard plant material at 900–1300 psi. The optimal pressure and the optimal distance between the gun and the target can be determined empirically for each experiment. For optimization, bombard with a visible marker such as *gus* A, to determine the efficiency of transformation. Note that optimal transformation occurs when many small spots of GUS activity are detected. Large sectors of GUS positive tissue are usually dead!

9. Transfer embryos to fresh osmoticum medium after bombardment and incubate in darkness at 24 °C.

Protocol 7

Regeneration of transgenic wheat plants

The following steps are depicted in *Figure 2*.

Reagents

- Murashige and Skoog medium
- 2,4-D
- 10 mg/litre Zeatin
- 2–4 mg/litre PPT

Method

1. Transfer bombarded wheat embryos to MS medium supplemented with 2 mg/litre 2,4-D, and incubate in darkness at 24 °C for two weeks.

2. Transfer embryos to MS medium supplemented with 10 mg/litre Zeatin and 2–4 mg/litre PPT (use higher concentrations of PPT for more healthy-looking material, this depends on the wheat variety used).

3. Transfer regenerated shoots onto half-strength MS medium supplemented with 2–4 mg/litre PPT, under strong light (130 mE), 18 h photoperiod, 24 °C. After two weeks, transfer elongated shoots to tubes containing the same medium.

4. Transfer to soil.

Figure 2: please see plate section between pages 226–227.

5 Screening regenerated plantlets for immunoglobulin chain production

Typically, capture ELISAs are used for screening potential transformed plants, followed by Western blotting to confirm the correct relative molecular mass of the recombinant product. The techniques are identical to those in routine standard use, however due to the ubiquitous nature of plant products in all diets and the consequent presence of plant protein reactive antibodies, it is important to use affinity purified antibodies at all stages. Similarly, if monoclonal antibodies are used it is often necessary to remove any contaminating antibodies that might be present, for example in fetal calf serum.

Protocol 8

Plant sample preparation for ELISA or Western analysis

Equipment and reagents

- Microcentrifuge tubes (1.6–2 ml) and microcentrifuge tube pestle
- Microcentrifuge

- Tris-buffered saline: 150 mM NaCl, 10 mM Tris base pH 8, containing 10 μg/ml leupeptin; chilled on ice

Method

Providing the samples are processed reasonably rapidly and kept as cold as possible throughout, there is no need for any further addition to this simple extraction buffer.

1. Plants may be assayed as soon as they are well established with two or three small leaves. Remove approx. 1 cm² leaf tissue and immediately grind in cold extraction buffer in a chilled microcentrifuge tube.

2. Centrifuge at 12 000 r.p.m. for 2 min.

3. Apply supernatant directly to an appropriate pre-coated ELISA plate or use in SDS–PAGE analysis in the usual way.

For detection of assembled antibodies, crude plant extracts can be prepared in a non-reducing buffer and analysed on SDS–PAGE followed by Western blotting. For IgG antibodies, 8–10% SDS–PAGE is used, for secretory IgA we routinely use 4% SDS–PAGE without a stacking gel. The latter is extremely fragile and requires careful manipulation when blotting onto nitrocellulose.

5.1 Self- and cross-fertilization of transgenic plants

The process of fertilization involves transfer of pollen from one transgenic plant flower to another. This is done to generate progeny that inherit the genetic characteristics of both parents (e.g. light chain expressing plants cross-fertilized with Ig heavy chain expressing plants will yield approximately 25% progeny that express both chains and assemble functional antibody). To ensure correct cross-fertilization, an immature recipient flower is used in which the anthers (pollen-producing organs) can be removed before they produce pollen, which eliminates the chance of self-fertilization. Donor pollen is then lightly dusted onto the stigma of the flower and the entire flower is covered with a light polythene bag, again to eliminate the possibility of airborne pollen contamination. In the case of tobacco, seeds will be ready for harvesting approximately six to eight weeks after fertilization, once the seed pod has fully dried.

In order to identify homozygous plants, these are 'back-crossed' with non-transgenic plants, and the progeny screened for Ig chain production. The off-spring of homozygous parent plants will be 100% positive for recombinant protein.

6 The overall advantages in expressing antibodies in plants

There are currently two approaches for introducing foreign genes into plant cells, both of which are versatile and can utilize a number of plant hosts. The use of viral-based vectors results in the rapid production of very large quantities of recombinant product, but at present appears to be less suitable for larger complex molecules such as full-length antibodies, but quite suitable for single antibody fragment, such as scFv. At present, the transgenic approach is favoured for full length antibodies, and expression levels of 1–5% of total soluble protein are achieved consistently.

There are a number of considerations to be taken into account for the development of any heterologous expression system. These include fidelity of the expressed recombinant protein in terms of function, which in turn depends on folding, structure and glycosylation, protein stability, ease of purification, the potential for scale-up production to produce sufficient quantities, cost, and safety issues. The strength of bioengineering in plants is that there are significant advantages over other expression systems in respect to many of these issues. With regards to protein folding and structure, small peptides, polypeptides, and even complex proteins can be expressed in plants that are fully assembled and functional. For larger molecules such as IgG antibodies, this is associated with the presence of endoplasmic reticulum (ER) resident chaperones that are homologous to those involved in protein assembly in mammalian cells.

Targeting recombinant proteins for secretion through the ER and Golgi apparatus is achieved using either native or plant leader sequences, and this also ensures that N-glycosylation take place. In plants, glycosylation differs from mammals in the complex glycans, but for the recombinant proteins expressed so far, this has not led to any loss of structure or function.

The storage of immunoglobin genes and gene products in plants can be very stable. Transgenic plants can be conveniently self-fertilized to produce stable, true breeding lines, propagated by conventional horticultural techniques, and stored and distributed as seeds. The expressed recombinant immunoglobins can be targeted to stable environments within the plant, for example the extracellular apoplastic space. Alternatively, tissue-specific promoters can be used to direct expression in storage organs such as seeds or tubers. Extraction and purification from these sites is generally simple.

One of the most obvious benefits of plants is the potential for scale-up production, in which virtually limitless amounts of recombinant antibody could be grown at minimal cost. Plants are easy to grow, unlike bacteria or animal cells their cultivation is straightforward, does not require specialist media or equipment, or involve toxic chemicals. The use of plants also avoids many of the potential safety issues associated with contaminating mammalian viruses, as well as ethical considerations involving the use of animals. Various estimates have been made of the commercial advantages of expressing IgG antibodies in plants. A strain of corn has been developed that would allow the production of 1.5 kg of pharmaceutical-quality antibodies per acre and estimates that antibodies could be produced at approximately US $1–3/gram have been made. It seems likely that even at current levels of expression, sufficient antibody could be 'grown' in plants for most medical applications, on only a few acres of land and at minimal cost, compared with alternative methods of production.

References

1. Ma, J. K.-C., Hiatt, A., Hein, M. B., Vine, N., Wang, F., Stabila, P., *et al.* (1995). *Science*, **268**, 716.
2. Hiatt, A. C., Cafferkey, R., and Bowdish, K. (1989). *Nature*, **342**, 76.
3. Tavladoraki, P., Benvenuto, E., Trinca, S., De Martinis, D., Cattaneo, A., and Galeffi, P. (1993). *Nature*, **366**, 469.
4. Benvenuto, E., Ordas, R. J., Tavazza, R., Ancora, G., Biocca, S., Cattaneo, A., *et al.* (1991). *Plant Mol. Biol.*, **17**, 865.
5. Owen, M., Gandecha, A., Cockburn, B., and Whitelam, G. (1992). *Bio/Technology*, **10**, 790.
6. De Neve, M., De Loose, M., Jacobs, A., Van Houdt, H., Kaluza, B., Weidle, U., *et al.* (1993). *Transgenic Res.*, **2**, 227.
7. Schouten, A., Roosien, J., van Engelen, F. A., de Jong, G. A. M., Borst-Vrenssen, A. W. M., Zilverentant, J. F., *et al.* (1996). *Plant Mol. Biol.*, **30**, 781.
8. Fiedler, U., Phillips, J., Artsaenko, O., and Conrad, U. (1997). *Immunotechnology*, **3**, 205.
9. During, K., Hippe, S., Kreuzaler, F., and Schell, J. (1990). *Plant Mol. Biol.*, **15**, 281.
10. Hein, M. B., Tang, Y., McLeod, D. A., Janda, K. D., and Hiatt, A. C. (1991). *Biotechnol. Prog.*, **7**, 455.
11. Fontes, E. B. P., Shank, B. B., Wrobel, R. L., Moose, S. P., O'Brian, G. R., Wurtzel, E. T., *et al.* (1991). *Plant Cell*, **3**, 483.

12. Denecke, J., Goldman, M. H., Demolder, J., Seurinck, J., and Botterman, J. (1991). *Plant Cell*, **3**, 1025.

13. Ma, J. K.-C., Lehner, T., Stabila, P., Fux, C. I., and Hiatt, A. (1994). *Eur. J. Immunol.*, **24**, 131.

14. van Engelen, F. A., Schouten, A., Molthoff, J. W., Roosien, J., Salinas, J., Dirkse, W., *et al.* (1994). *Plant Mol. Biol.*, **26**, 1701.

15. Mestecky, J. and McGhee, J. R. (1987). *Adv. Immunol.*, **40**, 153.

16. Ma, J. K.-C., Hikmat, B. Y., Wycoff, K., Vine, N., Chargelegue, D., Yu, L., *et al.* (1998). *Nature Med.*, **4**, 601.

17. Kumagai, M. H., Turpen, T. H., Weinzettl, N., della-Cioppa, G., Turpen, A. M., Hilf, M. E., *et al.* (1993). *Proc. Natl. Acad. Sci. USA*, **90**, 427.

18. Della-Cioppa, G. and Grill, L. K. (1996). *Ann. N. Y. Acad. Sci.*, **792**, 57.

19. Sturm, A., Van Kuik, J. A., Vliegenthart, J. F., and Chrispeels, M. J. (1987). *J. Biol. Chem.*, **262**, 13392.

20. Faye, L., Johnson, K. D., Sturm, A., and Chrispeels, M. J. (1989). *Physiol. Plant.*, **75**, 309.

21. Cabanes, M., Fitchette-Laine, A.-C., Bourkis, C., Vine, N. D., Ma, J. K.-C., Lerouge, P., *et al.* (1999). *Glycobiology*, **9**, 365.

22. von Schaewen, A., Sturm, A., O'Neill, J., and Chrispeels, M. J. (1993). *Plant Physiol.*, **102**, 1109.

23. Fraley, R. T., Rogers, S. G., Horsch, R. B., Sanders, P. R., Flick, J. S., Adams, S. P., *et al.* (1983). *Proc. Natl. Acad. Sci. USA*, **80**, 4803.

24. Wilmink, A. and Dons, J. J. M. (1993). *Plant Mol. Biol. Rep.*, **11**, 165.

25. Jefferson, R. A., Kavanagh, T. A., and Bevan, M. W. (1987). *EMBO J.*, **6**, 3901.

26. De Block, M., Herrera-Estrella, L., Van Montagu, M., Schell, J., and Zambryski, P. (1984). *EMBO J.*, **3**, 1681.

27. Ow, D. W., Wood, K. V., Deluca, M., DeWet, J. R., Helsinki, D. R., and Howell, S. H. (1986). *Science*, **234**, 856.

28. Koncz, C., Olsson, O., Langridge, W. H. R., Schell, J., and Szalay, A. A. (1987). *Proc. Natl. Acad. Sci. USA*, **84**, 131.

29. Sheen, J., Huang, S. B., Niwa, Y., Kobayashi, H., and Galbraith, D. W. (1995). *Plant J.*, **8**, 777.

30. Odell, J. T., Nagy, F., and Chua, N. H. (1985). *Nature*, **313**, 810.

31. Lawton, H. A., Tierney, M. A., Nakamura, I., Anderson, E., Komeda, Y., Dube, P., *et al.* (1987). *Plant Mol. Biol.*, **9**, 315.

32. Bevan, M., Barnes, W. M., and Chilton, M. D. (1983). *Nucleic Acids Res.*, **12**, 369.

33. De Greve, H., Dhaese, P., Seurinck, J., Lemmers, M., Van Montagu, M., and Schell, J. (1983). *J. Mol. Appl. Genet.*, **1**, 499.

34. DiRita, V. J. and Geluin, S. B. (1987). *Mol. Gen. Genet.*, **207**, 233.

35. Stoger, E., Williams, S., Kenn, D., and Christou, P. (1999). *Transgenic Res.*, in press.

36. Davey, M. R., Curtis, I. S., Gartland, K. M. A., and Power, J. B. (1994). In *Systematics association special* (ed. M. A. J. Williams), Vol. 49, p. 9. Clarendon Press, Oxford.

37. Petit, A., David, C., Dahl, G. A., Ellis, J. M., Guyon, P., Casse-Delbert, F., *et al.* (1983). *Mol. Gen. Genet.*, **190**, 204.

38. Nester, E. W., Gordon, M. P., Amasino, R. M., and Yanofsky, M. F. (1984). *Annu. Rev. Plant Physiol.*, **35**, 387.

39. Shaw, C. H., Ashby, A. M., Brown, A., Royal, C., and Loake, G. J. (1988). *Mol. Microbiol.*, **2**, 413.

40. Douglas, C. J., Staneloni, R. J., Rubin, R. A., and Nester, E. W. (1985). *J. Bacteriol.*, **161**, 850.

41. Hooykaas, P. J. J. and Schilperoort, R. A. (1992). *Plant Mol. Biol.*, **19**, 15.

42. Zambryski, P. (1992). *Annu. Rev. Plant Physiol. Plant Mol. Biol.*, **43**, 465.

43. Yanofski, M., Porter, S., Young, C., Albright, L., Gordon, M., and Nester, E. (1986). *Cell*, **47**, 471.
44. Stachel, S. and Nester, E. (1986). *EMBO J.*, **5**, 1445.
45. Howard, E. and Citovski, V. (1990). *Bio/Essays*, **12**, 103.
46. Yusibov, V. M., Steck, I. R., Gupta, V., and Gelvin, S. B. (1994). *Proc. Natl. Acad. Sci. USA*, **91**, 2994.
47. Gheysen, G., Van Montagu, M., and Zambryski, P. (1987). *Proc. Natl. Acad. Sci. USA*, **84**, 6169.
48. Zambryski, P., Joos, H., Genetello, C., Leemans, J., Van Montagu, M., and Schell, J. (1983). *EMBO J.*, **2**, 2143.
49. Hoekema, A., Hirsch, P. R., Hooykaas, P. J. J., and Schilperoort, R. A. (1983). *Nature*, **303**, 179.
50. Mersereau, M., Pazour, G. J., and Das, A. (1990). *Gene*, **90**, 149.
51. Walkerpeach, C. R. and Velten, J. (1994). In *Plant molecular biology manual B1* (2nd edn) (ed. S. B. Gelvin and R. A. Schilperoort), p. 1. Kluwer Academic Publishers, Dordrecht.
52. Bevan, M. (1984). *Nucleic Acids Res.*, **12**, 8711.
53. Horsch, R. B., Fry, J. E., Hoffman, N. C., Eichholtz, D., Rogers, S. G., and Fraley, R. T. (1985). *Science*, **227**, 1121.
54. Bechtold, N., Ellis, J., and Pelletier, G. (1993). *C. R. Acad. Sci. Paris /Life Sci.*, **316**, 1194.
55. Trick, H. N. and Finer, J. J. (1997). *Transgenic Res.*, **6**, 329.
56. Stachel, S. E., Hessen, E., Van Montagu, M., and Zambryski, P. (1985). *Nature*, **318**, 625.
57. Alt-Morbe, J., Kuhlmann, H., and Schroder, J. (1989). *Mol. Plant-Microbe Interact.*, **2**, 301.
58. Alt-Morbe, J., Neddermann, P., von Lintig, J., Weiler, E. W., and Schroder, J. (1988). *Mol. Gen. Genet.*, **213**, 1.
59. Veluthambi, K., Krishnan, M., Gould, J. H., Smith, R. H., and Gelvin, S. B. (1989). *J. Bacteriol.*, **171**, 3969.
60. Cangelosi, G. A., Ankenbauer, R. G., and Nester, E. W. (1989). *Proc. Natl. Acad. Sci. USA*, **87**, 6708.
61. Ankenbauer, R. G. and Nester, E. W. (1990). *J. Bacteriol.*, **172**, 6442.
62. Vernade, D., Herrera-Estrella, A., Wang, K., and Van Montagu, M. (1988). *J. Bacteriol.*, **170**, 5822.
63. James, D. J., Vratsu, S., Cheng, J., Negri, P., Viss, P., and Dandekar, A. M. (1993). *Plant Cell Rep.*, **12**, 559.
64. Hood, E. E., Helmer, G. L., Fraley, R. T., and Chilton, M. D. (1986). *J. Bacteriol.*, **168**, 1291.
65. Jin, S., Komari, T., Gordon, M. P., and Nester, E. W. (1987). *J. Bacteriol.*, **169**, 4417.
66. Curtis, I. S., Power, J. B., Blackhall, N. W., De Laat, A. M. M., and Davey, M. R. (1994). *J. Exp. Bot.*, **45**, 1441.
67. Sanford, J. C. (1988). *Trends Biotechnol.*, **6**, 299.
68. Christou, P., McCabe, D. E., and Swain, W. F. (1988). *Plant Physiol.*, **87**, 671.
69. Christou, P., Swain, W. F., Yang, N.-S., and McCabe, D. E. (1989). *Proc. Natl. Acad. Sci. USA*, **86**, 7500.
70. McCabe, D. E., Swain, W. F., Martinell, B. J., and Christou, P. (1988). *Bio/Technology*, **6**, 923.
71. Klein, T. M., Harper, E. C., Svab, Z., Sanford, J. C., Fromm, M. E., and Maliga, P. (1988). *Proc. Natl. Acad. Sci. USA*, **85**, 8502.
72. Klein, T. M., Gradziel, T., Fromm, M. E., and Sanford, J. C. (1988). *Bio/Technology*, **6**, 559.
73. Klein, T. M., Fromm, M., Weissinger, A., Tomes, D., Schaaf, S., Sletten, M., *et al.* (1988). *Proc. Natl. Acad. Sci. USA*, **85**, 4305.
74. Finer, J. J. and McMullen, M. D. (1991). *In Vitro*, **27**, 175.

75. Fitch, M. M. M., Manshardt, R. M., Gonsalves, D., Slightom, J. L., and Sanford, J. C. (1990). *PCR*, **9**, 189.

76. Gordon-Kamm, W. J., Spencer, T. M., Mangano, M. L., Adams, T. R., Daines, R. J., Start, W. G., *et al.* (1990). *Plant Cell*, **2**, 603.

77. Fromm, M. E., Morrish, F., Armstrong, C., Williams, R., Thomas, J., and Klein, T. M. (1990). *Bio/Technology*, **8**, 833.

78. Birch, R. G. and Bower, R. (1994). In *Particle bombardment technology for gene transfer* (ed. N.-Y. Yang and P. Christou), pp. 3–37. UWBC/Oxford University Press, New York, Oxford.

79. Sanford, J. C., Klein, T. M., Wolf, E. D., and Allen, N. J. (1987). *J. Part. Sci. Technol.*, **6**, 559.

80. Christou, P., McCabe, D. E., Martinell, B. J., and Swain, W. F. (1990). *Trends Biotech. Biotechnol.*, **8**, 145.

81. Sautter, C., Waldner, H., Neuhaus-Url, G., Galli, A., Neuhaus, G., and Potrykus, I. (1991). *Bio/Technology*, **9**, 1080.

82. Iida, A., Seki, M., Kamada, M., Yamada, Y., and Morikawa, H. (1990). *Theor. Appl. Genet.*, **80**, 813.

83. Takeuchi, Y., Dotson, M., and Keen, N. T. (1992). *Plant Mol. Biol.*, **18**, 835.

84. Finer, J. J., Vain, P., Jones, M. W., and McMullen, M. D. (1992). *Plant Cell Rep.*, **11**, 323.

85. Sanford, J. C., Devit, M. J., Russell, J. A., Smith, F. D., Harpending, P. R., Roy, M. K., *et al.* (1991). *Technique*, **3**, 3.

86. Kohli, A., Griffiths, S., Palacios, N., Twyman, R. M., Vain, P., Laurie, D. A., *et al.* (1999). *Plant J.*, **17**, 591.

87. Bano-Maqbool, S. and Christou, P. (1999). *Mol. Breed.*, in press.

88. Tisserat, B. (1985). In *Plant cell culture: a practical approach* (ed. R. A. Dixon), p. 79. IRL Press, Oxford.

89. Williams, E. G. and Maheshwaran G. (1986). *Ann. Bot.*, **57**, 443.

90. Dey, M., Kalia, S., Ghosh, S., and Guhanumukherjee, S. (1998). *Curr. Sci.*, **74**, 591.

91. Franklin, C. J. and Dixon, R. A. (1994). In *Plant cell culture: a practical approach* (2nd edn) (ed. R. A. Dixon and R. A. Gonzales p. 1). IRL Press, Oxford.

92. De Cleene, M. D. and De Ley, J. D. (1976). *Bot. Rev.*, **42**, 389.

93. Chan, M. T., Chang, H. H., Ho, S. L., Tong, W. F., and Yu, S. M. (1993). *Plant Mol. Biol.*, **22**, 491.

94. Ishida, Y., Saito, H., Ohta, S., Hiei, Y., Komari, T., and Kumashiro, T. (1996). *Nature Bio/Technol.*, **14**, 745.

95. Mahalakshmi, A. and Khurana, P. (1995). *J. Plant Biochem. Biotech.*, **4**, 55.

Chapter 9

Radiolabelling of monoclonal antibodies

Stephen J. Mather

Department of Nuclear Medicine, St. Bartholomew's Hospital,
London EC1A 7BE, U.K.

1 Introduction

Antibodies have been tagged with radionuclides since the 1940s when David Pressman, perhaps the first to recognize the potential of radiolabelled antibodies *in vivo*, radiolabelled polyclonal antibody preparations with iodine-131, a radionuclide first produced by artificial means only a decade or so previously (1). Today the researcher may wish to radiolabel an antibody for a wide variety of different applications, many of which are described in other chapters in this book.

2 The choice of radionuclide

The radionuclide employed and the technique used to incorporate it into the antibody will depend upon the application envisaged. For the purposes of this chapter applications can be divided into *in vitro* and *in vivo*.

3 *In vitro* applications of radiolabelled antibodies

The principle *in vitro* types of application are for radioimmuno- (or related) assay and pre-clinical development studies prior to future *in vivo* use, but antibodies may also be labelled for use as probes in immunocytochemistry, Western blots, or for immunoprecipitation. The choice of radionuclide will depend upon such factors as ease of tracer preparation, ease and efficiency of detection (counting or autoradiography), shelf-life of radiotracer, cost, and safety. A list of some of the most common radionuclides which might be employed for preparation of radiotracers with applications *in vitro*, together with their physical decay properties is shown in *Table 1*.

While the soft beta emitters, tritium, carbon-14, and sulfur-35 are widely used for synthesis of radiotracers by commercial sources, they are not commonly used for the preparation of tracers in academic laboratories. Although antibodies can be labelled with these isotopes, the main advantage of these radionuclides is their high resolution for autoradiographic studies, and, while

Table 1 Some radionuclides used commonly *in vitro*

Radionuclide	Type of decay	Energy (MeV)		Half-life
		$E_{\beta max}$	E_γ	
^3H	β^-	0.018		12.26 years
^{14}C	β^-	0.156		5736 years
^{125}I	EC		0.035 (7%)	60 days
^{131}I	β^-, γ	0.61 (86%)	0.364 (80%)	8.04 days
		0.33 (13%)	0.284 (6%)	
^{35}S	β^-	0.167		87.4 days
^{32}P	β^-	1.71		14.3 days

their use for more common applications such as radioimmunoassay has some potential advantages related to tracer stability and radiation safety issues, the more extensive sample preparation required for liquid scintillation counting means that their use is very limited. Accordingly, methods which can be used for labelling antibodies with tritium and carbon-14 will not be addressed in detail and the interested researcher is referred to another useful publication in the Practical Approach series (2). By far the most useful and widely used radionuclide for *in vitro* antibody studies is iodine-125. The two month half-life of this radionuclide means that the useful shelf-life of the isotope is at least a month and the shelf-life of the radiotracer of similar duration. The chemistry used to label antibodies is (normally) very simple, the starting material is readily available and quite inexpensive, and the low energy gamma rays and X-rays emitted can be counted with high efficiency on a gamma counter and will normally give an acceptable sensitivity and resolution on X-ray film for autoradiographic detection. The main drawback of radioiodine is its safety profile.

3.1 Iodine-125

Iodine-125 decays with a half-life of 60.2 days by electron capture to the excited state of tellurium-125. A gamma photon of 35 keV is emitted during 7% of subsequent de-excitations and the remaining 93% by the emission of 27–35 keV X-rays as external orbital electrons cascade down to fill vacancies in the inner orbitals. These gamma and X-rays also interact with circulating electrons resulting in expulsion of about 15 low energy internal conversion or Auger electrons per disintegration (3). Provided they are not absorbed by the walls of the container, the photons and X-rays will be detected with about 50% efficiency by a standard 2 inch NaI(Tl) scintillation detector and the low-energy electrons provide good resolution on X-ray film.

The predominant mechanism of radiolabelling of antibodies with radioiodine involves electrophilic substitution of the iodine into the activated phenolic ring of tyrosine side-chains as shown in *Scheme 1*. Provided that low substitution ratios are used it is unlikely that radiolabelling will take place at any other site in the large immunoglobulin molecule but in proteins lacking tyrosine residues

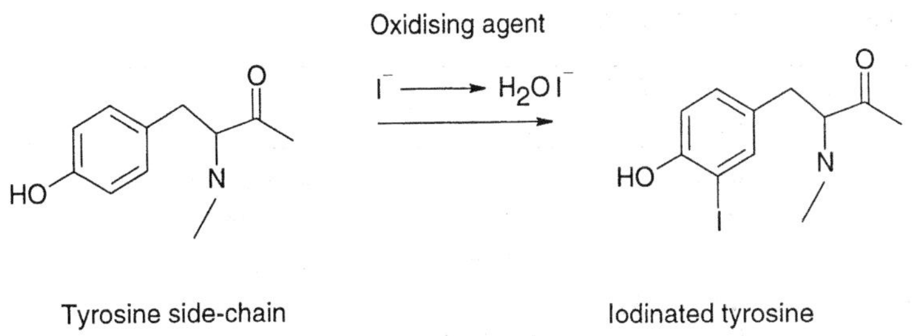

Scheme 1 Iodination of tyrosine side-chains by electrophilic substitution.

then substitution can also take place (albeit with a lower degree of stability) in histidine side-chains.

The radiolabelling grade of radioiodine purchased from commercial suppliers will be in the form of sodium iodide. It will be essentially carrier free (i.e. uncontaminated by stable iodine-127) and is normally dissolved in sodium hydroxide solution (typically 0.01–0.001 M) at concentrations of about 100 mCi (3.7 GBq)/ml. The function of the hydroxide ions are to maintain the chemical state of the iodide and prevent formation of volatile radioiodine molecules. Freshly prepared and purchased solutions of radioiodine will invariably give the best labelling efficiencies but with time, radiolysis produces free radicals which react with the iodide ions to produce non-reactive by-products. Nevertheless radioiodine solutions stored (at room temperature) for one to two months will normally give acceptable results.

A large number of techniques for labelling proteins with radioiodine have been developed over the past half-century but, for reasons of efficiency and ease of use, many of these are now only of historical interest, at least so far as antibody labelling is concerned. For a detailed review see Dewanjee (3). The most widely practised methods are those based upon the oxidation of radioiodine to a reactive intermediate species, the exact nature of which has not been identified, but is thought to be a positively charged species such as the hydrated iodinium ion H_2OI^+. This or related cations can be produced by reacting the radioiodide with a variety of oxidizing agents but two in particular have become the most popular—Chloramine-T (*N*-chloro-*p*-toluenesulfonamide) and Iodogen (diphenyl-glycoluril). Methods for labelling antibodies with radioiodine using these two methods are described in *Protocols 1* and *2*. Both methods have their own inherent advantages and disadvantages. The Iodogen method (4) is very simple and largely invariable. The concentration of this oxidant is controlled by its very sparing solubility in aqueous solvents and one therefore has very little control over the reaction other than by changing the incubation time and temperature. Nevertheless this method will normally produce very acceptable labelling efficiencies of the order of 80–95%. The Chloramine-T method (5), on the other hand, varies widely from laboratory to laboratory. Different researchers have their own favourite Chloramine-T concentrations, incubations times, choice and concentration of quenching reagents. This method has the (somewhat theoretical) advantage that it can be tailored for different antibodies, but has the dis-

209

advantage that the reagents have to be freshly prepared prior to use and over-enthusiastic attempts to improve labelling efficiencies with high concentrations of the oxidant can lead to antibody damage.

Protocol 1

Radioiodination with iodine-125 using Chloramine-T as oxidant

Reagents

- Antibody: 100 µg–1 mg/ml in 0.1 M phosphate buffer or phosphate-buffered saline pH 7.4
- Chloramine-T: 0.5 mg/ml in 0.1 M phosphate buffer pH 7.4 (freshly prepared)
- 0.5 M phosphate buffer pH 7.4
- Sodium metabisulfite: 50 µg/ml in 0.1 M phosphate buffer pH 7.4
- Radioiodine: Na^{125}I (100 mCi/ml, 3.7 Gbq/ml) (IMS30 Amersham Pharmacia Biotech or equivalent)

Method

1. In a small polypropylene tube mix 10–100 µg of the antibody, 50 µl of 0.5 M phosphate buffer pH 7.4, and the desired amount of iodine-125, typically 100 µCi–1 mCi.

2. Add 20 µl of Chloramine-T solution and vortex mix briefly.

3. After 5 min add 40 µl of sodium metabisulfite solution and mix.

4. If desired, check labelling efficiency by ITLC (see *Protocol 4*).

5. Separate labelled antibody from free iodine (see *Protocol 3*).

Protocol 2

Radioiodination with iodine-125 using Iodogen as oxidant

Reagents

- Antibody: 100 µg–1 mg/ml in 0.1 M phosphate buffer pH 7.4 or phosphate-buffered saline
- Iodogen™ (Pierce Chemical Company)
- Dichloromethane
- Radioiodine: Na^{125}I (see *Protocol 1*)
- 0.1 M phosphate buffer pH 7.4
- 0.5 M phosphate buffer pH 7.4

A. Preparation of Iodogen tubes

1. Dissolve 1 mg of Iodogen in 10 ml of dichloromethane in a glass or polypropylene container.

2. Pipette 50 µl into as many glass or polypropylene test-tubes as required.

3. Evaporate the solvent either in a Speed-Vac, with a stream of nitrogen or by leaving in a laminar-flow hood for 2–4 h with the lights turned off.

Protocol 2 continued

4. Cap the tubes and store in a closed container at $-20\,°C$ for up to a year until required.

B. Antibody labelling

1. In an Iodogen tube mix 10–100 µg of the antibody, 50 µl of 0.5 M phosphate buffer pH 7.4, and the desired amount of iodine-125, typically 100 µCi–1 mCi. Wait for 10 min.
2. Transfer the reaction mixture to a fresh test-tube, wash the Iodogen tube with 0.5 ml of 0.1 M phosphate buffer pH 7.4, and add to the mixture.
3. If desired, check labelling efficiency by ITLC (see *Protocol 4*).
4. Separate labelled antibody from free iodine (see *Protocol 3*).

Only in rare circumstances will the labelling procedure result in labelling efficiencies approaching 100%. This means that the reaction mixture will contain a certain proportion of unreacted 'free iodine'. Depending upon the application it may be desirable to remove this free iodine in order to provide a preparation with high radiochemical purity. The most widely used method for purification of labelled antibody preparations is gel filtration on a short Sephadex column as described in *Protocol 3* but (potentially more convenient) alternatives such as ion exchange chromatography or the use of spin columns exist. Both before and after purification it is useful to obtain a measure of the purity of the labelled antibody preparation, first to gain an idea of the efficiency of the labelling procedure and secondly to determine the purity of your reagent. A very simple means of measuring this purity, based on ascending thin-layer chromatography is described in *Protocol 4*.

Protocol 3

Separation of radiolabelled antibody from free iodide

Reagents

- Sephadex G50 fine grade or pre-packed PD-10 column (Amersham Pharmacia Biotech)
- Bovine (or human) serum albumin: 1% in phosphate-buffered saline pH 7.4 (1% BSA/PBS)

A. Preparation of column

If pre-packed columns are not available prepare one as follows:

1. Weigh out 1 g of Sephadex powder and add 15 ml of deionized water. Mix well and either leave overnight or heat in a boiling water-bath for 1 h to allow the gel to swell.
2. Remove the barrel from a 10 ml disposable syringe and cap the Luer tip. Plug the end of the syringe with a small circle of filter paper or lint dressing.

Protocol 3 continued

3. Clamp the syringe vertically in a retort stand. Swirl the swollen Sephadex gel and pour as much as possible into the syringe.

4. Allow the gel to settle for a few minutes. Then remove the Luer cap and allow the liquid supernatant to run through into a waste container. Gently layer 10 ml of deionized water on top of the gel and allow to run through to waste. Replace the Luer cap and use the prepared column as soon as possible.[a]

B. Purification procedure

1. Clamp either a pre-packed or home-made column vertically in a retort stand and remove the Luer cap.

2. Wash the column with 30 ml of 1% BSA/PBS.

3. Apply the labelled antibody reaction mixture to the surface of the column and allow it to run into the gel. Gently pipette 1 ml of 1% BSA/PBS onto the gel and collect the eluate in a test-tube.[a]

4. Repeatedly elute the column with ten 1 ml aliquots of 1% BSA/PBS and collect each 1 ml of eluate in a fresh, numbered test-tube.

5. Pipette 10 µl samples from each of the eluate fractions into counting tubes and count them in a gamma counter in order to identify the tubes containing the labelled antibody fractions (typically tubes 3–5). Use or store the contents as required.

[a] When home-made columns are used, care must be taken that the surface of the gel remains flat and does not run dry.

Protocol 4

Determination of radiochemical purity by TLC

Equipment and reagents

- Chromatographic support material: this can be Whatman 3MM chromatography paper, silica gel coated plastic TLC sheets, or silica gel impregnated glass fibre (ITLC) sheets (Gelman Sciences)

- Glass beaker or similar container 10–15 cm tall
- 85% methanol in water

Method

1. Cut a piece of the chromatographic support approx. 1×10 cm in size. Make a faint pencil mark 1.5 cm from one end.

2. Pour enough 85% methanol into the beaker until it is 0.5 cm deep. Cover the beaker with a Petri dish lid, aluminium foil, or similar.

3. Place a 1 µl spot of the sample to be analysed on the chromatographic strip level with the pencil mark and allow it to dry.

4. Using forceps gently place the strip in the beaker with the sample at the lower end. Cover the beaker and allow the solvent to run up the support material.

5. When the solvent is about 5 mm from the top of the strip, remove it from the beaker using forceps and lay it on a clean tissue to dry.

6. Cut the strip into upper and lower halves and place each half into counting tubes.

7. Count the tubes in a gamma counter. Calculate the radiochemical purity as follows:

$$\% \text{ labelled antibody} = \frac{\text{counts in lower half}}{\text{counts in lower half} + \text{counts in upper half}} \times 100\%$$

The main purpose in labelling an antibody is to obtain a radioactive molecule which binds to a specific recognition site. It is therefore essential that the antibody retains the ability to bind to its epitope throughout the labelling process. A number of tests have been developed to check the immunoreactivity of antibodies after conjugation procedures, many based on ELISA or related assays, but not all of these are entirely appropriate for use with radiolabelled antibodies. ELISAs essentially measure the binding of the entire population of antibody molecules in solution, not only the labelled or conjugated ones. Depending upon the specific activity to which the antibody is labelled the solution will contain a varying proportion of labelled and unlabelled molecules. It is quite likely that the former will be in the minority and will represent 10% or less of the entire population. If the labelling procedure were to entirely abolish the binding of the labelled molecules (but leave the unlabelled molecules unaffected), an ELISA would still return a result of 90% immunoreactivity. This result would be considered generally acceptable despite the fact that the labelled antibody would fail to deliver its primary function. The best types of assay for use with radiolabelling procedures are direct radioligand binding assays which measure only the binding of the labelled molecules. These fall into two categories—those intended to measure the antibody's binding affinity, and those intended to measure the proportion of labelled molecules which retain some ability to bind specifically to their epitope. For the purposes of a relatively simple check to see if the antibody remains functional after labelling the latter type of assay is the more appropriate and a protocol describing such a test can be found in *Protocol 5* (6).

There are two main mechanisms through which immunoreactivity can be compromised. The first is due to the effect of substitution of the large iodine atom into a critical tyrosine residue present in the binding site of the antibody. Despite the fact that the radioiodine can potentially be labelled to any of the (usually quite large number of) tyrosine side-chains scattered throughout the molecule, factors such as local charge distribution and physical access mean that one or more sites are preferentially labelled. If one of these sites happens to be in one of the important CDRs then a significant degree of immunoreactivity will be lost. In this case, the number of possibilities for resolving the situation

Protocol 5

Measurement of immunoreactive fraction of radiolabelled antibody[a]

Reagents

- 15×10^6 cells expressing the appropriate antigen
- 1% bovine serum albumin in phosphate-buffered saline (1% BSA/PBS)

A. Assay procedure

1. Harvest the cells from a sufficient number of tissue culture flasks. Wash the cells and resuspend in cold 1% BSA/PBS. If necessary break up clumps of cells by repeated pipetting or by syringing them through a 23G needle. Measure the cell concentration with a haemocytometer or automated cell counter and dilute to a final concentration of 4×10^6/ml in cold 1% BSA/PBS.

2. Prepare two rows of 7×1.5 ml microcentrifuge tubes. Label the tubes 1–7 and 8–14. Pipette 0.5 ml of cold 1% BSA/PBS into tubes 2–5 and 9–12.

3. Pipette 0.5 ml of the stock cell suspension into tubes 1, 2, 6, 8, 9, and 13.

4. Add 200 µg of unlabelled antibody to tubes 6 and 13. Mix well.

5. Briefly vortex tube 2 and transfer 0.5 ml of the contents to tube 3. Vortex tube 3 and transfer 0.5 ml to tube 4. Vortex tube 4 and transfer 0.5 ml to tube 5. Vortex tube 5 and discard 0.5 ml.

6. Repeat step 5 with tubes 9–12.

7. Dilute the labelled antibody to a concentration of 50 ng/ml with cold 1% BSA/PBS. At least 4 ml of the final dilution is required.

8. Pipette 250 µl of diluted radiolabelled antibody into tubes 1–14. Mix well.

9. Incubate the tubes for 2 h at a constant temperature (preferably 4 °C but alternatively at room temperature or 37 °C). Mix the tubes either constantly (on a mechanical shaker) or regularly (about every 15 min by hand) during the incubation.

10. After the incubation, put the tubes on ice for 10 min. Centrifuge tubes 1–6 and 8–13 at high speed (e.g. > 1000 r.p.m.) for 2 min. Carefully remove the supernatant, taking care not to disturb the pellet. Pipette 0.5 ml of cold 1% BSA in PBS into each tube and mix to resuspend the pellet. Re-centrifuge and again discard the supernatant.

11. Count all tubes in a gamma counter.

B. Data analysis

1. Divide the counts in tubes 1–5 and 8–12 (BOUND counts) by an average of the counts in tubes 7 and 14 (TOTAL counts) to get the fraction of counts bound to the cells for each cell concentration. Average the values for duplicate tubes (1 and 8, 2 and 9, 3 and 10, 4 and 11, 5 and 12).

2. Calculate the cell concentration in each tube (approx. 40, 20, 10, 5, and 2.5 × 10^5/ml).

3. Plot as *y* values the *reciprocal* of fraction bound (i.e. TOTAL/BOUND) against the *reciprocal* of cell concentrations as *x* values. A straight line plot should be obtained.[b] Determine the intercept on the *y* axis and calculate the reciprocal. This is the immunoreactive fraction.[c]

[a] This assay has been adapted from a method published by Lindmo *et al.* (6). It includes a number of 'short-cuts' which make the assay simpler but which could rightly be criticized if used out of context. The assay is, therefore only recommended as a 'quality control check', rather than as a way of determining the true immunoreactive fraction of the antibody for which the original published method is recommended.

[b] The most common problem experienced with this assay is that a straight line is not obtained when the data are plotted. This is nearly always caused by inaccuracies in diluting and losses in washing the cells. With practice and care, the problem usually goes away.

[c] This calculation excludes the contribution made by non-specific binding (NSB). A measure of NSB can be determined from the fraction of counts bound in tubes 6 and 13 (binding of the hot antibody in the presence of a large excess of cold antibody). This is generally very low. If NSB is less than 5% of specific binding it is reasonable to exclude it for this type of assay. If NSB is high, it is necessary to explore and eliminate the reasons for this.

are limited. It is possible that a change in pH during the labelling procedure may help. Although this would probably result in a lower labelling efficiency, the change in charge brought about may promote labelling at a different site. If this does not help then it will be necessary to use an entirely different chemistry for radiolabelling which substitutes the radioiodine at a different site such as the Bolton and Hunter method described in *Protocol 6*. The other possible mechanism responsible for loss of immunoreactivity is oxidation. As well as oxidizing the radioiodide to a reactive form, the oxidant used may also oxidize critical residues, particularly methionine, in the antibody molecule. One way to determine which of the two mechanisms is responsible is to perform the labelling procedure without the addition of the radioiodine and to perform an ELISA—if an oxidative mechanism is responsible then immunoreactivity will still be lost since all the antibody molecules will be affected, not only those substituted with radioiodine. If this is found to be the cause of immunoreactivity loss then either a Bolton and Hunter approach can be pursued or an electrophilic substitution method which does not subject the antibody to such strong oxidizing conditions can be used. Two approaches may work. The first is to use a milder oxidizing agent such as the lactoperoxidase system (7). The alternative is to use a modification of the Iodogen system in which the radioiodine is first oxidized in the Iodogen tube but then transferred from the oxidizing environment to another tube containing the antibody (8). Both of these procedures will result in a reduction in labelling efficiency but may overcome the problem of oxidative damage to the antibody.

Protocol 6

Iodination of antibody with 'Bolton and Hunter' reagent

Reagents

- 10–100 µg of antibody at a concentration of 2–5 mg/ml in 0.1 M Hepes, phosphate, or borate buffer pH 8.0[a]

- 0.2 M glycine in 0.1 M Hepes, phosphate, or borate buffer pH 8.0

- Bolton and Hunter reagent: N-succimidyl-3-(4-hydroxy-3-[^{125}I]iodophenyl)propionate (IM5861, Amersham Pharmacia Biotech, or equivalent)

- Phosphate-buffered saline pH 7.4 containing 0.05% polysorbate 20 (PBS/Tween)

- Pre-packed PD-10 or 10 ml Sephadex column

Method

1. Into a 1.5 ml microcentrifuge tube (e.g. Eppendorf) pipette the required radio-activity of Bolton and Hunter reagent. Evaporate the solvent, ideally with a 'Speed-Vac' or alternatively under a gentle stream of nitrogen.

2. Add the required amount of antibody to the vial. Mix briefly and incubate for 10 min at room temperature.

3. Add 0.5 ml of 0.2 M glycine solution. Mix and incubate for a further 5 min.

4. If desired, check labelling efficiency by TLC (see *Protocol 4* and use ITLC paper and 20% trichloroacetic acid as mobile phase).

5. Separate labelled antibody from free iodine (see *Protocol 3*) using PBS/Tween for pre-washing and elution buffer.

[a] The higher the antibody concentration, the greater is the labelling efficiency achieved with this method. Antibody concentrations lower than 1 mg/ml will produce labelling efficiencies below 30%. Buffers must contain no amino-containing compounds (e.g. Tris) which will compete with the antibody for the conjugation reaction.

For many radiotracers the issue of specific activity is important, but this is normally not the case with radioiodinated antibodies as most *in vitro* applications can be performed with preparations having low specific activities in the range of 40–400 MBq (1–10 mCi)/mg. This is equivalent to a substitution ratio of 0.1–1 atoms of iodine per antibody molecule and, in most instances will have little or no effect on antibody function. Although it may be possible to increase this specific activity by a factor of ten or more, the greater the substitution ratio, the greater the likelihood of a significant loss in immunoreactivity.

When stored for an appreciable length of time radioiodinated antibodies undergo radiolysis leading to a loss in purity and immunoreactivity. The rate of radiolysis can be reduced by the addition of carrier proteins or antioxidants such as ascorbic acid which scavenge the free radicals responsible (9). A concentration of 0.1–1% albumin or 0.5% ascorbic acid is commonly used and antibodies may be stored in these solutions either at 4 °C or −20 °C for at least a month without significant loss of quality. If stored frozen, then the preparation should be aliquoted to prevent repeated freezing and thawing which tends to produce aggregation. If stored at 4 °C then, provided it does not interfere with the application sodium azide can be added to a concentration of 0.05% to limit microbial growth.

4 *In vivo* applications of radiolabelled antibodies

Radiolabelled antibodies may be administered to experimental animals or to human recipients for the purposes of biodistribution and pharmacokinetic studies, radionuclide imaging, or therapeutic purposes. Again the radionuclide of choice will depend upon the application and a list of the most likely candidates can be found in *Table 2*.

Table 2 Some radionuclides used commonly *in vivo*

Radionuclide	Type of decay	Energy (MeV)		Half-life
		$E_{\beta max}$	E_γ	
^{125}I	EC		0.035 (7%)	60 days
^{131}I	β^-, γ	0.61 (86%) 0.33 (13%)	0.364 (80%) 0.284 (6%)	8.04 days
^{123}I	EC		0.159	13 hours
^{111}In	EC		0.173 (91%) 0.247 (94%)	2.8 days
^{99m}Tc	IT		0.141 (89%)	6.02 hours

In biodistribution, metabolic, and pharmacokinetic studies, the radiolabelled tracer is administered to the subject and samples of tissues are subsequently taken to be measured in a scintillation counter in order to quantify the amount of radioactivity present in that sample. For this type of application any of the radionuclides listed in *Table 1* may be used. In animal studies the amount of radioactivity which needs to be administered will be determined by the efficiency of the detection system and normal radiation safety considerations. For studies in human subjects the radiation dose to the recipient is likely to be paramount. For these reasons iodine-125 will probably be the radionuclide of choice for this type of study and the labelling procedure described above may be used.

For imaging biodistribution with a gamma camera, gamma emitting radionuclides must be used and the choice is determined by the gamma photon yield in relation to the radiation dose received by the patient. The three radioisotopes of choice are technetium-99m, iodine-123, and indium-111. All are 'pure' gamma emitters producing good quality images at the cost of an acceptable radiation dose to the patient. The decision on which to use will be made on the basis of cost, availability, and half-life.

4.1 Iodine-123

I-123 decays by electron capture with an intermediate half-life of 13 hours and its single gamma photon at an ideal energy of 159 keV produces images of the highest quality. A large number of internal conversion electrons are also produced. If sufficient radioactivity is administered, then images may be acquired up to 48 hours after injection. The disadvantage of this isotope is its cost and availability. It is prepared in a high-energy cyclotron and, due to its half-life must be shipped to the user on the day of use. This results in a high cost per patient dose of the order of £300–500.

The chemistry of iodine-123 is identical to that of I-125 and the same labelling methods may be employed. However, there are some differences between the radionuclide preparations which may necessitate some modifications of the procedures. First, owing to the much shorter half-life of I-123 compared to I-125 the specific activity is considerably higher and consequently the concentration of iodide (in terms of micrograms/ml) in the radionuclide solution is lower and can result in reduced labelling efficiencies. This can be overcome by the addition of a very small amount of cold iodide (about 50 ng/mCi) to the labelling reaction. Secondly, the quality of I-123 preparations is somewhat variable, both between and within manufacturers. This will result in variations in labelling efficiency due to inconsistencies in radiochemical purity and specific activity of the starting material and relates (presumably) to the rapid processing of the radionuclide. If problems are encountered with variability in labelling efficiency with iodine-123 then, in the first instance, the manufacturer should be contacted and asked to provide evidence of radiochemical purity. It is possible to check the radiochemical purity of the radioiodine in-house and to take steps to improve it if necessary (10), but the level of expertise and equipment required is likely to be restricted to specialized radiochemistry laboratories.

Although radioiodinated antibodies have acceptable stability for *in vitro* applications, their *in vivo* stability is less ideal. These tracers can act as substrates for the iodotransferase enzymes which are normally responsible for metabolism of the thyroid hormones thyroxine and tri-iodothyronine which bear structural similarities. Although not present in the vascular circulation, these enzymes are widely distributed in many tissues and their action results in the intracellular release of the radionuclide which then either diffuses or is actively transported out of the cell and subsequently excreted in the urine. In order to try and increase the *in vivo* stability of antibodies labelled with radioiodine, alternative synthetic strategies have been pursued which result in molecules which are not substrates for the iodotransferases (11, 12). All of these techniques are more complicated than the electrophilic substitution methods described above and require some synthetic chemistry in order to prepare the precursors which are not commercially available.

One of the potential advantages of radiometals as protein labels is their enhanced *in vivo* stability and tissue residence times, however unlike iodine, a direct attachment of the metal to the antibody cannot be easily achieved.

4.2 Indium-111

In-111 decays by electron capture with a half-life of 2.8 days to produce stable cadmium-111. In the process it emits a number of Auger and internal conversion electrons and two principal gamma photons with energies of 174 and 246 keV. Although these photons can be imaged with the gamma camera, their energy is not as ideal as that of iodine-123 or technetium-99m. The greater energy of the 246 keV photon requires the use of a thicker collimator which reduces the sensitivity of the detection and hence degrades the quality of the images obtained. However, the longer half-life of this radionuclide means that images can be acquired for a much longer period—up to about a week—after administration of the tracer. This allows a longer time for clearance of the radioactivity from the blood stream and may result in clearer delineation of target uptake. Like iodine-123, indium-111 is produced in a cyclotron and must be shipped on demand. It is also relatively expensive with a cost of the order of £300–400 per patient dose.

All strategies for labelling antibodies with indium rely on the use of bifunctional chelating agents (BFC). These are molecules which are able to bind metals with high affinity but also contain reactive functional groups which allow them to be conjugated to the antibody as illustrated in *Scheme 2A*. A considerable number of bifunctional chelating systems useful for indium and other radiometals have been published but a detailed review of them is outside the scope of this chapter. Most of these BFC are not commercially available and must be either synthesized in-house, or obtained by collaboration with other groups with the necessary synthetic expertise. The exception to this is a BFC based on the chelating agent diethylenetriaminepentaacetic acid (DTPA) (13). The cyclic dianhydride of DTPA is available from Sigma-Aldrich Chemical company and for this reason, despite some shortcomings, this BFC has become widely used. A method for labelling antibodies with indium-111 using this reagent is described in *Protocol 7*. Cyclic DTPA anhydride has two main disadvantages, the first somewhat theoretical the second a more practical drawback. As can be seen in *Figure 1a* the DTPA is functionalized through the carboxylic acid groups of the parent molecule and one of these groups subsequently becomes the point of attachment of the chelator to the antibody. This has the effect of removing some atoms from the co-ordination sphere of the radiometal and may influence the subsequent strength of chelation. Since the binding of indium to the chelator is reversible, once the radiolabelled antibody is administered, the potential exists for the indium to transfer from the chelate to other natural molecules which have high affinity for metallic ions. Perhaps the most important of these is transferrin which is present in large concentrations in serum. Indium shares many similar chemical characteristics with ferric iron and the affinity of transferrin for indium actually exceeds that of DTPA. Since the concentration of transferrin in the blood will be greater than that of the antibody–DTPA conjugate and transferrin has a greater stability constant then the indium will tend to transfer from the antibody to transferrin.

However, while thermodynamic stability issues decide the direction in which the transfer of the radiometal occurs, of greater practical importance is the kinetic stability which determines the rate at which it happens. Studies have shown that indium does in fact transfer from antibody–DTPA conjugates to transferrin *in vivo*, but that this occurs at a rate of only about 10% per day which is considered to be too slow to radically influence the quality of the images obtained. The fact that only a slow rate of dissociation is observed even when one of the carboxylic acid groups is employed for conjugation is explained by the fact that indium requires only seven atoms to complete its co-ordination. This is not the case for other radiometals such as yttrium which are discussed below.

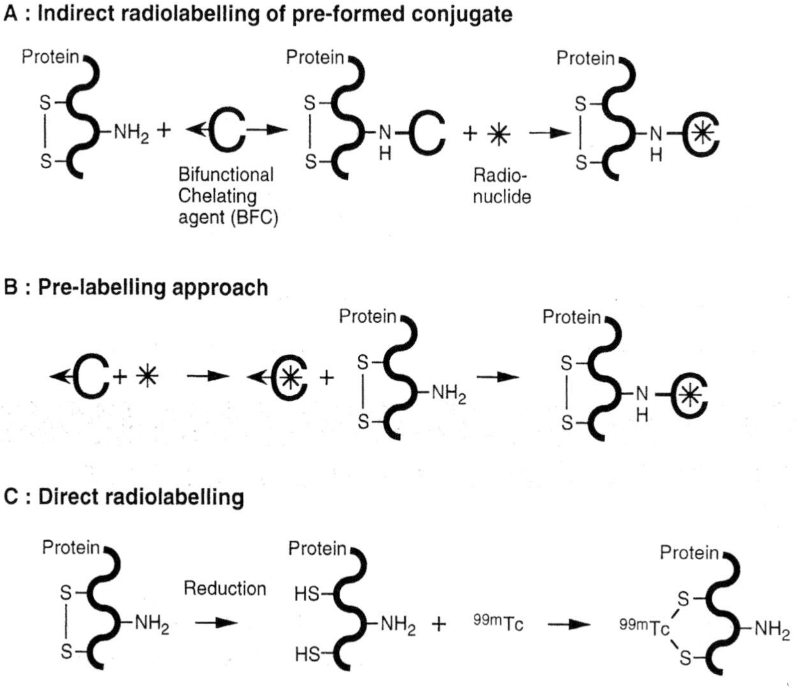

A : Indirect radiolabelling of pre-formed conjugate

B : Pre-labelling approach

C : Direct radiolabelling

Scheme 2 Indirect and direct approaches to radiolabelling of antibodies.

The other drawback of the cyclic DTPA anhydride is the fact that the molecule actually contains two chemically reactive anhydride groups, one at either end of the molecule. The desired chemistry is that one of these will react with amino groups present in the antibody while the other reacts with water to liberate the rest of the DTPA molecule. However, it is also possible for both anhydride groups to react with antibody causing intra- and intermolecular crosslinking. This results in the formation of antibody dimers and, if high DTPA:antibody ratios are used, higher oligomers during conjugation and necessitates the use of an efficient means of purification to remove them from the preparation.

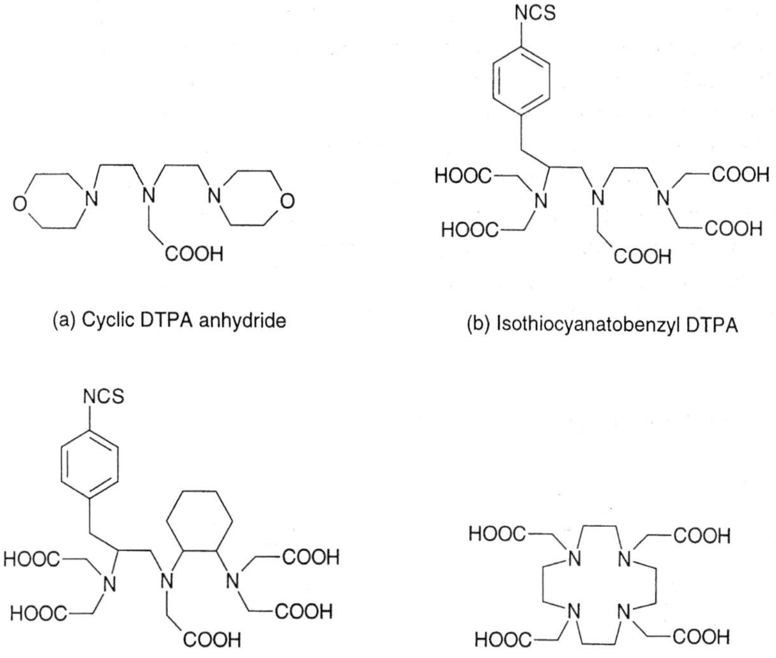

(a) Cyclic DTPA anhydride

(b) Isothiocyanatobenzyl DTPA

(c) Isothiocynatobenzyl-cyclohexyl DTPA

(d) DOTA

Figure 1 Some bifunctional chelating agents based on diethylenetriaminepentaacetic acid.

Protocol 7

Indium-111 labelling of antibodies using cyclic DTPA anhydride[a]

Equipment and reagents

- Size exclusion chromatographic equipment such as FPLC (Amersham Pharmacia Biotech)
- 10–20 mg of antibody at a concentration of 5–10 mg/ml in 0.1 M Hepes, phosphate, or bicarbonate buffer pH 8.0
- Cyclic DTPA anhydride (CDTPAA, Sigma-Aldrich Chemical Company)

- Anhydrous dimethyl sulfoxide
- 0.1 M ammonium acetate buffer pH 5
- 1 M sodium acetate buffer pH 5
- 0.1 M sodium acetate pH 5 containing 50 mM EDTA
- Indium-111 chloride (Nycomed-Amersham INS1P or equivalent)

A. Derivatization

1. Dissolve cyclic DTPA anhydride in dry DMSO to a concentration of 2–10 mg/ml.

2. Add sufficient CDTPAA solution to give an anhydride:antibody molar ratio of 2–4:1 to the stirred antibody solution dropwise. Stir the reaction mixture for a further 10 min.

Protocol 7 continued

3. Although not essential, it is useful at this stage to estimate of the success of the derivatization procedure as follows:

   (a) Adjust the pH of a small sample of the reaction mixture to pH 5–6 by the addition of 1 M acetate buffer pH 5.

   (b) Add a trace amount (approx. 1 MBq) of 111-In acetate and incubate for 5 min.

   (c) Separate the protein bound and unbound forms of indium either on a short gel filtration column (see *Protocol 3*) or by an appropriate TLC method (see *Protocol 4* and use ITLC paper developed with 0.1 M citrate buffer). The proportion of activity associated with the protein (Rf = 0) should be about 30–40%.

B. Purification

1. Separate the DTPA-conjugated antibody from unbound DTPA and antibody by passage down a suitable chromatographic column. Ideally this should be a plastic-based HPLC column such as the FPLC Superose columns. Alternatively a 30–60 cm Sephacryl or equivalent column may be used.

2. The column should be first equilibrated and the sample then eluted with 0.1 M ammonium acetate buffer pH 5. Fractions containing the monomeric antibody peak should be collected and pooled. After measurement of the protein concentration, the derivatized antibody can be aliquoted into suitable sized patient doses (typically 0.5 mg) and frozen for future use.

C. Radiolabelling

1. To a vial of 111-In chloride add sufficient 1 M acetate buffer pH 5 to produce a final concentration of 0.2 M acetate. Thaw a vial of antibody and add the contents dropwise to the indium acetate solution.

2. Incubate the mixture for 10–30 min and add 0.1 M acetate pH 5/EDTA 50 mM to a final concentration of 5 mM EDTA.

3. Measure the radiochemical purity of the preparation by TLC (see *Protocol 4* and use plastic-backed TLC strips developed in 0.1 M sodium acetate pH 5 containing 50 mM EDTA; labelled antibody has an Rf = 0). If further purification is required, this can be performed on a short Sephadex column (see *Protocol 3*).

[a] All reagents and materials should be as free of metal ion contamination as is practically possible. The highest analytical grade of chemicals, and in particular water, should be used. Disposable plastic containers are preferable to glassware which, if unavoidable, should be of borosilicate glass, washed with 1 M hydrochloric acid, and rinsed with deionized water until neutral to litmus.

Some other BFCs overcome both of these disadvantages of the DTPA anhydride (see *Figure 1b*). These have a single reactive group which is situated in the ethylene backbone of the DTPA molecule and thus leave all five carboxylic acid groups free for complexation of the metal (14). The synthesis of these BFCs is however, relatively complex and they are not currently commercially available.

Unlike radioiodine, during intracellular processing and metabolism of the antibody, the radioindium is not released but is metabolized to a non-freely diffusible form which remains trapped within the cell. This results in a longer residence time of the radioisotope in the tumour and may improve target to background ratios at later imaging times. It also results, however, in retention of the radiolabel is some normal tissues, notably the liver and bone marrow, and this can make visualization of tumour sites in these regions more difficult.

4.3 Technetium-99m

Tc-99m is by far the most widely used radionuclide in nuclear medicine. It decays by isomeric transition to technetium-99 with a half-life of six hours emitting a single gamma photon of 140 keV. This allows imaging up to 24 hours after administration. Most gamma cameras have been designed principally for use with technetium so that the image quality obtained with this isotope represents the best that can be achieved in single-photon radionuclide imaging. One of the main advantages of technetium-99m is its means of manufacture. Unlike iodine-123 and indium-111 which are both produced in a cyclotron, Tc-99m is produced in a small radionuclide generator which is operated in the hospital radiopharmacy.

The generator consists essentially of an ion exchange column to which is adsorbed between 5–50 GBq (about 100 mCi–1 Ci) of molybdenum-99 in the form of ammonium molybdate. Mo-99 decays with a half-life of 67 hours to produce Tc-99m which, since it binds less strongly to the ion exchange support, can easily be separated from the molybdenum with a solution of normal physiological saline. The molybdenum remains bound to the column and continues to produce technetium which can be eluted as desired. While it is possible to make a 'home-made' technetium generator, in practice these are made and supplied by specialist manufacturers. While the principle of the system is very simple, the design is complicated by the need to limit the radiation dose arising from the device and of ensuring that the technetium eluted therefrom is in a form suitable for injection, i.e. sterile, free from particles, and of a high radionuclidic and radiochemical purity. Thus the column is highly shielded with lead or, in the case of the biggest generators, depleted uranium, and a system of elution has been devised whereby the technetium can be eluted aseptically and with the minimum of human intervention. The shelf-life of the generator is effectively determined by the half-life of the parent radionuclide, molybdenum-99, by the amount of radioactivity loaded onto the column, and the amount of technetium required by the user. In practice generators are normally delivered to the radiopharmacy on a weekly basis and used for 10–14 days before they are returned to the supplier for recycling.

The continual availability of technetium-99m in the department greatly simplifies the logistics of supply of this short half-life radioisotope and the economy of scale associated with its use reduces its cost by an order of magnitude compared with I-123 and In-111. Because technetium-99m is so widely available, it is not surprising that a large number of strategies for radiolabelling antibodies with this radionuclide have been devised (15). The situation is further complicated by the fact that the chemistry of technetium is rather more complex than that of iodine or indium. In order to understand the mechanisms involved in the labelling methods used for the radioisotope, a brief description of technetium chemistry is appropriate.

4.3.1 Chemistry of technetium

Technetium is a transition element with the atomic number 43. It can exist in a number of different oxidation states ranging form -1 to $+7$ but the forms most commonly used in radiopharmaceutical chemistry are $+1$, $+3$, $+5$, and $+7$. Technetium is eluted from the generator in its most highly oxidized form as sodium pertechnetate. In this state it is very stable and remains chemically unchanged indefinitely under normal ambient conditions. In order for the technetium to form complexes with other compounds it must first be reduced by removing electrons from the outer orbital. Many different reducing agents can be used to achieve this, but the most widely used reagent by far is stannous ion, normally in the form of stannous chloride $SnCl_2$. Once some of the outer shell of electrons are removed, the technetium becomes more reactive and will form complexes with a wide variety of compounds which contain atoms, normally nitrogen, sulfur, phosphorus, or oxygen, able to share electrons with the radionuclide and thus form co-ordinate bonds. Although technetium will form complexes with many different types of ligands making compounds with a wide variety of physicochemical and biological properties, the methods used for radiolabelling antibody normally take advantage of only a small part of this chemistry. Under the right chemical conditions technetium will be reduced to oxidation state $+5$ by stannous ions and form simple tetrahedral complexes with many ligands which contain a sequence of four co-ordinating atoms, normally nitrogen or sulfur, as shown in *Figure 2a*. These complexes are very stable, even under biological conditions, and form the basis of many different approaches to technetium labelling of antibodies.

Antibody labelling methods are normally described either as 'direct' in which the technetium is complexed directly by the atoms present in the protein chain

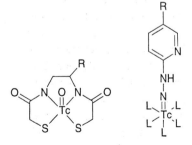

(a) 'Typical' Tc (V) - oxo core   (b) 'Tc-HYNIC core

Figure 2 Examples of technetium-99m co-ordination 'cores'.

or 'indirect' in which a synthetic ligand or 'bifunctional chelate' is used in a manner analogous to the methods used for indium labelling. However for technetium labelling the indirect methods have a further possible variation in that the chelator can be radiolabelled either before or after it is conjugated to the antibody. The three most commonly used approaches to antibody labelling are illustrated in *Scheme 2*.

4.3.2 Direct antibody labelling methods

These are the perhaps most simple of all the methods employed for antibody labelling and can be performed with a minimum of expertise and equipment. As described above, once the technetium is reduced it will form complexes with many different molecules and among these are sequences present in all proteins and polypeptides. However, the stability of this complex will depend upon the nature of the actual co-ordinating atoms employed and it is likely that in a large molecule such as an antibody, the technetium will take up a number of different configurations some of which will be more stable than others. Antibodies labelled directly in the most simple fashion will normally be quite stable *in vitro*, but when introduced into a biological environment, are likely to decompose as the technetium *trans*-chelates from weak co-ordinating sites on the antibody to more stable configurations presented by competing biological molecules *in vivo*. In order to achieve sufficient stability for the technetium complex to maintain its integrity *in vivo*, the antibody must be manipulated in such a way as to present co-ordination sites of the highest possible stability to the radionuclide. Technetium–sulfur bonds are particularly stable and if one or more sulfur molecules are present in the co-ordination sphere then the *in vivo* stability of the complex is greatly enhanced. Free thiols are not normally present in the antibody molecule but can be produced by a partial reduction of the disulfide bridges present in the hinge region. Unlike many smaller biological molecules, the disulfides present in immunoglobulins can be reduced without loss of biological activity since other, mainly hydrophobic, interactions, continue to maintain the configuration of the reduced protein. Technetium will bind very strongly to these reduced thiols and antibodies labelled in this way have been shown to be stable *in vivo* for many hours. The main variable among the reduction-mediated labelling methods is the choice of reducing agent. The most commonly used are the thiol-containing reductants such as mercaptoethanol and dithiothreitol. These have the advantage that they are stable to air and are simple and convenient to use. They have the disadvantage that, since they contain thiols themselves they will also complex technetium and will therefore compete with binding sites on the antibody for the radionuclide. For this reason as well as toxicological considerations they need to be removed from the antibody preparation prior to radiolabelling. A detailed protocol describing a method for labelling antibodies following mercaptoethanol reduction (16) is described in *Protocol 8*. In order to simplify the labelling method still further, attempts have been made to use non-toxic reducing systems which do not interfere with the radiolabelling and consequently remove the need for a purification step after reduction. Among the agents proposed have been stannous ions (17) and ascorbate/dithionite mixtures (18) but, perhaps due to problems caused by their instability to air, these methods have not become so widely employed. One novel approach which may well find more widespread future application is the use of a physical rather than a chemical reducing agent. Low wavelength ultra-violet radiation of 180–200 nm is able to induce a number of chemical reactions in proteins but increasing the wavelength to around 300 nm results in a more selective reactivity with the disulfide bridges resulting in their reduction to free thiols. Thus antibodies can be easily reduced by exposing the solution to UV light for a few minutes and the reducing agent is simply removed by switching off the light. In fact, it appears that this treatment results in an active inter-mediate which can be stabilized by the addition of a small amount of stannous ions, but no purification step is required and the antibody can be easily labelled by the simple addition of the required amount of technetium-99m (19). While greater intensity of irradiation (requiring shorter exposure times) can be achieved with a specialized commercial UV reactor, acceptable results can also be obtained with more widely available sources of UV light such as laboratory transilluminators.

Protocol 8

Reduction-mediated technetium-99m labelling of antibodies with 2-mercaptoethanol

Equipment and reagents

- Nitrogen cylinder
- 10–20 mg of antibody at a concentration of 5–10 mg/ml in 0.1 M phosphate buffer pH 7.4
- 2-Mercaptoethanol
- 20 ml Sephadex G50 column (see *Protocol 3*)
- Phosphate-buffered saline pH 7

A. Antibody reduction

1. Purge 100 ml of phosphate-buffered saline (PBS) with nitrogen on ice for 30 min.
2. Calculate the volume of 2-mercaptoethanol that must be added to the antibody to give a molar ratio of 2000:1 (2-ME:Ab) This translates to 0.47 µl of mercaptoethanol/mg of antibody.
3. Draw up the required volume of mercaptoethanol, add to the antibody, cap the tube, and mix well. (2-Mercaptoethanol emits a foul odour, this should be done quickly and ideally in a fume cupboard.)
4. Leave the tube to incubate at room temperature for 30 min.
5. Meanwhile, prepare a 20 ml Sephadex G50 column as described in *Protocol 3*. Equilibrate the column with three column volumes of nitrogen-purged PBS.
6. When the 30 min reduction period is complete, purify the reaction mixture on the column, eluting the column with nitrogen-purged cold PBS and collecting 1 ml

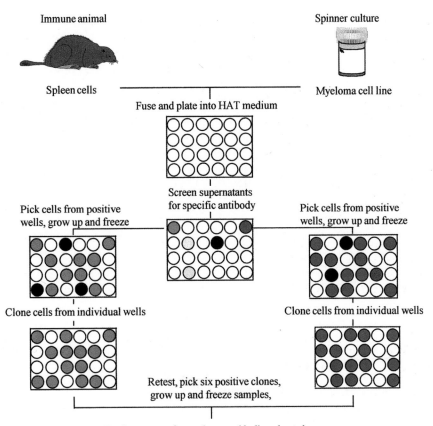

Chapter 1 Figure 1. Production of monoclonal antibodies.

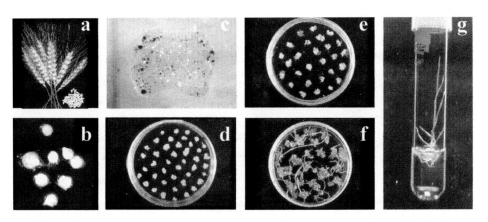

Chapter 8 Figure 2. The wheat transformation process. (a) Wheat spikes and seeds.
(b) Explants ready to be bombarded. (c) Transient GUS activity following transformation.
(d) Callus on selective medium. (e) Induction of shoots under selection. (f) Shoot elongation.
(g) Transfer to tubes for rooting.

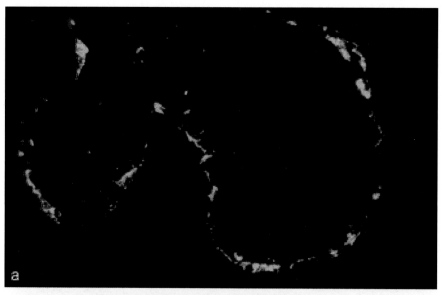

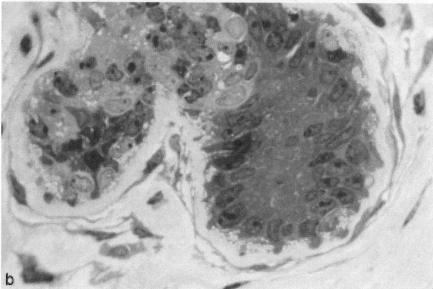

Chapter 11 Figure 1. Human breast lobules processed by PLT into Lowicryl HM20. The semi-thin section was labelled with anti-actin antibody followed by anti-mouse 5 nm colloidal gold conjugate. This was silver enhanced for 24 min. (a) The silver enhanced signal is imaged by epipolarized light microscopy showing the actin in the outer myoepithelial cell layer. (b) Combined epipolarized and transmitted light signals showing the labelling superimposed on the Toluidine blue stained section. Magnification: × 700.

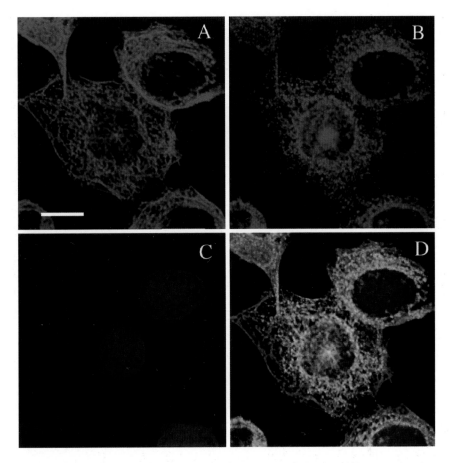

Chapter 16 Figure 1. Triple staining of cells. Triple fluorescent staining of (A) microtubules, (B) rab-11-GFP, and (C) nuclear DNA. A431 cells expressing rab-11-GFP were methanol fixed and immunostained for beta tubulin using a mouse anti-beta tubulin primary antibody and an Alaxa 546 anti-mouse secondary antibody (Molecular Probes). Hoechst 33258 DNA stain (1 μg/ml; Sigma) was added in the second last wash of the immunostaining procedure to visualize the nuclear DNA. Confocal images were taken on a Zeiss510 inverted microscope using a 63 × Planapo lens (1.3 NA) sampling at rates of 50 nm in the lateral and 150 nm in the axial direction, respectively. Images for the different channels were acquired simultaneously. Stacks for each channel were subjected to image restoration using the iterative maximum likelihood estimation algorithm (4) implemented in the Huygens software package for image deconvolution (Scientific Volume Imaging BV). (D) After maximum intensity projection images were merged to give a RGB coloured representation. Bar = 15 μm.

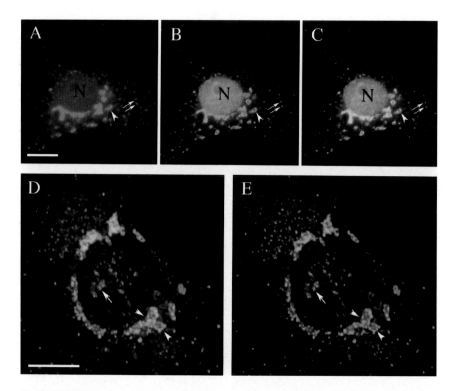

Chapter 16 Figure 2. Increasing resolution by deconvolution. Vero cells were incubated for 90 min with CY3 conjugated shiga-like toxin (SLT) at 37 °C. Then they were fixed with paraformaldehyde and permeabilized with Triton followed by incubation with a monoclonal anti-COPI antibody (5) and an Alexa 488 conjugated goat anti-mouse antibody. Confocal images were taken on a Zeiss510 inverted microscope using a 63 × Planapo lens (1.3 NA) sampling at rates of the lateral and 150 nm in the axial direction, respectively. Acquisition of the different colour channels was simultaneous. Stacks for each channel were subjected to image restoration using the iterative maximum likelihood estimation algorithm (4) implemented in the Huygens software package for image deconvolution (Scientific Volume Imaging BV). Maximum intensity projections before (A) and after (B) deconvolution are shown. Arrowheads point to Golgi structures with improved resolution in the deconvolved image. Bar = 15 μm.

fractions. Cap the tubes containing the fractions, and keep them on ice when they come off.

7. Measure the OD 280 nm of each fraction on a UV spectrophotometer. Pool all antibody fractions with a concentration > 1 mg/ml. (Absorbance of 1 mg/ml solution of IgG = 1.4/cm light path.) Measure the concentration of the pooled antibody. Cap the tube and keep on ice. Divide into suitable size aliquots (typically 250–500 μg) and freeze at the lowest temperature available.

B. Antibody labelling

1. Reconstitute a Medronate II MDP kit (Nycomed Amersham) or equivalent with 5 ml of 0.9% sodium chloride injection. Add 40 μl of this to 250–500 μg of the reduced antibody. Mix. Add the required amount of sodium [Tc^{99m}] pertechnetate (typically 50–1000 MBq depending on the application). Mix and wait 10 min.

2. Measure the labelling efficiency of the preparation by TLC (see *Protocol 4* and use ITLC-SG paper developed in 0.9% saline; labelled antibody has an Rf = 0).

Reduced disulfide bridges are readily re-oxidized by air and to maintain an acceptable shelf-life, oxygen must be rigorously removed from the preparation. The stability can also be enhanced by the use of low storage temperatures (ideally < −40 °C) and by the addition of chelating agents, such as EDTA, which complex the trace amounts of metals which can catalyse the re-oxidation. However all reduced antibody preparations will ultimately re-oxidize and this creates a problem. Since technetium will bind to non-reduced (or re-oxidized) antibodies in addition to reduced immunoglobulins, how can we tell if the radionuclide is complexed by high affinity, stable co-ordination sites or by low affinity, unstable sites which are likely to dissociate *in vivo*? This can be ensured in two ways, first by the use of a colorimetric thiol assay, for example using Ellman's reagent which permits calculation of the number of free thiols present (20). Secondly by the addition of a weak complexing agent to the labelling mixture. Compounds such as methylene diphosphonate and diethylenetri-aminepentaacetic acid are known to form weak complexes with reduced technetium. They are able to compete effectively with the low affinity technetium binding sites on non-reduced antibody, but not with the high affinity thiol-containing sites. Thus a non-reduced or re-oxidized molecule will fail to label satisfactorily in the presence of a small amount of such an agent while a high labelling efficiency will be maintained for reduced antibody.

4.3.3 Indirect technetium labelling methods

Antibodies labelled using methods similar to those described in the previous section have been widely used for radioimmunoscintigraphy for a number of years and, indeed, antibody preparations based on this technology are now registered and commercially available. In attempts to improve stability still further and to influence patterns of biodistribution *in vivo*, however, other labelling approaches based on the use of bifunctional chelates have been developed. In its simplest form, as described above for use with indium-111, a suitable BFC is first conjugated to the antibody. Following a purification step the conjugate is then radiolabelled with the radionuclide of choice. However, while such a 'pre-conjugation' approach is entirely suitable for indium-111 it has problems when applied to labelling with technetium. Since the technetium can form complexes directly with the protein side-chains, unless the chelator has a very high affinity for the technetium, it is possible for the radionuclide to bypass the chelator, and bind directly to low affinity sites on the protein with resulting stability problems *in vivo*. For this reason great care must be taken in the choice of chelating agent and appropriate experiments must be performed to ensure that radiolabelling does occur at the site of the chelator and not elsewhere on the antibody. Because of this potential drawback, this labelling approach is not widely applied to the labelling of intact antibodies with technetium, but there is one exception which has recently gained acceptance. This uses a novel chelating chemistry of high affinity which is able to overcome the possibility of adventitious direct labelling. While most approaches to technetium labelling utilize technetium (v)-oxo complexes as described above, hydrazinonicotinic acid (HYNIC) forms very stable technetium nitride complexes. Immunoconjugates labelled with technetium using this reagent have been shown to be more stable and have a different pattern of biodistribution *in vivo* (21). HYNIC is not commercially available but can be synthesized using a relatively simple two-step process. The *N*-hydroxysuccinimide ester of HYNIC is prepared and first conjugated to the antibody (22). After purification to remove excess HYNIC, the antibody is normally divided into smaller aliquots until required for radiolabelling. The labelling of HYNIC differs somewhat from most other chelating agents and in fact HYNIC is not really a chelator in the normally accepted sense of the word. The hydrazino nitrogen forms a single bond with the technetium, but atoms from other molecules are required to fill the remainder of the metal co-ordination sphere. These atoms are provided by 'co-ligands' as shown in *Figure 2b*. A number of different co-ligands have been employed (23) but the most widely used for antibody labelling are either glucoheptonate or tricine. Thus, radiolabelling the HYNIC immunoconjugate with technetium, in the presence of either glucoheptonate or tricine results in a binary complex.

An alternative means of overcoming the problems of direct labelling is provided by the last of the three generic approaches to technetium labelling, known as pre-labelling. In this technique instead of first conjugating the BFC to the antibody, the chelator is first radiolabelled and then subsequently attached to the immunoglobulin (24). Because all the radionuclide is pre-chelated it is unable to bind to low affinity sites on the antibody and high affinity binding is assured. The main disadvantages of this method are first that a suitable BFC needs to be synthesized, since none are commercially available, and secondly that the labelling efficiency is inevitably considerably lower than 100%. This means that a purification step is required to remove unbound chelate before the radiolabelled antibody can be administered. However, the advantage is that the BFC can be designed in such a way that the pattern of biodistribution of the antibody, and more specifically its metabolites, can be controlled. This labelling method forms the basis of the commercially available antibody preparation Verluma, used for imaging small-cell lung cancer, the chemistry of which is illustrated in *Scheme 3*.

The last of the three types of *in vivo* application of radiolabelled antibodies is targeted radioimmunotherapy. For such an application a cytotoxic radionuclide emitting particulate radiation is required. The added complexity of targeted radiotherapy and its earlier stage of current developments means that it is not possible to be prescriptive in identifying the most appropriate radionuclides. A list of candidate radioisotopes is shown in *Table 3*, these emit a variety of types of radiation including short-range internal conversion electrons and alpha particles, and medium- to long-range beta particles. The choice of radioisotope will be determined by both scientific and logistical considerations. In types of disease where it is possible to deliver the radionuclide uniformly to all targeted cells, the short-range particles have an advantage, these deliver their energy within a small sphere giving a locally intense radiation field, particularly in the case of alpha particles, and therefore spare surrounding, potentially 'normal' tissues. The extreme example of this approach is the use of internal conversion electrons which require internalization within the cell for a cytotoxic effect. These radioisotopes are therefore likely to be most effective in small volume

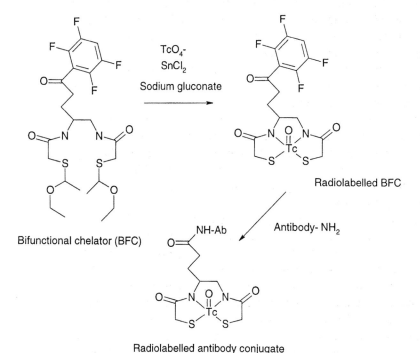

Scheme 3 Pre-labelling approach to labelling antibodies with technetium-99m.

Table 3 Some radionuclides for radioimmunotherapeutic application

Radionuclide	Type of decay	Energy (MeV)		Half-life
		$E_{\beta max}$	E_γ	
^{131}I	β^-, γ	0.61 (86%)	0.364 (80%)	8.04 days
		0.33 (13%)	0.284 (6%)	
^{32}P	β^-	1.71		14.3 days
^{90}Y	β^-	2.27		64 hours
^{188}Re	β^-, γ	2.1	155 (15)	17 hours
^{186}Re	β^-, γ	1.08	137 (9)	89 hours
^{47}Sc	β^-, γ	0.4–0.6	159 (68)	82 hours
^{67}Cu	β^-, γ	0.4–0.6	185 (49)	62 hours
^{211}At	α	5.8 (α)		7.2 hours
^{213}Bi	α, β^-, γ	0.2–0.4	440 (25)	45 minutes
		6–8 (α)		

disease, such as leukaemia, where access to antigen sites on the cell surface is good. However, for larger solid tumours, access is limited by poor blood supply and high intratumoural pressures and consequently antibody uptake is hetero-geneous. In order to be able to irradiate those areas of tumour not reached by the antibody, radionuclides emitting longer-range high energy beta particles must be employed. Often, however, uncertainty over which radionuclide might be most efficacious in treating a particular type of tumour is overcome by simple logistics of supply. Very few of these isotopes are widely available in the quantities required to run even moderate scale clinical trials. Reliance can only be placed on a regular supply of iodine-131, yttrium-90, and phosphorus-32. The availability of all the other radionuclides in *Table 3* is patchy and often limited only to a few months per year.

The chemistry employed for labelling antibodies with therapeutic isotopes is often very similar to that used for other radionuclides described in previous sections. Thus exactly the same methods can be used for iodine-131 as iodine-125, the chemistry of yttrium is very similar to that of indium, and that of rhenium almost identical to that of technetium. Some differences do exist, however, and they will be discussed below. However, often the most important consideration in developing a method for labelling antibodies for radio-immunotherapy is one of radiation safety. Not only are these isotopes poten-tially more toxic, they are also used in much greater quantity. A typical radioiodination of an antibody with I-131 might use 200–300 mCi (8–12 GBq) of radioactivity compared to 5–10 mCi (200–400 MBq) for a diagnostic study. Radiation safety issues are of particular importance if a purification step is re-quired as this normally requires much more hands-on manipulation, it increases the likelihood of accidents, and also generates a lot more radioactive waste. Both pharmaceutical and radiation safety issues must be considered in deciding where the labelling will be performed. While a shielded fume cupboard will normally provide adequate radiation protection, it does not provide an environment of adequate pharmaceutical quality. The reverse is true of most laminar-flow workstations. The best solution to this problem is the use of a closed glove-box or isolator such as those produced by Amercare. These are entirely closed systems swept with HEPA-filtered air which is subsequently ducted externally and this provides a high quality environment which protects both the product and the operator.

The labelling procedure itself should be kept as simple as possible and the use of remote handling or automated devices should be explored. In order to reduce the possibility of spillages an entirely closed labelling and purification system should be devised wherever possible.

4.4 Iodine-131

Two different approaches have been taken to labelling antibodies with therapeutic levels of I-131. The first approach is to use a very simple but efficient labelling method which removes the need for a purification step. N-bromo-succinimide is used as the oxidizing agent and this is effectively titrated with the antibody and radioiodine in order to achieve the highest possible labelling efficiency—normally greater than 90% (25). Significant amounts of free iodine do remain in the preparation and are administered to the patient when this method is used, but, since approximately 30% of the radiolabelled antibody will be metabolized to release free iodide and other low molecular weight catabolites in the first 24 hours after administration, the co-injection of small amounts of radioiodine has little effect on tumour and whole-body dosimetry.

The alternative approach is to semi-automate the labelling and purification process in order to reduce the radiation dose received by the operative. Although such a system is necessarily more expensive, time-consuming, and problem-prone it does produce a better quality product and, with modifications, can be used for radiolabelling antibodies with a variety of different radionuclides. The degree of automation to be applied will vary and will depend upon the local circumstances and available facilities. While it is clearly possible to fully auto-mate the process using electronically controlled valves, compressed air lines, and even robotics, such resources are likely to be beyond the reach of most researchers, however, an extremely robust 'low-tech' approach to automation of radioiodination was developed by Coulter Pharmaceutical for use in Phase I/II trials of Iodine-131 B1 antibody in Lymphoma. This method relies on the use of syringes, three-way taps, and evacuated vials to perform the liquid transfers. It has the added advantages that it provides a completely closed sterile system, and can therefore be performed in a non-aseptic environment such as a fume cupboard, and uses entirely disposable materials which can be discarded after the procedure. A diagrammatic outline of the method is shown in *Figure 3*.

4.5 Yttrium-90

Yttrium-90 is a relatively short-lived radionuclide which emits high energy beta radiation and has therefore been widely employed in clinical trials of solid

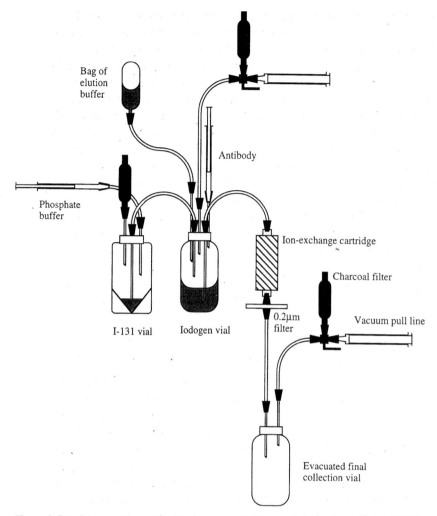

Figure 3 Possible arrangement for semi-automated therapeutic iodinations of monoclonal antibodies.

tumours. It is produced in a generator system following decay of strontium-90. While it is possible to set up a Sr-90/Y-90 generator in a hospital environment, radiation safety issues and difficulties in scale-up mean that commercial suppliers will be the source for most research groups. As indicated above, the chemistry of yttrium is similar to that of other trivalent metallic radionuclides such as indium-111, and the labelling methods employed use similar bifunctional chelate approaches based on DTPA. However, an important difference between indium and yttrium is the manner in which it is complexed by the DTPA. While indium requires seven co-ordinating atoms to provide a stable metal–chelate complex, yttrium requires eight. This means that DTPA derivatives in which one of the carboxylate groups is used to link the chelator to the antibody are not appropriate for yttrium-90 which needs all five carboxylates plus the three nitrogens for stability. For use with yttrium, therefore, the reactive group used for antibody conjugation must be substituted on the backbone of the chelator as illustrated in *Figure 1b*. This renders the synthesis of the molecule more complex and, since such compounds are not routinely available from commercial sources, means that a considerable degree of synthetic chemistry expertise is required. When therapeutic radionuclides are employed a high degree of *in vivo* stability is even more essential and this is particularly true when, as is the case with yttrium-90, any radioactivity which does dissociate is taken up by the bone and adds significantly to the radiation dose received by the bone marrow (26). In attempts to improve the therapeutic index of these immunoconjugates, chelators with higher stability constants have been employed. These are the macrocyclic chelators based on DOTA (see *Figure 1d*) which, by providing a cage within which the metal is protected from attack by competing cations, produces a complex with high thermodynamic and kinetic stability (27). The principal disadvantage of the macrocycles is that, while dissociation is very slow, the same is also true of the rate of association. Labelling efficiencies of DOTA conjugates can therefore be low unless extended incubation times and elevated temperatures are used.

4.6 Other therapeutic radionuclides

Many reports of, mostly pre-clinical, studies with other cytotoxic radionuclides have been published, but, since their use is largely restricted by their poor availability, they will not be discussed in detail here. Many of these radionuclides employ similar chemistry to that described in the previous sections, DTPA and DOTA provide stable co-ordination sites for many of these elements including scandium (28), lutetium (29), and samarium (30), while others such as rhenium (31) and copper (32) require more specific bifunctional chelators. A recent development in targeted radioimmunotherapy has been the clinical application of the alpha emitter bismuth-213 (33). For such toxic radioisotopes, a high degree of *in vivo* stability is paramount, but the short physical half-life of the radionuclide means that DOTA is not ideal. Accordingly, novel DTPA analogues such as cyclohexyl DTPA (see *Figure 1c*) which provide a pre-organized co-ordination site have been developed and these show promise as universal bifunctional chelators of the future. Not all therapeutic radionuclides are metals, however, and some candidates require a different type of chemistry. The alpha emitter astatine has properties very similar to those of iodine and can be incorporated using similar techniques to those described above. However, because of the reduced stability of the At-C bond and the high potential toxicity of the radionuclide, those approaches which offer enhanced *in vivo* stability are to be preferred (34). Phosphorus-32 is a radionuclide with a long history of use in nuclear medicine which also has potential for targeted radioimmunotherapy and methods of radiolabelling antibodies based on the use of phosphokinases have been described (35).

5 Antibody fragments and genetically engineered constructs

Because of the pharmacokinetic drawbacks of intact immunoglobulins, there is interest in the use of smaller antibody-based molecules such as conventional F(ab')₂ and Fab fragments and the newer engineered molecules such as sFv. Most of the radiolabelling methods described above are also directly applicable to these fragments, the exception being the direct technetium labelling methods which use reduced interchain disulfide bonds as their labelling site. This technique can only be used to label intact immunoglobulins and Fab' fragments. If used for F(ab')₂ fragments then it will result in cleavage of the bivalent molecule to the Fab' monomer. Other engineered constructs either lack redundant disulfide bridges or else they are required to maintain the configuration of the molecule and their reduction would result in a loss of stability or antigen binding. A possible solution to this problem is to engineer a direct labelling site into the antibody at a point where it will not interfere with antigen interactions (36). However this approach is not without its own problems and has achieved varying success in different laboratories.

6 Quality control of radiolabelled antibodies

The effective use of radiolabelled antibodies for any purpose, but especially for use in patients, requires that they are of the quality necessary for their intended application. Most importantly they should contain low levels of radioactive and non-radioactive impurities, the radiolabel should be bound to the antibody with a stability which ensures that it will not dissociate during the conditions of its

Table 4 Suggested quality control tests for radiolabelled antibodies

1. Radiolabelling development

Parameter	Method	Specification
Radiolabelling efficiency	TLC/HPLC	As high as routinely possible
Radiochemical purity	TLC/HPLC	Greater than 95%
Stability in solution	24 h storage in solution at 20 °C	< 5% loss in purity
Serum stability	4 h incubation in fresh serum at 37 °C	< 10% loss in purity
Antigen binding	Direct radioligand binding assay	> 70% immunoreactive fraction
Sterility	Pharmacopoeial sterility test	No growth
Apyrogenicity	Limulus test	< 7.5EU per dose

2. Final product

Parameter	Method	Specification
Radiochemical purity	TLC/HPLC	Greater than 95%
Antigen binding	Direct radioligand binding assay	> 70% immunoreactive fraction
Sterility	Pharmacopoeial sterility test	No growth

use, and the radiolabelling process should not compromise the ability of the antibody to interact with its respective antigen. Any antibodies to be administered to patients must in addition be pharmaceutically acceptable in terms of sterility, apyrogenicity, and freedom from particulate contamination. It is therefore recommended that a number of quality control tests be performed on the labelled antibody, either during development of the labelling process or on the final product before it is used. The tests listed in *Table 4* are suggested by the author as appropriate for a system of quality control suitable for radiolabelled antibodies intended for human use.

References

1. Pressman, D. and Keighley, G. (1948). *J. Immunol.*, **59**, 141.
2. Slater, R. J. (1990). In *Radioisotopes in Biology: a practical approach* (ed. D. Rickwood and B. D. Hames), p. 200. IRL Press, Oxford.
3. Dewanjee, M. K. (1992). *Radioiodination: theory, practice and biomedical applications.* Kluwer Academic Publishers, Dordrecht.
4. Fraker, P. J. and Speck, J. C. (1978). *Biochem. Biophys. Res. Commun.*, **80**, 849.
5. Hunter, W. M. and Greenwood, F. C. (1962). *Nature*, **194**, 495.
6. Lindmo, T., Boven, E., and Cuttita, F. (1984). *J. Immunol. Methods*, **27**, 77.
7. Morrison, M. and Bayse, G. S. (1970). *Biochemistry*, **9**, 2995.
8. van der Laken, C. J., Boerman, O. C., Oyen, W. J., van de Ven, M. T., Chizzonite, R., Corstens, F. H., *et al.* (1997). *J. Clin. Invest.*, **100**, 2970.
9. Chakrabarti, M. C., Le, N., Paik, C. H., De Graff, W. G., and Carrasquillo, J. A. (1996). *J. Nucl. Med.*, **37**, 1384.
10. Sartor, J., Guhlke, S., Tentler, M., and Biersack, H. J. (1998). *J. Nucl. Med.*, **39**, 143P.
11. Garg, P. K., Garg, S., and Zalutsky, M. R. (1993). *Nucl. Med. Biol.*, **20**, 379.
12. Badger, C. C., Wilbur, D. S., Hadlee, S. W., and Fritzberg, A. R. (1990). *Nucl. Med. Biol.*, **17**, 381.
13. Hnatowich, D. J., Childs, R. L., Lanteigne, D., and Najafi, A. (1983). *J. Immunol. Methods*, **65**, 147.
14. Gansow, O. A. (1991). *Nucl. Med. Biol.*, **18**, 369.
15. Eckelman, W. C. (1995). *Eur. J. Nucl. Med.*, **22**, 249.
16. Mather, S. J. and Ellison, D. (1990). *J. Nucl. Med.*, **31**, 692.
17. Rhodes, B. A. (1991). *Nucl. Med. Biol.*, **18**, 667.
18. Thakur, M. L., DeFulvio, J., Richard, M. D., and Park, C. H. (1991). *Nucl. Med. Biol.*, **18**, 227.
19. Stalteri, M. A. and Mather, S. J. (1996). *Eur. J. Nucl. Med.*, **23**, 178.
20. Ellman, G. L. (1958). *Arch. Biochem. Biophys.*, **74**, 443.
21. Claessens, R. A. M. J., Koenders, E. B., and Oyen, W. J. G. (1996). *Eur. J. Nucl. Med.*, **23**, 1536.
22. Abrams, M. J., Juweid, M., tenKate, C., Schwartz, D. A., Hauser, M. M., Gaul, F. E., *et al.* (1990). *J. Nucl. Med.*, **31**, 2022.
23. Babich, J. W. and Fischman, A. J. (1995). *Nucl. Med. Biol.*, **22**, 25.
24. Kasina, S., Sanderson, J. A., Fitzner, J. N., Srinivasan, A., Rao, T. N., Hobson, L. J., *et al.* (1998). *Bioconjug. Chem.*, **9**, 108.
25. Mather, S. J. and Ward, B. G. (1987). *J. Nucl. Med.*, **28**, 1034.
26. Brechbiel, M. W. and Gansow, O. A. (1991). *Bioconjug. Chem.*, **2**, 187.
27. Camera, L., Kinuya, S., and Garmestani, K. (1994). *Eur. J. Nucl. Med.*, **21**, 640.
28. Anderson, W. T. and Strand, M. (1985). *Cancer Res.*, **45**, 2154.

29. Schlom, J., Siler, K., Milenic, D. E., Eggensperger, D., Colche, R. D., Miller, L. S., *et al.* (1991). *Cancer Res.*, **51**, 2889.

30. Kraeber-Bodere, F., Mishra, A., Thedrez, P., Faivre-Chauvet, A., Bardies, M., Imai, S., *et al.* (1996). *Eur. J. Nucl. Med.*, **23**, 560.

31. van Gog, F. B., Visser, G. W., Stroomer, J. W., Roos, J. C., Snow, G. B., and van Dongen, G. A. (1997). *Cancer*, **80**, 2360.

32. Rogers, B. E., Anderson, C. J., Connett, J. M., Guo, L. W., Edwards, W. B., Sherman, E. L., *et al.* (1996). *Bioconjug. Chem.*, **7**, 511.

33. McDevitt, M. R., Sgouros, G., Finn, R. D., Humm, J. L., Jurcic, J. G., Larson, S. M., *et al.* (1998). *Eur. J. Nucl. Med.*, **25**, 1341.

34. Zalutsky, M. R., Stabin, M. G., Larsen, R. H., and Bigner, D. D. (1997). *Nucl. Med. Biol.*, **24**, 255.

35. Band, H. A., Creighton, A. M., and Britton, K. E. (1995). *Tumour Targeting*, **1**, 85.

36. Huston, J. S., George, A. J., Adams, G. P., Stafford, W. F., Jamar, F., Tai, M. S., *et al.* (1996). *Q. J. Nucl. Med.*, **40**, 320.

Chapter 10

Non-radioactive antibody probes

G. Brian Wisdom

School of Biology and Biochemistry, The Queen's University, Medical Biology
Centre, Belfast BT9 7BL, U.K. – Northern Ireland.

1 Introduction

The high specificity of monoclonal antibodies makes them valuable reagents,
however, to exploit them fully requires a means of detecting them at low concentration. This is usually done by labelling the antibody in one of various ways.
Radioisotopes are widely used (Chapter 9) as are non-isotopic labels such as
enzymes, fluorogenic and luminogenic molecules, as well as various low molecular weight groups that can themselves be easily detected by a secondary
reagent. The non-isotopic labels provide low detection limits in a wide range of
techniques without the complications of handling radioisotopes, although it
should be noted that all the non-isotopic labelling procedures have the hazards
associated with protein-modifying reagents.

Labelled monoclonal antibodies are employed in detecting, assaying, and
locating their complementary epitopes and are the key reagent in the techniques of immunoassay, immunocytochemistry, immunoblotting, and antibody-
based cell sorting.

In most of these applications one can use a direct or indirect approach to
labelling. In the direct approach the monoclonal antibody itself is labelled and
the label is detected after the antibody–antigen reaction has taken place. In the
indirect approach the primary, monoclonal antibody is unmodified and is detected by a labelled, secondary reagent. This secondary reagent can be an anti-
immunoglobulin (Ig) antibody, e.g. rabbit anti-mouse IgG, or one of the bacterial
Ig binding proteins, e.g. Protein G. The use of a secondary reagent adds an extra
step to the protocol and may increase the levels of background but the indirect
approach has advantages:

- It avoids damage to the primary antibody

- It diminishes the need to purify the primary antibody

- It increases the sensitivity of detection by allowing a greater concentration of
label at the site of reaction

- There are many labelled secondary reagents available from commercial sources

- A single secondary reagent can be used to detect primary antibodies of
different specificities

The labelling of antibodies with the low molecular weight molecules, biotin and digoxigenin (DIG), requires a secondary reagent for their detection and measurement. In the case of DIG a labelled anti-DIG antibody is used while for biotin, labelled forms of the biotin binding proteins streptavidin and avidin are employed.

This chapter provides protocols for the most convenient methods of labelling mouse IgG monoclonal antibodies with enzymes (alkaline phosphatase and peroxidase), a fluorescent molecule (fluorescein), biotin, and DIG. A general protocol for the evaluation of the labelled monoclonal antibody is also given.

2 Choice of label

The choice of label will depend on the application envisaged. Enzymes are widely applicable; they are used in assays, such as ELISAs, and for detection of antigen blotted or dotted on membranes or embedded in tissue sections. Enzyme labels have also be used for the location of antigen in electron microscope sections but gold labelled antibodies (Chapter 11) are now extensively employed for this purpose. Antibodies labelled with a fluorescent molecule are used in assays and for the detection of antigens in tissue sections and also for flow cytometry and fluorescence-activated cell sorting.

The small labels, biotin and DIG, have wide applicability; they are stable and provide a valuable alternative in cases where an enzyme label interferes with the binding ability of the antibody. As these labels are detected by a secondary reagent that can be labelled in various ways there is great flexibility in their use; for example, a biotin labelled monoclonal antibody can be detected with commercially available conjugates of streptavidin with various enzymes, fluorescent molecules, or gold.

3 General aspects of labelling

3.1 The labelling reactions

The protocols given in this chapter utilize the ε-amino groups of the antibody's lysine side-chains as targets for labelling. There are about 120 of these groups in IgG and a small fraction can be modified without detriment to the antibody's binding ability. In the cases of the low molecular weight labels, fluorescein, biotin, and DIG, N-hydroxysuccinimide ester derivatives are used to react with the amino groups and form amide bonds with the protein. The ε-amino group reacts in its unprotonated form (pK 10 approximately) however a sufficient proportion of this form is available at lower pHs thus allowing the labelling reaction to be performed under moderate conditions. The use of a pH in the 7.5–8.5 range also diminishes the hydrolyses of N-hydroxysuccinimide esters.

Other sites on IgG can be used for labelling. The oligosaccharides (which are chiefly in the Fc part of the molecule) can be oxidized and the resulting aldehyde groups reacted with hydrazide derivatives of, for example, biotin (1).

Another option is to use maleimide derivatives to label via a thiol group. Free thiol groups can be introduced by a variety of reagents (usually by modifying lysine side-chains) or they can be formed by the partial reduction of the IgG. This approach can be valuable when Fab or Fab' fragments are labelled.

The chemistry and procedures involved in labelling proteins are covered in detail in several texts (2–4).

3.2 The monoclonal antibody

The antibody should be highly purified to avoid complications due to the labelling of other proteins. (Methods for the purification of monoclonal antibodies are described in Chapter 7.) For most of the protocols the antibody should be in phosphate-buffered saline solution, pH 7.5. Buffer exchange can achieved by dialysis or ultrafiltration; there are various techniques for handling small volumes (5, 6), for example by placing a piece of sealed glass tubing in the dialysis bag. Commercial devices are also available for the efficient dialysis of small volumes. Gel filtration on a small column containing a matrix with a narrow pore size, such as Sephadex G25, is a rapid method of buffer exchange however it dilutes the sample. The concentration of the IgG can be estimated rapidly by measuring its absorbance at 280 nm; 1 mg/ml = 1.3 A approximately.

Because the labelling methods described exploit amino groups on the antibody it is vital that all interfering molecules, such as ammonium sulfate, sodium azide, Tris, and other amine-containing compounds, are absent from the antibody and other solutions.

3.3 Scale and ratios

The labelling protocols are based on 1–2 mg of the monoclonal antibody usually in a volume of 1 ml. The volume can be increased but there are advantages in keeping the antibody concentration high; this reduces losses due to inactivation and adsorption. The ratio of label to antibody is very important. It must be sufficient to provide a strong signal in the detection system but not so high that the antigen binding activity of the antibody is impaired. The ratios given in the protocols allow for the incorporation of a small number of label molecules and are generally suitable for the effective labelling of monoclonal IgG antibody. Optimization may be required in some cases and particularly if IgM antibodies are being labelled.

3.4 Purification and storage of the labelled antibody

After labelling it is necessary to terminate the reaction by the addition of an amine molecule, such as ethanolamine, and to remove any excess labelling reagent and by-products. (The presence of excess unlabelled antibody is generally not a problem.) Gel filtration, dialysis, and ultrafiltration may be used (see Section 3.2). An antimicrobial agent, e.g. sodium azide, and bovine serum albumin (BSA) are usually added to protect the labelled antibody and prevent losses; BSA is also sometimes added to the buffer used in the purification of the conjugate by gel filtration.

4 Labelling with an enzyme

Several enzymes have been used to label antibodies however alkaline phosphatase from bovine intestinal mucosa and horseradish peroxidase are the most widely exploited. They both can react with a range of substrates which permits detection and measurement by light absorbance or reflectance (via products that form insoluble complexes for example on blotting membranes) and by fluorescence and luminescence. An important attribute of enzyme labels is that their coloured products are detectable by eye.

The labelling of IgG with alkaline phosphatase using glutaraldehyde as a crosslinker (7) is not efficient in terms of the yield of antibody and enzyme activity, however, the procedure is easy to carry out and very stable conjugates are produced (see *Protocol 1*). The glutaraldehyde reacts, in an oligomeric form, primarily with the proteins' amino groups to give secondary amine bonds. It is not usually necessary to purify these conjugates as they are complexes of several molecules of each component with little or no free enzyme or antibody remaining when the recommended proportions are used.

Protocol 1

Labelling antibody with alkaline phosphatase

Reagents

- Monoclonal antibody
- Alkaline phosphatase from bovine intestinal mucosa:[a] 2000 U/mg or greater with 2-nitrophenyl phosphate as substrate (M_r 140 000) (Sigma, Boehringer)
- PBS: 20 mM sodium phosphate buffer pH 7.5, 0.15 M NaCl
- Glutaraldehyde
- 50 mM Tris–HCl buffer pH 7.5, 1 mM $MgCl_2$, 0.02% NaN_3, and 2% BSA

Method

1. Add 1.0 mg of antibody in 200 μl of PBS to 3.0 mg of alkaline phosphatase.

2. Add 5% glutaraldehyde in water to the antibody slowly and with stirring to give a final concentration of 0.2%.

3. Stir the mixture at room temperature for 2 h.

4. Dilute the mixture to 1 ml with PBS and dialyse it against the same buffer (2 litres) at 4 °C overnight.

5. Dilute the solution to 10 ml with 50 mM Tris–HCl buffer pH 7.5, containing 1 mM $MgCl_2$, 0.02% NaN_3, and 2% BSA, and store at 4 °C.

[a] The enzyme is available at a concentration of 10 mg/ml in 30 mM triethanolamine buffer pH 7.6. (This buffer does not interfere with the labelling reaction.)

Peroxidase labelling of antibody gives highly active conjugates (8). In the method described (*Protocol 2*) the oligosaccharide groups in the peroxidase are oxidized with periodate and the aldehyde groups produced are then allowed to react with amino groups on the IgG. (The peroxidase has few free amino groups so self-coupling is not a significant problem.) Finally, the Schiff bases formed are reduced to give a stable amide bond. These conjugates contain several molecules of each component and are purified by gel filtration in a matrix that fractionates in the 10–1000 kDa range.

Protocol 2

Labelling antibody with peroxidase

Equipment and reagents

- Fractionation collector and chromatography column (approx. 1 × 60 cm)
- Monoclonal antibody
- Horseradish peroxidase:[a] 100 U/mg or greater with 2,2-azinobis[ethylbenzthiazoline-6-sulfonic acid] as substrate (M_r 44 000) (Sigma, Boehringer)
- 0.1 M sodium periodate
- 1 mM sodium acetate buffer pH 4.4

- 10 mM sodium carbonate buffer pH 9.5
- 0.2 M sodium carbonate buffer pH 9.5
- 4 mg/ml sodium borohydride (freshly prepared)
- Gel filtration medium: e.g. Sepharose CL-6B (Pharmacia), Bio-Gel A 0.5m (Bio-Rad), or similar
- PBS (see *Protocol 1*)
- BSA

Method

1. Dissolve 1 mg of peroxidase in 250 μl of water.
2. Add 50 μl of freshly prepared 0.1 M sodium periodate and stir the solution at room temperature in the dark for 20 min.
3. Dialyse the modified enzyme against 1 mM sodium acetate buffer pH 4.4 (2 litres) at 4 °C overnight.
4. Dissolve 2 mg of antibody in 250 μl of 10 mM sodium carbonate buffer pH 9.5.
5. Adjust the pH of the dialysed enzyme solution to 9.0–9.5 by adding 10 μl of 0.2 M sodium carbonate buffer pH 9.5, and immediately add the IgG solution.
6. Sir the mixture at room temperature for 2 h.
7. Add 40 μl of sodium borohydride solution and stir the mixture at room temperature for 2 h.
8. Purify the conjugate by gel filtration in PBS. Monitor A_{280} and A_{403}. Pool the fractions where the absorbances coincide in the first peak.
9. Add BSA (final concentration 2%) to the pool and store the solution in aliquots at −20 °C or lower.

[a] Peroxidase is inhibited by sodium azide consequently this antimicrobial agent must be avoided when using this label.

5 Labelling with fluorescein

There are many fluorophores used in antibody labelling such as rhodamine, coumarin derivatives, phycobiliproteins, and rare earth chelates, however fluorescein is probably the most widely used. Individual choice may be determined by the optical filters available for the microscope or other detector. *Protocol 3* describes the use of fluorescein-*N*-hyroxysuccinimide ester to label IgG amino groups (3). Fluorescein isothiocyanate may also be used for the same purpose but it requires harsher conditions. The reaction is quenched by the addition of ethanolamine and excess labelling reagent is removed by gel filtration.

Protocol 3

Labelling antibody with fluorescein

Reagents

- Monoclonal antibody
- 5(6)-Carboxyfluorescein-*N*-hydroxysuccinimide ester (M_r 473) (Boehringer)[a]
- PBS (see *Protocol 1*)
- Dimethyl sulfoxide (DMSO)
- 0.1 M ethanolamine pH 8.5
- Sephadex G25 (Pharmacia)[b]
- BSA

Method

1. Dissolve 1 mg of the antibody in 1 ml of PBS.

2. Prepare 20 mM carboxyfluorescein-*N*-hydroxysuccinimide ester in DMSO immediately prior to use.

3. Add 20 µl of the fluorescein derivative to the antibody slowly with stirring at room temperature.

4. Incubate the mixture at room temperature for 1 h.

5. Terminate the reaction by adding 0.1 ml of 0.1 M ethanolamine and incubate at room temperature for 15 min.

6. Remove excess fluorescein derivative by gel filtration[b] in Sephadex G25 equilibrated with PBS containing 0.1% BSA. The IgG is in the first, coloured peak.

7. Store the conjugate in the dark at 4 °C with 0.05% NaN_3 or in aliquots at −20 °C or lower.

[a] Fluorescein isothiocyanate is an alternative reagent however it requires 0.1 M sodium carbonate buffer pH 9.0, and a reaction time of 8 h at 4 °C.

[b] Ready-made Sephadex G25 columns are available for use with gravity feed (PD-10) or syringe (HiTrap) from Pharmacia. The PD-10 columns with a bed volume of approx. 9 ml are very suitable for processing samples of 1–2 ml. Similar crosslinked dextran columns are available from Pierce (Presto and Kwik columns).

6 Labelling with biotin

Biotin labelled antibodies are effectively detected using the biotin binding proteins avidin and streptavidin. Streptavidin is usually preferred as the secondary reagent because avidin is glycosylated and has a high pI and these properties can give rise to non-specific binding in some situations.

A variety of reagents is available for the biotinylation of proteins through various functional groups including the N-hydroxysuccinimide ester for labelling via amino groups. There are several forms of this ester. There is the 'long chain' version which has a 6-aminocaproate spacer; this facilitates reaction with streptavidin. There are also water soluble derivatives containing the sulfonate group; they avoid the use of solvents such as DMSO and dimethylformamide which may damage the antibody. *Protocol 4* describes the use of the 'long chain' biotinyl-ε-aminocaproic acid-N-hydroxysuccinimide ester and its sulfo derivative.

Protocol 4

Labelling antibody with biotin

Reagents

- Monoclonal antibody
- Biotinyl-ε-aminocaproic acid-N-hydroxysuccinimide ester (Reagent I, M_r 455) or biotinyl-ε-aminocaproic acid-3-sulfo-N-hydroxysuccinimide ester (Reagent II, M_r 556) (Pierce, Sigma)

- PBS (see *Protocol 1*)
- DMSO
- 0.1 M ethanolamine pH 8.5
- BSA
- Sephadex G25 (Pharmacia)[a]

Method

1. Dissolve 1 mg of the antibody in 1 ml of PBS.

2. Prepare the biotinylating reagent immediately prior to use. Dissolve Reagent I in DMSO at a concentration of 2 mg/ml or Reagent II in water at a concentration of 2 mg/ml.

3. Add 15 μl of Reagent I or 19 μl of Reagent II to the antibody solution slowly with stirring.

4. Incubate at room temperature for 2 h.

5. Terminate the reaction by adding 0.1 ml of 0.1 M ethanolamine and incubate for 15 min.

6. Remove the excess biotinylating reagent by gel filtration[a] in Sephadex G25 equilibrated with PBS. The IgG is in the first A_{280} peak.

7. Store the labelled antibody at 4 °C with 0.05% NaN_3 or in aliquots at −20 °C or lower.

[a] See footnote *b* in *Protocol 3*.

7 Labelling with digoxigenin (DIG)

DIG is a plant steroid and provides a valuable alternative to biotin. It is particularly useful in situations where the biotin–streptavidin system gives high backgrounds (often due to the presence of biotin-containing enzymes). DIG is available as an N-hydroxysuccinimide ester for coupling with amino groups and *Protocol 5* describes its use in labelling IgG. There is a range of commercially available mouse and sheep anti-DIG Fab labelled with various enzymes and fluorescent molecules for the detection of the DIG labelled primary antibody.

Protocol 5

Labelling antibody with digoxigenin (DIG)

Reagents

- Monoclonal antibody
- Digoxigenin-3-0-succinyl-ε-aminocaproic acid-N-hydroxysuccinimide ester (M_r 659) (Boehringer)
- DMSO
- PBS (see *Protocol 1*)
- 0.1 M ethanolamine pH 8.5
- Sephadex G25 (Pharmacia)[a]

Method

1. Dissolve 1 mg of the antibody in 1 ml of PBS.

2. Prepare the labelling reagent immediately prior to use by dissolving it at a concentration of 2 mg/ml in DMSO.

3. Add 24 μl of the reagent to the antibody solution slowly with stirring.

4. Incubate at room temperature for 2 h.

5. Terminate the reaction by adding 0.1 ml of 0.1 M ethanolamine and incubate for 15 min.

6. Remove the excess biotinylating reagent by gel filtration[a] in Sephadex G25 equilibrated with PBS containing 0.1% BSA. The IgG is in the first A_{280} peak.

7. Store the labelled antibody at 4 °C with 0.05% NaN_3 or in aliquots at −20 °C or lower.

[a] See footnote *b* in *Protocol 3*.

8 Evaluation of labelled monoclonal antibodies

Conjugates of IgG and label may be evaluated in several ways but the essential criterion is their performance in the planned application. Comparisons can be made by titrating the conjugate in the chosen technique and confirming its specificity for the antigen by immunoblotting or immunocytochemistry using appropriate controls. A general method of conjugate evaluation and comparison is to titrate it in an ELISA as described in *Protocol 6*. This involves adding dilutions of the labelled antibody to antigen immobilized in the wells of a microtitre plate and measuring the amount of label present. In the case of enzyme labelled monoclonal antibody this can be done directly but the other labels require an indirect approach, i.e. the use of a secondary reagent. However, if a microtitre plate reader capable of measuring fluorescence is available then fluorescein can also be measured directly.

Protocol 6

Titration of labelled antibody

Equipment and reagents

- Polystyrene microtitre plates
- Labelled monoclonal antibody
- Antigen
- Coating buffer: 50 mM sodium carbonate pH 9.6
- Blocking solution: 2% BSA in 20 mM Tris–HCl buffer pH 7.4, 0.15 M NaCl (TBS)
- TBS-Tween: TBS containing 0.05% Tween 20
- Enzyme labelled secondary reagents: anti-fluoroscein IgG labelled with alkaline phosphatase or peroxidase (Boehringer) streptavidin labelled with alkaline phosphatase or peroxidase (Boehringer, Vector) anti-DIG IgG labelled with alkaline phosphatase or peroxidase (Boehringer)

- Alkaline phosphate substrate solution: 2-nitrophenyl phosphate (10 mg) in 10 ml of 10 mM diethanolamine buffer pH 9.5, 0.5 mM $MgCl_2$
- Peroxidase substrate solution: 3,3',5,5'-tetramethylbenzidine (100 μl of a 10 mg/ml solution in DMSO) and 7.5 μl of 6% hydrogen peroxide (freshly prepared from 30% stock solution) in 10 ml of 0.1 M sodium acetate/citrate buffer pH 6.0
- Alkaline phosphatase stopping solution: 2 M NaOH
- Peroxidase stopping solution: 2 M H_2SO_4

Method

All volumes are 200 μl[a] and all incubations are at 37 °C unless stated otherwise. Controls should include 'no antigen' as well as 'no antibody'.

1. Coat the wells of the microtitre plate with antigen at 10 μg/ml[a] in coating buffer and incubate for 3 h or at 4 °C overnight.

2. Remove the contents of the wells by inversion and tapping the plate on paper towelling. Add blocking solution and incubate for 1 h.

3. Wash the wells with TBS-Tween. If an automatic washer is not available wash the wells by adding TBS-Tween (about 300 μl), inverting the plate and tapping it on paper towelling; repeat this twice.

4. Add dilutions of the labelled antibody in TBS-Tween and incubate for 2–3 h.

5. Wash the wells as in step 3.

6.[b] Add the enzyme labelled secondary reagent in TBS-Tween (diluted as recommended by the supplier) and incubate for 2 h.

Protocol 6 continued

7. Wash the wells as in step 3.

8. Add substrate solution and incubate at room temperature until the colour in the wells with the highest antibody concentration is of moderate intensity, then add 50 μl of stopping solution.

9. Measure the absorbance of the solutions: at 405 nm for alkaline phosphatase and 450 nm for peroxidase.

10. Plot the titration curve and determine the titre of the labelled antibody.

[a] If the amounts available are restricted the volumes of the solutions may be reduced to 100 μl and the concentrations of the antigen to 1 μg/ml.

[b] When measuring enzyme labelled antibody directly steps 6 and 7 are left out. When measuring fluorescein labelled antibody directly steps 6, 7, 8, and 9 are omitted; fluorescein is excited at 495 nm and emits at 525 nm.

Measurement of the incorporation of the label into the IgG molecule is not practical in most cases as the labels do not have a characteristic physical feature that can be easily measured when conjugated to a large protein. The exceptions are peroxidase and fluorescein. The haem group of peroxidase absorbs light at 403 nm and the ratio of this absorbance to that of the protein peak at 280 nm is used to characterize peroxidase preparations and their conjugates. Fluorescein has an absorbance peak at 495 nm ($\varepsilon = 68\,000$ M^{-1} cm^{-1}) and this can be used to calculate the molar concentration of the label and relate it to the concentration of the IgG.

It is possible to measure the change in the number of amino groups in the IgG with 5,5′-dithiobis(2-nitrobenzoic acid) (9) but as the degree of modification is very small this approach will not give accurate results.

References

1. O'Shannessy, D. J., Doberson, M. J., and Quarles, R. H. (1984). *Immunol. Lett.*, **8**, 273.
2. Hermanson, G. T. (1996). *Bioconjugate techniques*. Academic Press, London.
3. Garman, A. (1997). *Non-radioactive labelling*. Academic Press, London.
4. Wong, S. S. (1991). *Chemistry of protein conjugation and cross-linking*. CRC Press, Baco Raton.
5. Overall, C. M. (1987). *Anal. Biochem.*, **165**, 208.
6. Cholewa, O. M. and Hornemann, U. (1998). *BioTechniques*, **25**, 212.
7. Engvall, E. and Perlmann, P. (1972). *J. Immunol.*, **109**, 129.
8. Wilson, M. B. and Nakane, P. P. (1978). In *Immunofluorescence and related staining techniques*. W. Knapp, K. Hulubar and G. Wicks Eds., p. 215. Elsevier, Amsterdam.
9. Ellman, G. L. (1959). *Arch. Biochem. Biophys.*, **82**, 70.

Chapter 11

Immunogold probes for light and electron microscopy

Paul Monaghan* and David Robertson[†]
*Institute for Animal Health, Pirbright Laboratory, Ash Road, Pirbright, Woking, Surrey GU24 0NF, U.K.
[†]The Institute of Cancer Research, Haddew Laboratories, 15 Cotswold Road, Belmont, Sutton, Surrey SM2 5NG, U.K.

1 Introduction

The characteristics of monoclonal antibodies make them particularly suitable for electron microscope (EM) immunolabelling studies. At the high magnification of the EM, any non-specific labelling can be a real problem, and some of the best antibodies we have used are monoclonal antibodies. The principles of immunocytochemistry are essentially the same for both light and electron microscopy. A primary antibody is persuaded to bind to its antigen and this is then detected by a suitable marker. For electron microscopy, a number of markers have been used including ferritin and horseradish peroxidase/DAB reaction product. However colloidal gold is an almost perfect marker in this context and alternatives are rarely used. Colloidal gold is particulate and electron dense making it readily visible in the EM. In addition, it does not obscure the underlying cellular structure, making it easy to localize antigens not just to specific organelles, but regions within these structures. Colloidal gold can be formed in defined sizes and this has implications for its use, particularly in multiple labelling experiments.

Colloidal gold has another interesting characteristic and that is its ability to act as a nucleus for the deposition of silver. In the right solutions, silver grows upon the gold particles and the longer the incubation, the greater the silver deposition until the particle is large enough to be visible in the light microscope. For light microscopy, if the signal is strong enough, the silver enhanced gold can be seen as a small black dot. However, low levels of labelling are not so easy to see, and the signal is detected more easily in an epipolarized light microscope. This silver enhancement technique can be used as a dedicated light microscope (LM) method, or as a useful preliminary step in electron microscope localizations. Methods will be described for using silver enhancement for resin or thin cryosections, but the principles are identical if used for wax or frozen sections for light microscopy (1).

Although colloidal gold is readily conjugated to proteins, preparation of direct antibody/gold reagents is not often done. The simplest approach to labelling with colloidal gold is an indirect one, where the primary antibody is detected with either species-specific gold conjugate or Protein A or G gold conjugates. Again, the simplest approach is not to make conjugates but to buy them. Commercial reagents are readily available, reproducible, and long-lasting.

Preparing cells or tissues for light microscope immunocytochemistry is relatively straightforward. Sadly, the routine methods of fixing and embedding samples for electron microscopy are generally bad for antigens and relatively few antibodies will recognize their antigen in glutaraldehyde fixed and epoxy resin embedded material. A considerable literature has grown around possible methods to process material for EM immunocytochemistry which claim to retain both good ultrastructure and antigenicity. There are many variants of the basic methods which will be described, but the concepts remain the same. It is a disappointing fact there is no one single method which is applicable to all antigens and all situations, and it is generally accepted that there are three main approaches to the problem (with one other showing potential). The three main methods are low temperature embedding by the progressive lowering of temperature technique (PLT) (2, 3), thawed cryosections (4), and rapid freezing followed by freeze-substitution (5). Which method to choose? Review of the following protocols will help to determine some of the advantages and disadvantages of the methods, but pragmatism also plays a role. All these methods rely to some extent on specialized (and in some cases very expensive) equipment. What is readily available may well affect the choice of approach. The one imponderable in the decision making is the antibody. The behaviour of the antigen/antibody combination is the overriding factor, and choice of method is ideally made to optimize the labelling with the antibody of choice.

A brief review of the methods may help to clarify the problem. Two methods embed the samples in a resin and then the antigen is localized on the resin sections. The alternative method is the preparation of thin frozen sections which are thawed and immmunolabelled (thawed cryosections). The simplest method is the PLT method. Here, almost all the steps used to prepare samples for routine EM are used, but the damage to antigens is reduced as far as possible. Thus, fixation times are minimized and avoid high concentrations of glutaraldehyde and osmium altogether. Dehydration is at low temperature to reduce the extraction of cellular components, and resin embedding and polymerization are also at low temperature. This method is designed around the Lowicryl resins and these allow dehydration and embedding at temperatures down to $-60\,°C$. Cellular morphology in samples embedded by PLT can be excellent and the method is applicable to a wide range of samples (2, 3). Many antigens have been detected in samples processed by PLT, but if the antigen is sensitive to fixation, then this could negate the PLT approach. The only alternative in this case is to replace the immersion fixation with rapid freezing. Specialist equipment is needed to freeze samples fast enough to prevent ice crystals forming, but the protocols below allow rapidly frozen samples to be freeze-substituted and embedded in low temperature resin for immunocytochemistry. An alternative to freeze-substitution may be freeze-drying, and recent reports (6, 7) are promising in this area. Some antigens have been shown to be resistant to aldehyde fixation but are sensitive to the resin embedding stages of the PLT and rapid freezing/substitution methods and thawed cryosections would be the way to go in this case. Here the sample is stabilized with a short fixation, cryoprotected to prevent ice crystal growth, frozen, sectioned, and the sections thawed and immunolabelled.

Whatever approach is chosen, it is vital to include adequate controls in all labelling experiments. These should include positive controls, negative controls (both cells that should be negative and checks to ensure the reagents do not contribute non-specifically to labelling), and preferably a Western blot to ensure the protein detected in the EM experiments really is the one your antibody is supposedly directed against.

The methods described here are suggested starting points. For many projects they will work perfectly well. However, many variations of these basic methods have been devised to encompass different samples and experimental situations. Whatever the approach, however, the same basic principles of antibody labelling apply. The antigen has to be 'exposed' to the antibody without damaging cell ultrastructure, the antibody must bind specifically to the antigen, and the primary antibody is detected by a suitable (gold) conjugate. Simple really!

2 Pre-embedding labelling for SEM and TEM

Where a cell surface antigen is of interest, and particularly where the cells are grown *in vitro*, a labelling method which avoids all the problems of fixation and embedding can be used. Antibody and conjugate incubations can all be completed prior to fixation and embedding for TEM or critical point drying for SEM.

Protocol 1

Surface labelling adherent of cell cultures or tissue samples

Reagents

- PBS
- PBS/0.5% BSA (PBS/BSA)
- Primary antibody
- Colloidal gold conjugate: working dilution will usually be 1:50 to 1:100, but different manufacturer's reagents may vary

- 2% glutaraldehyde in phosphate buffer
- 2% paraformaldehyde in PBS (optional)
- High quality distilled water (e.g. double glass distilled or 18 MΩ resistivity, UHQ)

Method

All incubations should be below 4 °C to avoid membrane redistribution and/or antigen internalization. A brief (< 10 min) fixation in low concentration of paraformaldehyde (e.g. 2%) could be used instead of cooling if indicated. This very short fixation is sufficient to temporarily stabilize the cells without markedly affecting antigenicity.

Protocol 1 continued

1. Remove culture medium and wash with PBS for 5 min.

2. Incubate with primary antibody diluted in PBS/BSA for 60 min.

3. Wash 3 × 5 min in PBS.

4. Incubate with gold conjugate diluted in PBS/BSA for 60 min.

5. Wash 3 × 5 min.

6. Fix with glutaraldehyde as for routine TEM or SEM.

If the colloidal gold is to be silver enhanced, the samples should be washed (3 × 5 min) in UHQ water and then silver enhanced. More details of the enhancement times will be given in *Protocol 10* for labelling resin sections.

Process for TEM as for routine samples, although osmium should be avoided if the conjugate has been silver enhanced. For SEM, avoid osmium, dehydrate, critical point dry, and coat with carbon. Coating SEM samples which have been labelled with colloidal gold with a thin layer of gold is clearly not a good idea. Image the gold in the backscattered electron mode. The size of gold conjugate used depends on the resolution of the SEM; 5 nm or larger is detectable in a FEG SEM. Silver enhanced 5 nm gold can be detected in tungsten or LaB6 gun SEMs.

Protocol 2

Surface labelling of suspension cells

Equipment and reagents

- Bench centrifuge
- PBS/0.5% BSA
- Primary antibody
- Gold conjugate
- EM fixative

Method

1. Pellet the cells by centrifugation (500 *g*) for 10 min.

2. Resuspend in primary antibody diluted in PBS/BSA for 60 min.

3. Centrifuge, remove antibody, and resuspend cells in PBS. Repeat.

4. Resuspend in gold conjugate diluted in PBS/BSA for 60 min.

5. Centrifuge, remove antibody, and resuspend cells in PBS. Repeat.

6. Centrifuge and remove PBS.

7. Resuspend in fixative for TEM or SEM.

8. Silver enhance if required as in *Protocol 1*.

For SEM, the cells will adhere to a poly-L-lysine coated coverslip sufficiently to withstand fixation, dehydration, and critical point drying.

3 Thawed cryosections

This technique is analogous to the preparation of frozen sections for light microscopy. It is, however, a more difficult technique to master. The method owes its development to pioneering work of Tokuyasu (4) who established many of the basic steps in the procedure and reviewed by Griffiths (8). Whilst the cryoultramicrotomes have improved enormously in recent years, the technique is still much as worked out by Tokuyasu.

The method begins with a short aldehyde fixation. This is necessary to allow cryoprotection without osmotic damage and to stabilize the sample after thawing. Two fixatives are in common use; 4% paraformaldehyde and 2% paraformaldehyde plus 0.05% glutaraldehyde. Both fixatives should be buffered as for routine electron microscopy. Fixatives containing glutaraldehyde may give better preservation with some tissues, but on the other hand may give reduced antigen labelling with some antibodies.

Ice crystal damage will be caused by the freezing process without cryoprotection and this damage will be readily apparent. Samples are routinely cryoprotected in 2.3 M sucrose in PBS prior to freezing, but as the cryoprotectant also acts as the sample support during sectioning, alternatives such as poly(vinylpyrrolidone) (PVP) have been proposed which are said to improve sectioning properties. Whilst improvements have been reported with some samples, 2.3 M sucrose is a good starting point. It is important that the sample is fully infused with sucrose and it will become translucent when this has taken place. Freezing is now easy—place the sample onto a sample pin from the cryomicrotome and drop the pin into liquid nitrogen. Samples can be stored indefinitely at this stage, assuming the nitrogen bank does not fail.

The sectioning process takes place inside the cryochamber of the microtome and does require some practice. The first step is to ensure the sample contains what you are looking for. Trimming (at −80 °C or so depending on the sample) removes the sucrose layer covering the sample and produces a square block. It is now possible to collect 0.5 μm sections for LM. These sections are collected on 2.3 M sucrose drops on a wire loop and placed on a glass microscope slide to thaw. They can be stained with Toluidine blue, or by immunocytochemistry to check the presence of antigen. The two best markers for this step are colloidal gold or fluorescence. The advantage of using suitably silver enhanced colloidal gold is that a picture of the labelling obtained will help understand the labelling that will be seen at the EM level.

Trimming for thin sections produces a very small square block and is eased considerably by the use of a diamond trimming tool. Thin sections are cut at around −100 °C and this stage requires a little patience. Both diamond and glass knives can be used, but for ease (making high quality glass knives is not easy) diamond knives are to be preferred. Static electricity can be a problem, particularly with diamond knives and this can be alleviated with an antistatic device (Diatome).

Without doubt the step which follows is the biggest advance in cryosectioning for many years. One of the problems with cryosections is that the morphology of the sections in the EM is variable and often 'fuzzy'. This seems to result from the rapid surface tension changes that take place when the thawed section is transferred from 2.3 M sucrose to buffer for labelling. Collection of the sections from the cryochamber on droplets of a 50:50 mix of 2.3 M sucrose: methyl cellulose prevents the sections from being stretched, and gives superb preservation of cellular detail (9). This method works best on thin (< 70 nm) sections.

Immunolabelling is straightforward, and the final stage is to support and contrast the sections. The simplest approach is to use uranyl acetate to contrast the sections and support them during the drying stage with methyl cellulose.

Protocol 3

Preparation of thawed cryosections for LM

Equipment and reagents

- Specimen pins
- Glass knife
- Subbed slides: dip racks of microscope slides into a mixture of 1% gelatin plus 1% formalin in distilled water; drain the slides and dry in a warm oven
- Methyl cellulose: heat 100 ml UHQ water to 95 °C, add 2.0 g of 25cp methyl cellulose, and heat for 5 min with stirring. Cool and stir for four to five days at 4 °C. Clarify by centrifugation (60 000 g for 60 min). Store at 4 °C.

- 3% uranyl acetate in UHQ: store at 4 °C
- Fixative: 4% paraformaldehyde or 2% paraformaldehyde plus 0.05% glutaraldehyde in PBS
- 2.3 M sucrose
- Liquid nitrogen
- Eyelash on holder
- 0.1% Toluidine blue in 0.1% borax

Method

1. Fix samples in 4% paraformaldehyde or 2.05 (see PLT protocol, *Protocol 7*) for 60 min.

2. Wash samples in PBS and use immediately or store for a short time at 4 °C.

3. For cell suspensions, prepare as a block of 4% gelatin as for PLT.

4. Cryoprotect in 2.3 M sucrose. This will take from 60 min to overnight in rare cases. The sample should be translucent when fully infused with the cryoprotectant.

5. Place cryoprotected sample on a specimen pin with a drop of 2.3 M sucrose.

6. Remove most of the sucrose with a piece of filter paper and drop the pin into LN_2.

7. Place frozen specimen into cryomicrotome set at -80 °C for trimming.

8. Advance knife holder and trim the front of the block in 1 μm sections. The initial trimmings will be sucrose and will be crumbly. It is usually easy to see when the tissue is being sectioned.

Protocol 3 continued

9. Cut and move a 0.5 μm section away from the knife edge with an eyelash and pick up on a droplet of sucrose.

10. Thaw the drop and touch onto a subbed slide.

11. Gently wash away the sucrose with PBS.

12. For routine LM, stain sections with Toluidine blue and carefully wash.

Protocol 4

Immunolabelling 0.5 μm sections for LM

Equipment and reagents

- Dako pen
- PBS/0.5% BSA
- PBS
- Primary antibody
- Gold conjugate or fluorescent conjugate

- UHQ water
- Silver enhancer
- 1% Toluidine blue in 1% borax

Method

1. Mark around the sections with a Dako pen or similar hydrophobic pen.

2. Block non-specific binding with PBS/0.5% BSA or similar for 15–30 min.

3. Incubate in primary antibody diluted in PBS/BSA for 1–2 h.

4. Wash 3 × 5 min in PBS.

5. Incubate in colloidal gold or fluorescent conjugate diluted in PBS/BSA for 1–2 h.

6. Wash 3 × 5 min in PBS.

7. Mount fluorescent labelled sections in aqueous mountant.

8. Wash with UHQ water and silver enhance the colloidal gold label (see *Protocol 10*).

9. Stain colloidal gold labelled sections with Toluidine blue, wash, and air dry.

Protocol 5

Preparation of thawed cryosections for EM immunolabelling

Equipment and reagents

- Glass or diamond knife
- Eyelash on holder
- 2.3 M sucrose:methyl cellulose 1:1 mixture

- Copper wire loop
- Carbon coated, formvar grids
- PBS

Protocol 5 continued

Method

1. Select a suitable area for thin sections and trim the sides of the block. This can be done on the sides of a glass knife or with a diamond trimming tool. The smaller the block the more likely it is to section well.

2. Cool down to cutting temperature ($-90\,°C$ to $-100\,°C$ is suitable for many samples).

3. Cut sections and remove from the knife-edge with an eyelash.

4. Collect sections on a drop of 50:50, 2.3 M sucrose:methyl cellulose.

5. Thaw and place onto a formvar coated, carbon coated grid.

6. Place grid onto a drop of PBS.

If the block being cut contains the area of interest and the immunolabelling is satisfactory, it is time to cut thin sections for EM immunolabelling

The grids of sections can be stored for a short time, but will usually be immunolabelled soon after sectioning. For labelling thawed cryosections, use 5 nm or 10 nm gold conjugates. 1 nm gold conjugates may give improved labelling on cryosections but will require the additional step of silver enhancement which can sometimes be unreliable on these sections.

Protocol 6

Immunolabelling thawed cryosections

Equipment and reagents

- Copper wire loops approx. 4.0 mm in diameter
- Filter paper
- Syringe needle
- PBS/0.5% BSA
- Primary antibody
- Gold conjugate
- UHQ water
- 3% uranyl acetate
- 2% methyl cellulose

Method

1. Incubate in PBS/0.5% BSA or similar for 15–30 min.

2. Incubate in primary antibody diluted in PBS/BSA for 1–2 h.

3. Wash 3 × 5 min in PBS.

4. Incubate in gold conjugate for 1–2 h.

5. Wash 3 × 5 min in PBS.

6. Add 100 µl 3% uranyl acetate to 900 µl of 2% methyl cellulose and mix gently.

7. Place 2 × 50 µl drops per section on Fuji film (or similar) on ice.

8. Place grids on the first drop. Leave for 30 sec. Transfer to second drop and leave (in dark) for 20 min.

Protocol 6 continued

9. Pick up grids in a film of methyl cellulose within the copper wire loops.

10. Drain off excess methyl cellulose on a filter paper and leave to dry.

11. Carefully remove grid from the loop with the sharp tip of a syringe needle.

4 Progressive lowering of temperature embedding (PLT)

The alternatives to preparing and labelling thawed cryosections are the two resin embedding methods; PLT and freeze-substitution. The simplest to do is the PLT method which can be completed without the necessity of expensive equipment.

Two fixatives are commonly used for the PLT method. The less antigenically damaging one is 4% paraformaldehyde. However with some samples the morphology of PLT embedded and sectioned material is not sufficiently good, and improved preservation can then be achieved with an alternative fixative which has a small amount of glutaraldehyde. A fixative containing 2% paraformaldehyde plus 0.05% glutaraldehyde (2.05) will usually give better preservation, but may adversely affect some antigens.

Solid tissues should be fixed as rapidly as possible in 4% paraformaldehyde or 2.05. Suspension samples (tissue culture cells, microbiological samples, or very small samples) need to be held in a matrix through the processing steps. Centrifuge the samples (500 g) and discard the supernatant. Resuspend in 4% gelatin in PBS and centrifuge again. Remove the supernatant and cool the centrifuge tube on ice. The gelatin will set and hold the cells in a pellet which can be removed into cold PBS and carefully cut up into suitably small pieces. Plastic centrifuge tubes can be cut off above the pellet to ease its removal.

Protocol 7

Embedding by the progressive lowering of temperature method

Reagents

- PBS or L15: Leibovitz L15 air-buffered tissue culture medium plus 4% FCS

- 4% paraformaldehyde or 2.05 fixatives

Method

1. Fix sample for 60 min.

2. Rinse two or three times in PBS or L15 until colour no longer changes.

3. Chill on ice and continue (or store overnight at 4°C in PBS or L15 if necessary).

Protocol 7 continued

4. Dehydrate in ethanol:
   - 30% ethanol for 30 min at 4 °C
   - 55% ethanol for 30 min at −15 °C
   - 70% ethanol for 30 min at −30 °C
   - 100% ethanol for 30 min at −50 °C
   - 100% ethanol for 30 min at −50 °C

5. Prepare Lowicryl HM20 resin to manufacturer's instructions, and mix by bubbling nitrogen gas through the solution.

6. Infiltrate with resin:
   - 25% resin/75% ethanol for 60 min
   - 50% resin/50% ethanol for 60 min
   - 75% resin/25% ethanol for 60 min
   - 100% resin for 60 min
   - 100% resin overnight
   - 100% resin for 60 min

7. Embed and UV polymerize for 48 h at −50 °C, warm to 20 °C at 5 °C per hour, and polymerize for a further 24 h.

The polymerization times will require optimization for any particular system depending upon the intensity of the lamp and the distance from the lamp. Depending on the system used, blocks may come out clear, pink, or even pink at the top and clear at the bottom. This appears to be unimportant as the pink coloration disappears with time. Labelling of Lowicryl blocks will be the same for PLT processed or rapidly frozen samples, and LM and EM labelling is described in *Protocols 10* and *11*.

5 Rapid freezing

Immersion fixation is slow and allows redistribution of ions and antigens. In contrast, rapid freezing is fast, allows visualization of rapidly occurring events, and does not allow movements of cellular components. So if it is so much better, why is it not the only method of sample stabilization? First it is tricky to get right, with the exception of high pressure freezing it will only freeze one cell layer without ice crystal damage, and finally, freeze-substitution and low temperature embedding cannot be considered to be routine techniques. None the less, the morphology which can be obtained with this method is excellent.

There are a number of methods described for rapid freezing, but for immuno-cytochemistry, plunge freezing, impact freezing, and high pressure freezing are probably the best methods. For the purposes of this chapter a method will be described for impact freezing. Techniques will require modification for high pressure freezing, and the interested reader is referred to a recent publication on immunocytochemical application of this freezing method (10).

Protocol 8

Rapid freezing of suspension samples

Suspension samples will freeze well without any form of support, but during freeze-substitution they will disperse into the substitution medium. A dilute gelatin solution surrounding the sample will prevent this.

Equipment and reagents

- Freezing device
- Bench centrifuge
- Small nylon washers
- 4% gelatin in PBS

Method

1. Centrifuge (500 g) the sample into a pellet.

2. Resuspend in 4% gelatin.

3. Re-centrifuge and remove almost all the supernatant. Resuspend in gelatin.

4. Introduce a small amount of the resulting suspension of sample plus gelatin into the nylon washer on the holder of the freezing device. The sample volume should be sufficient to form a meniscus above the level of the washer.

5. Freeze.

Frozen samples can be stored under liquid nitrogen indefinitely. The method for freezing solid tissues is essentially the same as described in *Protocol 8*. Only the surface of the sample which contacts the freezing surface will be well frozen, however, and even then only 10–15 μm of the sample will be free from ice crystal damage. It is important therefore that the surface to be frozen is as flat as possible, not covered by a layer of fluid, and at the other extreme not allowed to dry.

Precise details of freezing will be dictated by the particular freezing product in use. Following freezing, the samples are freeze-substituted and low temperature embedded. Many protocols have been described for freeze-substitution. Some will suit particular specimens better than others. In general, substitution with acetone containing 1% osmium tetroxide will give superb morphology, but freeze-substitution with fixatives in the substitution medium is not a good idea for immunocytochemical studies. Substitution with methanol is rapid (< 18 h) but leaves the sample looking extracted. Substitution in pure acetone is slow ($>$ five days) but gives good morphology. A protocol for freeze-substitution in acetone which has been tested on a variety of samples will be given as an example. Resin embedding is best undertaken in one of the Lowicryl resins, but no firm rules can be given about the choice of Lowicryl resin. The protocols will be given for HM20.

The acetone to be used for freeze-substitution must have the minimum water content possible. 1–2% water content will prevent substitution (11).

Protocol 9

Freeze-substitution in pure acetone and Lowicryl embedding for immunocytochemistry

Equipment and reagents

- Dried acetone: dry 5A molecular sieve in a container in dry oven at about 150 °C and allow to cool in a desiccator; add the acetone and leave for 24 h

- Dried ethanol
- Lowicryl HM20
- Nitrogen gas

Method

1. Pre-cool the dried acetone to −90 °C. There is some evidence that the transfer from liquid nitrogen temperature to substitution temperature should not be too rapid, otherwise ice crystal growth may occur.

2. Introduce sample to substitution medium at −90 °C for 24 h.

3. Warm samples to −80 °C at 3 °C per hour and leave for five days. Change acetone at least once.

4. Warm samples at 3 °C per hour to −50 °C.

5. Incubate in dried ethanol for 1 h.

6. Infiltrate the samples with Lowicryl HM20 and polymerize as described in *Protocol 7* for PLT.

6 Immunocytochemistry of resin sections

This step relies on the silver enhancement of colloidal gold for detection. To keep background to a minimum, sections are washed in water containing a minimum of ions—which can act as nucleation sites for the silver—before and after the enhancement.

Resin blocks prepared by PLT or freeze-substitution are treated the same for both LM and EM immunocytochemistry. In the early stages of an investigation, much of the work of defining antibody dilutions, and positive and negative controls can be undertaken on 0.5 μm sections for light microscopy. In addition, the larger sections for the LM give a better overview of the labelling given by the antibody. When all the preliminary investigations are complete, the area of the block of interest can be chosen for EM studies. The trick is to use for the LM studies the gold conjugate that will be used for EM studies (usually 5 nm) and silver enhance sufficiently for it to be seen in the light microscope (*Figure 1*). For labelling EM sections the same conjugate is used at the same concentration, but with little or no enhancement (*Figures 2a, b*).

Figure 1: please see plate section between pages 226–227.

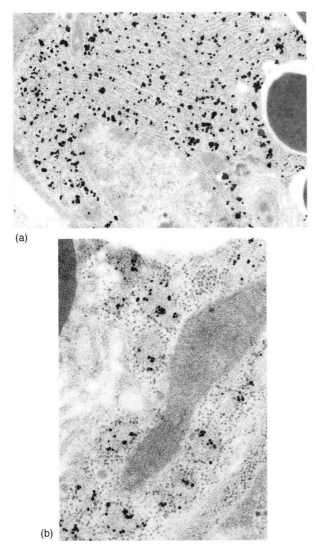

(a)

(b)

Figure 2 Mouse small intestine processed by PLT into Lowicryl HM20. The rough ER is labelled with an anti-PDI (protein disulfide isomerase). (a) Low power view: the 5 nm gold conjugate was silver enhanced for 12 min. Magnification: × 21 000. (b) Higher power view where the 5 nm gold was enhanced for 6 min. Magnification: × 62 000.

If the antigen is present in high concentration, the silver enhanced gold will appear as black particles on the section when viewed in a conventional microscope. Lower levels of labelling will be more difficult to detect and are best imaged by epipolarized light microscopy (12). In this relatively inexpensive method, the silver enhanced colloidal gold is seen as an intense light blue dot against a black background. The epipolarized light image is illustrated in *Figure 1a*, and the combined epipolarized and transmitted light image is seen in *Figure 1b*. Between them, these two images illustrate how the colloidal gold labelling pattern can be clearly seen (17), and compared with the cellular pattern of the tissue. When the correct dilutions and labelling distribution has been established, an area of the block used for the preliminary work can be chosen, and trimmed for electron microscopy. When well infiltrated and embedded, Lowicryl blocks section readily and the sections can be collected onto uncoated grids. If, as sometimes happens, the sections have some weaker areas, they should be collected on formvar coated grids. The disadvantage of using coated grids is that unlike uncoated grids, they can only be labelled on one side. A 50% reduction in labelling! Coated grids are therefore floated on drops of reagent, whilst naked grids of sections are sunk in the reagents. The thin sections collected on grids can be immunolabelled with the same reagents/concentrations as for LM. The only difference is to drastically reduce or omit the silver enhancement time.

Protocol 10

Protocol for labelling resin sections for light microscopy

Equipment and reagents

- Subbed slides: dip racks of microscope slides into a mixture of 1% gelatin plus 1% formalin in distilled water; drain the slides and dry in a warm oven
- Dako pen (or similar hydrophobic pen)
- Moist chamber
- Primary antibody
- Gold conjugate
- UHQ water
- Silver enhancer
- Washing buffer: PBS, 0.1% IGSS gelatin, 0.8% BSA
- Blocking buffer: 5% FCS in washing buffer

Method

1. Cut 0.5 μm sections and place on subbed slides.
2. Mark around the section with Dako (or similar) pen.
3. Incubate for 30 min in blocking buffer (5% FCS in washing buffer).
4. Incubate with first antibody diluted in PBS/BSA overnight in moist chamber.
5. Wash 3 × 5 min with washing buffer.
6. Incubate with gold conjugate diluted in blocking buffer for 90 min.
7. Wash 3 × 5 min with washing buffer.
8. Wash 3 × 2 min with UHQ water.
9. Silver enhance (approx. 1 × 24 min for Amersham IntenSE M).
10. Rinse × 3 with UHQ water.
11. Dry and stain with dilute Toluidine blue (0.1% in 0.1% borax).
12. Image using epipolarized light microscope.

Protocol 11

Electron microscope immunocytochemistry of resin sections

Equipment and reagents

- Moist chamber
- Primary antibody
- Gold conjugate
- UHQ water
- Silver enhancer

- Washing buffer: PBS, 0.1% IGSS gelatin, 0.8% BSA
- Blocking buffer: 5% FCS in washing buffer
- Uranyl acetate
- Lead citrate

Method

1. Cut 80–90 nm sections and collect on grids.
2. Incubate for 30 min in blocking buffer.
3. Incubate with first antibody overnight in moist chamber.
4. Wash 3 × 5 min with washing buffer.
5. Incubate with gold conjugate diluted in blocking buffer for 90 min.
6. Wash 3 × 5 min with washing buffer.
7. Wash 3 × 2 min with UHQ water.
8. Silver enhance 3–12 min (depending upon reagent used).
9. Rinse × 3 with UHQ water.
10. Dry and contrast with uranyl acetate and lead citrate.

7 Multiple labelling

The availability of gold conjugates in different non-overlapping sizes means that multiple labelling can be readily done. The method chosen depends to a large extent on the specific problem and the reagents to hand. If the two antigens are detected by two different species of primary antibody, then the sections can be incubated with both primary antibodies together, washed, and then labelled with the two species-specific conjugates (of different sizes, of course) also concurrently (*Figure 3*). Care must be taken when interpreting these experiments that comparisons of labelling density are not made without recognition of the fact that smaller gold conjugates have a higher labelling efficiency than larger gold markers.

If two species of primary antibody are not available, sequential labelling is an option, but care must be taken to ensure that cross-reactivity between the two primary antibodies is minimized. This is best achieved by a suitable blocking step after the first gold conjugate and the second primary antibody.

An alternative is to carefully float the grids on droplets of reagent and complete labelling for one antigen. The grid is dried and turned over. The second

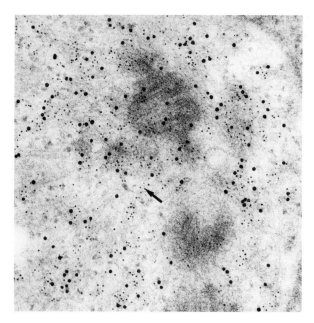

Figure 3 Double labelling of α- and β-tubulin components of microtubules labelled with 5 nm and 10 nm gold conjugates in a section of a PLT embedded cell. Whilst lines of gold marker can be seen occasionally following the profile of a microtubule (*arrow*) this figure illustrates the point that although the microtubules can be seen in the section, only those antigens accessible at the section surface can be labelled. Magnification: × 74 000.

antigen is then detected by floating the grids on the second set of reagents. Clearly this has the disadvantage of halving the sensitivity of the labelling and will only work for sections on uncoated grids.

Whatever the method chosen, additional controls should be introduced to check for cross-reactions occurring during the double labelling steps.

8 Conclusion

This chapter can only act as a starting point for immunogold labelling. The methods described will work for many studies, but they should be regarded to some extent as a statement of basic principles. If the methods described are not directly applicable, then as long as these principles are understood, modifications can be designed to tailor the method to the project.

No mention of labelling of freeze-fracture replicas has been made here and the interested reader could consult the excellent book on the topic by Severs and Shotton (13).

References

1. Holgate, V. S., Jackson, P., Lauder, I., Cowen, P. N., and Bird, C. C. (1983). *J. Clin. Pathol.*, **36**, 742.
2. Carlemalm, E., Garavito, R. M., and Villiger, W. (1982). *J. Microsc.*, **126**, 123.

3. Robertson, D., Monaghan, P., Clarke, C., and Atherton, A. J. (1992). *J. Microsc.*, **168**, 85.
4. Tokuyasu, T. (1986). *J. Microsc.*, **143**, 139.
5. Van Harreveld, A. and Crowell, J. (1964). *Anat. Rec.*, **149**, 381.
6. Edelman, L. and Rut, A. (1996). *Scanning Microsc.*, **10** (**Suppl**), 295.
7. Linner, J. G., Livesey, S. A., Harrison, D., and Steiner, A. L. (1986). *J. Histochem. Cytochem.*, **34**, 1123.
8. Griffiths, G. (1993). *Fine structure immunocytochemistry.* Springer–Verlag, Berlin.
9. Liou, W., Geuze, H., and Slot, H. (1996). *Histochem. Cell. Biol.*, **106**, 41.
10. Monaghan, P., Perusinghe, N. P., and Muller, M. (1998). *J. Microsc.*, **192**, 248.
11. Humbel, B. M. and Müller, M. (1986). In *The science of biological specimen preparation* (ed. M. Müller, R. P. Becker, A. Boyd, and J. J. Wolosewick), pp. 175–83. SEM Inc., AMF O'Hare, Chicago
12. De Waele, M., Renmans, W., Segers, E., Jochmans, L., and Van Camp, B. (1988). *J. Histochem. Cytochem.*, **36**, 679.
13. Severs, N. J. and Shotton, D. M. (1995). *Rapid freezing and deep etching.* Wiley Liss New York.

Chapter 12

Characterization of cellular antigens using monoclonal antibodies

Gillian Hynes

CRC Centre for Cell and Molecular Biology, The Institute of Cancer Research, Chester Beatty Laboratories, 237 Fulham Road, London SW3 6JB, U.K.

1 Introduction

Monoclonal antibodies are powerful biological tools due to their ability to distinguish an antigen molecule, typically a protein, from a complex mixture of macromolecules. A number of techniques have been developed to detect cellular antigens using monoclonal antibodies as specific probes. Consequently, a characterization study can be advanced considerably after the generation of a monoclonal antibody reagent to a target protein, and a possible strategy is outlined below:

(a) A useful starting point is to study the distribution and expression levels of an antigen protein in whole cell lysates (*Protocol 1*) prepared from a range of species and tissue types. The mixture of proteins is separated by sodium dodecyl sulfate–polyacrylamide gel electrophoresis (SDS–PAGE; *Protocol 4*) and the antigen polypeptide is identified by immunoblotting (*Protocols 5* and *7*).

(b) The electrophoretic properties, isoelectric point (pI) and relative molecular weight (M_r), of the antigen polypeptide are determined by resolution of a protein mixture by two-dimensional PAGE (2D PAGE; *Protocol 11*) followed by identification of the antigen by immunoblotting.

(c) The subcellular localization of the antigen is studied biochemically by differential extraction of cellular proteins (*Protocols 2* and *3*) or subcellular fractionation of cell homogenates on density gradients (Section 5); the antigen protein is detected *in situ* by immunofluorescence staining (Section 5).

(d) Electrophoresis of protein extracts on non-denaturing PAGE gels (*Protocol 12*) followed by immunoblotting is used to determine whether the antigen resolves as a component of an oligomeric complex.

(e) Intact oligomeric complexes are recovered by immunoprecipitation under non-disruptive conditions (*Protocol 14B*) and the subunit composition is determined by resolution on SDS–PAGE or 2D PAGE gels. The antigen protein is identified in the complex by subsequent immunoblotting or by performing immunoprecipitation under disruptive conditions (*Protocol 14C*).

This scheme can be regarded as a basis for a characterization study; however, it should be noted that some steps may not be universally relevant and that a number of the methods described in this chapter can be adapted and optimized depending on the nature of individual studies.

1.1 Initial characterization of monoclonal antibodies

The methods used to produce, purify, and characterize monoclonal antibodies are described in detail in Chapters 1–11, and in ref. 1. Initial characterization of a monoclonal antibody is essential before use in experiments and should include:

(a) Screening the monoclonal antibody, usually by an enzyme-linked immunosorbent assay (ELISA) or immunoblotting, to ensure that the target protein is recognized.

(b) Determination of the immunoglobulin class and subclass; important for immunoprecipitation experiments (see Section 7).

(c) Purification of the monoclonal antibody and determination of antibody concentration.

2 Preparation of cell lysates

Cells and tissues must be lysed to release the antigen protein for analysis by immunoblotting and immunoprecipitation. The method of extraction depends upon the subcellular localization of the antigen and the subsequent mode of analysis and detection. The methods described in this section are intended as guides, and can be adapted to optimize the extraction of individual proteins.

Protocols 1A and 1B use detergents to achieve whole cell lysis of tissue culture cells and result in solubilization of the plasma membrane, nuclear and organelle membranes, and the cytoskeleton. Tissue samples can also be homogenized in buffers containing detergents using a Polytron homogenizer.

Protocol 1A uses a harsh extraction buffer that enables the simultaneous analysis of soluble and insoluble proteins. Heating the proteins in the presence of the anionic detergent SDS and a reducing agent results in the disruption of secondary and tertiary protein structure, and physiological analysis is now precluded due to the extent of protein denaturation. However, the sample is suitable for analysis by SDS–PAGE (Protocol 4). Protease inhibitors are not included in Protocol 1A as any proteases will be denatured during sample preparation. In general, however, it is important to perform cell lysis at low temperatures in buffers that contain protease inhibitors (see Protocol 1B) to avoid protein degradation during sample preparation.

In Protocol 1B, a soluble whole cell extract is produced by lysis in RIPA, a mixed micelle detergent-containing buffer. RIPA buffer results in extraction of proteins without complete denaturation and cellular antigens are maintained in conformations that can be detected by immunoprecipitation (Protocol 14B). However, it should be noted that some physiological interactions are disrupted under these conditions.

Buffers containing SDS for cell lysis (*Protocols 1*A and *1*B) are incompatible with some subsequent methods of analysis, for example non-denaturing PAGE (*Protocol 12*) or immunoprecipitation under non-disruptive conditions (*Protocol 14*B). In order to study physiological interactions, it is necessary to perform cell lysis in buffers containing no detergents or mild non-ionic detergents, such as Triton X-100 or Nonidet P-40.

Protocol 1

Rigorous whole cell lysis to produce a soluble extract of proteins

Equipment and reagents

- Syringe and 19 gauge needle
- Cell scraper
- Dulbecco's PBS.A: 170 mM NaCl, 3.35 mM KCl, 10.6 mM KH_2PO_4, 1.76 mM Na_2HPO_4 pH 7.2
- Modified 2 × SDS–PAGE sample buffer: 80 mM Tris–HCl pH 6.8, 50 mM DTT, 2% (w/v) SDS
- Glycerol

- Bromophenol blue
- RIPA buffer: 50 mM Hepes, 100 mM NaCl, 1% (w/v) sodium deoxycholate, 1% (v/v) Triton X-100, 0.1% (w/v) SDS pH 7.2
- A cocktail of protease inhibitors used at the following concentrations: 1 mM AEBSF, 5 mg/ml chymostatin, 10 mg/ml leupeptin, 5 mg/ml antipain, 5 mg/ml pepstatin A, 0.3 U/ml aprotinin

A. Lysis of cells in modified 2 × SDS–PAGE sample buffer

1. Take one 90 mm dish of adherent tissue culture cells growing in monolayer culture (equivalent to approx. 10^7 cells).

2. Wash the cells twice in PBS.

3. Add 0.5 ml modified 2 × SDS–PAGE sample buffer and leave the cells at room temperature for 5 min with occasional swirling to ensure efficient extraction of proteins.

4. Remove the extract using a pipette, transfer to a polyethylene Eppendorf tube, and add glycerol to 10% (v/v) and bromophenol blue to 0.01% (w/v).

5. Sonicate the extract or pass through a 19 gauge needle to shear the glutinous DNA, which may cause smearing when the sample is subjected to electrophoresis.

6. Heat the sample at 100 °C for 5 min.

7. Centrifuge the sample at 13 000 g (top speed of benchtop microcentrifuge) for 10 min to remove any insoluble material, and remove the supernatant to a new Eppendorf tube; this sample is now ready for resolution by SDS–PAGE (see *Protocol 4*).

B. Lysis of cells in RIPA buffer

1. Take one 90 mm dish of adherent tissue culture cells growing in monolayer culture (equivalent to approx. 10^7 cells).

Protocol 1 continued

2. Wash the cells twice in ice-cold PBS.

3. Add 1 ml ice-cold RIPA buffer containing protease inhibitors.

4. Leave cells on ice for 15 min with occasional swirling to ensure efficient extraction of proteins.

5. Harvest the cells using a cell scraper and transfer to an Eppendorf tube.

6. Clear the lysate by centrifugation at 2000 g for 30 min at 4 °C to remove insoluble material, and remove the supernatant to a new Eppendorf tube.

Note: cells that grow in suspension can also by lysed by *Protocol 1A* or *1B*. Recover the cell pellet by centrifugation at 400 g for 10 min at 4 °C, wash the cells in PBS, re-pellet, and resuspend in the appropriate lysis buffer at a ratio of approx. 10^7 cells/ml in Eppendorf tubes. Proceed with cell lysis according to *Protocol 1A* or *1B*.

Protocol 2 describes a sequential extraction procedure. The first incubation in a hypotonic buffer results in cell swelling and lysis to produce an extract enriched in cytosolic proteins. This is followed by incubation in a high salt buffer to dissociate nuclear structures, including chromatin, and recovery of an extract enriched in nuclear proteins.

Protocol 2

Sequential extraction of cells to recover lysates enriched in cytosolic and nuclear proteins

Equipment and reagents

- Cell scraper
- PBS (see *Protocol 1*)
- Low salt lysis buffer: 5 mM NaCl, 1.5 mM MgCl$_2$, 0.2 mM EDTA, 0.1% Triton X-100, 1 mM DTT, 20 mM Hepes pH 7.9
- Protease inhibitors (see *Protocol 1*)

- High salt lysis buffer: 350 mM NaCl, 1.5 mM MgCl$_2$, 0.2 mM EDTA, 0.1% Triton X-100, 1 mM DTT, 20 mM Hepes pH 7.9
- No salt buffer: 1.5 mM MgCl$_2$, 0.2 mM EDTA, 0.1% Triton X-100, 1 mM DTT, 20 mM Hepes pH 7.9

Method

1. Take one 90 mm dish of tissue culture cells growing in monolayer culture (equivalent to approx. 10^7 cells).

2. Wash the cells once with 10 ml ice-cold PBS.

3. Add 200 μl low salt lysis buffer to the dish of cells.

4. Harvest the cells using a cell scraper and transfer to an Eppendorf tube.

5. Spin the cells at 13 000 g (top speed of a benchtop microcentrifuge) for 5 min at 4 °C.

6. Remove the supernatant, which should be enriched in cytosolic proteins.

7. Add 100 μl high salt lysis buffer to the pellet, vortex briefly, and place on a rotating wheel for 20 min at 4 °C.

8. Add 100 μl no salt buffer to the sample to adjust the concentration of NaCl to 175 mM, and place the tube on ice for 20 min.

9. Spin the sample at 13 000 g for 10 min at 4 °C.

10. Remove the supernatant, which should be enriched in nuclear proteins and proteins extracted from membrane compartments that have pelleted with the nuclei, such as endoplasmic reticulum.

Protocol 3 describes a controlled lysis procedure to recover a post-nuclear supernatant homogenate of cytoplasmic proteins. The cells are suspended in Harms buffer, which is an iso-osmotic buffer that causes the plasma membrane to become fragile. Cells are then lysed by the shearing forces due to Dounce homogenization or passing the cells through a homogenization chamber containing a steel ball-bearing. A centrifugation step removes nuclei and unlysed cells. This method facilitates extraction of cytosolic proteins whilst maintaining the integrity of subcellular organelle membranes and nuclear membranes, and can therefore be used to isolate nuclei, whole organelles, and membrane fractions. Physiological interactions should be preserved in the native state under these conditions, and since detergents are not used this method is compatible with enzymatic assays.

Protocol 3

Controlled cell lysis to produce a post-nuclear supernatant homogenate of cytoplasmic proteins

Equipment and reagents

- Dounce homogenizer or steel ball-bearing homogenization chamber (e.g. a number of different-sized chambers are available from H&Y Enterprise, Redwood City, California)
- PBS (see *Protocol 1*)

- Harms buffer: 250 mM sucrose, 10 mM triethanolamine, 10 mM acetic acid, 1 mM EDTA pH 7.5
- Protease inhibitors (see *Protocol 1*)

Method

1. Use approx. 10^8 cells for lysis; a suspension of cells, such as non-adherent tissue culture cells or testis germ cells, or adherent tissue culture cells growing in monolayer culture in 90 mm dishes may be used (10^8 cells is approximately equivalent to ten 90 mm dishes).

2. Wash the adherent tissue culture cells in ice-cold PBS and harvest using the cell scraper.

3. Recover the cell pellet by centrifugation at 400 g for 5 min at 4 °C.

4. Wash the cell pellet once with ice-cold Harms buffer, re-pellet the cells, and re-suspend into 2 ml Harms buffer containing protease inhibitors.

5. Lyse the cells using a steel ball-bearing homogenization chamber or a Dounce homogenizer until approximately 90% cell lysis is achieved without lysis of the nuclei. This step must be optimized for individual cell types and lysis may be assessed by phase microscopy.

6. Recover the post-nuclear supernatant homogenate (PNS) by centrifugation at 2000 g for 10 min at 4 °C.

3 Detection of cellular antigens by immunoblotting

Immunoblotting is a technique that enables the detection of cellular antigens immobilized on a membrane matrix using antibody probes. *Figure 1* outlines the steps involved in immunoblotting. Protein mixtures are separated by poly-acrylamide gel electrophoresis (PAGE) and electrotransferred to the membrane matrix where they are readily accessible to detection by antibodies. After block-ing the non-specific binding sites on the membrane, the immobilized proteins are incubated with a specific primary monoclonal antibody that distinguishes and binds to the target antigen. This is followed by incubation with a secondary anti-immunoglobulin antibody conjugated to a marker that facilitates detection. Depending on the sensitivity of the primary monoclonal antibody and the system of detection, 10 ng or less of protein may be readily detected by immunoblotting.

3.1 Separation of proteins by SDS–PAGE

SDS–PAGE is the most widely used method of separation on polyacrylamide gels prior to immunoblotting, and has been described in detail in other volumes of the Practical Approach series (2, 3). SDS–PAGE can resolve complex protein mixtures and is used for the estimation of the relative molecular weight (M_r) of a protein. The protein mixture is denatured by heating at 100 °C in the presence of the anionic detergent SDS and a reducing agent (to disrupt disulfide bonds). SDS coats the polypeptides with a uniform negative charge in a constant charge-to-mass ratio, and polypeptides are resolved on the basis of size due to the sieving effects of the polyacrylamide matrix. *Table 1* shows the protein molec-ular weight range resolved by different percentages of acrylamide during SDS–PAGE.

An advance in the SDS–PAGE separation of proteins has been the optim-ization of the mini gel system. Mini gel systems are available from a number of

Separation of proteins by PAGE

↓

Electrotransfer of proteins from gel to membrane

↓

Blocking the non-specific binding sites on the membrane

↓

Incubation with specific primary monoclonal antibody :

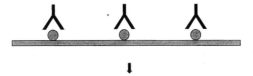

↓

Incubation with secondary antibody conjugated to marker :

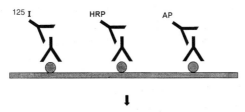

↓

Detection of signals :

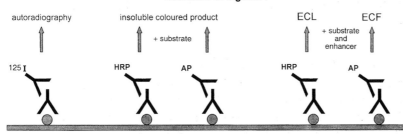

Figure 1 The steps involved in detection of cellular antigen by immunoblotting. Individual steps are described in the text.

commercial suppliers (e.g. Hoefer, Bio-Rad), and their major advantage is the reduced separation time. *Protocol 4* describes SDS–PAGE (using a mini gel apparatus) based on the discontinuous system developed by Laemmli (4). The potential disadvantages of mini gel systems are that resolution of polypeptides is sometimes inadequate and sample volume is limited; these may be overcome by running large preparative SDS–PAGE gels.

GILLIAN HYNES

Table 1 Protein molecular weight range resolved by SDS–PAGE
(using a mini gel system)

Polyacrylamide gel (% acrylamide)	M_r range
8	30 000–200 000
10	20 000–175 000
12	10 000–100 000

The distribution and expression levels of a protein antigen in lysates prepared from different species and tissue types can be analysed by immunoblotting following separation of the protein mixtures by SDS–PAGE. The
amount of protein in the lysates should be normalized before loading, and this
is achieved by quantitation of total protein levels using a protein assay (e.g.
Bradford assay; available as a kit from Bio-Rad, Pierce, Sigma). As a guide,
approximately 50 μg protein/lane should be loaded onto gels. A set of molecular
weight standards should be included during SDS–PAGE and pre-stained or [14]C-
radiolabelled markers are available (e.g. Rainbow Markers, Amersham).

Protocol 4

Separation of polypeptides by SDS–PAGE on mini gels

Equipment and reagents

- Mini gel apparatus (e.g. Hoefer SE 250 Mighty Small II or similar)
- Glass plates (8 × 10 cm), spacers, and Teflon combs
- Hamilton syringe
- 40% (w/v) acrylamide solution[a]
- 2% (w/v) bis-acrylamide solution[a]
- Ultrapure water
- 1 M Tris–HCl pH 8.8
- 1 M Tris–HCl pH 6.8
- 10% (w/v) SDS
- TEMED
- 10% (w/v) ammonium persulfate
- Water-saturated isobutanol
- 2 × SDS–PAGE sample buffer: 80 mM Tris–HCl pH 6.8, 50 mM DTT, 2% (w/v) SDS, 10% (v/v) glycerol, 0.01% (w/v) bromophenol blue
- SDS–PAGE running buffer: 192 mM glycine, 25 mM Tris, 0.1% (w/v) SDS

Method

1. Assemble the glass plates for casting mini gels.
2. Prepare the resolving gel solution (20 ml volume is sufficient for four 0.75 mm thick mini gels or two 1.5 mm thick mini gels):

Stock solution	Resolving gel (% acrylamide)		
	8%	10%	12%
• 40% acrylamide solution[a] (ml)	4	5	6
• 2% bis-acrylamide solution[a] (ml)	2	2.7	3.2
• Ultrapure water (ml)	6.1	4.4	2.9

• 1 M Tris–HCl pH 8.8 (ml)	7.5	7.5	7.5
• 10% SDS (μl)	200	200	200
• TEMED (μl)	20	20	20

Degas under vacuum for 10 min then add:

• 10% ammonium persulfate (μl)	200	200	200

3. Pipette the resolving gel solution into the space between the glass plates leaving sufficient space for the stacking gel.

4. Overlay the resolving gel with water-saturated isobutanol, and allow polymerization to proceed for 30 min at room temperature.

5. Prepare the stacking gel solution (10 ml volume):

Stock solution	Stacking gel (4% acrylamide)
• 40% acrylamide solution[a] (ml)	1
• 2% bis-acrylamide solution[a] (ml)	0.5
• Ultrapure water (ml)	7.05
• 1 M Tris–HCl pH 6.8 (ml)	1.25
• 10% SDS (μl)	100
• TEMED (μl)	10

Degas under vacuum for 10 min then add:

• 10% ammonium persulfate (μl)	100

6. Wash the top of the resolving gel with distilled water to remove the isobutanol overlay.

7. Pipette the stacking gel solution to the top of the glass plates, insert the Teflon comb to mould the sample wells, and allow polymerization to proceed for 30 min at room temperature.

8. Add an equal volume of 2 × SDS–PAGE sample buffer to the protein samples, mix well, and heat at 100 °C for 2 min.

9. Clamp the gel to the mini gel apparatus and fill the upper and lower chambers with SDS–PAGE running buffer.

10. Remove the Teflon comb, and load the samples using a Hamilton syringe.

11. Connect the upper chamber to the cathode and the lower chamber to the anode and carry out SDS–PAGE at constant current (typically 20 mA for one gel and 40 mA for two gels) until the dye front reaches the bottom of the gel.

[a] Acrylamide and bis-acrylamide are neurotoxins; appropriate safety measures should be taken.

3.2 Electrotransfer of proteins from gels to membranes

During electrotransfer, an electrophoretic field is generated to transfer proteins from a polyacrylamide matrix to a membrane matrix (also known as Western blotting). Two types of system are used for electrotransfer:

- Tank apparatus
- Semi-dry blotting apparatus

Protocol 5 describes electrotransfer using the Genie Blotter tank apparatus from Idea Scientific Company, however tank systems are available from a number of companies (e.g. Hoefer, Bio-Rad). An advantage of the Genie Blotter tank system is that it enables transfer in 40 minutes since it utilizes two plate electrodes instead of wire electrodes and can generate high fields. Electrotransfer in other tank systems is achieved in 2–16 hours, and for these longer transfer times it is necessary to pre-cool the transfer buffer or perform transfer at 4 °C. The transfer sandwich is assembled as shown in *Figure 2*. The polyacrylamide gel is placed on the membrane, which is tightly compressed in a transfer cassette or using sponge pads to ensure close contact of the gel and membrane. The transfer sandwich is then submerged in a tank of buffer for electrophoretic transfer of proteins from the gel on the cathode side to the membrane on the anode side as described by Towbin (5).

The composition of the transfer buffer may vary with different protocols and is dependent on the membrane used and the nature of the proteins to be transferred, however, the Tris/glycine buffer described in *Protocol 5* is useful as a general transfer buffer. Methanol is included to prevent swelling of the gel during transfer, but may rapidly remove SDS from some proteins. Therefore it may be necessary to reduce the concentration of methanol or add SDS to the transfer buffer to optimize the transfer of some proteins.

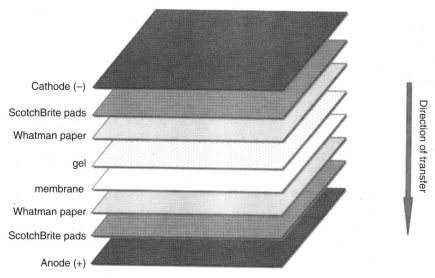

Figure 2 Composition of a transfer sandwich with plate electrodes at the top and bottom. The polyacrylamide gel is placed onto the membrane, and the assembly is tightly compressed using ScotchBrite™ sponge pads to ensure close contact of the gel and membrane. The transfer sandwich is then submerged in a tank of buffer for electrophoretic transfer of proteins from the gel on the cathode side to the membrane on the anode side.

Nitrocellulose and polyvinylidene difluoride (PVDF) are the types of membrane matrix most commonly used in immunoblotting. *Protocol 5* uses nitrocellulose membrane, which has high binding capacity for proteins (approx. 80–100 $\mu g/cm^2$) and, unlike PVDF, does not require pre-wetting in methanol. Supported nitrocellulose (e.g. Hybond-C extra, Amersham) is more robust and durable and the membrane blot can be readily re-probed (see *Protocol 10*).

The efficiency of transfer can be assessed by:

- The transfer of pre-stained marker proteins to the membrane
- Staining the gel after transfer with a protein stain such as Coomassie blue
- Staining the membrane blot after transfer

Ponceau S is used to stain the membrane before performing the blocking step to obtain a total protein pattern (*Protocol 6*). This is a relatively insensitive method (1 μg protein detected in a protein band), but has the advantage of being rapid, completely reversible, and compatible with subsequent immunoblotting since the immobilized proteins are not damaged.

Protocol 5

Electrotransfer of proteins from gels to nitrocellulose membranes

Equipment and reagents

- Transfer apparatus (e.g. Genie Blotter, Idea Scientific Co.)
- Sponge pads (e.g. ScotchBrite pads)
- Whatman 3MM paper
- Nitrocellulose membrane
- Transfer buffer: 20 mM Tris, 150 mM glycine, 20% (v/v) methanol

Method

1. Place the anode (+) electrode plate into the bottom of the Genie tray followed by two grids to dissipate the current evenly during transfer.
2. Wearing gloves, cut four pieces of Whatman paper and one piece of nitrocellulose membrane to the same size as the gel.
3. Assemble the transfer sandwich as shown in *Figure 2* taking care to remove any air bubbles at each stage by rolling with a plastic pipette.
4. Place two ScotchBrite pads pre-soaked in transfer buffer on the electrode plate followed by two sheets of Whatman paper pre-wet with transfer buffer.
5. Gradually immerse the nitrocellulose membrane into a tray of transfer buffer in order to wet it evenly without trapping air in the membrane, and place onto the Whatman paper.
6. Remove the gel from the glass plates. The stacking gel may be removed using a scalpel blade, however, it is sometimes useful to transfer the entire gel to assess the extent of protein precipitation in the sample wells or at the interface between the stacking and resolving gels.

275

Protocol 5 continued

7. Place the gel onto the nitrocellulose membrane, followed by two pre-wet sheets of Whatman paper and more ScotchBrite pads (enough to ensure that the gel is held in close contact with the nitrocellulose membrane), two grids, and the cathode (–) electrode.

8. Place the Perspex cover on top of the assembly and slide the tray into the vertical transfer tank.

9. Fill the tank with transfer buffer and connect the anode and cathode plates to the power supply. Limit the current to 900 mA and carry out electrotransfer at 50 V for 40 min at room temperature (current should be maintained at around 900 mA for the duration of the transfer).

10. After transfer is complete, remove the transfer sandwich. The gel can be stained in Coomassie blue to assess efficiency of transfer, and the nitrocellulose membrane may be stained with Ponceau S (*Protocol 6*) to visualize the total protein pattern prior to immunoblotting (*Protocol 7*).

Protocol 6

Staining membranes with Ponceau S after transfer to obtain the total protein pattern

Equipment and reagents

- Clean container
- Whatman 3MM paper
- Whatman blunt-edged forceps
- Ponceau S staining solution: 0.2% (w/v) Ponceau S in 0.5% acetic acid
- Transparent polythene sheeting
- Destaining solution: 0.5% (v/v) acetic acid in water
- PBS (see *Protocol 1*)

Method

1. Using Whatman forceps, place the membrane blot into the container and rinse in water.

2. Incubate the membrane in Ponceau S staining solution for 5 min with constant agitation.

3. Destain the membrane by incubation in destaining solution until the protein bands are visible and background staining is low. Note: this is a reversible stain and prolonged incubation in the destaining solution will result in loss of stain from protein bands.

4. Remove excess liquid by placing the membrane onto Whatman paper and place the stained membrane between transparent polythene sheets. Photocopy the membrane to obtain a permanent record of the staining pattern.

5. Rinse blot in PBS before the blocking stage (*Protocol 7A*).

3.3 Blocking non-specific binding sites on the membrane

Prior to incubation with the antibody solutions, the non-specific protein binding sites on the membrane blot are blocked by incubation with a quenching buffer, such as blocking buffer (*Protocol 7A*). Blocking buffer is a protein-rich solution that affords partial renaturation of antigens that have been denatured by electrophoresis. However, blocking buffer is not suitable for use with avidin–biotin conjugates or antibodies that recognize carbohydrate moieties due to cross-reactivity with components of the buffer. Other quenching buffers containing 5% BSA, 0.05% Tween 20, or 1% gelatin may be used. It is important to ensure that the membrane blot does not dry out at any stage of the immunoblotting procedure after performing the blocking step.

3.4 Incubation with the primary monoclonal antibody

Incubation with the primary monoclonal antibody is used for specific recognition of the target antigen. Some epitopes, such as non-linear epitopes, are only recognized when the protein is in a native conformation; non-linear epitopes will not be detected if the protein is denatured during electrophoresis, for example after SDS–PAGE (*Protocol 4*) or 2D PAGE (*Protocol 11*). Conversely, some epitopes may be inaccessible to detection when the protein is in a native conformation, for example after non-denaturing PAGE (*Protocol 12*). The reactivity of the antibody with native and denatured antigen will depend upon the conformation of the antigen used for immunization.

3.5 Detection of secondary antibodies and quantitation of signals

Antigen–primary monoclonal antibody complexes are recognized by a secondary anti-immunoglobulin antibody conjugated to a marker to provide a means of detection. Enhanced sensitivity is achieved in this signal amplification step as each primary monoclonal antibody is recognized by many molecules of the secondary antibody.

3.5.1 Radiolabelled secondary antibodies

Secondary antibodies labelled with radioisotopes (typically ^{125}I) are detected by autoradiography, and the signals are quantitated by densitometric scanning of the film. Alternatively, radiolabelled secondary antibodies result in excitation of a Phosphor storage screen, and the signals are detected and quantitated by PhosphorImaging.

3.5.2 Enzyme-conjugated secondary antibodies

Secondary antibodies conjugated to horseradish peroxidase (HRP) or alkaline phosphatase (AP) are detected in an enzymatic reaction:

(a) Colorimetric reactions. HRP- or AP-conjugated secondary antibodies are detected in reactions in which a substrate changes colour and precipitates to result in a coloured band (*Protocols 8A* and *8B*). Colorimetric signals are quantitated by densitometric scanning of the membrane. Colorimetric detection systems are useful for the analysis of immunoprecipitation reactions by immunoblotting. The cross-reactivity often observed with the heavy chains ($\sim$ 50 kDa) and light chains ($\sim$ 25 kDa) of the immunoglobulins used for immunoprecipitation is less pronounced due to the relatively low sensitivity of colorimetric reactions. Colorimetric detection is also useful when it is necessary to overlay a radiolabelled signal with an immunoblot signal.

(b) Enhanced chemiluminescence (ECL). HRP-conjugated secondary antibodies are detected in a reaction in which the oxidation of luminol by HRP, in the presence of chemical enhancers, leads to the production of luminol radicals which emit light as they decay (*Protocol 9*) (available as a kit from a number of commercial suppliers, e.g. Amersham, Pierce). The signals are detected on autoradiography film sensitive to the blue light produced by ECL (λ max = 425 nm), and signals are quantitated by densitometric scanning. Some specialized films exhibit a linear response to the light emitted by ECL (e.g. Hyperfilm-ECL, Amersham), and the range over which the film response is linear can be extended by pre-flashing the film prior to exposure. ECL signals are quantitated directly by specialized light-gathering and imaging systems. Incubation with ECL reagents does not result in any damage to the proteins on the blot, and re-probing with a different antibody is possible (*Protocol 10*). In general, ECL detection has superseded the colorimetric methods of detection due to the advantages of increased sensitivity, opportunity for multiple exposures, hard copy results, and option to re-probe the blots.

Protocol 7

Immunoblotting

Equipment and reagents

- Clean container
- Whatman forceps
- Polythene sheeting
- Thermal bag sealer
- Whatman 3MM paper
- Blocking buffer: 1 M glycine, 5% (w/v) dried skimmed milk powder, 1% (w/v) ovalbumin, 5% (v/v) newborn calf serum

- Primary monoclonal antibody stock solution
- Secondary antibody stock solution
- TBS: 150 mM NaCl, 10 mM Tris–HCl pH 7.4
- TBS-Tween: 150 mM NaCl, 10 mM Tris–HCl pH 7.4, 0.05% (v/v) Tween 20

A. Blocking non-specific sites on nitrocellulose membranes

1. After transfer or staining, submerge the membrane blot in blocking buffer in a container.

2. Incubate the membrane in blocking buffer for at least 2 h at room temperature or overnight at 4 °C with constant agitation on a shaker.

B. Incubation with the primary monoclonal antibody

1. Using Whatman forceps, place the membrane blot between sheets of polythene and seal three edges to form a bag using a thermal sealer.

2. Dilute the primary monoclonal antibody (mAb) stock solution in blocking buffer. The antibody dilution should be optimized, but dilution of a 1 mg/ml primary mAb stock solution in the range 1:200 to 1:5000 is customary.

3. Add the primary mAb solution to the blot in the polythene bag and ensure that large air bubbles are removed before sealing the bag with the thermal sealer. The advantage of this technique is that a small volume of antibody solution is required. Use 10 ml volume for a 15 cm^2 blot and 3 ml volume for a 5 cm^2 blot.

4. Incubate the blot in primary mAb solution for 1–4 h at room temperature or overnight at 4 °C with continuous agitation on a shaker. The blot may be placed between two glass plates to ensure an even distribution of the antibody solution.

5. Remove the blot from the polythene bag and transfer to a container. Wash the blot three times with TBS-Tween (10 min each wash). Rinse the blot with TBS.

C. Incubation with secondary antibody

1. Dilute the secondary antibody stock solution in blocking buffer according to the instructions of the supplier. Dilution in the range 1:2000 to 1:5000 is customary.

2. Seal the blot in a new polythene bag and add the secondary antibody solution. Incubate the blot for 1–4 h at room temperature or overnight at 4 °C with continuous agitation on a shaker.

3. Wash the blot as described in part B, step 5.

4. Place the blot protein side up on a sheet of Whatman paper to remove the excess wash solution (do not allow to dry). Proceed with the appropriate method of detection (see *Protocols 8* and *9*).

(c) Enhanced chemifluorescence (ECF). AP-conjugated secondary antibodies are detected in a reaction in which the alkaline phosphatase cleaves a phosphate group from a fluorogenic substrate to yield a highly fluorescent product (available as a kit; Vistra™ ECF, Amersham). An additional amplification step is possible, i.e. a fluorescein labelled secondary antibody detects the specific primary monoclonal antibody, and an alkaline phosphatase-conjugated anti-fluorescein tertiary antibody generates the fluorescent product in the ECF reaction. Fluorescent signals, such as those generated by ECF or fluorescein-conjugated secondary antibodies, are quantitated directly using a fluorescence imaging system which generates a digital image of the signals (e.g. Fluorimager, Molecular Dynamics; Storm, Molecular Dynamics; Fluor S Imager, Bio-Rad).

Protocol 8

Detection of signals by colorimetric reactions

Reagents

- TBS (see *Protocol 7*)
- 50 mM Na-glycinate pH 9.6
- 1 M MgCl$_2$
- Nitro blue tetrazolium (NBT) (Sigma product, N-6876): 1 mg/ml in 50 mM Na-glycinate pH 9.6
- 30% hydrogen peroxide solution
- 5-Bromo-4-chloro-3-indolyl phosphate (BCIP) (Sigma product, B-8503): 5 mg/ml in dimethylformamide
- 4-Chloro-1-naphthol solution (Sigma product, C-8302)
- PBS (see *Protocol 1*)

A. Detection of AP-conjugated secondary antibodies using BCIP/NBT substrate

1. Rinse the blot in TBS.

2. Wash the blot in 50 mM Na-glycinate pH 9.6 for 5 min.

3. Prepare 20 ml substrate solution as follows:
   - 1 M MgCl$_2$ 80 µl
   - 5 mg/ml BCIP   200 µl
   - 1 mg/ml NBT   2 ml
   - 50 mM Na-glycinate pH 9.6   17.8 ml

4. Incubate the blot with the substrate solution for about 10 min at room temperature with constant agitation. A purple signal should be observed.

5. Pour off the substrate solution and stop the reaction by rinsing the blot in distilled water.

B. Detection of HRP-conjugated secondary antibodies using chloronaphthol substrate

1. Rinse the blot in PBS.

2. Prepare the substrate solution by adding 5 µl of 30% hydrogen peroxide solution to 5 ml chloronaphthol solution.

3. Incubate the blot with the substrate solution for about 10 min at room temperature with constant agitation. A blue-black signal should be observed.

4. Pour off the substrate solution and stop the reaction by rinsing the blot in distilled water.

5. Store the blot in the dark to prevent fading.

Protocol 9

Detection of HRP-conjugated secondary antibodies by enhanced chemiluminescence (ECL)

Equipment and reagents

- Autoradiography film sensitive to the blue light produced by ECL (λ max = 425 nm) (e.g. Biomax MR, Kodak; Hyperfilm-ECL, Amersham; X-OMAT, Kodak)
- Dark-room facility
- X-ray film cassette
- X-ray film processor
- Transparent polythene sheeting
- Whatman forceps
- TBS
- ECL detection solution 1 (containing a peracid salt)
- ECL detection solution 2 (containing luminol and the enhancer)

Method

1. Rinse the blot in TBS.

2. Mix equal volumes of ECL detection solutions 1 and 2 to give sufficient working solution to cover the blot (allow 0.125 ml for each cm^2 of membrane).

3. Using Whatman forceps, briefly place the blot protein side up on a sheet of Whatman paper to remove the excess TBS, but do not allow the membrane to dry.

4. Place the blot protein side up into a clean container, add the ECL working solution to the surface of the blot, and incubate at room temperature for 1 min.

5. Briefly place the blot protein side up on a sheet of Whatman paper to remove the excess ECL working solution, but do not allow the membrane to dry.

6. Place the blot between sheets of transparent polythene and ensure that no air bubbles are trapped.

7. Place the blot protein side up into an X-ray film cassette and expose to film for 30 sec.

8. Develop the film and vary the timing of the second exposure according to the signal intensity. Multiple exposures may be taken at a variety of time points (2 sec–1 h) to obtain the optimum signal, as the ECL signal persists for over 1 h.

9. Following ECL detection, it is possible to re-probe the membrane several times (see *Protocol 10*).

Protocol 10

Re-probing blots after ECL detection

Equipment and reagents

- Container with lid for incubations
- Stripping buffer: 100 mM 2-mercaptoethanol, 2% (w/v) SDS, 62.5 mM Tris–HCl pH 6.7
- TBS (see *Protocol 7*)
- TBS-Tween (see *Protocol 7*)
- Blocking buffer (see *Protocol 7*)

Protocol 10 continued

A. Re-probing only (without stripping off previous antibodies)

1. Wash blot twice in 200 ml TBS-Tween for 10 min with constant agitation on a shaker.

2. Rinse blot in TBS and incubate the blot in blocking buffer as described in *Protocol 7A*.

3. Incubate the blot in primary and secondary antibody solutions as described in *Protocol 7*, and detect signals as described in *Protocols 8 or 9*.

B. Stripping and re-probing[a]

1. Submerge the blot in 500 ml of stripping buffer in a container with a lid.

2. Incubate at 50 °C for 30 min with occasional agitation.

3. Dispose of the stripping buffer by pouring down the sink and flushing with plenty of running water. This step should be performed in a fume hood since 2-mercapto-ethanol is pungent and toxic.

4. Wash the stripped blot twice in 200 ml TBS-Tween for 10 min with constant agitation on a shaker.

5. Rinse the blot in TBS and incubate in blocking buffer as described in *Protocol 7A*.

6. Incubate the blot in primary and secondary antibody solutions as described in *Protocol 7*, and detect signals as described in *Protocols 8 or 9*.

[a] Blots may be stripped and re-probed up to four times without loss of the immobilized proteins from the membrane.

3.6 Troubleshooting problems encountered during immunoblotting

3.6.1 Problems associated with electrotransfer from gel to membrane

(a) Distorted pattern of proteins on the membrane:

  (i) Ensure that no air bubbles are trapped during assembly of the transfer sandwich.

  (ii) Ensure that the Whatman paper and membrane are wet with transfer buffer.

  (iii) Ensure that close contact is maintained between the gel and membrane during electrotransfer.

(b) Inefficient transfer:

  (i) A longer transfer period may be necessary for thick gels and gels composed of high per cent acrylamide.

  (ii) Addition of 0.035–0.1% (w/v) SDS to the transfer buffer aids the transfer of large proteins (> 80 kDa) and promotes the efficiency of transfer from non-denaturing PAGE gels.

3.6.2 Problems associated with detection of signals

(a) High background signals.

    (i) Extend the duration of the blocking step.

    (ii) Re-wash the blot.

    (iii) Ensure that antibodies do not react with components of the quenching buffer.

(b) Weak signals.

    (i) Use increased amounts of antigen protein.

    (ii) Increase the concentration of primary monoclonal antibody.

3.6.3 Signals due to cross-reactivity of the primary or secondary antibodies with other proteins

Appropriate positive and negative controls should be included to confirm that the signals detected are due to specific recognition of the target antigen:

(a) Cross-reaction of the primary monoclonal antibody with proteins that share the same epitope (epitope-related proteins; see Section 8) may be observed. One way to determine which bands correspond to the target antigen is to perform immunoprecipitation with another monoclonal antibody to a different epitope on the target protein followed by detection of signals by immunoblotting with the first monoclonal antibody.

(b) Cross-reaction due to the secondary antibody reagent may be determined by incubating the blot with secondary antibody alone.

Optimization of the conditions during electrotransfer and immunoblotting is discussed in ref. 6.

4 Analysis of the electrophoretic properties of polypeptides by 2D PAGE

Two-dimensional polyacrylamide gel electrophoresis (2D PAGE) is a powerful approach to the analysis of complex protein mixtures as polypeptides are separated on the basis of two electrophoretic properties:

(a) Resolution in the first dimension by on the basis of charge by isoelectric focusing (IEF) or non-equilibrium pH gradient electrophoresis (NEPHGE).

(b) Resolution in the second dimension on the basis of size by SDS–PAGE.

In IEF, an amphoteric protein molecule migrates to the position in a pH gradient where it has no net charge. This is the characteristic pH at which the protein will not migrate in an electric field, known as the isoelectric point (pI). The pH gradient obtained by IEF does not usually extend beyond 7 due to cathodic drift; therefore, NEPHGE is used to resolve basic proteins (pI greater than 7) which would be poorly resolved by conventional IEF. The sample proteins are separated as cations and electrophoresis is carried out for short periods so that basic proteins do not migrate off the end of the gel. Equilibrium is not reached and therefore the position of a protein in NEPHGE will vary with the duration of focusing.

Resolution of a protein mixture by 2D PAGE followed by identification of the antigen by immunoblotting enables the pI and relative molecular weight (M_r) of the antigen polypeptide to be calculated. Detection of additional isoforms of the antigen polypeptide may indicate the presence of some post-translational modifications, such as phosphorylation. 2D PAGE is used for the study of proteomes (the proteins expressed by a genome or tissue), and several comprehensive proteome databases are available or in preparation. The apparent pI and M_r of cellular antigens may be compared with 2D PAGE reference maps, which show the pattern of proteins expressed in particular cells and tissues (some 2D PAGE reference maps have been established in the SWISS-2DPAGE computer database available through the ExPASy molecular biology server).

A number of 2D PAGE systems (both flat-bed and vertical systems) have been optimized for maximum separation of proteins using the large gel format. These specialized gel systems are essential for high-resolution studies that depend upon reproducibility, but they are generally expensive. Some mini gel systems (e.g. SE 250 Mighty Small II, Hoefer; mini-PROTEAN II, Bio-Rad) can be adapted to run IEF/NEPHGE separations in tube gels. This provides adequate resolution of proteins by 2D PAGE and is an inexpensive alternative that is available to most laboratories. *Protocol 11* describes the separation of polypeptides by analytical 2D PAGE (using a mini gel system), based on the method described by Corbett and Dunn (7), which is a modified version of that originally described by O'Farrell (8). As a guide, approximately 100 µg of a protein lysate is loaded onto IEF/NEPHGE gels. pI marker proteins can be run together with the sample of interest to calibrate IEF separations (e.g. carbamylated marker proteins; BDH, Pharmacia).

Protocol 11

Separation of polypeptides by analytical 2D PAGE

Equipment and reagents

- Equipment and reagents required for second dimension electrophoresis on SDS–PAGE mini gels (see *Protocol 4*)
- Mini gel apparatus with tube gel adaptor kit (e.g. Hoefer SE 250 Mighty Small II with SE 220 tube gel adaptor kit)
- Glass tubes (7.5 cm long, 1.5 mm inner diameter)
- 5 ml graduated cylinder
- Hamilton syringe
- Syringe to extrude tube gels
- Acrylamide stock solution:[a] 30% (w/v) acrylamide, 0.8% (w/v) *bis*-acrylamide; deionize with Amberlite MB-1 resin for 1 h, filter, and store in the dark at 4 °C

- Amberlite MB-1 resin (BDH)
- Ultrapure water
- 3-[(3-cholamidopropyl)-dimethylammonio]-1-propane sulfonate (CHAPS)
- Urea
- Dithiothreitol (DTT)
- 10% (w/v) ammonium persulfate
- TEMED
- 10 mM phosphoric acid
- 20 mM sodium hydroxide
- Equilibration buffer: 80 mM Tris–HCl pH 6.8, 3% (w/v) SDS, 32 mM DTT, 0.01% (w/v) bromophenol blue

Stock solutions for IEF

- 40% ampholytes pH 4–8 (e.g. Resolyte 4–8, BDH)
- IEF sample buffer: dissolve 5.4 g urea in 6.5 ml water and deionize with Amberlite MB-1 resin for 1 h to remove any cyanate ions resulting from urea degradation

which can result in protein carbamylation. Filter the solution before addition of 0.4 g CHAPS, 0.1 g DTT, 0.5 ml of 40% ampholytes pH 4–8, and 0.01% (w/v) bromophenol blue. Aliquot the IEF sample buffer and store at $-70\,°C$.

Stock solutions for NEPHGE

- 40% ampholytes pH 3–10 (e.g. Resolyte 3–10, BDH)
- 40% ampholytes pH 8–10.5 (e.g. Pharmalyte 8–10.5, Pharmacia)
- 4 M urea
- NEPHGE sample buffer: dissolve 5.4 g urea in 6.5 ml water and deionize with Amberlite MB-1 resin for 1 h to remove

any cyanate ions. Filter the solution before addition of 0.4 g CHAPS, 270 µl of 2-mercaptoethanol, 375 µl of 40% ampholytes pH 3–10, 125 µl of 40% ampholytes pH 8–10.5, and 0.01% (w/v) bromophenol blue. Aliquot the IEF sample buffer and store at $-70\,°C$.

A. Isoelectric focusing (IEF)

1. Rinse the glass tubes in HPLC grade water and ethanol, and dry in a hot air cabinet.

2. Mark the required length (7 cm) on the tubes and place in a clean 5 ml graduated cylinder.

3. To prepare 20 ml of IEF gel mixture, dissolve 10 g urea in 7.4 ml HPLC grade water and 3 ml acrylamide stock solution,[a] and deionize with Amberlite MB-1 resin for 1 h.

4. Filter the mixture and degas for 10 min before addition of 1 ml of 40% ampholytes (pH 4–8), 0.3 g CHAPS, 15 µl TEMED, and 30 µl 10% ammonium persulfate solution.

5. Pipette the IEF gel mixture into the graduated cylinder containing the glass tubes until the marked level is reached.

6. Allow the gels to polymerize for 1.5 h (the top 5 mm of gel solution does not polymerize and serves as an overlay).

7. Remove the tube gels from the cylinder, clean off the excess gel, and insert into the IEF chamber.

8. Fill the lower chamber with anode buffer (10 mM phosphoric acid) and any trapped air bubbles are removed from underneath the tube gels using a buffer-filled syringe.

9. Degas the cathode buffer (20 mM sodium hydroxide) for 10 min to remove CO_2 and add to the top chamber.

10. Remove any air remaining in the upper part of the tubes using a Hamilton syringe filled with cathode buffer ensuring that the unpolymerized gel solution is not disturbed.

11. Mix an equal volume of protein sample and IEF sample buffer and, without heating, load the sample (approx. 20 µl volume) onto the IEF gel using a Hamilton syringe.

12. Connect the upper chamber to the cathode and the lower chamber to the anode and carry out isoelectric focusing at a constant setting of 500 V for 4 h.

13. Store the IEF tube gels at $-20\,°C$ until the second dimension electrophoresis is performed.

B. Non-equilibrium pH gradient electrophoresis (NEPHGE)

1. Cast the NEPHGE tube gels as described above (part A) substituting 660 µl of 40% ampholytes (pH 3–10) and 330 µl 40% ampholytes (pH 8–10.5) in place of the 1 ml of 40% ampholytes (pH 4–8) used in isoelectric focusing.

2. Fill the lower chamber with degassed cathode buffer (20 mM sodium hydroxide), and insert the NEPHGE tube gels.

3. Mix an equal volume of protein sample and NEPHGE sample buffer and, without heating, load the sample (approx. 20 µl volume) onto the NEPHGE gel using a Hamilton syringe.

4. Overlay the sample with 5 µl of 4 M urea and add the anode buffer (10 mM phosphoric acid) to the top chamber.

5. Connect the upper chamber to the anode and the lower chamber to the cathode and carry out NEPHGE at 250 V for 1.5 h.

6. Store the NEPHGE tube gels at $-20\,°C$ until the second dimension electrophoresis is performed.

C. Second dimension electrophoresis on SDS–PAGE gels

1. Remove the IEF or NEPHGE tube gel from storage at $-20\,°C$ (freezing makes it easier to extrude the gels).

2. Extrude the tube gels onto lengths of Parafilm using a water-filled syringe.

3. Cast the SDS–PAGE slab gel with a Teflon preparative comb, which results in one large well along the top of the gel.

4. Lay the tube gel along the large well of the SDS–PAGE gel and incubate *in situ* with equilibration buffer (make fresh before use) for 5 min.

5. Drain off the equilibration buffer and perform the second dimension SDS–PAGE procedure as described in *Protocol 4*.

[a] Acrylamide and *bis*-acrylamide are neurotoxins; appropriate safety measures should be taken.

5 Analysis of subcellular localization of proteins

5.1 Immunofluorescence staining

Immunofluorescence staining permits detection of protein antigens *in situ*, in order to investigate the subcellular localization or cellular distribution within a tissue. The cells or tissue sections are fixed and incubated with the specific primary monoclonal antibody. The antigen–primary monoclonal antibody complex is bound by a second antibody conjugated to a fluorescent dye, such as rhodamine-β-isothiocyanate or fluorescein isothiocyanate, for detection by fluorescence microscopy. Immunofluorescence staining is described in more detail in Chapter 16.

5.2 Subcellular fractionation

The localization of a protein antigen is studied biochemically by subcellular fractionation of cell homogenates (9). Cell lysis is carried out under conditions that do not disrupt the nuclear and organelle membranes (see *Protocol 3*), and subcellular fractionation is achieved by centrifugation techniques such as:

• Differential centrifugation

• Sedimentation of organelles and oligomeric protein complexes during isopycnic separations on density gradients

Sucrose is most commonly used to prepare gradients but other materials, such as glycerol, Ficoll, Percoll, and Nycodenz, can also be used depending on the application. After centrifugation, fractions are collected for analysis by immunoblotting. The sedimentation profile of a protein antigen is compared with the sedimentation of native high molecular weight protein standards run on a parallel gradient; specific enzyme assays indicate fractions that contain intact organelles or membrane vesicles (i.e. the sedimentation of Golgi membranes, lysosomes, and peroxisomes can be determined by galactosyl transferase assay, β-hexose-aminidase assay, and horseradish peroxidase assay, respectively). As a result, gradient fractionation is a means of determining the subcellular localization of a protein antigen and whether it sediments as a component of a high molecular mass complex. Gradient fractionation may also be useful as an enrichment step in a purification strategy.

6 Resolution of oligomeric complexes by non-denaturing PAGE

Electrophoresis of protein extracts on non-denaturing PAGE gels followed by immunoblotting is used to determine whether the antigen resolves as a component of an oligomeric complex. Under non-denaturing conditions, separation is dependent on the size, net charge, and shape of the proteins analysed, and subunit interactions and the native conformation of proteins should be maintained. *Protocol 12* describes separation on non-denaturing slab mini gels based on the method described by Bollag and Edelstein (10). Migration in non-denaturing PAGE gels is influenced by the pH of the buffer system. The pH 8.8 buffer system described in *Protocol 12* is suitable for proteins that are negatively charged and stable at this pH; however, it should be noted any basic proteins present in the sample will migrate towards the cathode and will be lost from the gel. Cell lysis should be performed under non-disruptive conditions and a set of native molecular weight protein standards can be included on the gel (e.g. thyroglobulin, 669 kDa; apoferritin, 443 kDa; β-amylase, 200 kDa; bovine serum albumin, 69 kDa; chicken ovalbumin, 45 kDa).

Protocol 12

Non-denaturing PAGE

Equipment and reagents

- Mini gel apparatus (e.g. Hoefer SE 250 Mighty Small II)
- Glass plates (8 × 10 cm), spacers, and Teflon combs
- Hamilton syringe
- 40% (w/v) acrylamide solution[a]
- 2% (w/v) *bis*-acrylamide solution[a]
- Ultrapure water
- 1 M Tris–HCl pH 8.8

- TEMED
- 10% (w/v) ammonium persulfate
- Water-saturated isobutanol
- 10 × non-denaturing PAGE sample buffer: 125 mM Tris–HCl pH 8.8, 84% (v/v) glycerol, 0.1% (w/v) bromophenol blue
- Non-denaturing PAGE running buffer: 25 mM Tris, 192 mM glycine pH 8.8

Method

1. Prepare the non-denaturing resolving gel solution and adjust to pH 8.8 (20 ml volume is sufficient for four 0.75 mm thick mini gels or two 1.5 mm thick mini gels):

Stock solution	Resolving gel (% acrylamide)		
	6%	8%	12%
• 40% acrylamide solution[a] (ml)	3	4	6
• 2% *bis*-acrylamide solution[a] (ml)	1.5	1.5	1.5
• Ultrapure water (ml)	7.9	6.9	4.9
• 1 M Tris–HCl pH 8.8 (ml)	7.4	7.4	7.4
• TEMED (μl)	10	10	10
Degas under vacuum for 10 min then add:			
• 10% ammonium persulfate (μl)	200	200	200

2. Assemble the glass plates and cast the non-denaturing PAGE gels using the mini gel system as described in *Protocol 4*.

3. Prepare the stacking gel solution (10 ml volume) and adjust to pH 8.8:

Stock solution	Stacking gel (3.2% acrylamide)
• 40% acrylamide solution[a] (ml)	0.8
• 2% *bis*-acrylamide solution[a] (ml)	0.6
• Ultrapure water (ml)	7.9
• 1 M Tris–HCl pH 8.8 (ml)	0.57
• TEMED (μl)	10

Degas under vacuum for 10 min then add:

• 10% ammonium persulfate (μl)	150

6. Add 1 μl of 10 × non-denaturing PAGE sample buffer to 10 μl protein sample, mix well but do not heat.

7. Clamp the gel to the mini gel apparatus and fill the upper and lower chambers with non-denaturing PAGE running buffer pre-chilled to 4 °C.

8. Remove the Teflon comb, and load the samples using a Hamilton syringe.

9. Connect the upper chamber to the cathode and the lower chamber to the anode and carry out non-denaturing PAGE at 4 °C at constant voltage until the dye front reaches the bottom of the gel (typically 3 h at 90 V or overnight at 30 V).

[a] Acrylamide and *bis*-acrylamide are neurotoxins; appropriate safety measures should be taken.

7 Detection of cellular antigens by immunoprecipitation

During immunoprecipitation, the antibody–antigen complex is formed in solution and is captured and immobilized on a solid phase matrix (beads). Some of the applications of immunoprecipitation are:

- Detection and quantitation of antigens
- Analysis of the rate of synthesis or degradation of the antigen
- Analysis of the presence of certain post-translational modifications
- To study physiological interactions between the antigen protein and ligands, nucleic acids, and other proteins
- To study the enzymatic or ligand binding properties of the precipitated antigen
- To enrich a non-abundant antigen for subsequent detection by immunoblotting

  Immunoprecipitation is typically performed in five steps (illustrated in *Figure 3*):

(a) Radiolabelling of the antigen (optional). The antigen and other cellular proteins may be radiolabelled prior to immunoprecipitation to facilitate detection by autoradiography. *Protocol 13* describes metabolic labelling of cellular

Radiolabelling of cellular proteins

⇩

Cell lysis

⇩

Formation of antibody-antigen complexes

Under non-disruptive conditions :   **Under disruptive conditions :**

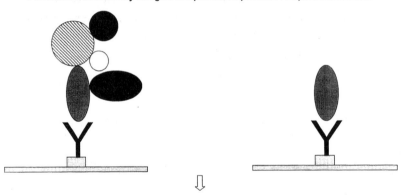

⇩

Purification of antibody-antigen complexes on protein A-Sepharose beads :

⇩

Analysis of antibody-antigen complexes by PAGE

Figure 3 The steps involved in detection of cellular antigens by immunoprecipitation. Individual steps are described in the text.

proteins with [^{35}S]methionine. Other radiolabels may be used to give an indication of post-translational modification, for example phosphorylated proteins can be identified by labelling with [^{32}P]orthophosphate and glyco-sylated proteins can be identified by labelling with [^{3}H]fucose. The detection limit is estimated at approximately 100 c.p.m./protein.

(b) Lysis of cells to release the antigen. Preparation of lysates that are com-patible with subsequent analysis by immunoprecipitation are described in *Protocols* 1B, 2, and 3.

(c) Formation of antibody–antigen complexes. The specific monoclonal antibody is added to the lysate and the antibody–antigen complexes are formed in solution. Unlike immunoblotting, where the antigen is presented in a highly concentrated state, immunoprecipitation relies on the formation of the antibody–antigen complex at relatively low concentrations. To confirm the specificity of the interaction, it is important to include an appropriate negative control. Ideally, this is a reaction using an immunoglobulin that is from the same species, class, and subclass as the specific monoclonal antibody, but does not recognize the target antigen. If this is not possible, incubation of the lysate with beads only under the same conditions as the immunoprecipitation reaction serves as a negative control.

(d) Purification and immobilization of the antibody–antigen complexes. Antibody–antigen complexes are purified by adding a solid phase matrix, usually beads containing *Staphylococcus* Protein A. Some immunoglobulin subclasses do not bind strongly to Protein A (*Table 2*). This is circumvented by a bridging step using a secondary anti-immunoglobulin antibody that exhibits strong binding to Protein A, or by capture of the antibody–antigen complex on *Streptococcus* Protein G beads (if the monoclonal antibody exhibits stronger binding to Protein G; see *Table 2*). Protein A and Protein G are bacterial cell wall proteins that bind to the Fc domain of the primary monoclonal antibodies, and thus do not interfere with the interaction with the antigen.

(e) Analysis of antibody–antigen complexes. The products of immunoprecipitation are analysed by SDS–PAGE or 2D PAGE, followed by:

   (i)   Silver staining or Coomassie blue staining of gels.
   (ii) Autoradiography to detect radiolabelled signals.
   (iii) Immunoblotting.

Table 2 Comparison of the binding of immunoglobulins to Protein A and Protein G[a]

Immunoglobulin		Affinity for Protein A	Affinity for Protein G
Species	**Subclass**		
Mouse	IgG1	−	++
	IgG2a	++	++
	IgG2b	++	++
	IgG3	+	+++
	IgM	−	−
	IgA	−	−
	IgE	−	−
Rabbit	IgG	+++	+++
Rat	IgG1	−	−
	IgG2a	−	+++
	IgG2b	−	++
	IgG2c	++	++

[a] − = very weak/no binding; + = weak binding; ++ = acceptable binding; +++ = strong binding.

7.1 Analysis of physiological interactions by immunoprecipitation

Immunoprecipitation is performed under non-disruptive conditions to maintain physiological interactions. *Protocol 14B* can be used to recover and examine the subunit composition of intact oligomeric complexes containing the antigen and other associated macromolecules. The antigen protein is identified in the complex by subsequent immunoblotting or by performing immunoprecipitation under disruptive conditions (*Protocol 14C*).

7.2 Immunoprecipitation under disruptive conditions

In *Protocol 14C*, immunoprecipitation is performed under disruptive conditions in a mixed micelle detergent buffer (RIPA). Most physiological interactions are dissociated under these stringent conditions, and usually only the antigen protein and very tightly-associated proteins are recovered by immunoprecipitation. This method is useful to quantitate levels of [^{35}S]methionine labelled protein antigen, and to study synthesis and degradation rates by pulse chase analysis.

7.3 Troubleshooting high background signals

The most common problem encountered in immunoprecipitation reactions is a high background due to non-specific binding of proteins to the antibody and/or the beads. The steps outlined below may remedy this problem:

(a) Pre-clear the protein sample prior to adding the specific monoclonal antibody. This is done by incubating the sample with an antibody that does not recognize the antigen followed by incubation with Protein A–Sepharose beads to remove the antibody. Rabbit immunoglobulins are recommended due to their high affinity for Protein A.

(b) Pre-incubate the beads with competitor proteins, such as BSA or a cell lysate, to block the sites of non-specific interaction (see *Protocol 14A*).

(c) Centrifuge the protein sample at 13 000 *g* for 30 min at 4 °C before adding the specific monoclonal antibody.

(d) Use more stringent wash conditions (e.g. increase the detergent concentration).

(e) Increase the number of washes.

Protocol 13

Metabolic labelling of cellular proteins with [^{35}S]methionine

Reagents

- [^{35}S]methionine
- PBS (see *Protocol 1*)

- Methionine-deficient growth medium
- Complete growth medium

Method

1. Take one 90 mm dish of adherent tissue culture cells growing in monolayer culture (equivalent to approx. 10^7 cells).

2. Remove the growth medium and rinse the cells once in pre-warmed PBS.

3. For pulse-labelling, add 0.5 mCi [^{35}S]methionine in 2 ml methionine-deficient growth medium and incubate the dish of cells at 37 °C (or appropriate temperature) for up to 2 h. For very short pulses (e.g. 5 min), pre-incubate the cells in methionine-deficient medium for 30 min to reduce the cellular pools of methionine.

4. If pulse-labelling is followed by a chase period, remove the radioactive medium and dispose of appropriately. Wash the cells twice with pre-warmed complete growth medium, add 2 ml complete medium, and incubate the cells further for the desired time.

5. For steady state labelling, add 1 mCi of [^{35}S]methionine in 2 ml complete growth medium and incubate the dish of cells at 37 °C for 16 h.

6. After radiolabelling, remove the growth medium from the cells and rinse twice with PBS prior to cell lysis.

7. To increase the total protein levels, the ^{35}S-labelled cell lysates can be mixed with non-labelled cell lysates in the ratio of 1:9 to increase the total protein levels.

Protocol 14

Immunoprecipitation

Equipment and reagents

- Rotating wheel
- PBS (see *Protocol 1*)
- Protein A–Sepharose beads
- Primary monoclonal antibody stock solution
- Secondary antibody stock solution
- 0.5% Triton X-100 in breaking buffer: 50 mM Hepes, 90 mM KCl pH 7.2

- 1% Triton X-100 in breaking buffer: 50 mM Hepes, 90 mM KCl pH 7.2
- 1 × RIPA buffer: 50 mM Hepes, 100 mM NaCl, 1% sodium deoxycholate, 1% Triton X-100, 0.1% SDS pH 7.2
- 2 × RIPA buffer: 100 mM Hepes, 200 mM NaCl, 2% sodium deoxycholate, 2% Triton X-100, 0.2% SDS pH 7.2

A. Preparation of Protein A–Sepharose beads

1. For Protein A–Sepharose beads supplied as a suspension in a liquid preservative such as 20% ethanol (e.g. Sigma product, P-9424): decant the slurry of beads into a 15 ml Falcon tube and centrifuge at 400 g for 5 min at 4 °C. Remove the liquid by aspiration and wash the beads five times in 15 ml immunoprecipitation buffer. Resuspend the beads in an equal volume of immunoprecipitation buffer to give a 1:1 slurry. Store the slurry of beads at 4 °C (sodium azide may be added to 0.02% to prevent microbial growth).

Protocol 14 continued

2. For Protein A–Sepharose beads supplied as a lyophilized powder (e.g. Sigma product, P-3391): add 20 ml PBS to the container of dried beads and place on a rotating wheel for 16 h at 4 °C to allow beads to swell. Proceed as described in part A, step 1.

3. Non-specific binding sites on the beads may be blocked by pre-incubation of the beads with BSA or a non-labelled lysate prepared from the same cell type as that used for immunoprecipitation. The beads pre-incubated with lysate are not suitable for use in immunoprecipitation reactions that will be subsequently analysed by immunoblotting, since the monoclonal antibody may react with target protein that has bound non-specifically to the beads. Incubate the beads with 1 mg/ml BSA or non-labelled cell lysate diluted in immunoprecipitation buffer for 1 h at 4 °C on a rotating wheel. Wash the beads as described in part A, step 1.

B. Immunoprecipitation in non-disruptive buffer

1. Add an equal volume of ice-cold 1% Triton X-100 in breaking buffer to 200 µl protein sample in an Eppendorf tube and mix gently.

2. Add 600 µl ice-cold 0.5% Triton X-100 in breaking buffer to the sample, mix, and place on ice.

3. Add 3 µg primary monoclonal antibody, mix, and place on ice for 30 min to allow the antibody–antigen complexes to form.

4. If the primary monoclonal antibody does not bind strongly to Protein A, add 6 µg of a secondary anti-immunoglobulin antibody that exhibits strong binding (e.g. rabbit anti-rat IgG) to form a bridge between the primary monoclonal antibody and the Protein A–Sepharose. Mix and place on ice for 30 min.

5. Add 200 µl of a 1:1 slurry of Protein A–Sepharose in 0.5% Triton X-100 in breaking buffer to the sample (equivalent to 100 µl packed volume of beads). Place the sample on a rotating wheel for 3 h at 4 °C to allow the antibody–antigen complex to be captured by the Protein A–Sepharose beads.

6. Pellet the beads by centrifugation at 13 000 g (top speed of benchtop microcentrifuge) for 2 min at 4 °C and remove the supernatant by aspiration.

7. Wash the beads three times with 1 ml of 0.5% TX-100 in breaking buffer. At the third wash, transfer the sample to a new Eppendorf tube.

8. Store the immunoprecipitates at −20 °C.

9. To elute the immunoprecipitates for analysis by SDS–PAGE, add an equal volume of 2 × SDS–PAGE sample buffer, mix well, and heat at 100 °C for 2 min. Pellet the beads by centrifugation at 13 000 g, and remove the supernatant.

10. To elute the immunoprecipitates for analysis by IEF or NEPHGE, add an equal volume of IEF or NEPHGE sample buffer, mix well but do not heat. Pellet the beads by centrifugation at 13 000 g, and remove the supernatant.

C. Stringent immunoprecipitation in RIPA buffer

1. Add an equal volume of ice-cold 2 × RIPA buffer to 200 μl volume of sample in an Eppendorf tube, and mix gently.

2. Add 600 μl ice-cold 1 × RIPA buffer to the sample, mix, and place on ice.

3. Add 3 μg primary antibody, mix, and place on ice for 30 min to allow the antibody–antigen complexes to form.

4. If necessary, add 6 μg of a secondary anti-immunoglobulin antibody that exhibits strong binding to Protein A. Mix and place on ice for 30 min.

5. Add 200 μl of a 1:1 slurry of Protein A–Sepharose in 1 × RIPA buffer to the sample (equivalent to 100 μl packed volume of beads). Place the sample on a rotating wheel for 3 h at 4 °C to allow the antibody–antigen complex to be captured by the Protein A–Sepharose beads.

6. Pellet the beads by centrifugation at 13 000 g for 2 min at 4 °C and remove the supernatant by aspiration.

7. Wash the beads three times with 1 ml of 1 × RIPA buffer. At the third wash, transfer the sample to a new Eppendorf tube.

8. Store the immunoprecipitates at −20 °C.

9. Proceed as described in part B, steps 9 or 10.

8 Epitope mapping

An advantage of using a monoclonal antibody for detection of an antigen is the specificity of the interaction due to the recognition of a single epitope. Epitope mapping is described in Chapter 1. Several approaches to map the epitope sequence are outlined below:

(a) Cleveland mapping, in which fragments produced by protease digestion of the antigen polypeptide are screened by immunoblotting to determine which fragment contains the epitope sequence.

(b) Epitopes are mapped by screening a peptide array immobilized on a solid phase matrix, such as activated membrane or polyethylene pins.

(c) The importance of individual amino acid residues in the epitope sequence is established in competition ELISA assays using soluble peptides.

Often the epitopes recognized by monoclonal antibodies are only four or five amino acid residues, and therefore, other proteins that share the epitope sequence are also detected by the monoclonal antibody. These epitope-related proteins are readily distinguished by PAGE separations, but not by immuno-fluorescence staining or ELISA. In some cases, a common epitope may indicate a structural or functional similarity of the epitope-related proteins, however, this should be established by further studies.

9 Further applications of monoclonal antibodies in protein characterization

After the initial characterization of protein antigens has been performed, more advanced studies can be undertaken with monoclonal antibodies:

(a) Structural studies.

   (i) Decoration of proteins with monoclonal antibodies and visualization in the electron microscope.

   (ii) Use of monoclonal antibodies as conformational probes; for example, ELISA competition assays or native band shift assays.

(b) Functional studies.

   (i) Immunodepletion of cell extracts using monoclonal antibodies to inactivate the antigen protein *in vitro*.

   (ii) Microinjection of monoclonal antibodies into cells to inactivate the antigen protein *in vivo*.

(c) Immunoaffinity purification under native conditions. Purification of oligomeric complexes containing the antigen protein, and identification of components of the complex by peptide mass mapping (11).

Acknowledgements

I would like to thank Prof. K. Willison, Prof. A. Ashworth, Dr J. Grantham, and Dr D. Bertwistle for helpful discussions.

References

1. Harlow, E. and Lane, D. (1988). In *Antibodies: a laboratory manual* (ed. E. Harlow and D. Lane), pp. 53–319. Cold Spring Harbor Laboratory Press, Cold Spring Harbor, New York.
2. Hames, B. B. (1990). In *Gel electrophoresis of proteins: a practical approach* (2nd edn) (ed. B. D. Hames and D. Rickwood), pp. 1–30. IRL Press, Oxford.
3. Makowski, G. S. and Ramsby, M. L. (1997). In *Protein structure: a practical approach* (2nd edn) (ed. T. E. Creighton), pp. 1–26. IRL Press, Oxford.
4. Laemmli, U. K. (1970). *Nature*, **277**, 680.
5. Towbin, H., Staehelin, T., and Gordon, J. (1979). *Proc. Natl. Acad. Sci. USA*, **76**, 4350.
6. Van Dam, A. (1994). In *Protein blotting: a practical approach* (ed. B. S. Dunbar), pp. 73–84. IRL Press, Oxford.
7. Corbett, J. and Dunn, M. J. (1992). In *Methods in molecular biology: biomembrane protocols* (ed. J. M. Graham and J. A. Higgins), Vol. 19, pp. 219–27. Humana Press, New Jersey.
8. O'Farrell, P. (1975). *J. Biol. Chem.*, **250**, 4007.
9. Hinton, R. H. and Mullock, B. M. (1996). In *Subcellular fractionation: a practical approach* (ed. J. M. Graham and D. Rickwood), pp. 31–69. IRL Press, Oxford.
10. Bollag, D. M. and Edelstein, S. J. (1991). In *Protein methods* (ed. D. M. Bollag and S. J. Edelstein), pp. 143–60. Wiley and Son Inc., New York.
11. Jensen, O. N., Shevchenko, A., and Mann, M. (1996). In *Protein structure: a practical approach* (2nd edn) (ed. T. E. Creighton), pp. 29–56. IRL Press, Oxford.

Chapter 13
Immunoassays

Jane V. Peppard

Novartis, Arthritis Biology Research, LSB 3293, Novartis Pharmaceutical Corporation, 556 Morris Avenue, Summit, New Jersey 07901, U.S.A.

1 Introduction

This chapter describes methods by which monoclonal and also polyclonal antibodies may be used in order to quantitate antigens. Since the choice of format is extensive, the information is divided so that options for each stage of the assay, e.g. choices of label or solid support, which can be 'mixed and matched' according to requirements, are discussed individually. Finally, the most frequently used methods of immunoassay and their set-ups are described.

2 General considerations

2.1 Selecting an antibody

Monoclonal antibodies are often considered superior to polyclonal antibodies simply because of their exquisite specificity and homogeneous structure. In the context of immunoassay, however, this is not always an advantage. For example, when the antibody is required to be bound to a solid support or be labelled and still retain activity, the loss through denaturation of a sensitive antibody species leaves many others in a polyclonal preparation; not so for a monoclonal. Monoclonal preparations are also less avid in binding than polyclonals, leading to lower performance in immunoassay (1). That said, successful assays are often set up with monoclonals, in competitive binding or by using their high specificity to 'capture' the antigen, followed by a second monoclonal or polyclonal antibody for detection purposes. Mixtures of monoclonal antibodies for either capture or detection can also be used effectively.

Since monoclonal antibodies are specific, if more than one is to be used, it is important to check that the epitopes that each recognize are not identical or overlapping. 'Epitope mapping' can be done by adsorbing a monoclonal to microtitre plate wells, binding antigen, and then checking other monoclonal antibodies for binding, or by competitive binding methods (e.g. ref. 2). This is iterative and time-consuming and can be a problem in terms of detecting the binding of the various antibodies which either have to be labelled or specifically detected by a further labelled antibody. However, it directly mimics an immunoassay. Antibody–antigen interactions can be more elegantly measured using surface plasmon resonance (3–5), where no labelling is required to visualize the protein–protein interactions, the measurements are in real time (a few minutes each), and association and dissociation constants for the different antibodies can be directly compared. This technique is especially suited to fast epitope mapping of many antibodies.

2.2 Selecting an assay standard

An alternative aspect of specificity to consider is that of the antigen, for the same epitope may be present on more than one antigen species. This is an important concept which is often overlooked in favour of the belief that since monoclonals are specific reagents, an assay employing them will be equally specific for the expected antigen.

The best antigen standard is one which is identical to the antigen contained in the samples which are to be assayed. A similar dilution series of standard antigen and 'unknown' antigen should, when plotted out as signal versus dilution factor, give lines which are parallel. If the lines are not parallel then reading antigen concentration against the standard curve will show 'trending' in the results over several different dilutions (see *Figure 1*).

For many proteins, a purified protein standard is ideal and straightforward to incorporate since the concentration is known. However, in reality such standard/sample accord is often not possible to obtain because the antigen contained in the sample may be a mixture of species. For example, it may come in a variety of molecular sizes (e.g. monomeric versus oligomeric forms of immunoglobulin A), or it may exist in complexes with other molecules (e.g. a protease bound to a macromolecular inhibitor, or a ligand bound to its soluble receptor), or as a

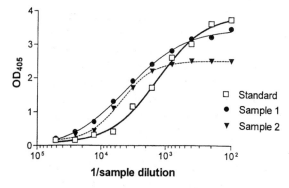

Figure 1 Comparison of readings from 1:2 dilution series of a standard and two samples in a 'sandwich' ELISA format. Sample 1 represents a sample exhibiting matrix interference: if interpolated from the standard curve, the lower dilutions of this sample would show a trend of increasing concentration with increasing dilution. Sample 2 represents a sample where antigen is non-identical with standard and only partially recognized by the antibodies used in the assay.

proform as well as the active form (e.g. proenzymes versus 'mature' form). These forms of non-identity will affect the quantitation if a purified and restricted protein standard is used, and are usually difficult or impossible to factor into the calculation later because of sample-to-sample variation.

In some cases it may be possible to use a single sample to provide the standard antigen, providing it contains a similar mix of molecular species. This can also be useful if a purified protein standard is not available or scarce. Samples may then be designated as '% standard sample', or in other units if the standard sample can be alternatively quantitated by some means (e.g. its functional activity).

2.3 Sample matrix

Assay interference by matrix, i.e. everything present in the reaction mixture other than the antigen and antibody, is common. As with other problems (see above), it is signalled by 'trending' in the results obtained from a series of dilutions of the sample read against the standard curve (*Figure 1*). Often it can be corrected simply by diluting the standard in the same matrix, e.g. diluted serum or tissue culture medium. Alternatively, if interference is removed by diluting samples sufficiently into assay buffer (see also Section 5.3), the assay may be set up at the sensitivity required to enable measurement of samples at such dilutions. Thus, samples from different origins, e.g. serum and tissue culture supernatants, should be compared using the standard and procedure adjusted as necessary to take account of matrix interference. This effect on quantitation can also be minimized if samples with similar matrix which are being compared are diluted to the same extent, e.g. everything is compared at dilutions of 1:100 and 1:200, and then normalized to a control.

2.4 Sample preparation and dilution

Special sample preparation may be necessary for certain antigens. For example, in assays for the eicosanoid PGE_2 in serum, it is often recommended that samples be extracted by protein removal before PGE_2 is assayed. If such extraction is necessary it should be carefully controlled for by including in every assay set samples of standard antigen 'spiked' into sample matrix, to check that recovery is uniform. Alternatively, it is often the case that sufficient dilution of sample may largely overcome the need for extraction. The errors produced by residual matrix interference may be less significant than those introduced by extraction methods, and also more uniform between samples.

Sample dilution is also an important consideration. The dilution procedure itself can introduce errors, and obviously very large dilutions can lead to proportionately large errors. This is particularly important with reference to the preparation of the standard curve between assays. Often, purified standard antigens are stored at relatively high concentrations to optimize stability. When devising an assay, it may be better to develop a less sensitive form of the assay rather than making an exquisitely sensitive assay where everything has to be diluted a thousand-fold. If large dilutions of any of the reagents are inevitable it is essential to devise a dilution protocol which is maintained for every assay where pipetting devices are kept calibrated to ensure interassay reproducibility. It is desirable that the standard antigen and samples are diluted using the same protocols, e.g. an initial 1:100 dilution and a set of serial 1:2 dilutions thereafter. Small errors in procedure or instrument calibration are thereby controlled by being introduced uniformly between standard and samples.

2.5 Assay turnaround time

When devising an assay it is important to consider how long is optimal for its completion once it becomes routine. This is something that is user-controllable during the setting up of a new protocol, but is best not changed once the assay is established. Unless very fast turnaround times are absolutely required, e.g. for samples in clinical laboratories, it may be more pragmatic for day-to-day operation to plan on a slower procedure. It is often the case that suppliers of commercial immunoassay 'kits' promote them on the brevity of their completion time when in reality the assay would be more user-friendly and robust (not to mention more sensitive) if the steps took longer.

Often, greater sensitivity will be obtained by longer incubations. As a rough rule of thumb, for most immunoassays a three to four hour incubation at around 20 °C (RT) will give near maximal binding for that step (*Figure 2*); equally effective is an 'overnight' incubation at 4 °C. Once equilibrium is reached, the assay may be held at that point for longer if necessary without significant change to the readout. An exception to this is the coating step for immunoassays which utilize solid supports to bind protein antigens or antibody which are applied in dilute solution. For most proteins, this binding occurs much faster, plus extended periods of binding may lead to unacceptable levels of inactivation of some antibodies. Incubations of two to three hours at RT are usually sufficient for this step, although it does depend on the protein, e.g. streptavidin or horseradish peroxidase require overnight binding to microtitre

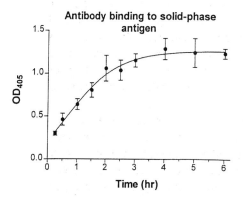

Figure 2 Binding of antibody to an antigen adsorbed to a microtitre well. Incubation was done at RT.

plates to achieve acceptable levels. While incubation at 37 °C will speed binding, if not properly controlled it may introduce artefacts, such as 'edge effects'
in 96-well plates, where the outer wells spend a disproportionate time at 37 °C
compared to the better-insulated inner wells. In general, if an assay is designed
where each step is allowed to come to equilibrium, interassay variation of the
standard curve will be minimized and, unless the process is entirely automated,
results in a more flexible protocol. However, as a caveat, sample stability may
occasionally influence choice of temperature and duration of assay. Finally,
especially where sample extraction is necessary or large sample numbers are
expected, it is also important to factor in a consideration of the time for sample
preparation, and wash steps.

3 Solid support and separation options

One of the first questions when considering what conformation a new assay is
to take is: how is the quantitative signal to be separated from the surrounding
matrix? Proteins bind tightly to plastic surfaces (6) and, if it is possible to adhere
an antibody or antigen to a solid support of some kind without harming its
activity, a variety of microtitre plates or microspheres are available for this
purpose. If the reaction needs to be done in solution and the separation carried
out afterwards, then, for technical ease of manipulation, a particle which can be
added to the reaction solution, or some other means of precipitating the required activity from solution, is necessary. This section describes some of the
more popular techniques.

3.1 Microtitre plates

Microtitre plates specifically intended for immunoassay (those for tissue culture
are not suitable) are available from many suppliers as a rigid polystyrene plate.
Peptides or proteins can be permanently adhered directly to the well surfaces.
The most common format is a 12 × 8 configuration, although 24 × 16-well
plates (and higher) are also available. If the wells need to be separated after the
assay, PVC plates, which can be cut with scissors, or snap-off polystyrene well
strips which can be fitted into a 12 × 8 frame, are available. Microtitre plates
can be purchased pre-coated with material such as streptavidin, Protein A,
antibodies, or wheat germ agglutinin, pre-activated for protein binding, or
impregnated with scintillant for use in scintillation proximity assays (see SPA in
Section 3.3).

Microtitre plates are limited since they restrict the volume of the reaction,
and thus the amount of sample that can be applied. Where this is a problem,
polystyrene particles (or other solid bead) to which antibody is adsorbed, can be
added to any reaction volume. Systems where analytes are absorbed to membranes (e.g. nitrocellulose) and then vacuum filtered through in sequence (e.g.
the Pierce ELIFA system) are also less limited with respect to volume, although
technically more difficult to use successfully.

3.2 Precipitation from solution

If a reaction involving antibodies has to be performed in solution, then a reagent which brings the antibody out of solution can be added at the end. The precipitate is then collected and washed, and subjected to a suitable detection step. This is most often used for radioactively labelled antigens.

3.2.1 Double antibody precipitation

This technique utilizes the propensity of polyclonal antibodies to form insoluble crosslinked complexes at antibody/antigen equivalence. The specific antibody (e.g. mouse IgG1 anti-Ag) is allowed to bind to antigen in solution and then a second, polyclonal, antibody (e.g. goat anti-mouse IgG) is added to precipitate the first antibody. Since the amount of the specific antibody (mouse IgG1) is quantitatively small, normal mouse IgG or mouse serum is added at the same time to increase the amount of precipitate that will form between it and the goat anti-mouse IgG, so that a visible pellet is made once the mixture is centrifuged. The precipitin reaction has to be pre-titrated carefully to get these proportions correct, and thorough mixing of analytes is key. It also takes time, usually overnight at 4 °C, for the precipitation to reach completion. The precipitate is removed by centrifugation and must be washed and re-centrifuged several times to reduce non-specific background. An advantage of this method is that none of the antibodies need be purified.

3.2.2 Polyethylene glycol

Addition of polyethylene glycol (PEG) 6000 in PBS to a final concentration of 14% is effective in precipitating immunoglobulin (including antibody–antigen complexes) from solution, in about one hour at RT. Precipitates can be conveniently harvested onto filters using a cell harvester apparatus, followed by a wash through with 15% PEG/PBS solution to reduce non-specific binding. Radiolabelled reagents can then be counted on the filters.

3.2.3 Magnetic particles

Magnetic particles with a variety of reagents can be obtained from suppliers such as Perseptive Biosystems or Accurate Chemical Co. Antibody (e.g. mouse IgG1) can be removed from solution by adding an excess of second antibody (e.g. rabbit anti-mouse IgG) coupled to magnetic particles and incubating for one to two hours. These are then attracted into a pellet using a strong magnet, highly speeding the washing steps. Centrifugation is also an option. One drawback with this method compared to the previous two is that samples (e.g. serum) containing amounts of IgG of the same type as the specific antibody, or cross-reactive with it as far as the 'magnetic antibody' is concerned, will lead to lower precipitation of the specific antibody. Addition of more particles or the adjustment of reagents may take care of this problem. For example, goat anti-mouse IgG is usually much more cross-reactive with human, rabbit, and mouse IgG than rabbit anti-mouse IgG is with human IgG.

3.3 Scintillation proximity

Scintillation proximity assays (SPA) may be used to measure ligands labelled with ^{125}I or other isotopes such as ^{14}C, ^{33}P, ^{32}P, ^{35}S, and ^{45}Ca, using a β-counter. In this technique, scintillant is incorporated into the plastic of microtitre plates (NEN or Wallac), or into fluoromicrospheres which are derivatized with molecules such as antibodies or wheat germ agglutinin (Amersham Pharmacia). Antibodies or antigens can be bound to the plate well or the bead, and assays employing radiolabelled antigen or antibody carried out. If a radioactive molecule is bound by the bead or well, it is brought into close proximity to stimulate the scintillant to emit light. Unbound radioactivity is too distant from the scintillant: the energy released is dissipated before reaching the solid support and is not detected. This eliminates the need for separation of bound and free radioligand (although in practice emptying the plates or spinning down the beads does improve the signal:background ratio), a major saving in time and effort.

4 Labelling and detection of antibody or antigen

There is a wide choice of reagents available today to label molecules for detection in immunoassay systems. It is important that structural changes to the labelled molecule, introduced either through addition of the labelled moiety itself or through the reagents used in the conjugation process, are minimal so that they do not lead to significantly diminished recognition of the antigen by the antibody. This aside, the choice of label largely depends on the sensitivity required and the local availability of the technology to recognize it. Several of the more frequently utilized techniques are described.

4.1 Enzyme labelling

The enzymes which are today most widely used for detection in quantitative immunoassay are horseradish peroxidase and alkaline phosphatase, and these are discussed in the following sections. Some discussion of the types of substrate currently available is included but these are continuously changing and improving. The various suppliers of the enzymes and substrates often provide very good overviews of their products and these should be consulted for the latest developments; ultimately however, the choice of system depends on the sensitivity required and the instrumentation available to measure the output. In general, chromogenic substrates can be very sensitive but the dynamic range (i.e. the range of linear readout of the assay) is limited by the absorbance cut-off of the spectrophotometer, plus reading above absorbances of 2.0 may not be linearly proportional. Luminometers or fluorometers usually claim a dynamic range of 4–5 logs, thus larger range standard curves can be set up, which is useful if samples contain a wide range of possible values.

4.1.1 Horseradish peroxidase

Horseradish peroxidase (HRP) (M_r 40 000) is one of the most commonly used enzymes in immunoassay techniques. Many antibodies are commercially available already labelled with HRP. HRP may be coupled with reasonable recovery to proteins via sodium borohydride (7) or glutaraldehyde coupling (see *Protocol 1*), or kits are available for conjugation of maleimide-activated HRP to antibody or other proteins from Pierce. HRP has a fast catalytic rate and thus development times are comparatively short. The choice of substrate affects both the sensitivity which will be obtained and the dynamic range of the assay. A variety of chromogenic substrates are available from various suppliers (e.g. Pierce, KPL, Sigma), including 3,3',5,5'-tetramethylbenzidine (TMB), *ortho*-phenylenediamine (OPD), and 2,2'-azino-di-(3-ethylbenzthiazoline-6-sulfonate) (ABTS). Chemiluminescent substrates, such as luminol-based products, are also available from the same manufacturers and others; these have a much wider dynamic range than chromogenic substrates. Disadvantages of HRP include substrate inhibition of enzyme activity so that development reactions are short by necessity rather than choice, the toxic hazard of some of the more common substrates, and, for some substrates, an extra step to halt the enzyme/substrate reaction through the addition of a 'stop' solution in order to gain the maximum sensitivity. HRP conjugates may also have a relatively short shelf-life.

4.1.2 Alkaline phosphatase

Alkaline phosphatase (AP) ($M_r \sim 140\,000$) is also commonly used as a detection enzyme for immunoassay. Many antibodies are available for purchase already labelled with AP. Compared to HRP, AP has a lower catalytic rate and development of substrate is thus slower. However, it has the advantage that it is not self-limiting due to substrate inhibition of enzyme and reaction rates remain linear over a long period of time (up to 18 hours at least if the absorbance remains < 2) (*Figure 3*). Therefore, signal:background ratios and sensitivity can be improved by allowing substrate development to proceed for longer periods (if necessary, substrate development can be stopped by adding a half-volume of 5 M NaOH). In addition, substrates available for AP are less toxic. A variety of substrates is available, both chromogenic, e.g. Sigma 104® , *para*-nitrophenyl phosphate, or BluePhos® (5-bromo-4-chloro-3-indolyl phosphate) (KPL) and chemiluminescent,

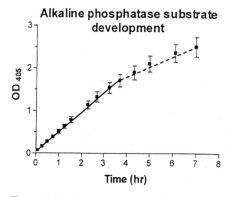

Figure 3 Development of Sigma 104® substrate with an alkaline phosphatase labelled second antibody.

e.g. LuciGLO™ (KPL) or CSPD® with enhancers such as Emerald-II™ from Tropix. Finally, the procedure for labelling antibody, or other proteins, with AP while still retaining its enzyme activity is straightforward, and the product is quite stable in long-term storage. Kits for labelling with AP can also be purchased (Pierce).

Protocol 1

Alkaline phosphatase labelling of IgG antibody[a]

Reagents

- PBS: 0.05 M phosphate-buffered saline pH 7.4
- Tris buffer: 0.05 M Tris pH 8.0, containing 1 mM $MgCl_2$
- Sodium azide
- Purified IgG antibody (preferably affinity purified)

- 25% glutaraldehyde (Grade 1, Sigma)
- Bovine intestinal alkaline phosphatase (e.g. Type XXX-T, Sigma, an affinity purified solution in NaCl)
- Bovine serum albumin (BSA) (e.g. RIA grade Fraction V powder, Sigma)

Method

1. Add 1 mg of IgG in 0.7 ml PBS to 3.5 mg ($\sim$ 3500 U) alkaline phosphatase. If alkaline phosphatase containing ammonium sulfate is used, dialyse against 500 ml PBS overnight at 4 °C.

2. Add 25% glutaraldehyde to make a final concentration of 0.2% (v/v) and mix gently.

3. Incubate 1–2 h at RT, then dialyse overnight at 4 °C against 500 ml PBS; change buffer once. Transfer dialysis to 500 ml Tris buffer and dialyse at 4 °C overnight; change buffer once. Note: for immunoassay, separation of free enzyme is usually not required, but can be done using a suitable gel filtration column.

4. Remove from dialysis and add BSA to 1% (w/v) and sodium azide to 0.05% (v/v). Store at 4 °C in the dark.

[a] This procedure can be adapted for HRP. Use 5 mg HRP (Type VI, Sigma) to 1 mg IgG and do not add azide as a preservative.

4.2 Biotin labelling

Biotin is a small (244 Da) vitamin found in tissue and blood which binds with an extremely high affinity (K_a = 10^{15} M^{-1}) for avidin. Once formed, this bond is extremely strong, being resistant to extremes of pH, organic solvents, and regular denaturing agents. It can however be released by 8 M guanidine or autoclaving. Avidin is a $\sim$ 68 kDa glycoprotein consisting of four subunits, each of which binds one biotin molecule. Avidin is found in egg white of birds and amphibia; a non-glycosylated form, streptavidin, is found in *Streptomyces*. Conjugation of biotin to a protein can be carried out using sulfonated N-hydroxysuccinimide esters of biotin which react with primary amines of the protein. The biotinylated protein can then be detected with a variety of reagents which are commercially available already bound to avidin or streptavidin, such as alkaline phosphatase or horseradish peroxidase (Section 3.2).

The small size of the biotin moiety is an asset since it is less likely to hinder antigen–antibody interactions. The availability of a variety of labels coupled to avidin is also an advantage. However, one drawback to the use of biotin–avidin is the propensity of avidin or streptavidin to bind non-specifically to other proteins or carbohydrates, giving rise to high backgrounds and consequent decrease in signal-to-noise ratio. Depending on the assay format, this is not always a problem, and may also be somewhat reduced by the use of higher ionic strength NaCl (0.3–0.5 M) or 200 mM α-methyl-D-mannoside in the buffer matrix. New derivatives of avidin, such as the NeutrAvidin marketed by Pierce, may in the future decrease this problem.

Protocol 2

Biotinylation of IgG

Equipment and reagents

- Centrifuge with fixed angle rotor capable of delivering 4000 g (e.g. Sorvall SS-34), set to 10 °C
- Centricon-30 (Amicon) 2 ml centrifugal microconcentrators (30 kDa M_r cut-off)
- PBS-A buffer: 0.05 M phosphate-buffered saline pH 7.3, 0.05% NaN_3

- 0.1 M Hepes buffer pH 8.0
- BSA
- Affinity purified IgG: 1.1 mg/ml in Hepes buffer
- Sulfo-NHS-biotin (Pierce Chemical Co.)

Method

1. Add 0.9 ml IgG solution into a Centricon-30. If a buffer exchange of the IgG is necessary this can be accomplished by repeated dilution into Hepes buffer and re-concentration to the starting volume, until pH 8.0 is achieved.

2. Make up Sulfo-NHS-biotin at 0.887 mg/ml in Hepes buffer and *immediately* add 0.1 ml to the IgG in the Centricon. Mix gently by tapping.

3. Incubate mixture for 30 min at RT.

4. Remove unconjugated biotin: add enough Hepes to fill the Centricon and centrifuge at 4000 g and 10 °C for 30 min or until only a small volume remains in the top compartment. Repeat dilution and re-concentration four more times. The unconjugated biotin will come through the filter leaving the IgG–biotin conjugate in the top compartment. Final volume should be around 0.2 ml.

5. Add PBS-A to give a 1 ml volume and BSA to 0.05%. Store at 4 °C.

4.3 Labelling with lanthanide fluorophores

Europium, samarium, and terbium conjugated antibodies can be prepared and used in immunoassays utilizing time-resolved fluorescence (8). The antibody–

lanthanide conjugate is essentially non-fluorescent and the lanthanide has to be released by the addition of a dissociating reagent to be detected. Compared to conventional fluorophores, lanthanide chelates have a 20–30 times longer fluorescence decay time and also a long Stokes shift so that the wavelength of the emitted light is 200–300 nm longer than that of the excitation light used. Time-resolved measurement starts only after background fluorescence from, for example, protein, has already decayed, thus leading to excellent signal-to-noise ratios compared to other fluorescent probes. The emission wavelength is also unique for each lanthanide fluorophore so that they can be used in multilabel assays. Kits for labelling proteins with Eu or Sm, or a small selection of Eu labelled reagents, are available from Wallac.

4.4 Radioactive labelling

The radioactive tag most frequently used in radiolabelling proteins is ^{125}I since it is easily conjugated and can be readily measured. It can be introduced by several methods. Highest specific activity is achievable using the Chloramine-T method which incorporates ^{125}I into tyrosines; however, the oxidizing and reducing reagents used tend to damage the protein and must be removed immediately using a separation column. Alternatively, Iodogen (1,3,4,6-tetrachloro-3α,6α-diphenylglycouril) (Pierce) also provides an efficient mechanism to incorporate ^{125}I into tyrosines (9). Since it is insoluble, it is used after being coated onto microcentrifuge tubes or polystyrene beads; the reaction is terminated by the removal of the protein from the Iodogen rather than through the addition of reducing reagents, which is less damaging to the protein. If the protein is particularly sensitive to these methods, or does not contain tyrosine, then [^{125}I]Bolton-Hunter reagent (N-succinimidyl 3-(4-hydroxy, 5-[^{125}I]iodophenyl)-propionate) may be used (Amersham Pharmacia) and then reacted under mild conditions via an active ester reaction with primary amines of the protein.

Protocol 3

Labelling of IgG with ^{125}I by the Chloramine-T method

Equipment and reagents

- Gamma counter
- 0.1 M phosphate buffer pH 7.0
- PBS-A buffer: 0.05 M phosphate-buffered saline pH 7.3, 0.05% NaN$_3$
- Chloramine-T: 25 mg/5 ml phosphate buffer
- Sodium metabisulfite: 48 mg/5 ml phosphate buffer

- 1% BSA in PBS-A
- Sephadex G25 or similar desalting column ($\sim$ 1 $\times$ 15 cm)
- ^{125}I (5 mCi, carrier-free) (Amersham Pharmacia)
- IgG at 1 mg/ml in phosphate buffer or PBS^a

Note: this protocol should be performed only by persons trained in the use of radio-isotopes. Perform the entire operation in a designated fume cabinet; use suitable safety precautions during the procedure, and approved methods of radioactive waste disposal.

Method

1. Unless it is purchased pre-blocked, prepare the desalting column by adding 1 ml of 1% BSA to the top of the bed, running it in under gravity, and then rinsing with a column volume of PBS-A. Run the buffer down to the top of the gel bed. In a rack, set up a row of about fifteen ~ 2 ml glass or polypropylene tubes for fraction collection.

2. Add 0.5 ml phosphate buffer to the ^{125}I solution. To a microcentrifuge tube, add 0.1 ml IgG solution (100 μg). Add 10–25 μl of the ^{125}I solution, depending on the desired level of activity. Mix gently by tapping the tube.

3. Dissolve the Chloramine-T, add 25 μl to the tube, and immediately close the lid. Mix gently.

4. Incubate 4 min at RT.

5. Make up the sodium metabisulfite solution and immediately add 25 μl to the iodination mix. Mix gently.

6. Immediately commence the separation of free ^{125}I from the bound: add the reaction mixture to the top of the separation column and allow it to run in. Rinse the reaction tube with 0.1 ml PBS-A and run in. Either set up the column to run from a buffer reservoir or add 0.5 ml PBS-A volumes sequentially to the top of the bed. Collect 0.5 ml fractions and count 10 μl of each in a gamma counter. The first peak of activity (eluting in the void volume) contains the labelled IgG, the second contains free ^{125}I; if the column was not overloaded, they should be well separated. Pool the fractions containing the labelled IgG and estimate the volume; count 10 μl and determine the c.p.m./ml.

7. Calculate the specific activity of the labelled material, using 1 μCi = 2.22 × 10^6 d.p.m. For example, assuming a 90% recovery of protein from the column in a volume of 3 ml at 10^7 c.p.m./ml, and a counter efficiency of 80%:

$$\text{Protein concentration} = (100 \times 0.9)/3 = 30 \text{ μg/ml.}$$

$$\text{Activity} = [10^7 \times (100/80)]/2.22 \times 10^6 = 5.6 \text{ μCi/ml.}$$

Specific activity = activity/protein concentration = 5.6/30 = 0.19 μCi/μg.

8. Bring the ^{125}I-labelled IgG to a final concentration of 1% (w/v) BSA. Store at 4°C. With time, the radiolabel will decay (^{125}I t$_{1/2}$ = 60 days) and the protein will also degrade through autoirradiation; depending on the antibody and specific activity, binding activity may be significantly reduced by one month after radiolabelling.

[a] NaN$_3$ is a very effective inhibitor of the Chloramine-T reaction (10). If antibodies were stored with azide, extensive dialysis will be necessary to remove it.

4.5 Mass spectrometry

Antibodies can be covalently coupled to a solid support and bound antigen eluted directly onto a mass spectrometer tip before analysis by time-of-flight mass spectrometry (11). By directly detecting the antigen at its characteristic mass-to-charge value, this novel type of immunoassay (MSIA) introduces an additional component of specificity over that derived from the specificity of the antibody, and can analyse multiple antigens in a single assay.

5 Setting up an assay

The type of assay selected will be governed by several factors. Availability of specific antibodies is the major consideration. If only one monoclonal antibody is available for a particular antigen, then the assay will take the form of a competitive inhibition assay. If more than one monoclonal and/or a polyclonal preparation is available then several formats are feasible. Another possibility is to exploit the binding of antigen to a non-antibody component—for example, assays for serum amyloid P (SAP) can be performed based on the detection by a single anti-SAP of the calcium-dependent binding of SAP to keyhole limpet haemocyanin (12). As already discussed in Section 4, the detection method selected will depend on the degree of sensitivity required as well as the instrumentation available. This section will describe the various forms of assay which can be set up using antibodies. For simplicity, the solid support described in these protocols is microtitre plates and the detecting label is alkaline phosphatase with a chromogenic substrate. However, other methods (see Sections 3 and 4) can be substituted as required and the general approach will remain the same.

5.1 Competitive binding immunoassay

As mentioned above, this is the assay format to consider when only one monoclonal (or oligoclonal) antibody species is available to the antigen of choice. It is often used for small molecular weight antigens where steric hindrance would preclude the binding to antigen of more than one antibody molecule, for example eicosanoids such as PGE_2. In a competitive assay, an increase in test antigen concentration leads to a corresponding decrease in the amount of labelled antigen bound to antibody, and so these assays are often called competitive inhibition immunoassays. For troubleshooting purposes, it is important to remember that falsely high, not low, readings will result from interference of matrix components with antigen–antibody binding, since, in this assay format, less binding reads out as more antigen.

5.1.1 Labelled antigen

Figure 4A outlines the basis of a simple competitive binding immunoassay where the antigen has been labelled. Antibody (preferably affinity purified, or an IgG fraction) is coated onto a solid support (e.g. a microtitre plate well), and a constant amount of labelled antigen is allowed to compete for the antibody binding sites with varying amounts of unlabelled standard or sample antigen.

Alternatively, the whole reaction may be carried out in solution (*Figure 4B*) and the antibody–antigen complexes subsequently precipitated from solution. The separation out of solution of antibody–antigen complexes can be achieved

A

Step 1: coat first antibody onto solid support; wash away unbound antibody.

Step 2: add labelled (fixed amount) and unlabelled antigen, which compete for first antibody. Wash away unbound antigen. Detect activity of bound labelled antigen.

B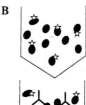

Step 1: mix labelled antigen with standard/sample unlabelled antigen.

Step 2: add specific antibody. Allow labelled and unlabelled antigen to compete for antibody binding sites.

Step 3: separate antibody and wash away unbound antigen. Detect activity of the labelled antigen.

Figure 4 Competitive binding inhibition assay—labelled antigen. (A) Antigen bound to microtitre plates. (B) Reaction in solution before separation of bound and free labelled antigen.

by several different methods (see Section 3.2). This method is most often used when radiolabelled antigen is employed.

5.1.2 Labelled antibody

In another format of competitive binding, purified antigen, or a portion of the antigen (e.g. a protein 'domain' or a peptide), may be coated onto a solid support and allowed to compete with soluble antigen (standard or sample antigen) for binding to a constant amount of labelled antibody (*Figure 5*). Alternatively, the first antibody (e.g. mouse IgG1 anti-Ag) need not be labelled but its binding to the antigen on the solid support can be detected by a further antibody which is labelled (e.g. rabbit anti-mouse IgG/HRP). The introduction of a second antibody should in theory lead to large amplification of signal but in practice seldom does, presumably due to steric hindrance considerations. It also adds an extra incubation step and adds to the cost of the assay. However, the user may prefer to purchase a labelled antibody rather than attempt to label the first antibody in-house.

Step 1: coat antigen onto solid support; wash away unbound antigen.

Step 2: add soluble antigen, then labelled antibody. Wash away unbound antibody/antigen. Detect activity of bound labelled antibody.

Figure 5 Competitive binding immunoassay—labelled antibody.

5.2 'Sandwich' immunoassay

This assay is one of the most commonly used immunoassays and depends on the availability of two different antibodies to the antigen to be measured. A scheme is shown in *Figure 6*.

5.2.1 Assay configuration

The two antibodies can be configured in a number of ways. First, two monoclonal antibodies which bind to non-overlapping epitopes of the antigen can be used (see Section 2.1). In this case, traditionally, the higher affinity antibody should be the best to coat onto the solid support, using the other to detect the bound antigen. However, not all monoclonal antibodies are the same in their reaction to being adhered to a solid support; loss of antigen binding capacity can result, and it is probably best to test each one before deciding which configuration to use. The second antibody can be labelled directly (Section 4) or, if there is a difference between the monoclonal antibodies used, e.g. in species or Ig isotype, it is possible to add a step and use a third (labelled) antibody. For example, IgG2a anti-Ag can be used to coat microtitre wells, IgG1 anti-Ag can be

Step 1: coat first antibody onto solid support; wash away unbound antibody.

Step 2: add antigen, which binds to first antibody; wash away unbound antigen.

Step 3: add an excess of labelled second antibody; wash away unbound second antibody. Detect activity of bound second antibody.

Figure 6 Sandwich immunoassay.

used as a second antibody to detect bound antigen, and a third antibody, rabbit anti-mouse IgG1/AP can be used to detect the second monoclonal. It is important that the third antibody not react significantly with the first (or vice versa), or high background binding will result.

Alternatively, one monoclonal antibody in conjunction with a polyclonal antibody may be used. It is unlikely that the epitope recognized by the monoclonal is not also recognized by some from the collection of antibodies present in the polyclonal preparation. Thus, the monoclonal is coated onto the support and the polyclonal, either labelled directly or indirectly through a labelled third antibody, is used subsequently to detect the binding of antigen to the first antibody.

In similar ways, it is also possible to use two polyclonal antibodies to the same antigen from different species in a sandwich assay, or even a single polyclonal preparation where a portion of it has been directly labelled for detection.

Finally, contaminants in the coating IgG preparation will also bind to the solid support and compete out the specific antibody. Thus, the first antibody should be (best case) affinity purified or, if the specific antibody content is high enough, the IgG fraction can be used for coating solid supports. Otherwise, consider coating purified antibody directed against the H chain of the specific antibody, and then building the assay from there. Also, the detecting antibody needs to be purified or semi-purified to get the best results during labelling. However, if a third antibody is employed for detection, the second need not be purified at all.

5.2.2 Titreing the antibodies

Several assumptions can be made to help in the initial phase of developing a new assay in sandwich format. If microtitre wells are to be used as the solid support, and the reaction volumes are 100 µl/well, the maximum amount of any protein, in this case IgG, that can adhere to the well is around 500 ng. The maximum amount of antigen which will bind to that antibody will be proportional—e.g. for a completely coated well containing IgG anti-human serum albumin (IgG = 155 kDa, HSA = 68 kDa), the maximum antigen that could possibly bind would be 2–3 µg. Thus, if the plate wells are first coated with a concentration of at least 10 µg/ml antibody, a titration of antigen starting from around 10 µg/ml is very likely the right range to achieve a standard curve. *Protocol 4* details the steps for titreing the antigen and the detecting antibody in a sandwich enzyme-linked immunosorbent assay (ELISA) format using a third antibody to detect binding of the second specific antibody. Two points in this initial titration are fixed: the coating antibody, and the third detecting antibody. Once the antigen and the first antibody have been titrated in, the other steps can be titrated also to adjust the assay for range and sensitivity. *Figure 7* shows the effects of varying the concentration of the third (AP labelled) antibody in a sandwich ELISA similar to the one in *Protocol 4*. With AP, where the reaction continues developing, it can be seen that, ultimately, similar signals can be achieved for very different antibody concentrations. For other labelling systems, the signal obtained is more dependent on the initial antibody concentration used.

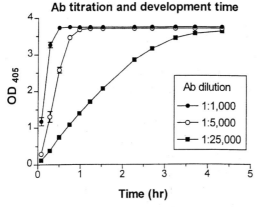

Figure 7 Development with time of alkaline phosphatase substrate (Sigma 104®) with an AP-conjugated antibody at three different dilutions.

Protocol 4

Titreing the antigen and the first detecting antibody for a sandwich ELISA

Equipment and reagents

- Microtitre 96-well immunoassay plate (e.g. Nunc Maxisorp plates)
- 12-channel multichannel pipette
- Spectrophotometric reader capable of accepting 96-well plates
- 'Damp box': box with well-fitting lid and lined with moist paper towels
- Coating buffer: 0.1 M borate buffer pH 8.6 + 0.1% NaN$_3$
- Wash buffer (PBS): 0.05 M phosphate-buffered saline pH 7.3

- PBS-A: PBS plus 0.01% NaN$_3$ (w/v)
- Assay buffer: 1% BSA (w/v) in PBS-A
- Substrate buffer: 10% ethanolamine (v/v), 0.01% MgCl$_2$ (w/v) pH 9.8
- Sigma 104® 5 mg substrate tablets
- Purified monoclonal anti-Ag
- Polyclonal rabbit anti-Ag
- Mouse (monoclonal) anti-rabbit IgG–alkaline phosphatase conjugate (Sigma)

Method

1. Prepare 12 ml of monoclonal anti-Ag at 10 µg/ml in coating buffer. Add 100 µl/well to all wells of the plate; incubate for 2–3 h at RT in a damp box.

2. Wash the plate with 200 µl wash buffer per well, three times. Slap the plates, wells down, onto absorbent paper towels.

3. Add 100 µl assay buffer to each well. Prepare 1 ml of antigen at 20 µg/ml in assay buffer. Add 100 µl of this solution to column 1 (wells A1 to H1) (8 wells). Using a multichannel pipette, make serial 1:2 dilutions of antigen from column 1 to column 11 (range 10 µg/ml to 10 ng/ml). Leave column 12 for a background control. Incubate plate at RT for 3 h (or overnight at 4 °C) in a damp box.

4. Wash the plate as before. Add 100 µl assay buffer to each well. Prepare 2 ml of

Protocol 4 continued

rabbit anti-Ag at 1:100 in assay buffer. Add 100 μl of this solution to wells in row A (wells A1 to A12) (12 wells). Make serial dilutions from row A to row G (range 1:200 to 1:12 800). Leave row H as a background control. Incubate as for step 3.

diluted   Ag: 1:2 1:4 1:8 etc. ⟶ 0 Ag

Ab:		1	2	3	4	5	6	7	8	9	10	11	12
1:2	A												
1:4	B												
1:8	C												
etc.	D												
	E												
	F												
	G												
0 Ab	H												

5. Wash the plate as before. Prepare 12 ml of 1:5000 mouse anti-rabbit IgG–alkaline phosphatase in assay buffer. Add 100 μl of this solution to all wells. Incubate as for step 3.

6. Wash the plate as before. Prepare 15 ml of 1 mg/ml Sigma 104® AP substrate in substrate buffer. Add 100 μl/well. Incubate in a damp box at RT and read in the plate reader (405 nm) at intervals until colour is sufficiently developed (reading of > 2 in wells A1–G1).

7. Plot the OD_{405} for antibody versus antigen. Select the dilution of rabbit anti-Ag which is just maximal, and the series of antigen concentrations suitable for a standard curve. Time of substrate colour development can also be selected during this titration. *Figure 8* shows the result of a similar titration, measured at a 5 h development time. For this system, background was progressively reduced by diluting out the detecting antibody, without significant loss of signal until the antibody was > 1:10 000. In this case, a 1:1000 dilution was selected for the final assay since a shorter development time was required.

8. Using the optimal antigen/antibody dilutions determined from the initial titration, check the assay using these conditions, with samples at least in duplicate, at this point standardizing the methods to be used for initial dilution of standard and detecting antibody. This is important since, as an example, making 1:2 serial dilutions of antibody into wells from 1:100 to get a 1:1600 dilution often gives a somewhat different result from making a 'bulk' solution of 1:1600 from dilutions of 1:10 and then 1:160.

9. Once these two reagents are fixed it is now possible, using similar methods, to perform a second titration to optimize the concentrations of third antibody (anti-rabbit IgG–AP) and the coating antibody. Use the optimal dilution of the second antibody determined in the first titration, and for the antigen, a fixed amount of half to three-quarters maximum (IC_{50}–IC_{75}) of the standard curve.

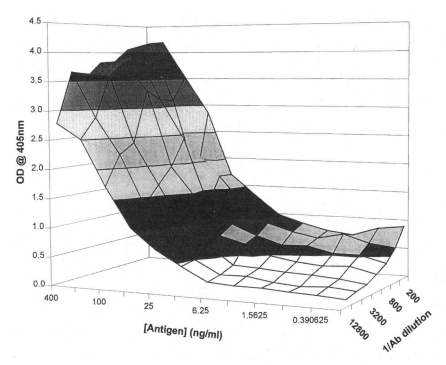

Figure 8 Two-dimensional titration (checkerboard) of antigen and second antibody in a 'sandwich' ELISA using a third (AP labelled) antibody to detect. First antibody (coated on the plate) and third antibody concentrations were kept constant. Detection was with Sigma 104® substrate at 5 h development.

5.3 Buffers

In *Protocol 4*, the coating buffer used was borate buffer containing azide, which can be stored at RT for extended periods. Many published methods use a bicarbonate buffer of a similar pH; this is equally effective but is not stable for long-term storage and should be prepared fresh. Some investigators recommend PBS, providing a more neutral pH. Detergents should not normally be present in a coating buffer (but see ref. 13).

Also, note that after the coating of the antibody, no discrete 'blocking step' was included to block vacant binding sites remaining on the solid support. The inclusion of a blocking step may sometimes be beneficial. However, 1% BSA or similar solutions used as assay buffers (e.g. 1% gelatin, 1% milk, or commercially available blocking buffers) block so fast, and the BSA concentration is in any case so high relative to the antigen, that it is often not necessary to wait before adding antigen. If a high background is observed, inclusion of 0.05% Tween 20 in the assay buffer may help.

In this protocol, PBS was used to wash the plates between steps. For certain antigens more stringent washing (e.g. 0.05% Tween 20/PBS) may be necessary to reduce background binding, but unless the background is found to be high, simpler is better, especially if automatic plate washers are used where foaming can be a problem. In many cases if the wash is fast, even distilled water is suitable. This is dependent on the individual assay and can be determined empirically.

5.4 Setting up an ELISA

Once the format of assay has been selected, the antibodies have been chosen, and titres determined, many of the procedures used for actually performing an ELISA will have become familiar. The remaining two protocols detail the setting up of two immunoassays using enzyme-linked antibodies as label. However, using analogous principles, these protocols can easily be reconfigured for other label types or assay formats.

Protocol 5

Basic sandwich ELISA

Equipment and reagents

- See *Protocol 4* for equipment and reagents
- Use the concentrations of antibodies and standard antigen determined using *Protocol 4*

- The detection method will be according to the nature of the label of the detecting antibody (see Section 4)

Method

1. Coat the first specific anti-Ag antibody onto microtitre plate wells (100 μl/well) in coating buffer. Incubate for 2–3 h at RT in a damp box.

2. Wash the plate. In duplicate, add 100 μl of assay buffer to the series of wells to be used for the standard curve. Perform serial dilutions (e.g. 1:2) of standard antigen from the first well to the last. Leave a pair of wells free of antigen to act as the background control. Add the samples, also in duplicate, and at suitable dilutions in assay buffer, to the other wells of the plate. Incubate the plate, in a damp box, for 3–4 h at RT or overnight at 4 °C.

3. Wash the plate. Add 100 μl/well of second anti-Ag antibody to every well. This can be directly labelled, or may be detected using a labelled third antibody which does not react with the plate coating antibody. Incubate as for step 2.

4. Wash the plate. Either add 100 μl/well of detecting (labelled) antibody and incubate as for step 2, or proceed directly to the detection step.

5. Plot a standard curve of antigen concentration (x axis) versus assay readout (y axis). The curve obtained by these methods is usually best fitted using a Four Parameter Logistic curve according to the equation:

$$y = (A - D)/(1 + (x/C)^B) + D$$

where A = background readout of the y axis, B is the slope of the curve, C is the IC_{50} of the antigen concentration, and D is the maximum readout of y.

Protocol 5 continued

6. Interpolate the sample values from the standard curve. It is best to run two or three dilutions of each sample in duplicate, and obtain the mean of all the values. Check for 'trending' of results according to dilution (Section 2.2).

Protocol 6

Competitive inhibition ELISA—labelled antibody

Equipment and reagents

- See *Protocol 4*
- Antibody labelled with biotin (*Protocol 2*)
- Purified antigen
- Streptavidin–alkaline phosphatase conjugate (e.g. Sigma, Pierce)

Note: for best results, individual reagents should be titrated before use. The quantities specified below, when possible, are for rough guidance only. The labelling suggested here is biotin but as an example only. See Section 4 for labelling choices.

Method

1. Coat the wells of a microtitre plate. Prepare antigen in coating buffer at 5 µg/ml. Add 100 µl/well of antigen and allow it to adhere for 2–3 h at RT. Note: at this point it may be possible to add 100 µl/well assay buffer, apply an adherent sealing cover (Corning or Pierce), and store the plate frozen.

2. In a separate microtitre plate (sample preparation plate), add 100 µl assay buffer to each well. Prepare antigen at 10 µg/ml in assay buffer. Make 1:2 serial dilutions of the standard antigen, from 5 µg/ml, in duplicate wells. Make suitable single or serial dilutions of samples to be quantitated, also in duplicate, into other wells.

3. Prepare labelled antibody at 100 ng/ml in assay buffer.

4. Wash the antigen coated plate with 3 × 200 µl/well wash buffer. Using a multichannel pipette, transfer 50 µl from the sample preparation plate to the appropriate wells of the antigen coated plate. Add 50 µl of labelled antibody to each well and mix gently, either by tapping the plate or briefly using a plate shaker. Place the plate in a damp box. Allow the competition for antibody between solid phase antigen and soluble antigen to proceed for 3 h at RT or overnight at 4 °C.

5. Wash the plate. Add 100 µl/well of Streptavidin-AP at 1:5000 in assay buffer; incubate as for step 4. Develop with Sigma 104® substrate (1 mg/ml in substrate buffer, 100 µl/well) until sufficient colour has developed. Analyse the results as for *Protocol 5*, remembering that, since this is competitive binding, the standard curve will be reversed compared with a binding assay.

References

1. Steward, M. W. and Lew, A. M. (1985). *J. Immunol. Methods*, **78**, 173.
2. Xuan, J. W., Wu, D., Guo, Y., Huang, C. L., Wright, G. L., and Chin, J. L. (1997). *Cell. Biochem.*, **65**, 186.
3. Fagerstam, L., Frostell, A., Karlsson, R., Kullman, M., Larsson, A., Malmqvist, M., *et al.* (1990). *J. Mol. Recognit.*, **3**, 208.
4. Johne, B., Gadnell, M., and Hansen, K. (1993). *J. Immunol. Methods*, **160**, 191.
5. Laricchia-Robbio, L., Liedberg, B., Platou-Vikinge, T., Rovero, P., Beffy, P., and Revoltella, R. (1996). *Hybridoma*, **15**, 343.
6. Catt, K. and Tregear, G. W. (1967). *Science*, **158**, 1570.
7. Nakane, P. K. and Kawaoi, A. (1974). *J. Histochem. Cytochem.*, **22**, 1084.
8. Hemmila, I., Dakubu, S., Mukkala, V., Siitari, H., and Lovgren, T. (1984). *Anal. Biochem.*, **137**, 335.
9. Fracker, P. J. and Speck, J. C. (1978). *Biochem. Biophys. Res. Commun.*, **80**, 849.
10. George, S. and Schenck, J. R. (1983). *Anal. Biochem.*, **130**, 416.
11. Nelson, R. W., Krone, J. R., Bieber, A. L., and Williams, P. (1995). *Anal. Chem.*, **67**, 1153.
12. Serban, D. and Rordorf-Adam, C. (1986). *J. Immunol. Methods*, **90**, 159.
13. Drexler, G., Eichinger, A., Wolf, C., and Sieghart, W. (1986). *J. Immunol. Methods*, **95**, 117.

Chapter 14

Immunoaffinity chromatography of macromolecules

Steve Hobbs

Institute for Cancer Research, Block F, 15 Cotswold Road, Belmont, Sutton, Surrey SM2 5NG, U.K.

1 Introduction

The epitope specificity of an antibody combining site allows the antibody to bind to its corresponding ligand in the presence of other complex molecules such as whole cell lysates or animal sera. If the antibody is immobilized to a solid surface the complex can easily be separated from the mixture, and the reversible nature of the antibody–antigen interaction allows the ligand to be recovered by dissociation once the unbound molecules have been washed away. This process is called immunoaffinity chromatography, a technique now in widespread use (1, 2). It can be used either to purify an antigen, or to remove a specific antigen from a solution in which case recovery of biologically active molecules is not an issue (3). The advent of monoclonal antibody technology (4) has enabled the preparation of homogeneous populations of antibody molecules with identical structure, epitope specificity, and affinity. These features are ideally suited to solid phase antigen isolation because the supply of antibody is reproducible, and elution of antigen from the immobilized antibody will take place under a narrow range of conditions.

The key to this technique is the ability to couple an antibody irreversibly to a solid support, often in the form of agarose beads, in such a way as to retain its biological activity and to prevent dissociation under denaturing conditions. For this to occur the solid matrix generally has to be activated chemically. Fortunately it is no longer necessary to carry out this sometimes hazardous step in the laboratory as many ready-activated matrices are now commercially available, either in bulk or pre-packed into a column, and come with full instructions for use. Coupling is achieved simply by mixing the two components together in a suitable buffer. This chapter will survey some of the types of matrices and coupling chemistries that are appropriate for use with immunoglobulins, and provide protocols for their use in antigen isolation. The antigen binding affinity will of course vary between different monoclonal antibodies so it is not possible to give precise instructions that will cover all cases; finding the optimum conditions for binding and elution may therefore involve a certain amount of trial and error. For most of these protocols it will be assumed that the monoclonal antibody in question has already been purified by one of the methods detailed in Chapters 1–11 and is present in solution in the absence of any other proteins which may become coupled to the matrix.

2 Choice of matrix

2.1 Physical structure

The choice of matrix is governed principally by the liquid handling techniques envisaged. For laboratory scale low pressure column chromatography or batch processing the most commonly used matrix is crosslinked beaded agarose (5–7). This material has good handling characteristics and is resistant to many salts, detergents, and denaturants. For higher-throughput techniques or medium to high pressure systems a more rigid structure is needed and various synthetic co-polymers are available for these applications.

2.2 Coupling methods

There are two methods by which an antibody molecule can be coupled to a solid phase support:

(a) Direct coupling: the immunoglobulin molecule is chemically linked through free amino acid or carbohydrate side-chains either directly to the matrix or via a spacer arm.

(b) Indirect coupling: an intermediate molecule that binds immunoglobulin via the Fc region (*Staphylococcal* Protein A or *Streptococcal* Protein G) is directly coupled to the matrix, the antibody molecule binds to this, and is subsequently chemically crosslinked to it.

The three most commonly used groups for direct coupling are side-chain amino acid $-NH_2$ from lysine, $-SH$ from cysteine liberated by partial reduction of interchain disulfide bonds, and oxidized $-OH$ groups from carbohydrate side-chains. Direct coupling through side-chain amino groups has the advantage of simplicity, the antibody molecule does not have to be pre-treated before link-ing, but the antibody may be coupled by any available lysine residue (*Figure 1A*). It will therefore be attached at more than one point and its orientation on the matrix will be random, so it may not be in the best conformation for ligand binding. For this reason the antigen binding capacity of such immuno-absorbents may be much lower than would be expected relative to the amount of coupled antibody (8). It is possible to control the orientation of the molecule by direct coupling through the carbohydrate side-chains located in the Fc C_H2 region (*Figure 1B*). In this way the antigen binding Fab' regions have free access to the moving phase containing the antigen which may improve the binding affinity and capacity of the matrix. Obviously this method is not applicable if an Fab' or F(ab')$_2$ fragment is being used, or with recombinant scFv fragments expressed in *E. coli* which are unable to carry out glycosylation. Coupling via free $-SH$ groups exposed following partial reduction of the interchain disulfide

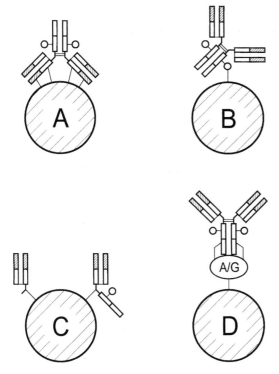

Figure 1 Different methods for coupling antibodies to solid matrices. (A) Random direct coupling via lysine ε-amino groups with multiple attachments. (B) Orientational coupling via C_H2 carbohydrate side-chains. (C) Orientational coupling via free hinge region –SH groups in partially reduced Fab' or half-molecule fragments. (D) Orientational indirect coupling via Protein A/G with DMP crosslinking. The VH/VL antigen binding domains are shaded.

bonds in the hinge region can also tether the antibody fragment in a better orientation for antigen binding (*Figure 1C*). Indirect coupling to a Protein A or G matrix involves extra steps, but the resultant bound antibody is again tethered by the Fc region allowing improved ligand access (*Figure 1D*).

2.3 Coupling chemistry

The main requirements for coupling an antibody to a solid phase are:

(a) The chemistry of the reaction should be relatively mild to prevent irreversible damage to the antibody.

(b) The bond formed should be stable to prevent leakage of the protein from the matrix especially under elution conditions.

(c) The coupling should cause minimum interference with the binding activity of the antibody.

(d) The linkage should not cause non-specific binding of other proteins either through ionic or hydrophobic interactions.

Table 1 Examples of commercially available matrices for spontaneous direct coupling

Product	Manufacturer	Activation	Groups	Matrix
CNBr–S4B	Pharmacia	CNBr	–NH$_2$	Crosslinked agarose
NHS–Superose	Pharmacia	NHS	–NH$_2$	Highly-crosslinked agarose
HiTrap-NHS ready packed	Pharmacia	NHS	–NH$_2$	Highly-crosslinked agarose
Affi-Gel 10	Bio-Rad	NHS	–NH$_2$	Crosslinked agarose
Affi-Prep 10	Bio-Rad	NHS	–NH$_2$	Pressure-stable acrylic polymer
Affi-Gel Hz	Bio-Rad	Hydrazide	–OH	Crosslinked agarose
Affi-Prep Hz	Bio-Rad	Hydrazide	–OH	Pressure-stable acrylic polymer
AminoLink	Pierce	Aldehyde	–NH$_2$	Crosslinked agarose
Reacti-Gel	Pierce	Carbodiimide	–NH$_2$	Crosslinked agarose
CarboLink	Pierce	Hydrazide	–OH	Crosslinked agarose
SulfoLink	Pierce	Iodoacetyl	–SH	Crosslinked agarose
TNB–Thiol	Pierce	TNB	–SH	Crosslinked agarose
UltraLink Iodoacetyl	Pierce	Iodoacetyl	–SH	Acrylamide/azlactone

Several different chemistries have been used for direct coupling, and a few examples of the many commercially available pre-activated matrices are shown in *Table 1*. The manufacturer's catalogues and Web sites should be consulted for further details.

2.3.1 Random direct coupling

(i) Cyanogen bromide

The most commonly used chemistry for coupling side-chain amino groups to agarose matrices is cyanogen bromide (9, 10). Ready-activated CNBr–Sepharose 4B can be purchased in freeze-dried form from Pharmacia; for small scale work this is preferable to handling hazardous cyanogen bromide in the laboratory (*Protocol 1*). Coupling of proteins to the CNBr-activated agarose proceeds spontaneously at alkaline pH via a complex chemistry involving a cyanate ester intermediate following reconstitution at acid pH (11), resulting in an isourea linkage. It is important to avoid buffers that contain free primary amino groups (Tris or glycine) that will compete for binding sites on the matrix. Once coupling of the antibody is complete, unreacted sites on the matrix can be blocked by incubation with excess free amino groups in the form of 1 M ethanolamine pH 8, 0.1 M glycine pH 8, or 0.1 M Tris–HCl pH 8. The disadvantages of this linkage are its potential instability (12, 13), multipoint attachment of the antibody leading to steric hindrance of antigen binding, and the fact that the resultant isourea bond is charged which may lead to non-specific binding of unwanted proteins through ion exchange effects.

Protocol 1

Coupling of antibodies to CNBr–Sepharose

Equipment and reagents

- Small grade 3 sintered funnel
- Rotating mixer
- Suitable affinity column: Bio-Rad Econocolumn, Pharmacia C column
- CNBr–Sepharose 4B, freeze-dried powder (Pharmacia)
- Purified antibody in 0.2 M bicarbonate or borate buffer pH 8.5 plus 0.5 M NaCl
- 1 mM HCl

- Coupling buffer: 0.2 M $NaHCO_3$ pH 8.5 plus 0.5 M NaCl
- Blocking buffer: 1 M ethanolamine pH 8 or 0.2 M glycine pH 8
- Washing buffer: 0.1 M acetate buffer pH 4 plus 0.5 M NaCl
- Storage buffer: phosphate-buffered saline (PBS) plus preservative (0.02% sodium azide or 0.02% thimerosal)

Method

1. Weigh out the required amount of CNBr–S4B, assuming the 1 g of freeze-dried powder reconstitutes to approx. 3.5 ml of hydrated bed.

2. Swell the gel in a suitable container by adding 1 mM HCl and swirling gently to ensure complete dispersion of the powder.

3. Transfer the mixture to a small grade 3 sintered funnel fitted to a side-arm flask and allow the liquid to drain through and top up until all the bed material has been transferred. Continue to wash the bed with several volumes of 1 mM HCl to remove the preservatives. A *gentle* vacuum may be applied but the bed must not be allowed to dry out. Drain the bed under gravity to remove the excess HCl.

4. Transfer the moist gel to a container and add the antibody solution at a loading of 5 mg antibody/ml of gel. The volume should be approximately twice the packed bed volume to allow the gel to mix freely, and should fill the container without leaving an air gap. Seal the container and mix the gel by inversion on a rotating mixer, do not use a magnetic stirrer as the flea can disrupt the agarose. Incubate for a few hours at room temperature, or overnight at 4 °C.

5. Separate the unbound material from the gel by *gentle* centrifugation, the bed will settle under gravity and does not require excessive compaction. Wash the bed once with coupling buffer, retain the supernatants for assaying of unbound antibody to calculate column loading.

6. Add 1 M ethanolamine or 0.2 M glycine pH 8 and incubate for a further 2 h at room temperature or overnight at 4 °C to block any unreacted groups.

7. Transfer the mixture to the sintered funnel and wash away the blocking solution with coupling buffer.

8. Wash the bed twice alternately with washing buffer and coupling buffer to remove unbound antibody.

9. Wash the bed in storage buffer (PBS plus preservative) and transfer to a suitable column ready for use.

(ii) N-hydroxysuccinimide ester

An alternative is to use *N*-hydroxysuccinimide activation (14, 15). The agarose is derivatized with a spacer arm terminating with an *N*-hydroxysuccinimide ester that reacts with side-chain $-NH_2$ groups spontaneously at alkaline pH resulting in an amide bond and liberating free NHS during the reaction. This link is generally more stable than the isourea bond generated by CNBr. Free $-SH$ groups can also compete for this reaction. The NHS active group has a short half-life in aqueous solution so addition of the antibody to be coupled must take place as soon as the matrix is prepared. Some forms of NHS-activated Sepharose are available as a freeze-dried powder (activated CH–Sepharose 4B from Pharmacia) and must be swollen and washed in 1 mM HCl before coupling at alkaline pH. Other forms are available ready-swollen and are stored in isopropanol to prevent hydrolysis of the active groups (e.g. HiTrap NHS-activated, Pharmacia; Affi-Gel 10, Bio-Rad), this must be washed away before use. Affi-Gel 10 contains a spacer arm that is attached to the agarose by a more stable link than that of activated CH–Sepharose. Again unreacted NHS groups can be quenched after the coupling step with ethanolamine or glycine, or left to hydrolyse overnight at alkaline pH. A variant of this utilizing a two-stage coupling system that does not require blocking of unreacted groups has just become available from Invitrogen, and comes as a kit complete with reagents, columns, and buffers (Linx Affinity Purification System). This relies on the interaction between salicylhydroxamic acid (SHA) and phenyldiboronic acid (PDBA). The SHA is conjugated to Sepharose beads and the PDBA is linked to *N*-hydroxysuccinimide. The PDBA–NHS conjugate is first mixed with the antibody to react with side-chain $-NH_2$ groups, then the derivatized antibody is mixed with the SHA-conjugated Sepharose and coupling takes place, the entire process taking just over one hour according to the manufacturer's claims.

(iii) Carbodiimide coupling

Carbodiimides are bifunctional crosslinking agents that promote coupling of ligands containing free $-NH_2$ groups to those containing free $-COOH$ groups by formation of a peptide bond. Activated Sepharose derivatives are available that have been derivatized to express $-NH_2$ groups (EAH Sepharose 4B, Pharmacia; Affi-Gel 102, Bio-Rad) or $-COOH$ groups (ECH Sepharose 4B, Pharmacia; CM BioGel A, Bio-Rad). These are used in conjunction with a water soluble carbodiimide such as EDAC to couple the corresponding ligand to a spacer arm connected to the gel. A pre-activated form of crosslinked agarose that will undergo spontaneous coupling with free $-NH_2$ groups is commercially available as Reacti-Gel from Pierce. The agarose has been treated with 1,1′-carbonyldiimidazole to give a matrix that is more stable to hydrolysis than NHS-activated agarose with a half-life of many hours at pH 9 for easier handling characteristics. It reacts with amino groups at alkaline pH to couple via a stable *N*-alkylcarbamate linkage which has the advantage of being non-charged so non-specific ionic interactions with the column are reduced (16). Reacti-Gel comes as a slurry suspended in acetone, this is washed away with water in a sintered funnel before coupling of the antibody in 0.1 M borate buffer pH 8.5. The coupling reaction releases free imidazole which absorbs UV light at 280 nm, interfering with spectroscopic estimation of protein coupling.

2.3.2 Orientational direct coupling

Multipoint attachment of the antibody to the matrix by linkage through lysine ε-amino groups can interfere with the ability of the antibody to bind antigen, reducing the ligand capacity of the column. Tethering the antibody through the Fc region alone may allow the antigen binding Fab' arms better access to the ligand. Two features of the Fc region make this possible; the complex carbohydrate chains located in the C_H2 domain and the interchain disulfide bonds in the hinge region, each of which can be used as an attachment point (17, 18). However, the actual increase in antigen binding capacity of immunoabsorbents made by these methods varies between antibodies, and may not result in the anticipated improvements (19). Antibodies coupled randomly through $-NH_2$ groups via spacer arms (e.g. to Affi-Gel 10, Bio-Rad) may have enough flexibility to bind small ligands efficiently without orientational immobilization (20). The consensus seems to be that orientation immobilization may be of value when purifying large ligands but for small ligands the steric effects of random immobilization may be less important.

(i) Carbohydrate coupling

Matrices derivatized with a hydrazide moiety will react with oxidized carbohydrate residues to form a stable hydrazone linkage (21). Vicinal sugar hydroxyl groups of immunoglobulin carbohydrate side-chains can be oxidized to aldehydes with sodium m-periodate without interfering with the antigen binding capacity (Protocol 2), treatment under acid conditions favours subsequent formation of the hydrazone linkage to the gel and reduces the tendency of the oxidized immunoglobulins to react with each other via side-chain amino groups to form oligomers (17). Suitable hydrazide-activated matrices include CarboLink (Pierce) and Affi-Gel Hz (Bio-Rad). This method may not be generally applicable as some monoclonal antibodies have carbohydrate side-chains in the Fab' regions which may compromise the Fc-specific carbohydrate linking (22).

(ii) SH coupling

The two immunoglobulin heavy chains are held together by interchain disulfide bonds in the hinge region. It is possible to separate the molecule into two H/L pairs by partial reduction of the hinge region bonds with 2-mercaptoethylamine hydrochloride without disturbing the S–S bonds holding the light and heavy chains together (Protocol 3). In this way the antigen binding activity of the half-molecule is retained. If intact IgG is used a half-molecule with a full-length heavy chain will be produced, a pepsin digested $F(ab')_2$ fragment will give an extended Fab' on partial reduction (18, 23). The free –SH groups can then be used for orientational coupling to a thiol-activated gel such as SulfoLink Gel or TNB-Thiol Gel (Pierce). TNB-Thiol gel couples the reduced antibody fragment through a reducible disulfide bond whereas SulfoLink gel produces an irreversible bond via an iodoacetyl group. A gel activated with maleimide will also form a stable thioether bond that is resistant to subsequent reduction by thiols (17). Excess activated groups can be blocked with cysteine. Bound antigen can be released from these columns by washing at low pH, or the antibody–antigen complex can be completely released from disulfide-coupled gels by reduction with dithiothreitol. A disadvantage of this approach is that the antibody fragment coupled to the gel only contains a single antigen binding site. If the antibody is of low affinity to start with the binding of antigen to the column may therefore be weakened, so the washing step will have to be monitored carefully to ensure that the antigen is retained on the column until elution.

Protocol 2

Linking of antibodies via carbohydrate side-chains[a]

Equipment and reagents

- Small desalting columns: e.g. PD-10 (Pharmacia); Econo-Pac 10DG (Bio-Rad); D-Salt (Pierce)
- Hydrazide-activated agarose: e.g. Affi-Gel Hz (Bio-Rad) or CarboLink (Pierce), both these matrices are also available as complete kits with all necessary components and instructions
- Coupling buffer: 0.1 M sodium acetate buffer pH 5.5

- Washing buffer : 10 mM phosphate buffer plus 1 M NaCl pH 7
- Purified antibody (5–10 mg/ml) in coupling buffer (desalt or dialyse into this buffer if necessary); keep the volume small to avoid overloading the desalting column
- Freshly made 0.1 M sodium m-periodate (e.g. Sigma, S1878)

Method

1. The antibody is first oxidized by adding sodium m-periodate to a final concentration of 10 mM and incubating for 1 h at room temperature. This oxidation step is light sensitive, so should be carried out in a brown vial or similar light-proof container.

2. Pass the antibody solution over a small desalting column equilibrated with coupling buffer to remove any unreacted periodate.

3. Remove any preservative in the gel by washing with coupling buffer. Add the oxidized antibody to the gel, a loading of 5 mg antibody/ml of gel is suitable. Mix by inversion overnight at room temperature.

4. Pack the gel into a small affinity column and wash with PBS containing 1 M NaCl pH 7 to remove any unbound antibody. Retain the washings for calculation of column loading. Store the column in PBS plus preservative (sodium azide).

[a] Procedure adapted from ref. 17.

Protocol 3

Coupling of partially reduced antibodies via –SH bonds[a]

Equipment and reagents

- Small desalting column equilibrated in coupling buffer
- Thiol-activated agarose gel: e.g. SulfoLink (Pierce)
- 2-Mercaptoethylamine HCl (Pierce, 20408)
- Monoclonal antibody or pepsin-digested F(ab')$_2$ fragments in 0.1 M phosphate buffer pH 6 plus 5 mM EDTA-Na

- L-Cysteine HCl
- Coupling buffer: 50 mM Tris pH 8.5 plus 5 mM EDTA-Na
- Washing buffer: 10 mM phosphate buffer plus 1 M NaCl pH 7
- Storage buffer: PBS plus 0.02% sodium azide

Method

1. Reduce the antibody by incubating in 50 mM 2-mercaptoethylamine for 1.5 h at 37 °C. MEA obtained from Pierce (product 20408) comes in 6 mg lots, 50 mM equates to 6 mg/ml and the antibody can be added directly to the MEA. The total volume should be kept low to allow efficient separation on the desalting column.

2. Separate the reduced antibody from excess MEA by passing the mixture over a small desalting column pre-equilibrated in coupling buffer. Pool the protein-containing fractions as measured by absorbance at 280 nm. Wash the required amount of SulfoLink gel with coupling buffer in a small sintered funnel or column. A loading of 5 mg antibody/ml of gel is suitable.

3. Transfer the gel to the reduced antibody solution in a suitable container and mix by inversion for 1 h at room temperature. Alternatively if the gel is packed into a column it can be sealed at the base and the antibody solution added. The column is then sealed at the top and mixed *in situ* by inversion. Allow the column to stand in the vertical position for 30 min after mixing to allow the bed to settle.

4. Remove the supernatant containing any unbound antibody and wash the bed with coupling buffer to remove any residual material. Retain the washings for estimation of coupling efficiency. Block any excess unreacted groups on the gel by incubating in 50 mM L-cysteine dissolved in coupling buffer for 30 min at room temperature.

5. Wash the gel with excess washing buffer (1 M NaCl), then with storage buffer. Pack into a suitable column.

[a] Procedure adapted from ref. 18.

2.3.3 Indirect coupling via Protein A/G

Staphylococcal Protein A or *Streptococcal* Protein G bind to immunoglobulin molecules in the Fc region. If an immunoglobulin-containing solution is passed over a column of immobilized Protein A or G the Ig will bind to the column, indeed this is one method for purifying monoclonal antibodies. Once binding has occurred and any non-specifically bound material has been washed away the

Table 2 Binding of rodent IgGs to Proteins A and G

Species	Isotype	Protein A	Protein G
Mouse	IgG1	±	+
	IgG2a	++	++
	IgG2b	++	++
	IgG3	++	++
Rat	IgG1	+	+
	IgG2a	–	++
	IgG2b	–	+
	IgG2c	++	++

Protein A/G–Ig complex can be stabilized with the bifunctional crosslinker dimethyl pimelimidate (24, 25). In this form the antibody on the column is oriented with the Fab' arms free from the matrix and available for antigen binding (*Protocol 4*). Complete kits for immobilization of antibody on Protein A or G are marketed by Pierce (IgG Orientation Kit), but any source of Protein A- or Protein G-agarose will do. The differences between Proteins A and G lie in the specificity for different immunoglobulin isotypes and species, Protein G in particular binds to rat IgG isotypes that Protein A does not. The majority of monoclonal antibodies made by conventional cell fusion will be IgGs of mouse or rat origin, so *Table 2* shows the relative affinities of these two proteins for different rodent IgG isotypes.

Protocol 4

Covalent coupling of antibodies to Protein A/G[a]

Equipment and reagents

- Protein A- or G-agarose gel from various suppliers (alternatively an ImmunoPure Protein A or G Orientation Kit from Pierce which contains all the necessary components and instructions)
- Binding buffer: 0.1 M borate buffer pH 8.2
- Crosslinking buffer: 0.2 M triethanolamine pH 8.2

- Blocking buffer: 0.1 M ethanolamine pH 8.2
- Elution buffer: 0.1 M glycine–HCl pH 2.8
- Dimethyl pimelimidate dihydrochloride (Pierce, 21666), dissolved at 25 mM in crosslinking buffer
- Monoclonal antibody in binding buffer (avoid buffers containing Tris or glycine)

Method

1. Wash away any preservative solution from the Protein A/G gel with binding buffer, either in a small column or sintered filter. Drain the bed and transfer to a suitable container (if not already in a column).

2. Add antibody solution to form a mobile slurry. Schneider and co-workers (24) obtained optimum binding performance from a column loaded with antibody to

Protocol 4 continued

50% saturation of the Protein A binding capacity. Mix by inversion at room temperature for 30 min.

3. Recover the gel by gentle centrifugation and wash with binding buffer. Retain the supernatants for estimation of column loading.

4. Wash the bed with crosslinking buffer, then resuspend in crosslinking buffer containing 25 mM DMP. Mix by inversion for 45 min at room temperature. Recover the gel by gentle centrifugation.

5. Block any unreacted DMP sites by resuspending the gel in an equal volume of blocking buffer and mixing by inversion for 10 min at room temperature.

6. Wash the gel with binding buffer then with elution buffer to remove any bound antibody that is not crosslinked with DMP. Store the column in PBS plus preservative.

[a] Procedure adapted from ref. 24.

2.4 Monitoring of coupling efficiency

If the antibody solution is purified the degree of coupling can be estimated by measuring the absorbance at 280 nm of the solution before addition to the beads, and the supernatant after removal of the beads by centrifugation. An extinction coefficient $E^{1\%}_{2\,80} = 13.5$ is appropriate for IgG. This technique cannot be used directly with NHS-activated matrices as the free NHS released during coupling also absorbs UV light at 280 nm in neutral or alkaline solution. This may be removed by passing the eluate down a small desalting column (Sephadex G25 in a Pharmacia PD-10 column) before measurement, alternatively protein concentration can be estimated by Coomassie blue dye binding (Bio-Rad). A similar approach may be used for CDI-activated supports which release imidazole on coupling that absorbs in the UV range. Coupling via –SH groups to TNB-Thiol agarose (Pierce) releases free TNB which absorbs at 412 nm, this can be used to monitor the progress of the reaction.

2.5 Crosslink stability

The performance of an immunoaffinity column will degrade slowly with use, the coupled antibody can be progressively degraded through the cycles of washing and elution, and can also leach from the column. Minimizing leakage is important for applications where the purified antigen is to be used for therapy *in vivo*, or where the column is being used to remove substances from the circulation. The rate of leaching will depend on the chemistry of coupling, the elution methods employed, and to some extent the antibody or fragment used (12, 26–28), and is difficult to prevent. NHS linking appears to be more stable than CNBr, although comparison of the various chemistries available varies depending on the system in use (1). Stabilization of the antibody on the support by crosslinking with glutaraldehyde (29) or bismaleimide (30) has been tried, although leakage may be reduced the antigen capacity of the column may also suffer as a consequence of the treatment (31). Antibodies crosslinked to Protein A/G matrices with DMP should similarly show improved stability. Free antibody or complexes leached from the column may be separated from the antigen by size exclusion chromatography after elution, the eluate can also be monitored for the presence of leached antibody by immunoassay (32, 33). Pre-washing or pre-cycling of affinity columns before use is also recommended, especially for those that have been stored for some time.

2.6 Column storage

The columns should be stored in a neutral or slightly alkaline buffer (PBS or borate buffers are suitable) and sealed to prevent drying out. Storage at 4 °C in the presence of a preservative such as 0.02% sodium azide or 0.02% Thimerosal (Sigma) will prevent bacterial growth. It is important to protect the column from freezing as this will fracture agarose matrices.

3 Sample preparation

As with all column chromatographic procedures the principal requirement of the sample is that it does not contain particulates that could clog the matrix and impede the flow, or that it might precipitate *in situ* with the same result. Any particles should be removed by centrifugation and/or by filtration through a 0.45 μm filter. Convenient syringe-mounted disposable filter units are available from Gelman, Sartorius, or Whatman for this purpose. If the sample consists of a cell membrane extract prepared by detergent lysis, detergent should also be added to the loading and washing buffers to maintain the antigens in solution. For soluble intracellular antigens extracted by detergent lysis the detergent should be removed before loading the sample. This can be difficult although detergents with a low molecular weight such as octyl glucoside may be removed by dialysis followed by centrifugation to remove any material that precipitates once the detergent has been removed. Proprietary pre-packed columns are available from Pierce (Extracti-Gel D) that will remove a variety of detergents; SDS, TX-100, CHAPS, Brij-35, Tween 20. An alternative is to use Triton TX-114 to extract the sample at 4 °C. This detergent has a cloud point around room temperature and undergoes phase separation when warmed above this point. The detergent and any proteins with hydrophobic regions such as membrane proteins that partition into the detergent phase can then be removed by centrifugation (34) before loading the soluble fraction onto the column. Care must be used with this approach to ensure that the antigen of interest although soluble does not tend to partition into the detergent-rich phase.

Attention must also be paid to the presence of any material in the sample that may compete with the desired antigen for binding sites on the column and reduce the yield of antigen on elution. For example if the column contains anti-immunoglobulin antibodies directed against the Fab' region or anti-idiotype and is being used to isolate a monoclonal antibody prepared with a light chain secreting myeloma (35), the culture supernatant may also contain free light chain which can bind to the column. In this case a preliminary gel permeation chromatography step to remove the smaller contaminant will increase the column yield. It is also possible for proteins in the sample to interact with the immobilized antibody at sites other than the antigen binding site. A cell membrane extract may contain cell surface immunogloblin Fc receptors that may bind to the Fc region of the monoclonal antibody; a column made from a pepsin digested F(ab')$_2$ would avoid this problem. Columns using the immunoglobulin-binding Proteins A or G to tether the monoclonal antibody may also bind other antibodies in serum samples if these sites are not saturated. Such samples would have to be absorbed on a pre-column of Protein A/G alone before exposure to the antibody-coupled matrix.

4 Binding of antigen

Binding of antigen to the conjugated matrix can be carried out in batch mode or on a column. Batch processing may be useful when using multiple small samples as several batches can be processed at the same time. The samples are mixed with an aliquot of conjugated matrix and mixed by inversion to allow ligand binding. The bed material must then be pelleted by gentle centrifugation and washed thoroughly before elution. An alternative to centrifugation is to use derivatized magnetic beads that can be recovered from suspension with a magnet (36). Washing becomes less efficient as the bed volume increases, and it is also not easy to monitor the progress of the purification. For these and other reasons it is generally preferable to carry out the purification steps in a column (*Protocol 5*). Many suitable types of column are available, those types with an open bed can be used for gravity flow elution or pumped downward flow from the bottom, a frit can be fitted on top of the bed to protect the matrix from disturbance during sample loading. If a flow adapter is fitted to the top of the column buffer can be pumped through in either direction and the elution step can be carried out in the reverse flow direction to prevent the eluted material rebinding to the column. Suitable columns include Bio-Rad Econo-Column (glass) with or without upper flow adapter or Econo-Pac (polypropylene) columns, Pharmacia C Columns, and others. For affinity chromatography a short wide column is preferable to a long narrow one to allow high flow rates without excessive pressures and to prevent the eluted sample peak from spreading too much. It is also possible to buy columns ready-packed with activated media such as the Pharmacia HiTrap affinity columns. These can be used for pumped flow or attached to a syringe for ease of use, and can be used as an alternative to batch processing. The ligand is coupled to the matrix *in situ* by loading with a syringe or recycling through the column with a pump. Other types of column are available for higher pressure systems from several manufacturers again with in-column linking protocols, but the principles of operation are similar whichever system is used. For pumped systems an in-line UV monitor connected to a chart recorder or computer logging system will provide a record of the purification. A fraction collector need only be used at the elution stage, the sample wash-through during loading can be collected in bulk, and washings can be diverted to waste. In the absence of in-line UV monitoring equipment the eluate may be collected in a fraction collector or by hand and the UV absorbance of individual fractions measured subsequently with a conventional UV spectrophotometer. If the column is run under gravity flow, allowing the buffer to drip directly from the end of the column without extra tubing will enable the flow to stop when the liquid reaches the top of the bed by surface tension. In this way the column can be left to load and wash unattended, and only connected to pump, monitor, and fraction collector for the elution step to provide a steady flow rate for equal fraction size. It is preferable to run the stages in a cold room or cold cabinet if the sample is labile and the run will take some time; syringe-based columns can be used at room temperature as the process is relatively fast.

Protocol 5

A typical immunoaffinity chromatography run

A typical protocol for purifying an antigen using an immunoaffinity matrix packed in a Bio-Rad Econocolumn, with a low pH or chaotrope elution method.

Method

1. Connect the upper column inlet to a buffer reservoir and wash under gravity with a high salt (1 M NaCl) solution or elution buffer (3 M KSCN, 0.1 M glycine–HCl pH 2.8, or 0.1 M citric acid pH 3.5) followed by PBS to remove any unbound material or leached antibody. Allow the buffer to drain to the top of the bed by gravity flow.

2. Apply the antigen-containing solution carefully to the top of the column and collect the flow-through in a clean container for recycling. If the dead space above the bed is insufficient to contain all the antigen the top of the column can be refitted and connected to a reservoir to allow the rest of the solution to drain through the column.

3. Wash the bed extensively with PBS plus 0.5–1 M NaCl to remove non-specifically bound material. Collect the washings separately in case the antigen is eluted at this stage.

4. Connect the column to pump, UV monitor, and fraction collector. Allow the wash buffer to drain to the top of the bed and carefully apply a layer of elution buffer without disturbing the matrix. Reconnect the column to a supply of elution buffer and collect 1 ml fractions. For elution with low pH buffers (0.1 M glycine–HCl pH 2.8, 0.1 M citric acid pH 3.5) neutralize the eluate by adding 100 μl of 1 M Tris–HCl pH 8 to each 1 ml fraction, this can be added to the collecting tubes before fractionation begins.

Protocol 5 continued

5. Pool fractions containing the eluted protein and pass over a desalting column equilibrated with PBS, and/or dialyse extensively. Alternatively the outlet from the pump can be connected directly to a desalting column fitted with an upper flow adapter for on-line desalting; this is useful for KSCN elutions where the separation between protein and eluting agent can be monitored by UV absorbance. If the antigen is to be radioiodinated additional dialysis is recommended.

6. Wash the column extensively with PBS and reapply the absorbed antigen solution. Repeat the cycle until no more antigen is extracted.

7. A gel permeation chromatography step will remove any aggregates or degraded material from the final antigen preparation

Before the column is used it must be washed through with a high salt buffer (0.5–1 M NaCl) or put through a complete elution cycle. This is to remove any antibody ligand that may have leached from the bed or antigen left over from a previous run, especially important if the column has been stored for some time. The antigen sample should be in a physiological buffer such as PBS or TBS; buffers of low ionic strength should be avoided as these can lead to problems with non-specific binding. A salt concentration of 0.1–0.2 M is suitable. Too high a salt concentration may lead to non-specific hydrophobic binding, especially if the bed has a hydrophobic linker between antibody and matrix. If the antigen-containing sample is a detergent extract then the detergent should be included in the column equilibration and washing buffers to prevent precipitation of the antigen on the column. Attention should be paid to the size of the column: too small a column in relation to the amount of antigen to be bound will result in overloading and pass-through of unbound antigen into the eluate. A column that is too large will bind all the antigen at one end, if it is eluted in the same flow direction the antigen will be able to interact with further ligand down the column on elution, spreading the small amount of antigen over a large peak. Retention of unwanted proteins by non-specific binding to the matrix will also increase as the column size gets larger. In any case it is advisable to retain the wash-through fraction during antigen loading as this can then be recycled until no more antigen is recovered.

Following sample loading the column should be washed in loading buffer until the unbound material has passed from the column as monitored by UV absorbance. It is then necessary to wash the column in a high salt solution (0.5–1 M NaCl is usual) to remove any material that may have bound non-specifically. Care must be taken at this stage as it is not possible to be precise about the conditions, it will depend on the affinity of the individual antibody–antigen interaction. A 1 M salt wash may well elute the antigen as well as non-specific material. If in doubt monitoring and collecting of fractions for assay at this stage may help. The Golden Rule is not to throw anything away until you are sure that the antigen is retained on the column.

5 Elution

To elute the bound antigen from the immobilized antibody the association between the two must be disrupted as gently as possible to avoid irreversible
denaturation or aggregation of either partner. The affinity of binding of antibody to antigen will vary between antibodies so the optimum conditions for
elution will have to be determined empirically for each combination. The main
methods used are to wash with a buffer of different pH, either high or low, a
high concentration of chaotropic ions (KSCN, NaSCN, NaI, KI), or a buffer of
high ionic strength. Denaturants such as 6–8 M urea or 3–4 M guanidine–HCl
are probably more powerful than is necessary for most monoclonal antibody-
antigen combinations and may cause irreversible damage to the antigen so tend
not to be used frequently. Other techniques include ethylene glycol and related
polarity-reducing agents, or biospecific elution with a competing hapten or
peptide although this method is unlikely to be generally applicable. Elution
may be possible under mild conditions using detergent to desorb the antigen,
this may be useful if membrane proteins are being purified: Herrmann and
Mescher (37) eluted the mouse MHC H-2K^k antigen from a monoclonal antibody
column using 0.5% deoxycholate in TBS pH 8. Whether such conditions may be
used will depend on the affinity of the antibody, and it may generally be
expected that stronger conditions will be needed for elution to occur. Pierce
supply a series of proprietary 'Gentle Ab/Ag' binding and elution buffers that
have been optimized for antibody production at neutral pH from immobilized
Protein A/G but are also compatible with antibody-based immunoaffinity
columns. It is also possible to elute the antigen electrophoretically under mild
conditions.

5.1 Extremes of pH

For low pH elution the column can be washed with 0.1 M glycine hydrochloride
buffer of pH 2.2–2.8, 0.1 M citric acid buffer pH 3.5–4, 0.5 M acetic acid/
ammonium acetate buffer pH 3.4, or propionic acid. Disadvantages of this
approach can include poor yield of product, or denaturation and precipitation
(28). A high pH elution may be more effective, especially if membrane proteins
are being purified as some membrane antigens tend to be more stable at
alkaline pH (38). Against this some mouse monoclonal antibodies may be
unstable at alkaline pH even when immobilized to a matrix (1). Buffers include
glycine–NaOH pH 10.5 (28), 0.05 M diethylamine pH 11.5 (24, 38), sodium borate
pH 10. Again the stability of the eluted antigen under these conditions must be
taken into account.

5.2 Chaotropic ions

High concentrations of chaotropic ions such as SCN$^-$, I$^-$, Cl$^-$ disrupt the water
structure around the antibody–antigen complex and reduce hydrophobic inter-

actions leading to desorbtion of the antigen. The efficiency of disruption can be ordered approximately:

- $SCN^- > ClO_4^- > I^- > NO_3^- > Br^- > Cl^-$
- $Ba^{2+} > Ca^{2+} > Mg^{2+} > Li^+ > Na^+ > K^+ > Cs^+ > NH_4^+$

Thiocyanate ions are the most commonly used chaotrope and elution can be tried with 3 M KSCN or 3.5 M NaSCN. One potential disadvantage of SCN^- ions is that they absorb at 280 nm which may interfere with UV monitoring of small amounts of released antigen, although generally the eluted protein can be seen as a sharp peak followed by a lower tail of chaotropic ion. Care must also be taken to ensure that the chaotrope is fully removed following elution (see next section).

5.3 Ionic strength

Washing the column with a high salt solution (e.g. 1 M NaCl) can remove material bound with low affinity and is usually used to remove non-specifically bound proteins. In general it will not be strong enough to elute the cognate antigen, but salts of divalent cations such as Mg^{2+} and Ca^{2+} have a higher ionic strength at equivalent molarities. $MgCl_2$ at 3–4 M can be used to elute some antigens, the two ions also having a weak chaotropic effect. Tsang and Wilkins (39) reported the most effective elution of polyclonal goat anti-human IgG antibodies from immobilized human IgG with a combination of 3 M $MgCl_2$ in 75 mM Hepes–NaOH pH 7.2 and 25% ethylene glycol.

5.4 Electroelution

A technique not as commonly used as chemical methods it can none the less provide good yields of eluted antigen in a small volume without subjecting it to extremes of pH or denaturants (40, 41). It requires the use of additional apparatus, either home-made or commercial (e.g. Bio-Rad Model 422 electro-eluter). The washed matrix carrying the adsorbed antigen has to be transferred from the column to an electroelution tube for the process to be carried out, so is more suited to self-packed or batch processed material rather than ready-packed columns. The antigen is eluted by the electric field into a small chamber sealed with a dialysis membrane and can be recovered in a small volume (400–600 µl) of suitable buffer. The membrane enclosed chamber can be obtained with membranes of different molecular weight cut-off to suit the size of the protein being eluted.

6 Antigen reconstitution

Once the antigen has been eluted from the column it is necessary to return it to physiological conditions; progressive irreversible denaturation can occur with prolonged exposure to an adverse environment. Samples eluted at high or low pH can be neutralized by addition of a small amount of concentrated buffer, e.g.

1 M Tris–HCl pH 7.5–8.0. This may be added to each sample as it is collected, or it can be added to each collection tube before fractionation starts, although mixing throughout the sample will be less effective. Denaturants and chaotropes are best removed by column desalting or dialysis; slower removal of denaturants by dialysis may allow time for the proteins to refold but this will depend on the protein in question and it is best initially to remove the denaturants rapidly (42). Samples containing the eluted antigen are pooled and loaded onto a column of Sephadex G25 or equivalent gel filtration medium pre-equilibrated in PBS or TBS pH 7–8 with additional detergents if membrane proteins are being purified. Pre-packed columns from various suppliers are available for this purpose. The easiest way is on-line desalting in which the affinity column undergoing elution is connected via a peristaltic pump to a desalting column fitted with a flow adapter and pre-equilibrated with PBS (e.g. a 2.2 × 30 cm column filled with Sephadex G25). The effluent from the gel filtration column is connected to the UV monitor and fraction collector. As the sample is eluted it is immediately loaded onto the desalting column and separated from the denaturant. If the eluting agent absorbs in the UV (SCN⁻ for example) the separation of eluted protein and denaturant can be clearly followed, the system can also be left to run unattended. Even after removal of denaturant ions by gel filtration the sample will need to be dialysed extensively to remove all traces, especially if the protein has been eluted with SCN⁻ and it is intended subsequently to label the protein with radioactive iodine as thiocyanate ions are potent inhibitors of the Chloramine-T or Iodogen-type reactions (43). Sodium azide as a preservative should also be avoided for the same reason at this stage. Whichever method is chosen the sample will generally need to be concentrated by Centricon concentrator (Amicon) or pressure stirred cells if the volume is larger. A final size fractionation step by gel filtration is recommended to ensure that the product does not contain aggregates or degraded material. This may also reduce the level of any contaminating antibody that may have leached from the bed and formed complexes in the eluate after removal of the denaturant.

7 Analysis of purified antigen

After elution and reconstitution of the antigen it is important to check that it has been recovered in as native a state as possible following disruption of the antibody–antigen bond, and free of any unwanted contaminants. Analysis by gel permeation HPLC will reveal any aggregation or fragmentation under non-denaturing conditions. The presence of any contaminating antibody leached from the column can be detected by immunoassay with a suitable anti-immunoglobulin antibody (32, 33), similarly for any unwanted Ig bound by non-saturated Protein A or G columns. Subunit molecular weight information for protein antigens can be obtained from SDS-PAGE, and a Western blot with the antibody used on the column will reveal the size of proteins carrying the antigenic epitopes. However, not all monoclonal antibodies will recognize antigen once it has been denatured by SDS and reducing agents so it may be necessary to run non-denaturing gels or to use a different antibody if one is to hand. By the nature of the technique the immobilized antibody will recognize and bind to molecules carrying a particular epitope, if this epitope is shared with molecules in the starting mixture other than the antigen of interest they too will be isolated. It may be possible to separate these subsequently by size fractionation, but careful analysis by 1D or 2D gels/blots may be needed to ensure a clean product.

If the antigen is to be used for immunization it may be sufficient to test that the required epitopes are present, but often it will be necessary to assay for recovery of biological activity. The antigen may be denatured or misfolded in some way that would not show up by these techniques. It is useful therefore to have some form of bioassay by means of which the purified ligand and the original starting material can be compared, this will of course depend on the nature of the antigen itself.

8 Future developments

The previous sections have described techniques for using monoclonal antibodies derived from conventional cell fusion and cloning methods. With the advent of recombinant DNA technology it is now possible to engineer immunoglobulin fragments, Fv, and single chain Fv (scFv), derived from hybridoma cDNA that can be used directly on immunoaffinity columns (44). In addition they can be modified to contain additional characteristics not found in the native molecule. These can include tags to facilitate detection or linking to affinity matrices: Protein A fusion proteins will bind to ready-made IgG columns (IgG Sepharose Fast Flow, Pharmacia), a C-terminal cysteine can be added to facilitate linking to sulfydryl gels without preliminary reduction of disulfide bonds. Another example is described by Kleymann and co-workers (45) in which a scFv was engineered to contain a C-terminal affinity tag to allow binding to a streptavidin–Sepharose column. In this indirect form of immunoaffinity chromatography the initial antibody–antigen interaction can be carried out in solution, in this case by mixing the antigen-containing preparation with a periplasmic fraction of E. coli bearing the bacterially-expressed scFv fragment. This may improve antibody–antigen binding as the antibody fragment is not constrained by crosslinking to a solid surface. The affinity chromatography step was then performed on streptavidin–Sepharose thus combining the antibody-matrix linking and antigen loading stages into a single step. Elution was performed with diaminobiotin to act as free ligand for the streptavidin, releasing the antigen–scFv as a complex. In a similar way to the indirect linking to Protein A–Sepharose described earlier purification of the antibody fragment to homogeneity is not required as this takes place automatically as the mixture is applied to the column. Similar fusion protein approaches can be used with other tags such as hexahistidine to couple the scFv-antigen complex to nickel-charged chelating Sepharose, or cellulose binding domains (46, 47) contained in plasmid vectors (e.g. pET-34(b) from Novagen) which would bind the scFv–

antigen complex to a cellulose-containing column bed directly avoiding the need to purchase expensive ready-conjugated matrices. An intermediate recombinant bifunctional linker constructed along similar lines with a cellulose binding domain connected to Protein A has been used to generate an immunoaffinity column by coupling an intact IgG antibody to a non-derivatized cellulose matrix (48). This could be used with any monoclonal antibody that will bind Protein A or an analogous Protein G or L linker. Again the initial antibody–antigen inter-action can be carried out in solution to improve binding characteristics and to obviate the need for preliminary antibody purification. Immunoaffinity can be combined with other affinity steps such as IMAC using an antibody to the His_6 tag to purify recombinant His-tagged proteins eluted from the metal–chelate column (49), further separating the protein of interest from contaminants.

References

1. Jack, G. W. (1994). *Mol. Biotechnol.*, **1**, 59.
2. Goding, J. W. (1996). In *Monoclonal antibodies, principles, and practice*, pp. 327–51. Academic Press Ltd., London, UK.
3. Fung, B., Lieberman, B., and Lee, R. (1992). *J. Biol. Chem.*, **267**, 24783.
4. Köhler, G. and Milstein, C. (1975). *Nature*, **256**, 495.
5. Hjertén, S. (1964). *Biochim. Biophys. Acta*, **79**, 393.
6. Cuatrecasas, P. (1970). *Nature*, **228**, 1327.
7. Porath, J., Janson, J.-C., and Låås, T. (1971). *J. Chromatogr.*, **60**, 167.
8. Subramanian, A. and Velander, W. H. (1996). *J. Mol. Recognit.*, **9**, 528.
9. Cuatrecasas, P. (1970). *J. Biol. Chem.*, **245**, 3059.
10. Kohn, J. and Wilchek, M. (1982). *Enzyme Microb. Technol.*, **4**, 161.
11. Jennissen, H. P. (1995). *J. Mol. Recognit.*, **8**, 116.
12. Tesser, G. I., Fisch, H. U., and Schwyzer, R. (1974). *Helv. Chim. Acta*, **57**, 1718.
13. Lasch, J. and Koelsch, R. (1978). *Eur. J. Biochem.*, **82**, 181.
14. Cuatrecasas, P. and Parikh, I. (1972). *Biochemistry*, **11**, 2291.
15. Frost, R. G., Monthony, J. F., Engelhorn, S. C., and Siebert, C. J. (1981). *Biochim. Biophys. Acta*, **670**, 163.
16. Bethell, G. S., Ayers, J. S., Hancock, W. S., and Hearn, M. T. W. (1979). *J. Biol. Chem.*, **254**, 2572.
17. Prisyazhnoy, V., Fusek, M., and Alakhov, Y. (1988). *J. Chromatogr.*, **424**, 243.
18. Domen, P. L., Nevens, J. R., Mallia, A. K., Hermanson, G. T., and Klenk, D. C. (1990). *J. Chromatogr.*, **510**, 293.
19. Highsmith, F., Regan, T., Clark, D., Drohan, W., and Tharakan, J. (1992). *Biotechniques*, **12**, 418.
20. Vankova, R., Gaudinova, A., Sussenbekova, H., Dobrev, P., Strnad, M., Holik, J., *et al.* (1998). *J. Chromatogr. A*, **811**, 77.
21. Hoffman, W. L. and O'Shannessy, D. J. (1988). *J. Immunol. Methods*, **112**, 113.
22. Orthner, C. L., Highsmith, F. A., Tharakan, J., Madurawe, R. D., Morcol, T., and Velander, W. H. (1991). *J. Chromatogr.*, **558**, 55.
23. DeSilva, B. S. and Wilson, G. S. (1995). *J. Immunol. Methods*, **188**, 9.
24. Schneider, C., Newman, R. A., Sutherland, D. R., Asser, U., and Greaves, M. F. (1982). *J. Biol. Chem.*, **257**, 10766.
25. Sisson, T. H. and Castor, C. W. (1990). *Immunol. Methods*, **127**, 215.
26. Lasch, J. and Janowski, F. (1988). *Enzyme Microb. Technol.*, **6**, 31.

27. Ubrich, N., Hubert, P., Regnault, V., Dellacherie, E., and Rivat, C. (1992). *J. Chromatogr.*, **584**, 17.

28. Riedstra, S., Ferreira, J. P., and Costa, P. M. (1998). *J. Chromatogr. B. Biomed. Sci. Appl.*, **705**, 213.

29. Kowal, R. and Parsons, R. C. (1980). *Anal. Biochem.*, **102**, 72.

30. Goldstein, M. A., Takagi, M., Hashida, S., Shoseyov, O., Doi, R. H., and Segel, I. H. (1993). *J. Bacteriol.*, **175**, 5762.

31. Sato, H., Kidaka, T., and Hori, M. (1987). *Appl. Biochem. Biotechnol.*, **15**, 145.

32. Beer, D. J., Yates, A. M., and Jack, G. W. (1994). *J. Immunol. Methods*, **173**, 103.

33. Beer, D. J., Yates, A. M., Randles, S. C., and Jack, G. W. (1995). *Bioseparation*, **5**, 214.

34. Bordier, C. (1981). *J. Biol. Chem.*, **256**, 1604.

35. Galfre, G., Milstein, C., and Wright, B. (1979). *Nature*, **277**, 131.

36. Quitadamo, I. J. and Schelling, M. E. (1998). *Hybridoma*, **17**, 199.

37. Herrmann, S. H. and Mescher, M. F. (1979). *J. Biol. Chem.*, **254**, 8713.

38. Parham, P. (1979). *J. Biol. Chem.*, **254**, 8709.

39. Tsang, V. C. and Wilkins, P. P. (1991). *J. Immunol. Methods*, **138**, 291.

40. Schulze-Osthoff, K., Michels, E., Overwien, B., and Sorg, C. (1989). *Anal. Biochem.*, **177**, 314.

41. Williamson. K. C., Duffy, P. E., and Kaslow, D. C. (1992). *Anal. Biochem.*, **206**, 359.

42. Hager, D. A. and Burgess, R. R. (1980). *Anal. Biochem.*, **109**, 76.

43. George, S. and Schenck, J. R. (1983). *Anal. Biochem.*, **130**, 416.

44. Berry, M. J. and Pierce, J. J. (1993). *J. Chromatogr.*, **629**, 161.

45. Kleymann, G., Ostermeier, C., Ludwig, B., Skerra, A., and Michel, H. (1995). *Biotechnology*, **13**, 155.

46. Greenwood, J. M., Ong, E., Gilkes, N. R., Warren, R. A., Miller, R. C., and Kilburn, D. G. (1992). *Protein Eng.*, **5**, 361.

47. Goldberg, M., Knudsen, K. L., Platt, D., Kohen, F., Bayer, E. A., and Wilchek, M. (1991). *Bioconjugate Chem.*, **2**, 275.

48. Ramirez, C., Fung, J., Miller, R. C., Antony, R., Warren, J., and Kilburn, D. G. (1993). *Biotechnology*, **11**, 1570.

49. Muller, K. M., Arndt, K. M., Bauer, K., and Pluckthun, A. (1998). *Anal. Biochem.*, **259**, 54.

Chapter 15

Immunochemical detection of BrdUrd labelled nuclei for monitoring cell kinetics

George D. Wilson

CRC Gray Laboratory, Mount Vernon Hospital, Nothwood, Middlesex HA6, U.K.

1 Introduction

Cell kinetics is defined as the measurement of time parameters in biological systems. This chapter describes the use of BrdUrd/DNA flow cytometry to study cell kinetics of cells in culture. The technique involves pulse labelling with BrdUrd after which samples are then taken at time intervals thereafter. The cells are stained with a monoclonal antibody against BrdUrd and visualized by a secondary antibody conjugated to a fluorochrome (usually fluoroscein isothiocyanate, FITC) and counterstained with propidium iodide (PI) to measure the DNA content, and analysed on the flow cytometer. The simultaneous measurement of BrdUrd incorporation and DNA content confers sensitivity and versatility in detection of cell cycle perturbations in response to drugs or radiation and tracing the lineage of a cell within the cell cycle. The speed and quantitative power of the flow cytometer, in conjunction with the specificity and sensitivity of monoclonal antibody techniques, provide the basis for the adoption and success of the BrdUrd technique in experimental and clinical investigations. The use of computer-generated windows facilitates the analysis of any population of cells within any phase of the cell cycle and is not restricted to the 'mitotic window', as previously was the case with ^{3}HTdR/autoradiography.

2 Basic concepts in cell kinetics

Readers are referred to the comprehensive treatise by Gordon Steel (1) for the basic theory of growing cell populations and a detailed review of cell kinetics in experimental systems and human tumours. Cell kinetics is essentially the quantitation of rates of cell production and cell loss, which balance the overall rate of growth of cells, tissues, and tumours. This can vary from the simple concept of exponential growth for cells in culture to the complex growth characteristics of human tumours. The cell cycle drives cell production and its duration (Tc) is defined as the time interval within which a cell completes a mitotic cycle, i.e. from birth at mitosis to division at the next mitosis to form two progeny. The cell cycle is made up of discrete sub-phases termed G1, S, G2, and M (mitosis). Each phase of the cell cycle is co-ordinated and accomplishes a sequence of biochemical events to ensure successful production of viable daughter cells. Events in G1 prepare and control the entry of cells into DNA synthesis. During S phase, DNA is correctly duplicated, whilst in G2 phase the events that allow entry into mitosis are accomplished. During mitosis the mitotic spindle is assembled, chromosomes become condensed, and the nuclear envelope breaks down in preparation for cell division. The decision to divide is an important one for any cell and the events required for this process are under feedback controls and scrutiny to ensure that correct replication takes place. Each phase of the cell cycle will have a finite time for its completion and this will vary from one cell type to another. Thus, the number of cells found in any particular phase will depend on the time they spend in that phase relative to the cell cycle time. In exponential growth, the growth rate of the population can be described by the equation:

$$N_t = N_0 \exp{(\beta t)} \quad \text{where } \beta = \frac{\log_e 2}{T_c}$$

Therefore, knowing the cell cycle time (T_c) the number of cells at any time (N_t) can be extrapolated from the starting number (N_0). In exponential growth the duration of each phase of the cycle is related to the T_c by the equation described below for the duration of S phase (T_s) and the number of cells in S phase (LI):

$$LI = \log_e 2 \cdot \frac{T_s}{T_c}$$

In more complex systems such as tissues and tumours not all cells will be involved in an active cell cycle. This introduces two further concepts of growth fraction and cell loss. The term G0 has been adopted to refer to those cells that are out of cycle. However, this rather nondescript term is used to describe cells that may be out of cycle for vastly different reasons. Cells may be out of cycle because they have started the process of differentiation and maturation or they may have reached their life span and are undergoing cell death. They may be in a poorly vascularized area of the tissue and cannot continue their cycle due to nutrient deprivation; this may leave them in a state of temporary hypoxia in which they may be able to re-enter the cycle if the microenviron-ment improves or they may die if the situation becomes chronic. Cell loss also describes a diversity of underlying processes from exfoliation in mucosal tissues to programmed cell death or apoptosis, necrosis, or even metastatic loss in tumours.

In cell kinetic terms, the concept of turnover time or potential doubling time has to be introduced to describe the time taken for the actively dividing cells to double the total number of cells. At each cell division, instead of two pro-liferating cells being formed there is a proportion of new proliferating cells (α)

and thus the growth fraction becomes $\alpha - 1$. The doubling time (T_d) would now be:

$$T_d = \frac{\log_e{}^2}{\log_e{}^\alpha} \cdot T_c$$

In a cell system with a growth fraction, the duration of individual phases of the cell cycle are related to the T_c by the equation outlined below T_s and LI:

$$LI = \frac{\log_e\alpha}{\alpha - 1} \cdot \frac{T_s}{T_c}$$

In tumours, the term potential doubling time (T_{pot}) is used to describe the shortest time a cell population can double its number taking into account growth fraction but in the absence of cell loss using the equation:

$$T_{pot} = \lambda \cdot \frac{T_s}{LI}$$

The term λ describes the age distribution of the population, it is a term which varies slowly with population doubling time with a minimum value of 0.693 in steady state tissues and a maximum value of 1.38 if the S phase occupied the complete cell cycle. In tumours a value of 0.8–1 is common. The volume doubling time represents the time interval in which a tumour doubles its volume and will be subject to cell loss and the growth fraction, as well as the influence of any host cells contained within the tumour mass, and is related to T_{pot} by the equation:

$$\theta = 1 - \frac{T_{pot}}{T_d}$$

where θ is the cell loss factor, that is the rate of cell loss divided by the rate of cell production (MI/T_m or LI/T_s).

Each of these cell kinetic parameters can be measured by different techniques although some, like θ, are difficult to quantitate and are usually extrapolated from other measurements. The BrdUrd/DNA staining technique can provide quantitative information on cell cycle phases and their duration, T_{pot}, and growth fraction.

3 Background to the BrdUrd flow cytometry technique

The technique became possible due to the development of monoclonal antibodies that recognize halogenated pyrimidines incorporated into DNA (2). Dolbeare and colleagues (3) then developed a simultaneous staining method, using FCM, to study the incorporation of BrdUrd relative to DNA content measured by propidium iodide (PI). The general approach to measuring cell kinetics is to identify a 'window' in the cell cycle and measure the movement of a cohort of labelled cells through the window. With ^{3}HTdR, the only identifiable window is mitosis; consequently, the per cent labelled mitosis (plm) analysis was developed. However, with the BrdUrd/PI method windows can be set in any phase of the cell cycle (see *Figure 1*) looking at either the BrdUrd labelled or unlabelled population using appropriate computer-generated regions.

The essence of the procedure is to pulse label with BrdUrd by a short incubation *in vitro* or by a single injection *in vivo*, samples are then taken at time intervals thereafter, and stained after fixation in ethanol. The cells are stained using a monoclonal antibody against BrdUrd that can either be directly conjugated to a fluorochrome (usually FITC) or alternatively bound to a second antibody conjugated with FITC. The cells are then counterstained with PI to measure DNA content, and analysed on the flow cytometer for red (DNA) and green (BrdUrd) fluorescence. The results are displayed as red (*x* axis) versus green (*y* axis) bivariate distributions.

Figure 1 shows a series of pseudo 3D distributions of DNA (*x* axis), BrdUrd incorporation (*y* axis), and cell number (*z* axis) obtained at hourly intervals after pulse labelling V79 Chinese hamster fibroblasts with 10 μM BrdUrd for 20 minutes *in vitro*. In the profile recorded after one hour the BrdUrd labelled S phase cells are clearly identified by their green fluorescence and lie between the G1 and G2 populations. The latter two populations can also be separately identified by virtue of their lack of BrdUrd uptake and the difference in their DNA content; note that the majority of cells reside in G1. The BrdUrd labelling is classically crescent-shaped with lower levels of uptake in cells that have just entered S phase from G1, and those found in late S phase about to enter G2. With time, the profiles show changing patterns. At two hours, the labelled cohort has become slightly skewed to the right as the cells have made more

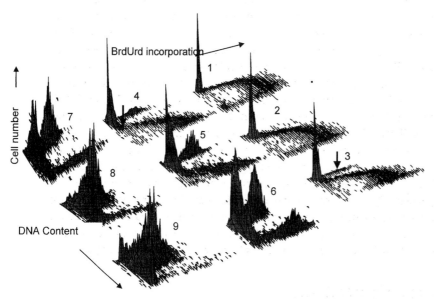

Figure 1 Isometric displays of V79 cells pulse labelled with BrdUrd and harvested at time intervals. Each display represents DNA content (*x* axis) versus BrdUrd uptake (*y* axis) versus cell number (*z* axis).

DNA. One hour later there is the appearance of BrdUrd labelled cells in G1 (see *arrow*), these represent cells in late S at the time of labelling which have finished S phase, gone through G2, divided, and become two daughter G1 cells. It should also be noted that the original G2 population has also disappeared, as these cells will divide prior to the labelled cells. As time progresses the BrdUrd labelled cohort moves through S phase and more cells divide. It is also evident that the original G1 population moves into S phase; these cells can be identified by virtue of their lack of BrdUrd but they have S phase DNA (see *arrow* in the four hour profile). This is another attribute of the technique as each population can be followed as it moves through its cell cycle whether it is BrdUrd labelled or not. Whilst BrdUrd cells reside in G1 they are not making DNA, therefore the number of labelled cells in G1 increases with time until, at six hours, some of the labelled cells in G1 begin to move into S phase for the second time as they complete their transit of the G1 phase. This is more evident in the subsequent profiles. Eventually the staining profile will revert to that seen in the one hour profile when all cells have completed their cycle; this would probably take 12 hours in this particular example.

4 *In vitro* BrdUrd incorporation

Incorporation into cell cultures is usually achieved by incubation with 10–20 μM BrdUrd for 20–30 minutes; this will achieve a pulse label. The drug would normally be made up as a stock solution of 1 mM in the medium used for the particular cell line. The BrdUrd solution should always be sonicated or shaken in warmed medium to ensure that it is fully dissolved. The stock can be diluted to the required final concentration by addition to the cell culture. It is important to make sure everything is kept at 37 °C as small fluctuations in temperature can cause cell cycle perturbations. This is particularly required at the next stage where the BrdUrd is washed out after the labelling period. At least two washes in warm medium are required. It is essential that there is no contaminating BrdUrd left in as it will continue to be incorporated as cells move from G1 into S. It is imperative that control cultures should be sham-treated in experiments involving cell cycle blocking agents. After removal of the DNA precursor, fresh medium is added and the cells allowed to progress through their cycle until the required sampling time. In continuous labelling studies, the BrdUrd is usually incubated for a complete cell cycle generation until all cells become labelled. This is particularly useful for looking at recruitment of cells into the cycle. Care must be taken to ensure that the concentration of BrdUrd used does not have a cytostatic effect although up to 100 μM BrdUrd has been used for continuous labelling in studies using the Hoechst 33342/BrdUrd quenching technique.

5 Fixation procedures

After harvesting by the appropriate method, cell suspensions can be fixed by a number of different procedures depending on the application (*Protocol 1*). The simplest option is to fix in ice-cold 70% ethanol after prior resuspension in a small volume of PBS. However, if preservation of other nuclear, cytoplasmic, or surface epitopes is required, then other fixation techniques might be preferable as outlined below. Alcohol fixation is preferable for optimal DNA staining but paraformaldehyde is superior for retaining surface and cytoplasmic epitopes. Long-term storage after paraformaldehyde is not advisable.

Protocol 1

Fixation of cells

A. Ethanol

1. Resuspend harvested cell pellet ($1-2 \times 10^6$) in 200 μl PBS.

2. Add 3-5 ml of ice-cold 70% ethanol dropwise whilst vortexing.

3. Store at 4 °C until required.

B. Methanol

1. Resuspend harvested cell pellet ($1-2 \times 10^6$) in 200 μl PBS.

2. Add 3 ml of absolute methanol, cooled to −20 °C, dropwise whilst vortexing.

3. Incubate for 20 min at −20 °C.

4. Dilute to 70% with distilled water and store at 4 °C until required.

5. Centrifuge cells for 5 min at 200 g.

6. Add 0.1% Triton X-100 in PBS and incubate for 5 min on ice prior to staining.

C. Paraformaldehyde

1. Resuspend harvested cell pellet ($1-2 \times 10^6$) in 100 μl PBS.

2. Add 1 ml of 0.5% paraformaldehyde in PBS whilst vortexing.

3. Incubate for 30 min on ice.

4. Centrifuge cells for 10 min at 200 g.

5. Add 0.1% Triton X-100 in PBS and incubate for 5 min on ice prior to staining.

6 Staining procedures

Several procedures have been described for identifying cells that have incorporated BrdUrd. The basic immunochemical staining with monoclonal antibody, either directly conjugated to FITC or indirectly through a second antibody, varies little between procedures. The diversity of techniques arises from the requirement for partial DNA denaturation to permit access of the monoclonal antibody to its epitope on the incorporated BrdUrd.

6.1 DNA denaturation

The DNA denaturation must be rigorously controlled to achieve sensitive detection of BrdUrd without disruption of the stoichiometry of binding of the intercalating dye, propidium iodide, which requires double-stranded DNA. The techniques that will be described include the use of acid, the use of heat, and the use of endonuclease/exonuclease enzymes. Each procedure has its advantages and disadvantages. The HCl method is relatively mild and applicable to all cell types; its disadvantage is that the level of denaturation achieved without disruption of the PI staining may not be enough to permit sensitive detection of BrdUrd incorporated at low levels. This has been improved by Schutte *et al.* (4) who combined acid denaturation with pepsin enucleation of cell suspensions.

The thermal denaturation procedures in formamide or water greatly increase the sensitivity of BrdUrd detection. Beisker (5) reported that the fluorescence ratio of BrdUrd labelled S phase cells to unlabelled G1 cells was fivefold using standard 2 M HCl at 25 °C but could be increased to 150-fold by extracting histones with 0.1 M HCl plus 0.7% Triton X-100 followed by denaturation in water for 10 min at 100 °C. The major drawback of thermal denaturation techniques is that substantial cell loss can occur; up to 90% of haematopoetic cells can be lost during the process. This renders the technique unsuitable for studies where the sample is at a premium such as those with human tumours. In experimental systems where material is not limiting, the thermal techniques would be particularly applicable for studies of drug resistance, unscheduled DNA synthesis, etc.

Combining BrdUrd staining with surface markers or other labile antigens has proven difficult due to the nature of the denaturation procedures. Perhaps the most promising method to obtain sensitive detection of BrdUrd without severe cell loss, morphological change, or protein loss is to use a restriction endonuclease and exonuclease (6). The disadvantages of this method are the extra costs and different cell types might require different endonucleases for optimal detection conditions.

Protocol 2

Denaturation procedures for BrdUrd detection

A. HCl acid alone

1. 1–2 × 10⁶ fixed cells are resuspended in 2.5 ml of 2 M HCl at room temperature. Cells are incubated for 20–30 min (this varies according to the cell type) with occasional mixing. Add 5 ml PBS.
2. Centrifuge at 200 *g* for 5 min.
3. Decant and repeat the washing to get rid of all the acid.
4. Alternatively, resuspend pellet in 5 ml of 0.1 M sodium tetraborate to neutralize the acid and wash twice in PBS.

B. Pepsin/HCl combined enucleation and denaturation

1. 1–2 × 10⁶ fixed cells are resuspended in 2.5 ml of 2 M HCl containing 0.2 mg/ml of pepsin. The pepsin should be dissolved in a small amount of water or PBS first. Incubate at room temperate for 20 min (this may vary according to the cell type).

Protocol 2 continued

2. Wash twice with 5 ml PBS, centrifuging at 300 g between washes.

C. Thermal denaturation methods

1. Incubate 1–2 × 10⁶ ethanol-fixed cells in 1.5 ml of PBS containing 1 mg/ml RNase for 20 min at 37 °C.

2. Centrifuge at 200 g for 5 min.

3. Resuspend pellet in 2 ml of 0.1 M HCl containing 0.7% Triton X-100 for 10 min at 0 °C.

4. Centrifuge at 200 g for 5 min and wash with 5 ml of cold distilled water.

5. Resuspend in 1.5 ml of 50% formamide for 30 min at 80 °C or 2 ml of distilled water for 10 min at 100 °C.

6. Plunge tubes into ice-bath to stop the reaction.

7. Wash with 5 ml PBS.

D. Endonuclease/exonuclease denaturation

1. Incubate 1–2 × 10⁶ ethanol-fixed cells in 1.5 ml of PBS containing 1 mg/ml RNase for 30 min at 37 °C.

2. Centrifuge at 200 g for 5 min and wash in 5 ml PBS.

3. Centrifuge at 200 g for 5 min and resuspend pellet in 1 ml of 0.1 M citric acid containing 0.5% Triton X-100 and incubate for 10 min on ice.

4. Centrifuge at 200 g for 5 min and wash in 3 ml of Tris buffer (0.1 M Tris–HCl, 50 mM NaCl, and 10 mM MgCl₂ pH 7.5).

5. Centrifuge at 200 g for 5 min and resuspend pellet in 100 μl of BamHI buffer (10 mM Tris–HCl, 5 mM MgCl₂, 1 mM mercaptoethanol pH 8.0) with 60 U of BamHI (Sigma). Incubate for 30 min at 37 °C.

6. Centrifuge at 200 g for 5 min and wash with 3 ml of Tris buffer.

7. Resuspend pellet in 100 μl of exonuclease III buffer (66 mM Tris–HCl, 0.66 mM MgCl₂, 1 mM mercaptoethanol pH 7.6) with 100 U of exonuclease III (Sigma). Incubate for 30 min at 37 °C.

8. Wash with Tris buffer and then with PBS.

6.2 Antibody staining

There are many commercial sources of monoclonal antibodies recognizing the halogenated pyrimidines. The choice of antibody often depends on individual preference to one or other company. However, several points should be taken into account as the quality of the measurement will depend on the purity, affinity, and specificity of the antibody. For instance high affinity antibodies such as IU-4 (Caltag) will be required if low levels of BrdUrd are to be detected. Most antibodies are divalent and their binding is limited when BrdUrd substitution is less than 1%. In some instances it may be preferable to use a rat-derived monoclonal such as BU/175 (Sera-Lab) to reduce non-specific staining if working with mouse tissue or to combine with another mouse-derived mono-clonal. Different antibodies have different specificities for the various halo-genated pyrimidines and this can be exploited in double labelling techniques. As with all antibody staining the optimal working concentration should be assessed by dilution analysis and this may vary from batch to batch.

The following protocol is our standard immunochemical detection system for BrdUrd utilizing the rat anti-BrdUrd antibody (Sera-Lab), if mouse antibodies are used the second antibody should be substituted with a goat anti-mouse FITC. We routinely use indirect fluorescence to increase the signal, as often the BrdUrd levels are low in human tumours. However, most of the major mono-clonals are now available as directly labelled conjugates and these work well in experimental systems and of course, are particularly useful in double labelling experiments. Again, the choice of the second antibody source is not crucial; we routinely use Sigma antibodies.

Protocol 3

Immunochemical detection of BrdUrd

1. After denaturation and washing, resuspend the pellet in 0.5 ml of PBS containing 0.5% Tween 20, 0.5% normal goat serum (NGS—this blocking agent can be sub-stituted with 1% BSA), and 20 µl of rat anti-BrdUrd antibody. Incubate for 1 h at room temperature with occasional mixing.

2. Add 5 ml PBS and centrifuge at 200 g for 5 min.

3. Resuspend pellet in 0.5 ml PBS/Tween/NGS containing 20 µl of goat anti-rat IgG (whole molecule) FITC conjugate (Sigma Chemical Co.). Incubate for 30 min at room temperature.

4. Add 5 ml of PBS and spin at 200 g for 5 min.

5. Resuspend pellet in 1 ml of PBS containing 10 µg/ml propidium iodide.

6. The preparation can be analysed immediately on the FCM.

7 Flow cytometry

This type of staining can be analysed on any of the modern flow cytometers with the proviso that the machine is equipped with a pulse processing facility to enable the discrimination of cell doublets. When using the Becton Dickinson FACScan, the propidium iodide should be collected in FL3 rather than FL2 to overcome any crossover with the FITC in FL1. Routinely the FL3 detector is set to linear amplification should, be set at around 400, whilst FL1 can be collected with log or linear amplification, and the PMT voltage is usually around 500 (linear) or 300 (log). Controls, either without BrdUrd or the monoclonal anti-body, should be included wherever possible to determine the lower limit of detection of the DNA precursor in the bivariate profile.

8 Examples of data and analysis

Figure 2A shows the type of data that can be obtained from cells grown in culture and incubated with BrdUrd in a pulse labelling experiment. Many different experiments can be designed using this type of approach to study progression through the cell cycle. In this example primary human fibroblasts cells have been irradiated with 2 Gy of X-rays. The figure shows the cell cycle distribution of irradiated and sham-treated cells at 10 hours after pulse labelling with BrdUrd. In the control cells, the BrdUrd labelled cohort have either divided and reside in G1 or are in G2 ready to divide. In contrast, the irradiated cells show an almost complete absence of divided BrdUrd labelled cells in G1 because radiation has caused them to accumulate in G2. The most appropriate procedure to analyse this effect would be to generate the DNA histogram of the BrdUrd labelled cells only by setting a region around labelled population. *Figure 2B* shows the DNA histogram of the BrdUrd labelled population in untreated cells (solid) overlaid with that obtained from the irradiated cells. Regions can then be set to extract quantitative information from the data. Region 1 defines the number of BrdUrd labelled cells in G1, i.e. those S phase cells which have transited G2 and successfully divided; this was used to generate the graph (*Figure 2C*) showing the time course of entry into G1. The irradiated cell curve is

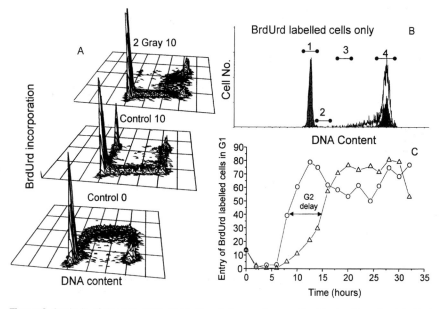

Figure 2 Analysis of the cell cycle effect of radiation on human fibroblast cells pulse labelled with BrdUrd. (A) Representative isometric displays of DNA content (*x* axis) versus BrdUrd uptake (*y* axis) versus cell number (*z* axis) are presented for irradiated and control cell cultures. (B) The DNA profile of the BrdUrd labelled cells at 10 hours has been gated. The solid fill represents the control cells which has been overlaid with the histogram from the irradiated cells. (C) A plot of entry of labelled cells into G1 (region 1 in B) displays the division delay effect of radiation (Δ) compared to control cultures (O).

shifted to the right due to division delay and the extent of this can be estimated using an iso-effective level (usually the point at which 50% of the maximum number of cells that enter G1).

The regions of analysis can be varied depending on the effect being measured. For example to measure G1/S delay it would be appropriate to gate a DNA histogram of the non-BrdUrd labelled cells and to use a region such as 2 in *Figure 2*. This region would then be monitoring the original G1 population as it moved into S phase; this is the population that would be first to experience a G1/S delay. The G2 delay of the original G2 population could be measured by region 4 also in the non-BrdUrd labelled DNA histogram.

The basic cell cycle parameters can be calculated by a variety of analyses. The most common method to measure the Tc is to set two narrow windows in mid-S (determined by measuring the mean DNA fluorescence of the G1 and G2 populations). One widow is set in the BrdUrd labelled population only, whilst the other spans both labelled and unlabelled populations (*Figure 3A*). The data is expressed as the ratio of labelled to total cells in mid-S (see *Figure 3B*). Initially the ratio should be 1 as all cells should be in the labelled window. This however

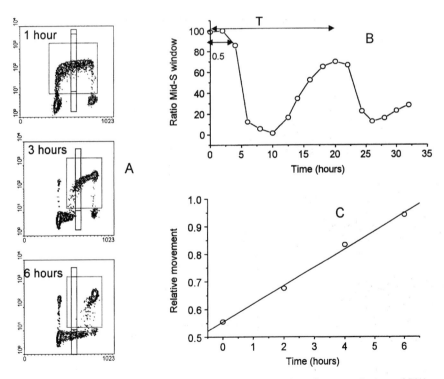

Figure 3 Calculation of basic cell cycle parameters. (A) A series of contour displays of DNA content (*x* axis) versus BrdUrd incorporation (*y* axis) for human fibroblast control cultures. (B) The graph has been generated from two narrow windows in mid-S and is used to calculate the cell cycle time (Tc). (C) The relative movement method has been used to calculate the duration of S phase by measuring the mean DNA content of the BrdUrd labelled cohort (using the dotted region in A).

may not be the case in experimental tumours, as some cells in S phase do not always incorporate the DNA precursor. This ratio should remain maximum for a period equal to 0.5 Ts, i.e. when the cells in early S phase at the time of labelling reach the window. The ratio will then fall as the labelled cells clear the region followed by the original (unlabelled) G1 population reaching mid-S. Thus the ratio remains low until the BrdUrd population complete G2, G1, and re-enters S phase for the second time. At which time the ratio rises and forms a second peak. The mid-point of that second peak is used to determine Tc.

The duration of G2 can be measured from the entry of labelled cells into G1 (*Figure 2*). This can be achieved by extrapolating a line back to zero which has been fitted to time points after the first two hours (because this region may contain cells which are still in early S but are included due to the CV of the G1 population) up until the entry plateaus when all labelled cells are in G1.

The duration of S phase can be calculated using a technique called 'relative movement' (RM) (7). The procedure is based on a measurement of the mean DNA content of the BrdUrd labelled cells defined by the dotted region in *Figure 3A*. Immediately after labelling, the mean of the BrdUrd labelled population is approximately midway between G1 and G2, as there is a uniform distribution of cells throughout S phase. To quantitate the function it is also necessary to measure the mean DNA contents of the G1 and G2 populations (this can be done from the single parameter DNA profile). The RM is calculated by subtracting the mean DNA of the G1 population from that of the labelled cells and dividing it by G1 subtracted from G2. If there is a uniform distribution at time zero, it should give a value of 0.5. With time the mean value of the labelled cohort (which remain undivided) will increase as it progresses through S to G2 (*Figure 3C*). If we make the assumption that the progression through S phase is linear, a point will be reached when all the labelled cells which remain undivided are in G2; the RM value will equal 1.0 and this time will be equivalent to the Ts as the cells in G2 will have been those in early S at the time of injection. Thus from single or multiple observations, made at a known time greater than G2 but less than Ts + TG2, the Ts can be computed assuming the value is 0.5 at time zero and 1.0 at Ts.

9 Conclusions

The development of antibodies to the halogenated pyrimidines combined with the attributes of FCM have led to a revolution in the study of cell kinetics and a resurgence of interest in their clinical application. The techniques are reproducible and relatively straightforward and can be tailored to individual experimental requirements.

References

1. Steel, G. G. (1977). *Growth kinetics of tumours*. Clarendon Press, Oxford.
2. Gratzner, H. G. (1982). *Science*, **218**, 474.
3. Dolbeare, F. A., Gratzner, H. G., Pallavicini, M. G., and Gray, J. W. (1983). *Proc. Natl. Acad. Sci. USA*, **80**, 5573.

4. Schutte, B., Reynders, M. M. J., van Assche, C., Hupperts, P. S. G., Bosman, F. T., and Blijham, G. H. (1987). *Cytometry*, **8**, 372.
5. Beisker, W., Dolbeare, F. A., and Gray, J. W. (1987). *Cytometry*, **8**, 235.
6. Dolbeare, F. A. and Gray, J. W. (1988). *Cytometry*, **9**, 631.
7. Begg, A. C., McNally, N. J., Shrieve, D. C., and Karcher, H. (1985). *Cytometry*, **6**, 620.

Chapter 16

Immunofluorescence microscopy

Rainer Pepperkok, Andreas Girod, Jeremy Simpson, and Jens Rietdorf

Advanced Light Microscopy Faculty, EMBL Heidelberg, Cell Biology/Cell Biophysics Program, Meyerhofstr. 1.69117 Heidelberg, Germany

1 Introduction

Antibodies provide a powerful tool to localize antigens in cells or tissues by immunocytochemistry at the light or electron microscope level. The development of efficient fluorescent dyes which can be coupled to antibodies for their visualization by fluorescence microscopy has pioneered a technology which is known as immunofluorescence microscopy (IFM). IFM is easy to apply to many biological and medical questions, the protocols involved are short, and the development of sophisticated imaging equipment has even made possible the acquisition of quantitative data on the 3D distribution of several antigens in the same specimen.

This chapter discusses protocols for fluorescent labelling of antibodies, optimized fixation, and permeabilization of cells, and describes several immunostaining procedures with the aim to quantitatively analyse the localization of antigens in fixed and living cells. Various aspects of the imaging equipment presently available, and the ways how to most efficiently use it in order to quantify and document results will be discussed. Labelling in living cells by microinjection of fluorescently labelled antibodies and subsequent time-lapse microscopy will also be discussed.

2 Immunofluorescence

At least two principal methods to label cellular antigens using monoclonal antibodies exist. First, the antibody, which specifically recognizes the cellular antigen is directly conjugated with a fluorescent dye. Upon incubation of fixed and permeabilized cells with such antibody the location of the antigen will be highlighted by the fluorescence of the dye conjugated to the antibody. By conjugating different antibodies with different fluorescent dyes this approach can be used for the labelling of a sample with several distinct antibodies each recognizing a different antigen. The only limitation in the number of antigens detectable by this approach is the availability of fluorescent dyes which can be spectrally discriminated by the fluorescence microscope. However, in some cases this staining method may not always provide sufficient sensitivity, in particular when antigens of low abundance have to be localized. Also, the antibodies used may not always be available in quantities which are required for their conjugation to fluorescent dyes. In order to circumvent these problems, indirect immunofluorescence staining protocols have been developed. They use a non-labelled antigen-specific antibody, which is allowed to bind to the antigen (first antibody). Then a secondary, fluorescently labelled, polyclonal, and species-specific antibody (e.g. anti-mouse or anti-rabbit IgG) is used to fluorescently label the first antibody. Such a staining protocol has the main advantage that several fluorescently labelled secondary antibodies can bind to the first antibody and thus amplify the fluorescent signal.

Regardless of which of the above approaches is used to fluorescently stain the cellular antigens under investigation, conjugation of antibodies to fluorescent dyes is required. As this is a critical step for the performance of high quality immunofluorescence microscopy we will discuss in more detail some protocols for the conjugation of antibodies to fluorescent dyes in the following section.

2.1 Fluorescent labelling of antibodies

The antibodies used for conjugation to fluorescent dyes should be purified and highly specific since contaminating molecules or inactive antibodies in the labelling solution will also be fluorescently labelled and may bind to the sample in a non-specific way causing a disturbing background fluorescence.

Several bright and photostable fluorescent dye derivatives, which are reactive with free amine or sulfhydryl groups (as found for example in lysines or cysteines, respectively) are available. Typical dye derivatives used for the labelling of amines or sulfhydryls are the isothiocyanates and carboxy-succinimidyl esters or iodoacetamides and maleimides, respectively. Amine containing side-chains are usually localized on the surface of a protein and thus they are very

Table 1 Amine group-specific dye derivatives and their characteristics

Derivative[a]	Ex./Em. (nm)	Solvent
Alexa 365 succinimidyl ester	365/440	Water
Fluorescein isothiocyanate (FITC)	494/520	DMSO
5-Carboxyfluorescein succinimidyl ester	491/518	DMSO
Alexa 488 succinimidyl ester	495/519	Water
Tetramethylrhodamine-5-isothiocyanate (TRITC)	544/570	DMSO
5-Carboxytetramethyl-rhodamine succinimidyl ester	546/579	DMSO
Cy3 succinimidyl ester[b]	550/570	Water
Cy5 succinimidyl ester[b]	649/670	Water

[a] All dye derivatives listed in the table, and many others with variations in excitation and emission maxima, are available from Molecular Probes (/www.probes.com/) except where indicated.

[b] Available from Amersham Pharmacia Biotech (/www.apbiotech.com/).

356

easy to derivatize. In contrast, the sulfhydryl groups in cysteines can usually be found in the core of a protein and are thus often inaccessible. Therefore, conjugation of antibodies with amine-specific dye derivatives is preferable.

Some amine-specific dye derivatives and their excitation and emission maxima are listed in *Table 1*.

Below we describe a protocol for the fluorescent conjugation of antibodies which we routinely use in our laboratories. A more general discussion of fluorescent dye derivatives and their applications to the labelling of marker proteins can be found in ref. 1.

Protocol 1

Labelling of antibodies with tetramethylrhodamine-5-isothiocyanate (TRITC)

The following method is used to fluorescently label amine groups (present for example in lysines) with the rhodamine derivative tetramethylrhodamine-5-isothiocyanate (TRITC; excitation = 546 nm, emission = 579 nm). Following this protocol a labelling efficiency of two to four moles of dye per mole of antibody is usually achieved.

Equipment and reagents

- Sephadex G25 (Pharmacia) or similar
- 0.1 M sodium carbonate buffer pH 8.5
- TRITC (Pierce)
- Microinjection buffer: 48 mM K_2HPO_4, 4.5 mM KH_2PO_4, 14 mM NaH_2PO_4 pH 7.2

Method

1. Dissolve the antibodies to be labelled in 0.1 M sodium carbonate buffer pH 8.5. The protein concentration should be at least 1 mg/ml. Similar buffers within the pH range 8.0–9.0 may be used as long as they do not contain amines (for example Tris buffer) which would compete with the amines to be labelled within the protein resulting in a very limited labelling efficiency.

2. Dissolve TRITC in dry DMSO at a concentration of 0.5 mg/ml. This solution should always be made up freshly, as breakdown of the thiocyanate group over time may drastically decrease the coupling efficiency. Protect the solution from exposure to daylight.

3. Slowly add 50 μl of TRITC solution/ml of protein solution. Mix gently.

4. React for at least 8 h at 4 °C in the dark. Occasionally mix the sample gently.

5. Stop the reaction by adding glycine at pH 8.5 (final concentration 1 mM). Incubate for 2 h at 4 °C to block remaining free TRITC which has not reacted with the antibodies.

6. Separate labelled antibodies from free TRITC by gel filtration using Sephadex G25 or similar matrices. To obtain optimum separation the column size should be at least 15 times the sample volume. As separation buffer we usually use microinjection buffer (48 mM K_2HPO_4, 4.5 mM KH_2PO_4, 14 mM NaH_2PO_4 pH 7.2). As

Protocol 1 continued

labelled proteins often stick to the gel filtration support, the column should not be reused. Alternatively, the sample may be dialysed against an appropriate buffer for at least 24 h in the dark.

7. Measure the OD of the fractions containing the labelled protein at 280 nm and 546 nm.

8. The ratio of OD_{546}/OD_{280} should be between 0.15–0.5 corresponding to approx. two to six dye molecules per antibody molecule.

Labelled antibodies should be divided into small aliquots (e.g. 10 μl), rapidly frozen in liquid nitrogen, and stored at −80 °C in the dark. Storage for longer than six months is not recommended as this often results in the loss of labelling efficiency. Due to over-modification by the labelling procedure a variable fraction of the antibodies may lose their activity or over-modification might be a source of unspecific background staining. In order to remove inactive antibodies, the labelled antibodies may be subjected to a further round of affinity purification using immobilized antigen. Alternatively, over-modified antibodies may be removed by ion exchange chromatography.

2.2 Fixation, permeabilization, and staining of cells

The quality of immunofluorescence of cellular components (and immunocytochemistry in general) often presents conflicting demands to the investigator: the cellular structures should be adequately conserved yet made readily available for antibody labelling, and the reactivity of the cellular epitopes should not be affected. To achieve this, an optimal fixation and permeabilization method should be selected, with the conditions optimized for:

(a) The structure under study (for example actin filaments or microtubules).

(b) The specific cell under study (some cells will not stay attached following light fixation).

(c) The specific antibody used.

A positive reaction in the immunolabelling may depend critically on the procedure used for fixation of the cells. The choice of a fixation protocol should also be made according to the subcellular localization of the antigen. Structures like microtubules, for example, which are in equilibrium with a soluble pool of tubulin are better visualized after a brief extraction of the cells with a buffer containing detergent. This eliminates the soluble 'background'. On the other hand, antigens associated with membranes often cannot be visualized after pre-extraction with detergent.

In this section, we will describe several procedures including fixation with paraformaldehyde, glutaraldehyde, methanol/acetone, and EGS.

For all the protocols described below, cells should be grown on glass coverslips (Ø = 15 mm).

2.2.1 Paraformaldehyde fixation

Fixation of cells by paraformaldehyde is used when cell surface staining is required. It is also a good choice for labelling of microfilaments, although microtubules are not fixed well by it.

Protocol 2

Cell surface labelling

A. Preparation of paraformaldehyde

1. Heat about 80 ml of D-PBS to 80 °C and add 3 g paraformaldehyde while stirring. Mix until clear.

2. Add 100 µl of 100 mM $CaCl_2$ and 100 µl of 100 mM $MgCl_2$ (to give a final concentration of 0.1 mM) with stirring whilst the solution is warm, to prevent precipitation.

3. Allow the solution to cool. Make up to 100 ml with D-PBS and adjust to pH 7.4.

4. Store at −20 °C in aliquots and use fresh each time.

B. Preparation of Dulbecco's phosphate-buffered saline (D-PBS) (minus calcium and magnesium)

1. Composition   (mg/litre)
   - NaCl   8000
   - KCl   200
   - Na_2HPO_4 (anhydrous)   1150
   - KH_2PO_4 200

2. Made up in double distilled water (dH_2O). The pH should be 7.4, and therefore pH adjustment of the solution ought not to be necessary.

C. Preparation of Mowiol mounting medium

1. Composition
   - Mowiol 4-88   2.4 g
   - Glycerol   6 g
   - DW   6 ml
   - 0.2 M Tris pH 8.5   12 ml

2. Place glycerol in 50 ml disposable conical centrifuge tube.

3. Add Mowiol and stir thoroughly. Add dH_2O and leave for 2 h at RT.

4. Add Tris and incubate at approx. 53 °C until the Mowiol has dissolved. Stir occasionally.

5. Clarify by centrifugation at 4000–5000 r.p.m. for 20 min.

6. Transfer the supernatant into glass vials with screw caps, about 1 ml in each.

7. Store at −20 °C. The solution is stable at this temperature for up to 12 months. Once it is defrosted, it is stable at RT for at least one month.

Protocol 2 continued

D. Cell surface labelling

1. Remove glass coverslips containing cells from a Petri dish and place (cells up) in 3% paraformaldehyde (see part A for its preparation) at RT for 20 min.

2. Aspirate the paraformaldehyde and wash with three changes of D-PBS (see part B for its preparation) at RT over ~ 15 min (when adding the D-PBS be careful not to pipette directly onto the coverslips). The second wash is with 30 mM NH_4Cl in D-PBS for 10 min to quench the residual paraformaldehyde crosslinking activity.

3. Tape a piece of Parafilm to the bench and pipette 50 μl drops of D-PBS containing the first antibodies onto it. Remove the coverslips from the D-PBS wash buffer and dry off excess liquid by touching the edge of the coverslips on filter paper. Place the coverslips, cells downward, onto the drops and leave for 20 min at RT.

4. Gently squirt a little D-PBS at the edge of each coverslip to raise them up from the Parafilm. Transfer back into D-PBS and repeat washings (two times, without glycine-quenching).

5. Place coverslips on Parafilm on 50 μl drops of D-PBS containing the secondary antibody (e.g. TRITC-conjugated anti-mouse). Incubate for 20 min.

6. Repeat step 4.

7. Remove the coverslips, dry off excess D-PBS with filter paper, and mount on glass slides by placing cell-side down on a 5 μl drop of Mowiol (see part C for its preparation). Allow to dry (> 30 min) and then observe under the fluorescence microscope.

As cells have not been permeabilized by this protocol, only antigen domains accessible from the outer side of the plasma membrane will be stained. When a directly fluorescently conjugated first antibody is used, *Protocol* 2D, steps 4 and 5 are omitted.

Protocol 3

Labelling of cytoplasmic or nuclear structures

Reagents
- See *Protocol 2*
- 1 M glycine pH 8.5
- 0.1% Triton X-100 (in D-PBS)

Method

1. Remove coverslips from Petri dish and place (cells up) in 3% paraformaldehyde (see *Protocol* 2A) at RT for 20 min.

2. Aspirate the paraformaldehyde and wash with three changes of D-PBS (see *Protocol* 2B) at RT over ~ 15 min (when adding the D-PBS be careful not to pipette directly to the coverslips). To the second D-PBS wash add approx. two drops of 1 M glycine pH 8.5 to quench the residual paraformaldehyde crosslinking activity.

3. Permeabilize the cells by incubation in 0.1% Triton X-100 (in D-PBS) for 4 min at RT.

4. Wash with D-PBS (two times, without glycine-quenching).

5. Continue as described in *Protocol* 2D, step 3 onwards.

2.2.2 Glutaraldehyde fixation

Using glutaraldehyde for fixation usually preserves subcellular structures better than other fixation methods using for example paraformaldehyde. It is difficult, however, to quench the autofluorescence of the glutaraldehyde completely which results in a high background fluorescence, as a consequence antigens of low abundancy will yield only weak fluorescent staining signals. In addition quite a few antigens lose their antigenic reactivity following glutaraldehyde fixation.

Labelling of antigens in fixed and permeabilized cells is as described in *Protocol* 2.

Protocol 4

Staining after glutaraldehyde fixation

Reagents

- See *Protocol* 2
- 1% glutaraldehyde (8% stock; Polyscience, EM grade) in D-PBS
- 0.5 mg/ml NaBH$_4$ (always made up fresh in D-PBS)
- 1 M glycine pH 8.5

Method

1. Remove coverslips from Petri dish and place (cells up) in 1% glutaraldehyde in D-PBS (see *Protocol* 2B) at RT for 20 min.

2. Aspirate the glutaraldehyde and wash with three changes of 0.5 mg/ml NaBH$_4$ (always made up fresh in D-PBS) for 4 min each to reduce the Schiff bases (which are fluorescent) to secondary amines.

3. Wash three times with D-PBS (include two drops of 1 M glycine pH 8.5 in the second wash).

4. Continue with permeabilization and staining of cells as described in *Protocol* 2D, step 3 onwards.

2.2.3 Methanol and methanol/acetone fixation

These two protocols cannot be applied for surface staining since methanol permeabilizes the cell membrane. They give good results for nuclear antigens, microtubule and intermediate filament labelling, and they may also be used for labelling of intracellular organelles like the Golgi apparatus or endoplasmic reticulum. Very often antigen accessibility is better with this fixation method, however it should be noted that the 3D architecture is slightly flattened by it.

361

Protocol 5

Staining after methanol/acetone fixation

Reagents

- See *Protocol 2*
- Methanol

- Acetone

Method

1. Remove the coverslip from the Petri dish, dry off excess medium, and place in methanol at $-20\,^{\circ}$C for 4 min.

2. Remove coverslips and immediately dip in acetone at $-20\,^{\circ}$C. After 4 min transfer into D-PBS (see *Protocol 2B*) or air dry on filter paper at RT (these coverslips can be stored for several days at $-20\,^{\circ}$C). Extraction in acetone is not always necessary, therefore, *this step may be omitted*.

3. Dry off excess D-PBS and invert coverslips onto 50 μl drops of first antibodies on Parafilm.

4. For the labelling of fixed cells with antibodies continue as described in *Protocol 2D*.

2.2.4 EGS fixation

Fixation with ethyleneglycol-bis-succinimidyl-succinate (EGS; Pierce) is comparable to glutaraldehyde fixation with respect to the preservation of cellular structures. There exists only little autofluorescence, however, resulting in a much lower background autofluorescence.

Protocol 6

Staining after EGS fixation

Reagents

- See *Protocol 2*
- DMSO

- EGS (ethyleneglycol-bis-succinimidyl-succinate; Pierce)

Method

1. Pre-warm D-PBS to $37\,^{\circ}$C on a thermostatic heating plate. For large glass coverslips, use 1040 μl D-PBS per 35 mm dish or per well in a 6-well tray. For small coverslips, pre-warm 217 μl D-PBS per well in a 4-well tray (well diameter 16 mm). (See *Protocol 2B* for the preparation of D-PBS.)

2. Add 160 μl of a 75 mM EGS stock solution (dissolve 34 mg of EGS in 1 ml of dry DMSO) to larger wells, or 33 μl of the same stock solution to smaller wells. Mix immediately with a pipette. Note: the EGS-DMSO stock solution should be diluted in D-PBS as quickly as possible (by vortexing during the dilution). EGS is highly unstable in water. Therefore, the dilution should be prepared just before use.

3. Wash the glass coverslips containing cells to be fixed by dipping in D-PBS, then dry off excess D-PBS by touching one edge on filter paper. Transfer immediately into the EGS solution; less than 1 min should elapse between diluting the EGS stock solution in D-PBS and putting the coverslips in the well.

4. Incubate (covered) at 37 °C for 10 min.

5. Wash three times in D-PBS (to the second D-PBS wash add two drops of 1 M glycine pH 8.5).

6. For the labelling of antigens with antibodies follow the permeabilization and staining protocols (*Protocol* 2D).

3 Microscopy and imaging equipment

The microscope and associated hardware components are the most important components in immunofluorescence microscopy as they are the tools to collect, analyse, and quantify the data. A good labelling of structures may be completely useless if the experiment is destroyed by using a microscope set-up where the emitted fluorescence cannot be collected with the appropriate sensitivity or resolution. Therefore, the microscope set-up has to be carefully chosen to carry out the experiments in mind, and the final composition may vary from case to case. Although a detailed discussion of individual microscope techniques is beyond the scope of this chapter, we will briefly discuss the considerations to be made for the setting up of such equipment to meet the demands of immuno-fluorescence microscopy. A more basic and detailed description of the micro-scopy techniques discussed here is given in ref. 2.

3.1 Microscopes and associated hardware

The performance of an imaging system for IFM depends on several parameters including the temporal, spatial, and spectral resolution, the imaging detector sensitivity (signal-to-noise ratio), and the image analysis hard and software used to quantify and document the results.

Unfortunately, some of these components have features which exclude each other. For example, the imaging detectors with the best spatial resolution are not very sensitive. In contrast, sensitive imaging detectors, such as intensified CCD cameras, require only a fraction of the exposure light compared to high resolution cameras in order to detect the fluorescent signals above background. However, these cameras usually suffer from a very low image resolution and details of labelled structures are very often missed. Thus the ideal microscope set-up and imaging detector for IFM does in our opinion not exist, and the suitability of the experimental set-up is therefore always dependent on the experiments in mind.

There are two major types of fluorescence microscopes, which are currently most widely used for immunofluorescence microscopy. One is wide field 3D sectioning microscopy and the other one is laser scanning confocal microscopy.

3.1.1 '3D' sectioning microscopes

A good 3D sectioning fluorescence microscope is capable of acquiring images with different excitation wavelengths and different focal planes of the stained specimen under investigation. After image acquisition the 3D data are re-constructed and possibly deconvolved using specialized, commercially available algorithms (see for example Fay, Agard, etc.) in order to remove the out-of-focus information contained in each individual image acquired at different focal planes. Many suppliers, who provide a complete set-up and software packages suitable for this purpose, do exist. For the reason of reproducibility, image acquisition is automated in these systems. Excitation and emission filter wheels equipped with the appropriate excitation and emission filters, which can be controlled by computer, are available and allow automatic image acquisition of samples stained with different fluorescent antibodies emitting at different wavelengths.

In epifluorescence mode, in addition to the excitation and emission filters, a dichroid mirror is used for every fluorophore to be detected in order to suppress the stray light of the excitation light. For the reliable collection of data it is of superior importance not to move this part of the microscope during the sequential acquisition at different wavelengths, as this mostly results in the distortion of the spatial relationship between the differently labelled structures (pixel shifts) in the acquired images. Some image analysis and documentation systems exist in order to correct such pixel shift in the acquired images. Altern-atively, dichroid mirrors, which are manufactured for the simultaneous acquisi-tion at two or more colours can be used. In this case images are acquired sequentially at different wavelengths by changing the excitation and emission filters only and not moving the dichroid mirror. Therefore, co-localization of differently labelled antibodies is possible with high precision. Unfortunately, the spectral characteristics of these dichroid mirrors, necessary to separate fluorescence of the different fluorophores properly, are at the price of a signifi-cant loss in the detected emitted fluorescence at each individual colour.

Automated control of the microscope z-focus drive with a submicron resolu-tion is usually used for the 3D sectioning of the sample.

3.1.2 Confocal laser scanning microscopes

Commercially available confocal laser scanning microscopes clearly overcome some of the limitations of 3D sectioning fluorescence microscopes. Simultane-ous image acquisition at up to four different colours is possible. Also, the image data acquired at each focal plane only contain a minimum of out-of-focus back-ground fluorescence and 3D image reconstruction does not require time-consuming image deconvolution as is the case in 3D sectioning microscopy.

However, the light budget of laser scanning microscopes appears to be poor as most of the emitted fluorescent light is discarded in order to obtain confocal images. Therefore, weakly labelled structures, detectable with a 3D sectioning fluorescence microscope, may not be detected with a laser scanning confocal microscope.

3.2 Imaging detectors for 3D sectioning microscopy

The camera used to acquire the images in 3D sectioning microscopy is of superior importance, since it is the device which collects the raw data. Although there has been considerable progress in camera design, it should be noted that the ideal camera capable of covering all applications still does not exist. In the following section we will discuss some features of the camera systems most frequently used. These considerations should help to decide which camera to implement into a microscope system with respect to the experiments in mind.

3.2.1 CCD cameras

The image detector in CCD cameras consists of a CCD semi-conductor chip, which comprises an array of distinct wells, which can be addressed individually. The wells have a specific size and capacity (*Table 2*). Each well corresponds to one pixel in the acquired image and thus the size and density of the wells determine the resolution of the camera. In contrast to imaging detectors using image intensifiers (see below) or image tubes, CCD cameras offer an excellent linearity over a broad range of incoming signal, and they show little geometric distortions at the camera edges. Therefore, they are very well suited for quantitative imaging.

Many CCD cameras which are presently commercially available allow the integration of the signals generated by the incoming fluorescent light on the CCD chip before it is read out and the signals digitized. Thus they offer the flexibility to image both very bright and dim signals by varying the sample exposure and camera integration times accordingly. As these CCD cameras usually also provide very good image resolution they are widely used for high quality imaging. In comparison with video rate CCD cameras, such on chip integrating CCD cameras achieve very high detection sensitivities by using extended exposure times. They also offer a feature which is called 'binning'. Using the binning function, wells are grouped and the corresponding signal added together on the CCD chip. In this way the sensitivity of the camera is increased by the binning factor, although this is at the price of reduced image resolution. Cooling of the CCD chip, usually below 0 °C, is extremely important for the image quality obtained with such cameras. This is because a major source of noise generated by the CCD chip, is thermal noise, which is also integrated together with the fluorescent signal of interest. Thus the signal above background is not very significantly improved with such cameras by using longer exposure times when they are operated with the CCD chip at room temperature. Cooling the CCD chip down to −20 °C can easily be achieved using peltier elements combined with air cooling. This is sufficient to reduce the thermal noise such that high quality imaging is possible.

Further parameters, which have to be considered when deciding which

Table 2 Typical features of a high quality slow scan cooled CCD camera[a]

Camera feature	Description	Typical values[a]
Chip size	Number of wells (pixels) in x, y direction, correspond to the number of image pixels.	1317 × 1035 wells (pixels)
Well size	Size of an individual well which corresponds to one pixel, determines the image resolution.	6.9 μm
Well capacity	Number of electrons, which can be integrated by a single well before saturation is reached. This sets an upper limit to how many grey values can be distinguished after A/D conversion.	45 000 electrons
Quantum efficiency (QE)	Percentage of incoming photons which are converted into an electronic signal.	For λ = 400 nm; QE = 5% For λ = 550 nm; QE = 40% For λ = 700 nm; QE = 45%
Dynamic range of A/D conversion	Corresponds to the number of distinct grey values which can be obtained for a fully saturated image.	12 bit, equals 4096 distinct grey values
Readout speed	Speed at which the data of the entire chip are read out.	1 MHZ, equals 1.4 sec for the readout of a full size image
Cooling temperature	Temperature at which the CCD chip is operated.	−20 °C
Readout noise	Noise which is generated by readout of the CCD chip.	10 electrons

[a] Princeton Instruments Micromax 1400/1.

camera to use for an experimental set-up are the quantum efficiency (QE), well capacity, and the overall geometry of the CCD chip (see *Table 2*).

The *quantum efficiency* (QE) of the CCD target is defined by the percentage of incoming photons, which are converted by it into a useful electronic signal. The QE is strongly dependent on the wavelength of the incoming light and usually does not exceed 40% for wavelengths in the visible range (400–700 nm) of the spectrum. As the QE is very poor below 400 nm for most CCD chips, several camera suppliers offer special coating of the chips which achieves quantum efficiencies up to 10% in the UV range. Also, 'back illuminated cameras' now offer up to 80% QE. However, they are more expensive than 'front illuminated cameras' and have an increased well size that results in a decreased image resolution.

The *size and capacity of the CCD wells* are two parameters, which determine the image resolution and dynamic range, respectively. The size of one CCD well corresponds to the physical size of one image pixel. Cameras with CCD chips, which easily cover the resolution of the microscope, are available. CCD chips with bigger well sizes offer less image resolution depending on the magnification used in the individual experiment.

The well capacity determines the amount of incoming signal that can be integrated before the well is physically saturated. When this occurs part of the signal will 'spill over' to the neighbouring ones and will thus distort the image and the linearity and the resolution of the camera will be lost. Therefore, cameras, which are manufactured to operate within a high dynamic range (e.g. 16 bit that corresponds to 65 536 distinguishable grey values), use larger well sizes to overcome this problem. It should also be kept in mind when deciding which camera to use, that although some camera systems use a 12 or 16 bit data digitization protocol, the well capacity of the used CCD chip may be too low to give a true 12 or 16 bit capacity.

3.2.2 Intensified CCD cameras

As described above, on chip integrating cooled CCD cameras achieve very good sensitivities by using longer sample exposure times and integration of the weak incoming signals. More sensitive cameras are intensified CCD cameras, which typically consist of a type IV image intensifier coupled to a CCD chip via a lens or optical fibres. These cameras are orders of magnitude more sensitive than normal CCD cameras when comparing identical exposure times. Therefore, with these cameras, the exposure light can be attenuated to a minimum, which causes less photobleaching. However, the image resolution, dynamic range, and geometric distortion at the edges of these cameras is determined by the image intensifier, whose performance with respect to these parameters is usually considerably worse compared to respective non-intensified CCD cameras. Thus intensified CCD cameras are used in experiments where less spatial resolution is required but the fluorescent signals to be detected are very weak.

3.2.3 Colour cameras

So far we have discussed black and white cameras capable of imaging single wavelengths at a time. Alternatively, three colour on chip integrating CCD cameras (red, green, and blue detectors) may be used. In combination with the appropriate markers (they should emit red, green, and blue light), they allow the simultaneous visualization of up to three differently labelled antibodies simultaneously. However, some fluorescent dyes, which can be very efficiently used for the labelling of antibodies in IF microscopy (for example Cy5) do not emit fluorescent light, which can be well separated into the spectral components of a colour camera. Thus quantitative association of the detected signals to the individual markers in use is impossible.

3.3 Data analysis, image processing, and data presentation

In recent years computer systems have become powerful enough to even perform 3D reconstructions involving time-consuming deconvolution algorithms at a reasonable speed. This has considerably increased the number of powerful and easy to use software packages commercially available. One of the most popular public domain software packages for scientific image analysis and documentation is the NIH image programme (available from: `http://rsb.info.nih.gov/nih-image/`). It allows the generation of simple macros, without the user having a background in computer programming, and thus allows the automated acquisition and evaluation of image data. It covers most basic features in image analysis, including the control of microscope periphery. It is also available as a source code, which allows users to implement their own extensions covering more sophisticated demands.

In general, the software package of choice for the more sophisticated analysis and documentation of image data should allows the possibility to automatically analyse geometric and densitometric parameters on multiple colour images in 3D.

4 Immunolabelling and visualization in living cells

Immunolabelling of organelles or structures by fluorescent antibodies in living cells has the advantage, compared to work in fixed cells, that the dynamics of the labelled structures or molecules can be studied by for example time-lapse microscopy. Also, labelling in living cells does not suffer from possible problems induced by the fixation or permeabilization of the cells as is necessary for immunofluorescence or electron microscopy.

4.1 Labelling by microinjection of fluorescent antibodies

A number of approaches to fluorescently label cytoplasmic and nuclear structures or organelles in living cells are presently available. They vary considerably in terms of ease of use, how specifically they label the target molecules, and in how much they interfere with cellular function. Direct microinjection of fluorescently labelled antibodies into living cells is a universal and efficient way of labelling cellular targets *in vivo*. In order to avoid interference with cellular function or non-specific cross-reactions the fluorescent antibodies should be prepared as described in Section 2.1 and be first characterized *in vitro* before they are introduced into cells.

At least two points have to be considered in order to obtain an optimal labelling by microinjection. First, antibodies are usually diluted10- to 20-fold in the cell upon microinjection. Therefore, the concentration of antibodies in the injection pipette will always have to be higher compared to the optimal concentration used for immunofluorescence using the same labelled antibodies. Equally, microinjecting too much fluorescent material into the cells may cause a high background due to the antibodies which have not bound to their cellular target.

Secondly, the injected labelled antibodies should not interfere with cellular function, which could significantly change the distribution of the antibodies' antigens in the living cells.

It is therefore most useful to microinject the fluorescent antibodies over a whole range of concentrations in order to achieve the optimal labelling conditions, but which does not interfere with cell function.

A detailed discussion of suitable microinjection equipment and technical aspects is beyond the scope of this chapter and are described in ref. 3. The glass micropipettes used for penetration of cells and delivery of the sample are available from Eppendorf (Eppendorf, Germany). The manipulators and the micro-injector, controlling the pressure applied to the microinjection pipette, used in our laboratory, are also from Eppendorf.

For microinjection and subsequent analysis (e.g. by time-lapse microscopy), cells are plated on 35 mm dishes with coverglass bottoms (MatTek Corp., Ashford, MA, USA), that allow for the use of high numerical aperture, short distance objectives. During injection and time-lapse microscopy, cells are incubated in tissue culture media preferably without phenol red and only low concentrations of serum (1%) to keep background fluorescence from the culture medium to a minimum. To keep the pH stable, carbonate-free culture medium (Gibco) buffered with 20 mM Hepes pH 7.4 is used. Temperature is kept at 37 °C, for example by enclosing the entire microscope and microinjection manipulators in a temperature controlled Perspex box.

In order to reduce phototoxicity due to free radicals forming by photobleaching of the fluorophores during time-lapse microscopy, ascorbic acid (1 µg/ml) or oxyrase (30 mM) may be added to the medium as free radical scavengers. For long-term observations (longer than two hours) the culture medium should be covered with mineral oil (from Sigma) to avoid the culture medium drying out.

4.2 Microscope set-up for live cell IFM

All basic aspects of the microscopy and imaging equipment described above for IFM are also applicable for imaging in living cells. In our laboratory we use for single label time-lapse microscopy an inverted fluorescence microscope equipped with a filter wheel/shutter for the controlled illumination of cells. Images are acquired with a slow scan cooled CCD camera. All moving parts of the microscope are controlled by computer. The entire set-up is enclosed in a Perspex box, which is temperature controlled via a heating fan and electronic control unit (plans describing the design of this temperature control box can be obtained from the Mechanical and Electronic Workshops at the EMBL in Heidelberg, Germany). The advantage of such imaging system for live cell observation is that up to 20 images can be acquired per second providing excellent temporal resolution. However it is less suitable for double labelling experiments, as image acquisition at different wavelengths is sequential and structures may move during image acquisition, thus distorting the spatial relationship of the fluorescently labelled antigens. Therefore, we prefer to use a confocal microscope for multi-labelling experiments. As imaging at two or three colours can be performed simultaneously with this microscope, the spatial relationship between the different fluorescently labelled antigens is preserved and is with confocal resolution. However, usually only up to two images (e.g. 512 × 512 image pixels) can be acquired per second with such microscope which sets an upper limit for the temporal resolution in time-lapse microscopy.

5 Examples of IFM in fixed cells

Figure 1 and *Figure 2* describe two examples of IFM in fixed cells to demonstrate the possibility of multicolour staining (*Figure 1*) and improvement in resolution by image deconvolution (*Figure 2*).

Figures 1 and 2: please see plate section between pages 226–227.

References

1. Hermanson, G. T. (ed.) (1996). *Bioconjugate techniques*. Academic Press.
2. Pawley, J. (ed.) (1995). *Handbook of Biological Confocal Microscopy*, 2nd edn., Plenum Press, New York.
3. Celis, J. (ed.) Pepperkok, R., Saffrich, R., and Ansorge, W. (1998). In *Cell biology: a laboratory handbook*, pp. 23–30. Academic Press.
4. Snyder, D. L., Schulz, T. J., and O'Sullivan, J. A. (1992). *IEEE Trans. Sign. Proc.*, **40**, 1143.
5. Palmer, D. J., Helms, J. B., Beckers, C. J., Orci, L., and Rothman, J. E. (1993). *J. Biol. Chem.*, **268**, 12083.

Chapter 17

FACS analysis of clinical haematological samples in transplantation for cancer

Barbara C. Millar

Department of Haemato-Oncology, Royal Marsden NHS Trust, Sutton, Surrey, U.K.

1 Introduction

The systematic approach to the naming of leucocyte antigens using the cluster of differentiation (CD) designation (1) was made using monoclonal antibodies (mAbs). Direct conjugation of these antibodies or indirectly, via a two-step procedure, to fluorochromes permits visualization of individual cell types. The development of automated fluorescence activated cell analysis (FACS) to assess leucocytes in peripheral blood (PB) and bone marrow (BM) samples by their CD designation has provided an accurate and reproducible method, which is relatively easy, quick to perform, and permits far greater numbers of cells to be analysed than can be done using microscopy.

2 Fluorescence activated cell analysis

2.1 General considerations

Detailed theoretical considerations for the use of FACS have been published elsewhere (2). Forward (FR-SC) and right-hand (RT-SC) scatter measurements are used to identify the white blood components after removal of red blood cells (RBC) by lysis (*Protocol 1*).

Protocol 1

Estimation of surface antigens in peripheral blood samples

Equipment and reagents

- Coulter Counter (Beckman-Coulter)
- LP3 tubes
- Ready-conjugated CD45-FITC/CD14-PE, test ready-conjugated mAbs[a]
- PBS
- Lysis solution: Ortho-mune™ Lysing Reagent (Ortho-Diagnostics)
- Heparinized blood
- Isotonic buffer: Isoton (Beckman-Coulter)
- Zap-oglobin (Beckman-Coulter)

Protocol 1 continued

Method

1. Count 40 μl of whole blood in 20 ml isotonic buffer containing six drops zap-oglobin using a Coulter Counter.

2. Dispense 10 μl of anti-CD45-FITC/anti-CD14-PE and 10 μl of each fluorochrome-conjugated mAb either singly or in pairs to LP3 tubes. For dual labelling use a separate pipette tip for the addition of each second antibody.

3. Add 50 μl of whole blood to the bottom of each tube.

4. Agitate briefly to mix.

5. Incubate for 10 min at 4 °C.

6. Add 1.0 ml lysis solution and incubate for approx. 12 min[b] at room temperature, then rest tubes in ice.

7. Process through the analyser using dual fluorescence with gating (A) for MNC (for assessing stem cells) or lymphocytes.

8. Calculate the concentration of stem cells or lymphocyte subset /ml, e.g. for total CD34$^+$ cells:

$$\text{Number of MNC in A} = \text{WBC} \times \%\text{A} \times (\%\text{CD45}^+ + \text{CD45}^+/\text{CD14}^+) = \text{B}$$
$$\text{Total CD34}^+ \text{ cells/ml} = \text{B} \times (\%\text{CD34}^+).$$

[a] Test mAbs as in *Table 1*.

[b] The precise time required for total lysis of RBC is dependent on the ambient temperature. The solution should be clear when lysis is complete. Any opacity will affect the scatter properties of MNC in the analyser.

The three major cell types are lymphocytes and monocytes, which constitute the mononuclear cell fraction (MNC), and granulocytes. Haemopoietic tumour cells are located within the MNC fraction. Cell debris is excluded from the analysis by setting a threshold below which particles are excluded from the calculations predetermined by the computer software. FR-SC is determined by the size of the cell; the larger the cell, the greater the FR-SC. RT-SC is determined by irregularities within the cell such as the shape of the nucleus and granules. Lymphocytes exhibit the least FR-SC and RT-SC, granulocytes the most. Monocytes have scatter properties intermediate between lymphocytes and granulocytes. Specific gating is employed to include only those cellular components of interest. This is done by setting a boundary around this population, defined by the scattergram (*Figure 1*), and assessing fluorescence in cells within this boundary. Although FACS analysis can employ more than one gate, for most operations a single gate is sufficient. Data can be accumulated as a matrix or in list mode file according to the software provided by the manufacturer. For clinical purposes list mode file has the advantage that the gate and statistical parameters can be changed easily. This is useful because of variation in the scatter properties of cells from different patients and different sites (BM or PBSC; *Figure 1*).

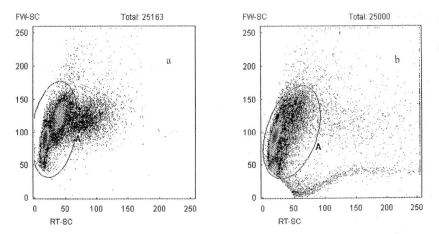

Figure 1 Forward scatter (FSC) and right-hand scatter (RHS) from (a) peripheral blood stem cell, and (b) bone marrow harvests.

2.2 Choice of antibodies, reagents, and general maintenance of the FACS analyser

Ready-conjugated mAbs developed specifically for FACS and quality controlled by the manufacturer should be used at all times. The most common fluoro-chromes used to assess clinical samples are fluorescein isothiocyanate (FITC) (green fluorescence, GR-FL) and phycoerythrin (PE) (orange fluorescence, OR-FL). When three-colour FACS is used, mAbs directly conjugated to Cy-Chrome™ (Pharmingen) (red fluorescence, RD-FL) provide reproducible results used in conjunction with FITC- and PE-conjugated mAbs. Since manufacturers develop their own clones to specific antigens once a suitable mAb, directed against a chosen antigen, has been purchased a similar mAb should not be purchased from a different supplier with a view to cost cutting. This can result in less effective measurements of some antigens due to differences in concentration and binding characteristics of mAbs from different manufacturers. Consistency in the supply of mAbs is important for clinical trials especially those which in-volve more than one centre, for which agreement, regarding the supplier, should be reached before commencing a trial so that internal and external quality control is maintained. *Table 1* shows a list of mAbs used typically for assessing lymphocyte subsets and CD34$^+$ cells. Paired ready-conjugated mAbs are available for some common lymphocyte subsets (e.g. LeucoGATE, Beckton Dickinson) and can reduce preparative procedures when a panel of mAbs is under investigation.

Commercially available lysis solution (e.g. Ortho Lysis, Ortho Diagnostics), supplied anhydrous, is best used to remove RBC because it is quality controlled by the manufacturer. Measured aliquots which reconstitute to 100 ml are made in 'clean' water (Elga, Reverse Osmosis water purifier) and used within two weeks.

Although some manufacturers recommend saline for the sheath buffer of

Table 1 Immunophenotype of lymphocyte subsets and stem cells, visualized by FITC or PE ready-conjugated mAbs[a] and analysed by FACS

Target cell	Antigenic determinant (positivity)–fluorochrome
Pan MNC	$CD45^+$–FITC/$CD14^+$–PE
Pan lymphocyte	$CD45^+$–FITC/$CD14^-$–PE
Pan T cell	$CD3^+$–FITC
Pan B cell	$CD19^+$–PE
T helper cell	$CD3^+$–FITC/$CD4^+$–PE
T suppresser cell	$CD3^+$–FITC/$CD8^+$–PE
Activated T cell	$CD3^+$–FITC/$CD25^+$–PE
NK[b] cell	$CD3^-$–FITC/$CD16^+$–PE
NK cell	$CD3^-$–FITC/$CD56^+$–PE
Pan stem cell	$CD34^+$–FITC
Committed stem cell	$CD34^+$–FITC/$CD33^+$–PE
Activated stem cell	$CD34^+$–FITC/$CD38^+$–PE

[a] All ready-conjugated antibodies are supplied from Beckton Dickinson.
[b] Natural killer.

FACS analysers, 'clean' water is preferable because it does not cause blockages in the ancillary tubing due to deposition of salts including calcium.

Extreme care should be taken to prevent contamination of the analyser with cell debris. This can arise because of the lysis of granulocytes in samples examined after thawing from cryopreservation or after T cell depletion. Blockages can be removed by flushing the system with 20–50% (v/v) bleach followed by at least six washes with 'clean' water. If this fails the system can be washed through with 0.1% sterile trypsin (Sigma Aldrich) followed by further washing with bleach and 'clean' water.

2.3 Dual fluorescence

Most analyses are made using dual fluorescence, since the presence of a single determinant is often insufficient to identify subsets of lymphocytes or progenitor cells (see *Table 1*). The gains and compensation factors for dual fluorescence are shown in *Table 2* for cells analysed with an Ortho Cytoron (Ortho Diagnostics). These settings may need adjustment when the analyser has been serviced. Histograms and cytograms can be acquired using either linear or logarithmic amplification of the fluorescence signal. Although linear amplification provides direct information about the relative fluorescence intensities, the range is limited and tends to emphasize the negative cell population. Therefore, for most purposes data is accumulated logarithmically, with negative cells being accumulated in the first decade of the display (*Figure 2*). For analysis of $CD34^+$ cells in buffy coat or lymphocytes in whole blood it is particularly important to fix the gate carefully, since damaged cells and granulocytes, identified outside the gate in the scattergram (*Figure 1b*), can cause autofluorescence (see the top right-hand quadrant of *Figure 3*).

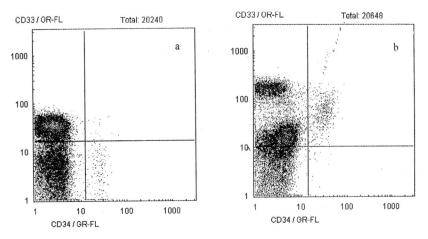

Figure 2 Cytograms of (a) peripheral blood stem cell (PBSC) and (b) bone marrow (BM) harvests stained with anti-CD34-FITC and anti-CD33-PE, gated for the MNC fraction.

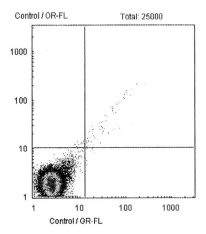

Figure 3 Cytogram of ungated bone marrow (BM) harvest processed to MNC without antibody.

All samples should be analysed in parallel with both a positive and negative control. The positive control is usually stained with ready-conjugated anti-CD45-FITC/anti-CD14-PE. Since monocytes and macrophages express both CD45 and CD14 whereas lymphocytes express only CD45 this procedure allows visualization of the MNC fraction and permits quantitation of both cell populations within the gate. Cells which do not stain with either marker are usually erythrocytes and nucleated RBC. Although a negative control should consist of an irrelevant mAb of the same isotype, it is acceptable to use cells which have not been exposed to antibody as a control for assessing surface antigens. This is not the case when fixed cells are used for determining levels of intracellular antigens. Since there is a wider variety of proteins intracellularly than at the cell surface, it is essential to keep fluorochrome-conjugated isotype controls for any

Table 2 Parameters, compensation factors, and gains for dual fluorescence using an Ortho-Cytoron to assess the composition of peripheral blood and bone marrow

Parameter name	Gain	Compensation A	Compensation B	Antibody	Range
FSC: FW-SC	13	N/A	N/A	N/A	N/A
RHS: RT-SC	45	N/A	N/A	N/A	N/A
FL 1: GR-FL	105	16 (FL 2)	0 (FL 3)	mAb–FITC	1–100%
FL 2: OR-FL	90	0 (FL 3)	6 (FL 1)	mAb–PE	1–100%
FL 3: RD-FL	0	0 (FL 3)	0 (FL 2)	None	1–100%

mAb directed against internal antigens. This avoids spurious results from non-specific binding.

2.4 Preparation of clinical samples

All clinical samples should be treated as infectious. Preparative procedures to make MNC or buffy coat are done aseptically, using sterile reagents and other consumables. Care should be taken to avoid contact with the mouth and eyes during handling and gloves must be worn if there is known infection with hepatitis A–E or HIV viruses. Samples should be discarded first into bleach and then sent for incineration after analysis. Peripheral blood stem cell (PBSC) harvests can be refrigerated overnight before FACS, however, BM samples, prepared as buffy coat, and whole blood should be examined as soon as possible after receipt. The half-life of granulocytes is approximately eight to nine hours. Refrigeration can result in cell death, the release of cell debris, and a reduction in the total cell count. If samples of BM or PBSC harvests are assessed after cryopreservation, no attempt is made to remove cell debris by centrifugation before recounting and FACS analysis. An aliquot of the thawed suspension is removed, avoiding gelatinous deposits, diluted 1:5, counted, and processed as in *Protocol 1*. Stem cells are resistant to cryopreservation and recovery is likely to be in excess of 95% (B. C. Millar, unpublished observation).

2.5 Quality control

Results from a particular marker used more than once with the same sample should be consistent. For example the total number of T cells measured by CD3 positivity should be equal to the sum of $CD3^+/CD4^+$ (T helper) and $CD3^+/CD8^+$ (T suppressor) cells. When the results do not corroborate the source of the discrepancy should be investigated. This may be trivial, such as improper gating or alignment of the statistical regions. If the results are significantly different the samples should be re-evaluated for those markers in case there was inadequate staining with the antibodies.

When two different mAbs are used simultaneously in defining a cell sub-population, controls should be included that define the four possible cell populations, i.e. cells that are negative for both fluorochromes, cells which are positive for either FITC or PE, and cells which are positive for both FITC and PE.

Table 3 Normal values for cellular components of human blood

	Cells/μl	Approx. normal range	% of total white cell count
Total WBC	9000	4000–11 000	
Granulocytes			
Neutrophils	5400	3000–6000	50–70
Eosinophils	275	150–300	1–4
Basophils	35	0–100	0.4
Lymphocytes	2750	1500–4000	20–40
Monocytes	540	300–600	2–8
Erythrocytes			
Females	4.8×10^6		
Males	5.4×10^6		
Platelets	300 000	200 000–500 000	

Also, since there may be more than one staining characteristic (e.g. 'dim' and 'bright') within one or more of these populations, statistical regions may need to be adjusted for accurate assessment of populations which exhibit dual fluorescence.

The percentage of different WBC in peripheral blood is shown in *Table 3*. Since the lymphocyte component is approximately 20–40% of the nucleated cell count, FACS analysis of lymphocyte subsets is done with approximately 10 000 cells yielding an appropriate number of lymphocytes to analyse. In contrast, since the number of CD34$^+$ cells is often less that 1% of WBC in blood, BM, and PBSC harvests, quantitation requires at least 20 000 cell to be analysed.

2.6 Calibration of the analyser

Before examining blood, BM, or PBSC harvests with mAbs, the scatter properties of whole blood are used to standardize the analyser that has been assessed for the differential count by other methods (e.g. Coulter Counter: ACT-8)). Computer software is provided by most FACS manufacturers to assist with the calibration. This procedure enables the mononuclear cell (MNC) and lymphocyte gates to be identified, so that there is > 95% agreement between the two methods.

3 Assessment of the progenitor cell content in bone marrow (BM) and peripheral blood stem cell (PBSC) harvests

3.1 Rationale for measuring the progenitor cell content of BM and PBSC harvests for transplantation after intensive therapy for cancer

The rationale for giving high-dose therapy for the treatment of cancer was based on the premise that larger doses of radiation or chemotherapeutic agents would result in greater tumour cell kill and produce a state of complete remission (3).

Since most normal tissue toxicity, particularly myelosuppression, is dose limiting, autologous BM rescue or allogeneic BM transplantation was given to assist haematological recovery (3). During the past decade PBSC rescue has superseded BM transplantation and resulted in rapid engraftment and low morbidity (4). In some instances PBSC rescue has permitted intensive therapy to be given to patients with a high tumour burden who would be ineligible for autologous BM rescue. Furthermore, since no general anaesthetic is required, PBSC can be taken at out-patient clinics, thus saving theatre time.

Unlike BM harvests the success of PBSC harvesting depends on mobilization of progenitor cells into the peripheral blood with cytotoxic agents and/or cytokines. This is influenced by the age of the patient and the duration and type of previous chemotherapy. Among the elderly (> 65 years) and those who have previous treatment with alkylating agents or radiation, mobilization can be severely impaired (5).

Until the identification of progenitor cells by their specific antigen (6) was possible the quality of harvests was assessed by clonogenic assay for granulocyte-macrophage colonies (GM-CFC) (7). This assay has two limitations. First the assay takes two weeks for colony formation. This means that there is no quality control for the stem cell content of harvests which might be returned to patients who have intensive therapy during this incubation period. Secondly, GM-CFC do not assess the potential for platelet recovery, which can be significantly delayed in heavily pre-treated patients.

3.2 The CD34 determinant

The recognition that a specific subset of lymphocytes which have the CD34 determinant could give rise to all of the elements in peripheral blood (8, 9) was of particular importance for transplantation. These cells are currently regarded as 'stem cells'. The ability to measure CD34$^+$ cells by FACS, within one hour after harvesting, enables quality control to be maintained for those harvests which are needed for immediate infusion. Although CD34$^+$ cells are heterogeneous with respect to the expression of other markers of differentiation and/or activation, quantitation of the total number of CD34$^+$ cells in PBSC and BM harvests has become the standard method for assessing the quality and efficacy of these procedures. As more is learnt about the role of CD34$^+$ cells in affecting long-term haemopoietic recovery the co-expression of other antigens is employed to improve assessment of the quality of harvests (Table 1). For example, recent evidence shows that a high numbers of CD34$^+$/CD33$^-$ cells (> 1.4×10^6/kg infused) is a better indicator of platelet recovery than simply the total number of CD34$^+$ cells/kg infused (10).

3.3 Collecting peripheral blood stem cell (PBSC) harvests for CD34$^+$ quantitation

A target dose of 2×10^6 CD34$^+$ cells/kg infused is set for all patients undergoing autologous PBSC rescue for the first time (10). In contrast, no target dose is set for allogeneic PBSC or BM transplantation or autologous BM rescue. For effective stem cell harvesting, a growth factor, usually granulocyte-colony stimulating factor (G-CSF), or chemotherapy followed by a growth factor is given to patients to stimulate migration of CD34$^+$ cells from the BM to the peripheral blood (PB). Normal donors are given growth factor alone. The time course of mobilization can be assessed by measuring the total CD34$^+$ cell content in PB (*Protocol 1*) at day 0, +2, +4, and thereafter daily for patients who receive growth factor alone, and at day 0, then day +9, and daily thereafter for patients who receive chemotherapy followed by growth factor. Commercially prepared mAbs enable lysis solution to be added directly to stained cells without the need for prior washing. A threshold dose of 10 CD34$^+$ cells/μl should be detectable before harvesting in started. Since mobilization is usually optimal by day +4 after growth factor alone, the collection of stem cells from normal donors is commenced at this time. Most patients and all normal donors undergo two or three separate leukaphoreses after a single mobilization. Mononuclear cells (MNC) are collected and concentrated using a cell separator (Cobe International) into a transfer bag. RBC and granulocytes are returned to the donor. No further processing of PBSC harvests is required before stem cell analysis. The protocol for analysing stem cell harvests is similar to that for whole blood (*Protocol 1*). The cell count/ml of the harvest, the weight of the recipient in kg, and the volume of the harvest in ml is recorded. Before FACS analysis the harvest is diluted 1:10 with PBSA. If subsets of CD34 cells are examined, a fresh pipette tip should be used for each mAb pair to avoid contamination of stock mAb solutions. The results are expressed as the total number of CD34$^+$ cells/kg infused in the patient or recipient. This is calculated from the product of the CD34$^+$ cell count/ml and the volume of the harvest divided by the weight of the recipient. Although most stem cells are small lymphocytes, gating is usually set on the MNC fraction (*Figure 1*). Among normal donors and patients in remission gating on lymphocytes or MNC results in similar values for the yield of CD34$^+$ cells. However, in some leukaemic patients, comparison between the yield of CD34$^+$ cells in the lymphocyte and MNC gates can identify leukaemic CD34$^+$ blasts which have scatter properties similar to monocytes.

Patients who fail to achieve the target dose for autologous rescue can be remobilized with the same procedure no sooner than one month later, or have BM taken to provide an additional reserve.

3.4 Collecting bone marrow (BM) harvests for CD34$^+$ quantitation

All BM harvests are processed to buffy coat to reduce the volume (*Protocol 2*) before FACS analysis (*Protocol 1*).

The apparatus for large scale preparation of BM is essentially a centrifuge in which a processing bag (P) occupies the entire bowl when full of liquid. A central line from P passes through a hole in the lid and is attached by a T-junction to five colour-coded lines. These lines connect the junction to transfer (T) and a waste (W) bag. Clamps are used to control the flow of BM, saline, and/or Ficoll-Hypaque into and out of the centrifuge. P rests on a flexible membrane which is attached to a hydraulic system. This allows separation of the components of the harvest after centrifugation, at a controlled rate whilst spinning. All of the component bags and lines are supplied sterile as part of the processing kit. Because the system is enclosed during most of the procedure, sterility is preserved. There are usually two or three bags of BM taken in theatre from an adult and one bag from a child. Each bag holds 200–400 ml. The total cell count from each bag is assessed as in *Protocol 1* before preparation of buffy coat or MNC to allow estimation of cell recovery after processing.

Protocol 2

Preparation of buffy coat from bone marrow harvests (BM) using a COBE 2991 cell processing machine

Equipment and reagents

- COBE 2991 cell processor, ancillary bags, and tubing (Cobe International)
- Transfer bags (Baxter Health Care Corporation)
- Clamps
- 50 ml syringes and 25 gauge needles
- PBS (Baxter Health Care Corporation)
- Heparinized bone marrow harvest

Method

1. Calculate the volume of the harvest in ml using the density of blood as 1.03:

$$\frac{\text{Total weight} - \text{weight of collection bag}}{1.03}$$

2. Take 0.3 ml from each bag for cell count in Coulter Counter (*Protocol 1*).

3. Attach each collection bag to one of the lines. Load harvest into the processing bag (P) in the cell processor through the lines one at a time, keeping all other lines clamped off. Use clamps throughout the procedure to direct the flow of liquids into the appropriate bags.

4. Start rotor to expel air in P through the line into the empty bag,[a] then clamp at the T-junction.

5. Spin at 1000 g for 8 min.

6. If there is a third bag of BM harvest attach it to a third line.

7. After centrifugation open the line to allow release of plasma towards the next bag of BM due for preparation. When the plasma runs clear, clamp this line, open the line attached to the waste bag, and collect the plasma.[b]

8. Pump in the buffy coat at 100 ml/min.

9. Collect RBC into a fourth bag, then clamp off the line.

10. Allow second and/or third bags of BM into the processor and repeat steps 3–9.

11. Take 0.4 ml of buffy coat. Determine the cell count and calculate the volume as in step 1. Take aliquot for FACS analysis (*Protocol 1*) and a second for sterility testing.

[a] The total volume in P will be 620 ml after expulsion of air.

[b] The centrifuge bag empties under centrifugal force up to a point when the hydraulic system is required. When most of the plasma has been pumped out, buffy coat will be visible above the RBC in P. Clamp the waste line.

BM which is ABO incompatible must be depleted of RBC and prepared as MNC (*Protocol 3*).

Protocol 3

Preparation of mononuclear cell fraction (MNC) from bone marrow (BM) harvests using a COBE 2991 cell processor

Equipment and reagents

- COBE 2991 cell processor and ancillary bags (Cobe International)
- Clamps
- 50 ml syringe with 16 gauge needle
- Ficoll-Hypaque (density 1.077: Nycomed AS Pharma)

- PBS pH 7.4 (Baxter Health Care Corporation)
- 4.5% human albumen (Zanalb: BPL Bio Products)
- BM harvest processed to buffy coat

Method

1. Measure 150 ml Ficoll-Hypaque into a 600 ml transfer pack and allow to run into processing bag (P) through one line. Use clamps throughout the procedure to direct the flow of liquids to the appropriate bags.

2. Expel air from P into the bag which contained the Ficoll-Hypaque.

3. Clamp off the line to P below the T-junction.

4. Adjust volume of buffy coat to 400 ml with PBS.

5. Attach buffy coat bag to a second line, via a transfer line and sterile tubing attached to a peristaltic pump. Return about 1/3 of the Ficoll-Hypaque to fill the tubing attached to the buffy coat bag.

6. Expel air in this line into the buffy coat by setting the peristaltic pump in reverse and pumping at 40 ml/min.[a]

7. Clamp all the lines except that attached to the buffy coat bag.

8. Start spin at 1000 g. Load buffy coat onto Ficoll-Hypaque using the peristaltic pump and spin for 30 min. Stop the pump briefly to add PBS to transfer bag to wash out residual cells.

9. After centrifugation, collect most of the plasma into a waste bag.

10. Collect the remainder of the plasma, mononuclear cells, and Ficoll-Hypaque into the original buffy coat bag.

11. Transfer MNC and Ficoll-Hypaque to a fresh processing bag and top up with PBS.

12. Spin at 1000 g for 7 min to sediment MNC.

13. Collect supernatant into the waste bag.

14. Remove P from the processor and transfer MNC to a new transfer pack by syringe and 16 gauge needle.

15. Rinse P with donor plasma (if ABO compatible) or albumen.

16. Make up the volume in the transfer pack to approx. 200 ml with additional plasma or albumen. Remove 0.4 ml for counting sterility testing, counting, and FACS analysis (Protocol 1).

[a] Ficoll-Hypaque is drawn along the line drawn towards the buffy coat, so that buffy coat is layered onto the Ficoll-Hypaque during centrifugation.

The dose of CD34$^+$ cells/kg infused is calculated as previously (Protocol 1).

3.5 Quantitation of purified CD34$^+$ cells from PBSC harvests from patients with chronic lymphocytic leukaemia

Patients with chronic lymphocytic leukaemia (CLL) who have tumour cells (CD5$^+$/CD19$^+$) (LeucoGATE, Beckton Dickinson) in their PBSC harvests taken for autologous rescue are eligible for reinfusion with CD34$^+$ cells that have been purified by positive selection. This is suitable if the total yield of stem cells from three separate PBSC harvests, after a single mobilization, is $> 3 \times 10^6$/kg. Procedures for large scale stem cell purification use kits, in which mAbs which target the CD34 antigen (e.g. CellPro) are mixed with PBSC harvests. The antibody-bound stem cells are collected by binding to a second antibody attached to a column or matrix through which the harvest is passed and are released subsequently by mechanical forces. Procedures should be followed according to individual manufacturer's guidelines The yield of CD34$^+$ and contamination with CD5$^+$/CD19$^+$ cells is assessed in each harvest (Protocol 1). One PBSC harvest is cryopreserved, for back-up, and the remainder and subjected to purification. FACS analysis is used as in Protocol 1 to assess the recovery, purity, yield, and tumour cell contamination of the purified population and in all washings that are collected during the preparation. Recovery of CD34$^+$ cells is approximately 50% of the original harvests, but purity is increased 50–100-fold, and tumour cell contamination should be less than 0.5%.

4 Analysis of lymphocyte subsets in peripheral blood and bone marrow harvests from unrelated donors

4.1 T cell depletion of paediatric matched (MUD) and unmatched (U-UD) bone marrow (BM) harvests from unrelated donors

Among paediatric patients who have no near relative, matched BM from an unrelated donor (MUD) or unmatched at one or more major histocompatibility locus (MHC) (U-UD) is used with T cell depletion. Bone marrow is supplied with serum from the same donor. MNC are prepared (*Protocols 2* and *3*) and assessed by FACS for lymphocyte subsets and CD34$^+$ cell content before removal of T cells with Campath-1M (11) *in vitro* (*Protocol 4*).

Protocol 4

T cell depletion of bone marrow (BM) harvests using Campath-1M in paediatric haemopoietic transplantation

Equipment and reagents

- COBE 2991 cell processing machine with ancillary processing, transfer and infusion bags (Cobe International)
- One vial (25 mg) Campath-1M (Therapeutic A-L)
- 60 ml clotted blood to produce 30 ml of serum from the same BM donor, collected in 10 ml vacu-tubes
- Universal containers
- RPMI-1640 (Flow Laboratories)
- Mononuclear cells (MNC) resuspended in RPMI-1640
- PBS (Baxter Health Care Corporation)
- 4.5% albumen (Zanalb: BPL Bio Products)

Method

1. Allow clotted blood to retract from sides of collection tubes.
2. Centrifuge at 300 g for 10 min at 18 °C.
3. Collect serum into Universal containers and use within 2 h.
4. Reserve 1.0 ml.
5. Count 40 μl of MNC as in *Protocol 1*.
6. Take 0.5 ml of MNC for assessment of lymphocyte subsets and stem cells (*Protocol 1*). Calculate total number of CD3$^+$/CD4$^+$, CD3$^+$/CD8$^+$, and CD34$^+$ cells/ml.
7. Add one vial of Campath-1M to the bag containing the remainder of MNC with a syringe and 22 gauge needle.
8. Incubate at room temperature for 10 min.
9. Add donor complement and note volume added.

10. Mix thoroughly.

11. Incubate for 45 min at 37 °C.

12. Wash cells in COBE 2991 using 50:50 PBS/albumen as in *Protocol 3*.

13. Resuspend cells and place in transfer pack. Measure the volume.

14. Take 0.4 ml. Count and analyse for T cell subsets and CD34$^+$ cells by FACS (*Protocol 1*).

Recipients are given a small dose of T cells (10^4 to 10^6/kg infused) calculated from the sum of CD3$^+$/CD4$^+$ and CD3$^+$/CD8$^+$, from the unmanipulated BM, before infusion of the treated BM. This is designed to minimize graft versus host disease (GVHD) by inducing some degree of tolerance to the remainder of the graft. FACS analysis of residual lymphocytes and CD34$^+$ cells is performed after T cell depletion. The CD34$^+$ cell dose/kg infused is also re-examined by FACS, since there is evidence that stem cells may be depleted from some harvests with Campath-1M (B. C. Millar, unpublished observation). In some instances lysis of T cells after the addition of serum is incomplete, resulting in a loss of fluorescence intensity from CD3, CD4, and CD8 antigens. This can lead to no apparent change in the total number of T cells. The percentage of damaged T cells can be calculated by adjusting the statistical regions after the samples have been through the analyser and calculating both the total number of each subset and the number which have reduced fluorescence intensity. The addition of insufficient serum may be responsible for incomplete lysis, however, there is no evidence that this decreases the time at which symptoms of GVHD appear in the recipient, suggesting that lysis continues *in vivo*.

4.2 Analysis of lymphocyte subsets after transplantation or autologous rescue

The WBC in patients after intensive therapy remains $< 2 \times 10^6$/ml for at least seven days after transplant or autologous rescue. Lymphocyte recovery can be assessed from day +1 after therapy (*Protocol 1*). Since T cells are the major component of the lymphocyte population, the statistical regions for identifying some subsets such as activated T cells (CD25$^+$) by dual fluorescence may require adjustment to ensure that only those cells which are positive for both antigens are assessed.

4.3 Measurement of intracellular cytokines in mononuclear cells (MNC) and T cells after transplantation or autologous rescue

Studies to examine cytokine production in T cells or monocytes after transplantation, using internal antigens, such as the production of interleukin-2 (IL-2) in T cells or interleukin-12 (IL-12) in monocytes, should not be done until the WBC is approximately 3×10^6/ml because there will be insufficient cells, in a routine blood sample, to provide adequate controls if the WBC is very low. Before assessing cytokine production in monocytes or T cells, subsets are quantitated in whole blood as in *Protocol 1* with gating for lymphocytes. This permits quantitation of cytokine production from stimulated cells to be calculated on the basis of their numbers in whole blood. MNC are prepared (*Protocol 5*).

Protocol 5

Small scale preparation of mononuclear cells (MNC) by Ficoll-Hypaque gradient centrifugation

Equipment and reagents

- 22 gauge needles
- 10 ml syringes
- 30 ml 'Universal' containers
- 5 ml Bijoux bottles
- 10–20 ml heparinized peripheral blood or bone marrow
- PBS pH 7.4 (Flow Laboratories)

- Ficoll-Hypaque solution (Lymphoprep, density 1.077 g/litre; Nycomed AS Pharma)
- Growth medium: complete RPMI-1640 supplemented with 10% FCS, 100 IU penicillin, and 100 μg/ml streptomycin (Flow Laboratories)

Method

1. If blood plasma is required for other purposes centrifuge whole blood for 10 min at 1000 g. Remove the supernatant into a Bijoux bottle. Disperse any lumps of cells in BM by pipetting gently through a syringe attached to a 22 gauge needle.

2. Dilute whole or plasma depleted blood or bone marrow to twice the original volume with PBS.

3. Add 5 ml of Ficoll-Hypaque to requisite number of Universal containers.

4. Layer diluted blood onto Ficoll-Hypaque keeping the container at an angle of 45° to maintain the Ficoll-Hypaque/blood interface.

5. Centrifuge for 20 min at 800 g at 18 °C in a swing-out rotor centrifuge, with no brake.

6. Remove the aqueous phase from above the cells and discard.

7. Remove the cells from the interface with the Ficoll-Hypaque and add to fresh Universal containers.

8. Make up the volume to approx. 30 ml with PBS.

9. Centrifuge at 1000 g for 10 min at 18 °C.

10. Discard the supernatant and resuspend the cells in a small volume (2–3 ml) of growth medium (for culture) or PBS (for FACS).

11. Count cells in Coulter Counter (*Protocol 1*).

12. Adjust cell concentration to 2–5 × 10^6/ml.

For the assessment of intracellular cytokines in T cells, MNC are depleted of macrophages and monocytes and stimulated with mitogens (*Protocol 6*) in the presence of Brefeldin A, which inhibits newly synthesized protein from traversing the cell membrane. This allows the fluorescence signal form the chosen intracellular antibody–fluorochrome conjugate to be amplified.

Protocol 6

Preparation of lymphocytes for analysis of intracellular cytokines *in vitro*

Equipment and reagents

- LP3 tubes
- 10 ml conical centrifuge tubes
- Magnet
- MNC in PBS
- 1.0 mg/ml carbonyl iron in PBSA
- 1.0 mg/ml PHA (Sigma)
- 100 μg/ml PMA (Sigma)

- 25 μg/ml calcium ionophore (A23187) (Sigma)
- 1 mg/ml Brefeldin A (Flow Laboratories)
- Growth medium: RPMI-1640, supplemented with 10% FCS, 100 IU penicillin, and 100 μg/ml streptomycin (Flow Laboratories)

Method

1. Add 100 μl carbonyl iron to each 5 ml of MNC in 10 ml centrifuge tubes.

2. Stir gently on a rotor for 30 min at room temperature.

3. Remove iron by adherence to wall of test-tube by a magnet.

4. Transfer cells to a fresh centrifuge tube and spin for 10 min at 1000 g at 18 °C.

5. Discard supernatant, resuspend cells in 2.0 ml of growth medium, and count in Coulter Counter (*Protocol 1*). Adjust cell concentration to $2–5 \times 10^6$/ml. Add 5 μg/ml PHA, 1 μg/ml PMA, 250 ng/ml calcium ionophore, and 10 μg/ml Brefeldin A (final concentrations).

6. Incubate cells at 37 °C for 4 h.

After activation, aliquots of cell suspension are stained for surface markers (*Protocols 1* and *6*). Suitable ready-conjugated antibody combinations to surface antigens on T cells and monocytes are shown in *Table 4*. The cells are then fixed before permeabilization and staining with PE-conjugated mAb to the internal antigen (*Protocol 7*). Fixation is not essential for assessing internal antigens, but is done because the total procedure takes in excess of eight hours. Treatment with paraformaldehyde allows samples to be stored at 4 °C overnight, after staining for surface antigens, without loss of the fluorescence signal or cellular integrity. Saponin in the permeabilization buffer allows access to the interior of the cells by large molecules such as PE.

Protocol 7

Staining for intracellular cytokines in activated mononuclear cells *in vitro*

Reagents

- 4% paraformaldehyde in PBS pH 7.4
- Permeabilization buffer pH 7.4: 0.1% saponin, 0.1% bovine serum albumin (BSA), 0.05% sodium azide (Sigma)
- Ready-conjugated mAbs (e.g. anti-IL-2-PE, Pharmingen) and appropriate isotype control
- Lysis buffer (Ortho-Lysis, Ortho-Diagnostics)
- Lymphocytes or MNC pre-stained with ready-conjugated mAbs to surface antigens (*Protocol 1*)
- PBS pH 7.4 (Flow Laboratories)

Method

1. Add 2.0 ml PBS to each sample pre-stained with mAbs to surface antigens and centrifuge for 10 min at 1000 *g* at 18 °C.

2. Discard supernatant, resuspend the cells, and add 100 μl paraformaldehyde to each tube.

3. Incubate at 4 °C for 10 min.

4. Add 2.0 ml PBS to each tube and centrifuge at 1000 *g* for 10 min at 18 °C.

5. Discard supernatant, resuspend the cells, and add 100 μl permeabilization buffer to each tube.

6. Add 5 μl mAb-PE, isotype control-PE, or no antibody as indicated below (*Table 4*).

7. Incubate for 30 min at 4 °C.

8. Add 2.0 ml permeabilization buffer to each tube and spin at 1000 *g* for 10 min at 18 °C.

9. Discard supernatant, resuspend the cells, and add 0.5 ml lysis buffer.

10. Process through analyser using three-colour or dual fluorescence, depending on choice of mAbs.

11. Calculate the number of cytokine positive cells based on the original WBC.

The incubation time for binding of intracellular antibody is longer than that required for surface antigens to allow entrance of the mAb and selective binding with the target. It is essential to wash cells with permeabilization buffer after exposure to allow unbound mAb to be released from the cells. The analysis is done using dual or three-colour fluorescence depending on the choice of fluorochrome and the number of antigens in the study (*Protocol 7*). Typical compensation factors and gains for three-colour fluorescence in an Ortho-Cytoron are shown in *Table 5*, for which a primary gate has been set on red fluorescence (RD-FL) and a second gate around cells within the gate which are positive for

Table 4 Staining for surface and intracellular antigens with ready-conjugated monoclonal antibodies (mAbs) before and after fixation[a]

Primary antibody (pre-fixation)	Secondary antibody	Target cell
CD3–CyChrome/CD4–FITC	IL2–PE	T helper cells
CD3–FITC/CD8–CyChrome	IL2–PE	T suppressor cells
CD3–FITC	IL2–PE	Total T cells
CD3–CyChrome/CD45RO–FITC	IL2–PE	Memory T cells
CD14–FITC	IL12–PE	Monocytes

[a] All primary antibody combinations should be done in duplicate so that relevant isotype controls can be made.

Table 5 Parameters, compensation factors, and gains for triple fluorescence using an Ortho-Cytoron to assess intracellular cytokines in peripheral blood lymphocytes

Parameter name	Gain	Compensation A	Compensation B	Antibody	Range
FSC: FW-SC	13	N/A	N/A	N/A	N/A
RHS: RT-SC	38	N/A	N/A	N/A	N/A
FL 1: GR-FL	90	3 (FL 2)	1 (FL 3)	mAb–FITC	1–100%
FL 2: OR-FL	90	17 (FL 3)	80 (FL 1)	mAb–PE	1–100%
FL 3: RD-FL	165	5 (FL 1)	80 (FL 2)	mAb–CyChrome	1–100%

this fluorochrome. The compensation factors and gains for dual fluorescence using FITC-conjugated and PE-conjugated are the same as in *Table 2*.

Assessment of cytokine production in monocytes is measured in cells pre-treated with interferon α before activation with mitogen compared with cells treated with mitogen alone (*Protocol 8*). Before preparation of an MNC fraction (*Protocol 5*) monocytes are quantitated in peripheral blood as in *Protocol 1* using anti-CD14-FITC. Assessment of intracellular cytokine production in monocytes is done as in *Protocol 8*.

Protocol 8

Preparation of monocytes for measurement of intracellular cytokines *in vitro*

Equipment and reagents

- LP3 tubes
- MNC prepared in growth medium as in *Protocol 5*
- 10^4 U/ml natural interferon α (Sigma)
- 1.0 mg/ml Brefeldin A (Sigma)
- 2.0 μg/ml LPS (Sigma)
- 10 ml conical centrifuge tubes
- PBS pH 7.4 (Flow Laboratories)

Method

1. Dispense cells at a concentration of $2\text{-}5 \times 10^6$/ml to two centrifuge tubes and add interferon α (100 U/ml) to one tube.

Protocol 8 continued

2. Incubate for 2 h at 37 °C.

3. Add 100 ng/ml LPS and 10 μg/ml Brefeldin A (final concentrations).

4. Incubate for 4 h at 37 °C.

References

1. Barclay, A. N., Birkeland, M. L., Brown, M. H., Beyers, A. D., Simon, S. J., Chamorro, S., *et al.* (ed.) (1993). *The leucocyte antigen: facts book*. Academic Press, London.

2. Ormerod, M. E. (ed.) (1990). In *Flow cytometry: a practical approach*, p. 29. IRL Press, Oxford.

3. McElwain, T. J. and Powles P. L. (1983). *Lancet*, **11**, 822.

4. Ho, T. B., Haylock, D. N., Kimber, R. J., and Juttner, C. A. (1984). *Br. J. Haematol.*, **58**, 399.

5. Tricot, G., Jagganath, S., Vesole, D., Nelson, J., Tindle, S., Miller, L., *et al.* (1995). *Blood*, **85**, 588.

6. Simmons, D. L., Satterthwaite, A. B., Tenen, G., and Seed, B. (1992). *J. Immunol.*, **148**, 267.

7. Bradley, T. R., Hodgson, G. S., and Rosendaal, M. (1978). *J. Cell. Phisiol.*, **97**, 517.

8. Bender, A. H., Unverzagt, K., Walker, D. E., Lee, W., Smith, S., Williams, S. (1994). *Clin. Immunol. Immunopathol.*, **70**, 10.

9. Galy, A. H., Cen, D., Travis, M., Chen, S., and Chen, B. P. (1995). *Blood*, **85**, 2770.

10. Millar, B. C., Millar, J. L., Shepherd, V., Blackwell, P., Porter, H., Cunningham, D., *et al.* (1998). *Bone Marrow Transplant.*, **22**, 469.

11. Waldman, H., Polliak, A., Hale, G., Or, R., Cividalli, G., Weiss, L., *et al.* (1984). *Lancet*, **2**, 483.

Chapter 18

Immunocytochemical staining of cells and tissues for diagnostic applications

Andrew R. Dodson and John P. Sloane

Department of Pathology, The University of Liverpool, Royal Liverpool University Hospital, Duncan Building, Liverpool L69 3GA, U.K.

1 Introduction

Immunocytochemistry can be defined as the *in situ* demonstration of cellular constituents by specific antibody–antigen reactions. It has its origins in the 1940s with the pioneering work of Coons *et al.* (1). Since that time many workers have developed and refined the technique, so that today it is an essential part of diagnostic pathology.

Due to limits on space this chapter will deal only with those issues and methods that are germane to immunocytochemistry as it is practised in most diagnostic pathology laboratories. The reader is directed to the excellent texts cited in the Further Reading section for more detailed information.

2 Tissue and cell substrates

2.1 Tissue sections

There are a number of methods of immobilizing antigens and preparing sections from tissues. The choices will depend on factors such as the ability of the antigen to survive fixation and paraffin processing, and how quickly the result is needed. In practise, however, relatively few methodologies are used, with sections from formalin-fixed, paraffin wax embedded material being usually employed for routine histological diagnosis.

2.2 Cytological specimens

Thin fluid specimens (e.g. fine needle aspirations, ascitic fluids, bronchial lavages) may be concentrated by centrifugation, or diluted if very cellular, before being spread onto glass slides. Cytospin® preparations (Cytospin®, Shandon Scientific Ltd.) may also be prepared; these have the advantage of localizing the cells to an area of approximately 1 cm in diameter.

Mucoid fluids such as sputa require treatment with a mucolytic agent (e.g.

Sputasol, Oxoid Ltd.) to remove the mucous, which may non-specifically bind antibodies.

Specimens can either be pre-fixed by suspension in fixatives such as formalin, or Saccomanno's Fluid (2% (w/v) polyethylene glycol, PEG 1500, BDH Ltd., in 50% ethanol). Alternatively the slides may be post-fixed, usually in alcohol, acetone, or methanol for 10–20 minutes at room temperature. In all cases glass slides coated with some form of adhesive should be used (see Section 4.1 for more details of suitable slide adhesives).

One of the major problems, when dealing with non-gynaecological specimens, lies in obtaining and maintaining adequate stocks of control material, due to the instability of unfixed cytological material. One solution is to prepare large numbers of slides from known positive cases and to store them at $-80\,°C$. Another is to prepare cell blocks from centrifuged material, usually suspended in agar, which may be fixed and processed to paraffin wax, and from which sections may be cut.

3 Fixation and processing for paraffin wax embedded tissues

3.1 Fixation

The type and duration of fixation have a profound effect on the results of immunocytochemical staining (2, 3). In general the simple formaldehyde-based fixatives (e.g. 10% formal saline), often used at an acid pH, produce the best results. However, satisfactory results can be achieved with most formaldehyde-based fixatives and several less common types (e.g. Carnoy's). Optimum fixation is essential, and it is standard practice in good histopathology departments to have protocols to ensure that all specimens are adequately fixed. In routine practice over-fixation is rarely encountered because turnaround times ensure that specimens are processed as quickly as possible. Under-fixation, however, causes problems. All specimens should be immersed in at least three times their volume of fixative. The rate of penetration is comparatively slow, so it is necessary to ensure it reaches all parts of the specimen before autolysis begins. This is usually achieved by slicing solid organs, opening hollow ones, and inflating lungs.

3.2 Decalcification

Calcified tissues must be decalcified before they are processed to paraffin wax, otherwise it will be impossible to produce good quality sections.

A number of decalcifying agents are in common use, and range from strong acids (e.g. nitric and formic acids) to calcium chelating agents (e.g. ethelyenedia-minetetraacetic acid, EDTA). Mineral acids have an extremely detrimental effect on subsequent immunostaining and should be avoided. Many antigens will survive decalcification with formic acid, but the best results are achieved when a calcium chelating agent is employed after fixation (4) (e.g. 15% (w/v) EDTA disodium salt, 1.4% (w/v) sodium hydroxide in deionized water, mix until completely dissolved then adjust to pH 7.6). The disadvantage of these agents is that they are comparatively slow (a bone marrow trephine will take between 12–36 hours to decalcify), and they may not be powerful enough to decalcify large pieces of cortical bone completely, at least in reasonable time. Mechanical agitation (e.g. a rotating stirrer) may be used to speed up the process.

A variety of surface decalcification agents effective on wax embedded tissues are available, but they have a detrimental effect on immunostaining.

3.3 Processing

While the reagents and procedures involved in processing wet fixed tissue into a paraffin block have an impact on immunocytochemistry it is far less profound than that of fixation. In one study it was found that 'increasing the processing temperature from ambient to 45 °C for all stages up to the wax infiltration step and increasing the duration of both the dehydration and wax infiltration stages improved the immunostaining of most of the antibodies tested'. None of the other factors tested had any effect (3).

4 Section preparation

Obtaining thin, flat, high quality sections from the wide range of tissues encountered in routine diagnostic pathology is not easy, but is crucial as section quality influences the ability to adhere to the glass slide. Poor quality sections are far more likely to be lost during the long incubation steps involved in most immunocytochemistry methods, especially if high temperature antigen retrieval (HTAR) is employed. Sections from paraffin embedded blocks are usually cut at 3–4 μm thick, floated onto warm water to flatten them, and picked up onto glass slides that have been coated with a chemical to improve section adherence (see Section 4.1). They are then dried to allow the section to adhere firmly. This can be done at various temperatures, but in general higher temperatures are associated with weaker immunostaining (5). A slide drying oven set at, or just above, the melting point of the wax (usually 56–60 °C) can be used to dry slides between two hours to overnight, with excellent immunostaining results. Some, however, recommend a lower temperature (e.g. overnight at 37 °C), but lower temperatures reduce section adhesion. The authors' personal preference is to dry slides overnight, well spaced in wooden racks, at 45 °C. This gives good section adherence without compromising staining. Hot plates should not be used to dry slides that are going to be subsequently immunostained, the cycling action of the heating element means that even if the temperature is set quite low, the slides can reach high temperatures and local hot spots often occur, leading to inconsistent and patchy staining.

4.1 Section adhesives

Immunocytochemical techniques involve long incubation steps that can result in section detachment. The use of HTAR techniques during which sections are immersed in a boiling aqueous solution has greatly exacerbated the problem. If a section adhesive is not employed 95–100% of sections will detach if subjected to HTAR (unpublished personal observation).

Several adhesives are available, including poly-L-lysine, 3-aminopropyltri-ethoxysilane (APES), Vectabond (Vector Laboratories Ltd.). Treated slides are also available commercially, as are electrostatically charged slides that are especially effective with fresh cell preparations. The most cost-effective method is to prepare adhesive coated slides 'in-house'. APES (5) is the section adhesive of choice; large numbers of slides can be prepared in batches, dried, and stored indefinitely without deterioration. The adhesive qualities are excellent and are vastly superior to those of poly-L-lysine, for example.

Protocol 1

The preparation of 3-aminopropyltriethoxysilane coated microscope slides

Reagent

- 3% (v/v) 3-aminopropyltriethoxysilane[a] (Sigma Chemical Co.) in acetone

Method

1. Load glass microscope slides[b] into metal slide carrier racks.

2. Immerse in fresh acetone for 5 min, agitate the rack occasionally.

3. Transfer to freshly made APES solution and allow to stand for 5 min, again with occasional agitation.[c]

4. Wash the slides for 1 min each in two changes of deionized water.

5. Drain off excess water and allow the slides to dry thoroughly before returning them to their original boxes for storage. The drying can be done overnight at room temperature, or more rapidly in a hot oven (e.g. at 60 °C). The treated slides have an indefinite shelf-life.

[a] The APES stock solution does not keep well once opened, volumes of reagent concomitant with usage within two or three weeks should be purchased.

[b] Slides from different manufacturers may vary in their 'stickiness' after APES coating.

[c] Four racks of slides are treated, then the APES solution is discarded, and replaced with fresh, the first deionized water is also discarded and replaced by the second one, which is in turn replaced with fresh.

4.2 Section storage

Proteins (and hence their epitopes) seem to survive indefinitely in an unchanged state once they are fixed and preserved in paraffin wax. This means that archived blocks are available for retrospective immunocytochemical investigations.

There is less certainty about paraffin sections. It seems that most epitopes remain unaffected in stored sections (6), although some workers have reported reduced, or lost staining for BCL-2, Ki-67, CD15, CD30, and others (7–10). The authors have occasionally experienced this phenomenon with regard to oestrogen receptor staining, but it appears to be highly variable, with sections from some blocks showing quite rapid diminution of staining (within two weeks) while others show no alteration.

5 Primary antibodies

Monoclonal antibody technology (11) has revolutionized diagnostic immunocytochemistry. Reagents with a vast range of specificities are now available, and with the advent of HTAR many of them can be made to stain formalin-fixed, paraffin processed material. Most are obtainable commercially. Many companies have invested significant resources in producing and evaluating high quality, dependable reagents, often in collaboration with diagnostic laboratories.

All companies produce data sheets on their antibodies, including useful information on expected dilution range, any required pre-treatments, suggested control tissues, expected staining patterns, and recommended storage conditions. The majority of monoclonal antibodies are stable when stored at 4 °C with the addition of sodium azide (15 mM) to prevent bacterial growth, however a small number deteriorate quite rapidly (within weeks) when stored in this way, and it is necessary to divide them into aliquots (the size depending on working dilution and volume used per test) and freeze usually at −20 °C to −30 °C (avoid 'frost-free' freezers which have a freeze–thaw–freeze cycle).

New antibodies should be evaluated on the laboratory's own material with regard to working dilution and pre-treatment requirements. As previously mentioned commercial antibodies are usually accompanied by data sheets that may give guidance, but even so a range of dilutions about the suggested one should be tried, as the sensitivity of different detection methods, or even of the same ones employed in different laboratories, can vary.

6 Immunocytochemical methodology

It is advisable to minimize the amounts of primary antibodies used to reduce cost. Staining sections flat on an enclosed humid incubating tray will necessitate 100–200 μl of reagent per slide. Alternatively a system such as the Sequenza™ Immunostaining System (Shandon Scientific Ltd.) can be used, in which a disposable Coverplate™ is closely attached to the slide forming a well over the entire area of the slide with a volume of approximately 100 μl, into which all necessary solutions can be introduced. The system is efficient and simple to use. Automated immunostainers are available from many companies in the field.

To produce consistent results it is necessary to pay attention to all details of the method(s) chosen, in particular timings, composition and pH of solutions, and storage conditions of antibodies. Immunocytochemistry is a craft as well as a science; time spent observing an experienced worker is well spent, and practise will make perfect.

6.1 Antigen retrieval

6.1.1 Proteolytic enzyme digestion

Fixation and processing sequesters a significant number of antigens, which in some cases can be made available for reaction by the use of various section pre-treatments. This was first noted in the 1970s (12, 13). Controlled proteolytic digestion of sections was performed before immunostaining, and positive results were obtained with many antibodies that had previously been effective only on frozen sections. Many different proteolytic enzymes have since been tried, including trypsin, protease XXIV, chymotrypsin, and pepsin. Success with the technique requires careful attention to the concentration of the enzyme, and to the time and temperature of incubation. Enzymes from different batches or suppliers will vary in their activity. The correct conditions for each laboratory's material should be established, as length and type of fixation and processing will all influence the result. Under-digestion of the section will lead to weak or negative staining while over-digestion often leads to spurious staining or loss of cytological detail.

Protocol 2

Proteolytic enzyme pre-treatment

Equipment and reagents

- Large water-bath or oven, set at 37 °C
- TBS: 0.05 M Tris–HCl pH 7.4, 0.15 M NaCl
- One of the following enzyme solutions:[a]
  0.05% (w/v) chymotrypsin (Sigma Chemical Co.), 0.1% (w/v) CaCl$_2$ in TBS at 37 °C

0.05% (w/v) trypsin-250 (Difco), 0.1% (w/v) CaCl$_2$ in TBS at 37 °C

0.05% (w/v) protease type XXIV (Sigma Chemical Co.), in TBS at 37 °C

Method

1. Pre-warm two containers of TBS to 37 °C (if using a water-bath this will take only 15–30 min, if using an oven leave overnight). A 50 ml Coplin jar can be used for small numbers of slides, a 400 ml trough for larger numbers.

2. Pre-warm the slides in one of the containers of TBS for 10 min.

3. Immediately before use make the enzyme solution and transfer the warmed slides quickly into it, incubate for 10–20 min with occasional agitation.

4. When the incubation time is finished quickly transfer the slides to running tap-water and wash them well in this for 5 min to remove all traces of the enzyme. Discard the enzyme solution.

[a] These are enzymes that have been found to work well in the authors' hands, it is not an exhaustive list, other enzymes and the same enzymes from other suppliers may be found to work equally well.

6.1.2 High temperature antigen retrieval

The introduction of HTAR methods has had a major impact on immunocyto-chemistry, allowing reliable staining of a plethora of epitopes previously not or barely demonstrable in formalin-fixed paraffin processed material.

Since the ground-breaking paper of Shi *et al.* in 1991 (14), many workers have developed and refined the technique (15–18). The basic requirements for successful HTAR are the input of large amounts of energy and some form of aqueous antigen retrieval (AR) solution. In current routine diagnostic practice energy is usually in the form of heat generated by a microwave oven or pressure cooker, while the AR solution is usually either sodium citrate or EDTA (19). The mechanism of action of HTAR is not clear but one theory suggests that it removes bound Ca^{2+} ions (20).

Microwave ovens should have turntables to ensure even heating, and digital timers for accuracy. An 800–850 watt model is preferable to a lower powered one. If a pressure cooker is used it should be of stainless steel (aluminium is corroded by some AR solutions). In most cases the results obtained from the two methods of heating are indistinguishable, however slightly better staining of nuclear antigens such as oestrogen receptor and MIB-1 (Ki 67) may be achieved after the use of the pressure cooker (Peter Jackson, personal communication).

Protocol 3

High temperature antigen retrieval

Equipment and reagents

- Microwave oven (with turntable and digital timer)

or

- Pressure cooker (6 litre, stainless steel model), and a hot-plate

- Antigen retrieval solution: 10 mM EDTA (Sigma Chemical Co.) pH 7.0[a]

A. Microwave oven method

1. Place slides in alternate slots in a plastic slide carrier,[b] and immerse them in 500 ml of AR solution in a 1 litre plastic container, loosely cover the top with several layers of absorbent paper to prevent excessive evaporation.

2. Place two containers[c] in the microwave oven and heat at full power for 25 min, after which remove them carefully from the oven and allow to stand for a further 5 min.

3. Remove the carriers quickly to running tap-water, taking care that the still hot slides do not dry out during the transfer. Wash them for 2 min in running tap-water, then rinse them in deionized water, and continue with the immunostaining method. Discard the used EDTA solution.

Protocol 3 continued

B. Pressure cooker method

1. Place 2 litres of AR solution in the pressure cooker, cover loosely with the lid, and bring the solution to the boil.

2. Place slides in alternate slots in metal slide carriers, immerse into the boiling AR solution, replace the lid tightly.

3. When full pressure is reached time 2 min, then cool the pressure cooker quickly by immersing in a sink of cold water.

4. As soon as the pressure cooker has depressurized remove the slide carriers to running tap-water. Wash them for 2 min, then rinse in deionized water, and proceed with the immunostaining method. Discard the used AR solution.

[a] It is convenient to make the EDTA solution in 10 litre batches as it keeps on the bench very well, in which case use the following recipe. Add together 37.2 g of EDTA, and 3.2 g of sodium hydroxide to 10 litres of deionized water, stir for 2–3 h, then pH to 7.0 with 1 M sodium hydroxide solution.

[b] Use alternate slots in the carrier to increase the gap between slides, this prevents any bubbles formed during the heating from becoming trapped between the slides.

[c] Always place the same 'load' into the oven for heating, if only one container is needed for slides the other may be filled with 500 ml of tap-water.

A recent innovation has been the plastic 'pressure cooker' that is heated in the microwave (A. Menarini Diagnostics), thus combining the superior staining of pressure cooking with the convenience of using the microwave oven.

6.2 Detection methodology

Methods for immunocytochemistry have shown an evolutionary progression from the simplest direct method used in pioneering work through two-step or indirect methods to the sensitive and robust multistep methods that are in common use today. The multistep methods have the great advantage of considerably improving sensitivity. High quality detection systems are available from several commercial suppliers.

6.2.1 Avidin biotin complex method

The detection methods that are currently the most popular are variations on the avidin biotin complex (ABC) method (21, 22). The technique relies on the extremely high affinity between avidin (a polypeptide originally derived from egg white), and biotin (a low molecular weight vitamin). Avidin has four binding sites for biotin and hence can be used as a linking reagent to couple other biotin-linked reagents together. Biotin is an ideal labelling molecule, as it does not interfere with the reactivity of the reagent it is bound to because of its small size. In the ABC method a complex of three biotin-conjugated horseradish peroxidase molecules per avidin molecule is formed, the remaining biotin binding site being free to bind to a biotin molecule conjugated to a secondary antibody.

When used to detect mouse monoclonal antibodies this secondary antibody would have anti-mouse immunoglobulin specificity. The method can be adapted for a number of reporter molecules, for example other enzyme labels such as alkaline phosphatase, and fluorescent molecules such as fluorescein isothiocyanate (FITC).

The avidin that was originally used carried an electrical charge; this caused some non-specific background staining due to electrostatic binding. The introduction of streptavidin, a homologous molecule derived from the *Streptomyces avidinii* bacterium that carries no charge, has eliminated the problem. However, most manufacturers now supply avidin that has been chemically modified to remove the charge. So in practice today there is little to choose between the commercially available avidin- and streptavidin-based detection reagents.

Protocol 4

The streptavidin biotin complex immunostaining method

Reagents

- TBS: 0.05 M Tris–HCl pH 7.4, 0.15 M NaCl
- Methanol/H_2O_2: 3% (v/v) 100-volume H_2O_2 in methanol[a]
- BSA/TBS: 5% (w/v) bovine serum albumin in TBS
- Primary monoclonal antibody (optimally diluted in BSA/TBS)

- Biotinylated anti-mouse Ig secondary antibody (optimally diluted in BSA/TBS, e.g. Amersham RPN1001 at 1:100)
- Streptavidin biotin complex Kit (used according to manufacturer's instructions, e.g. K0377 from Dako Ltd., and Elite from Vector Laboratories Ltd.)

Method

1. Rack the slides into metal slide carrier and dewax by immersing in two changes of xylene for 3 min in each.
2. Rinse in two changes of IMS for 30 sec in each.
3. Transfer to freshly made methanol/H_2O_2 to block the endogenous peroxidase activity, incubate for 12 min with occasional agitation.[a]
3. Quickly transfer to running tap-water and rinse for 30 sec, then rinse in deionized water.
4. At this point the slides should be subjected to any pre-treatment(s) that may be required (see *Protocols 2* and *3*).
5. Transfer to an incubating tray taking care not to allow the slides to dry out, cover with TBS, and allow to equilibrate for 5–10 min.
6. Apply 100 µl of primary antibody solution and incubate for 40 min (apply 100 µl of BSA/TBS to all negative controls).
7. Wash the slides well with at least two changes of TBS for 5 min each.
8. Apply 100 µl of biotinylated secondary antibody solution and incubate for 30 min (make the streptavidin biotin complex–HRP solution at this point because it needs to stand for 30 min before use).

Protocol 4 continued

9. Wash the slides well with two changes of TBS for 5 min each.

10. Apply 100 μl of streptavidin biotin complex–HRP solution and incubate for 30 min.

11. Wash the slides well with two changes of TBS for 5 min each.

12. Remove the slides from the incubating tray to metal slide carrier and immerse them in a trough of TBS.

13. Demonstrate the enzyme label (see *Protocols 6* and *7* for HRP, and *Protocol 8* for alkaline phosphatase).

^{a} The use of methanol/H_2O_2 is only necessary if the label is HRP, otherwise omit this step.

One major disadvantage that may be encountered is the presence of endogenous biotin in a variety of cell types, but fortunately it is 'masked' or removed during fixation and processing. However, the use of HTAR causes the retrieval of 'masked' biotin and can lead to false positive results (23), especially in liver, kidney, and pancreas that often contain significant amounts of biotin and hence produce very high background staining. If one is aware of the potential pitfall and the appropriate negative controls are carried out and carefully examined, endogenous biotin very rarely presents a serious problem (24).

6.2.2 Enhanced polymer methods

A recent development (25) has been the use of long chain dextran molecules as carriers for multiple molecules of antibody and reporter enzyme. Originally developed as a one-step method (EPOS™, Dako Ltd.), it shows acceptable sensitivity because of the presence of multiple enzyme molecules while being extremely simple and rapid to carry out. More flexibility has been achieved in the development of the indirect labelled polymer method (EnVision™, Dako Ltd.) (26, 27). This uses similar technology, with the dextran polymer carrying multiple enzyme molecules and secondary antibody molecules. This method shows great promise as it has a sensitivity at least equal to that of ABC and in some cases considerably better. It also eliminates any problems of endogenous biotin.

6.2.3 Alkaline phosphatase anti-alkaline phosphatase method

The alkaline phosphatase anti-alkaline phosphatase (APAAP) (28) method has gained great favour for use on cytological preparation and frozen sections, mainly because it avoids the necessity for blocking endogenous peroxidase, a procedure which can be disruptive to unfixed materials (an aqueous H_2O_2 solution rather than a methanolic one is better in these situations). The APAAP method is a three-step technique, the first being the specific antibody. The second is an antibody with specificity for the immunoglobulin species of the primary antibody, and is used at a comparatively low dilution to ensure some antibody molecules will be monovalently bound, leaving binding sites available to bind to the third stage, the APAAP complex, formed of molecules of alkaline phosphatase immunologically bound to antibody of the same species as the primary antibody.

Protocol 5

The alkaline phosphatase anti-alkaline phosphatase immunostaining method

Reagents

- TBS: 0.05 M Tris–HCl pH 7.4, 0.15 M NaCl
- BSA/TBS: 5% (w/v) BSA in TBS
- Primary monoclonal antibody (optimally diluted in BSA/TBS)
- Anti-mouse Ig secondary antibody (optimally diluted in BSA/TBS, e.g. Z0142, Dako Ltd. at 1:100)
- Alkaline phosphatase anti-alkaline phosphatase complex, mouse (optimally diluted in BSA/TBS, e.g. D0651, Dako Ltd. at 1:25)

Method

1. Rack the slides into metal slide carrier and dewax by immersing in two changes of xylene for 3 min in each.
2. Rinse in two changes of IMS for 30 sec in each.
3. Quickly transfer to running tap-water, and rinse for 30 sec, then rinse in deionized water.
4. At this point the slides should be subjected to any pre-treatment(s) that may be required (see *Protocols 2* and *3*).
5. Transfer to an incubating tray taking care not to allow the slides to dry out, cover with TBS, and allow to equilibrate for 5–10 min.
6. Apply 100 μl of primary antibody solution and incubate for 40 min (apply 100 μl of BSA/TBS to all negative controls).
7. Wash the slides well with at least two changes of TBS for 5 min each.
8. Apply 100 μl of secondary antibody solution and incubate for 30 min.
9. Wash the slides well with two changes of TBS for 5 min each.
10. Apply 100 μl of alkaline phosphatase anti-alkaline phosphatase complex solution and incubate for 30 min.
11. Wash the slides well with two changes of TBS for 5 min each.

    Steps 8–11 may be repeated if required, with antibody incubation time of 10 min in order to increase the sensitivity of the method.
12. Remove the slides from the incubating tray to metal slide carrier, immerse them in a trough of TBS.
13. Demonstrate the enzyme label (see *Protocol 8*).

The APAAP method is capable of extreme sensitivity if the second layer is reapplied followed by a further reapplication of the APAAP reagent. In practise these reapplication stages are often of shorter duration than the first with little detriment, otherwise the method becomes long and unwieldy.

7 Reporter molecules

Also called a label, the reporter is the reagent that produces the colour seen in the finished slide. It is usually an enzyme, and in routine diagnostic practice one of two enzymes is generally used: horseradish peroxidase (29, 30) or alkaline phosphatase (31).

7.1 Horseradish peroxidase

This is the enzyme reporter of choice in most situations. When used with the chromogen 3,3'-diaminobenzidine (DAB) (32), it yields a brown reaction product that is extremely well localized and insoluble in all the commonly used dehydrating and clearing agents. Slides can thus be permanently mounted in resinous mountant allowing very good optical resolution under the microscope.

Protocol 6

The 3,3'-diaminobenzidine chromogen for use with HRP (33)

Reagents

- 3,3'-diaminobenzidine tetrahydrochloride (Sigma Chemical Co.)[a]
- TBS: 0.05 M Tris–HCl pH 7.4, 0.15 M NaCl
- Mayer's Haematoxylin solution (Diachem International Ltd.)
- 100-volume (30%) H_2O_2
- Scott's tap-water substitute: 2% (w/v) magnesium sulfate, 0.35% (w/v) sodium hydrogen carbonate, 0.01% (w/v) thymol in tap-water

A. Preparing the stock DAB solution

1. In a fume cupboard, open the bottle of DAB powder and tip the contents into sufficient deionized water to give a final concentration of 10% (w/v).

2. Stir, covered until completely dissolved.

3. Divide into 2 ml aliquots in plastic tubes, and store frozen at −20 °C, where it will keep indefinitely.

B. Use of the DAB solution

1. Remove a tube of stock DAB solution from the freezer and thaw into 400 ml of TBS.

2. Immediately before use add 400 μl of 100-volume H_2O_2,[a] mix briefly then immerse slides in the solution, incubate for 8 min with occasional gentle agitation.[b]

3. Remove the slides to running tap-water and wash them well.

4. Counterstain the nuclei in the Mayer's Haematoxylin, the time will vary according to the age of the solution.

5. Rinse in running tap-water, and 'blue' in the Scott's tap-water substitute for 1 min, after which rinse well in running tap-water to remove all traces of the Scott's.

6. Dehydrate through graded alcohols, clear in xylene, and mount in a resinous mountant.

[a] Do not pipette directly from the stock H_2O_2 bottle, pour a little out, pipette from this, and discard the excess. Pipette tips can cause the H_2O_2 to break down, thus reducing the strength of the stock.

[b] The DAB solution may be used to stain two or three lots of slides; if however it begins to turn 'cloudy' and darken, discard and make fresh. After use discard the DAB solution into a large container and add 10–20 ml of household bleach. The solution will darken and may effervesce vigorously, the 'neutralized' DAB may then be run to waste with plenty of running tap-water.

The only significant disadvantages of DAB are:

(a) It has been reported as a potential carcinogen (the use of practical precautions reduce this to an acceptably low risk).

(b) It may be confused with some endogenous brown pigments such as lipofuscin especially in the liver and melanin in the skin and eye (careful scrutiny of negative controls can avoid confusion here).

Other chromogens that can be used with horseradish peroxidase include 3-amino-9-ethylcarbazole that produces a red colour and 4-chloro-1-naphthol (blue). Unfortunately both these yield reaction products that are soluble in alcohol and xylene and so require an aqueous mountant (commercial ones are available, e.g. Aquamount, BDH Ltd. and Glycergel, Dako Ltd.).

Protocol 7

The 3-amino-9-ethylcarbazole chromogen for use with HRP

Reagents

- 3-Amino-9-ethylcarbazole chromogen (AEC, Sigma Chemical Co.)[a]
- 0.05 M acetate buffer pH 5.0
- N,N-dimethylformamide (BDH Ltd.)
- 100-volume (30%) H_2O_2
- Mayer's Haematoxylin solution (Diachem International Ltd.)
- Scott's tap-water substitute (see *Protocol 6*)

Method

1. Pre-incubate the slides in acetate buffer for 10 min.

2. Meanwhile, dissolve 10–20 mg of AEC in 0.5 ml of N,N-dimethylformamide in a glass container, then gradually add 50 ml of acetate buffer with constant mixing, add 25 ml of 100-volume H_2O_2, mix, and filter onto the slides. Incubate for 5–15 min under microscopic control.

3. Rinse the sections with deionized water and then in running tap-water.

4. Counterstain the nuclei in the Mayer's Haematoxylin, the time will vary according to the age of the solution.[a]

5. Rinse in running tap-water, and 'blue' in the Scott's tap-water substitute for 1 min, after which rinse well in running tap-water to remove all traces of the Scott's, and mount from deionized water using an aqueous mountant.

[a] Do not differentiate with acid–alcohol, as the reaction product is alcohol soluble.

7.2 Alkaline phosphatase

This reporter is obviously encountered in the APAAP technique, but it can also be conjugated to biotin and used successfully in the ABC methods. The enzyme is sourced from calf intestine, this form being resistant to the blocking effect of levamisole, which is added to the substrate buffer and blocks all other types of endogenous alkaline phosphatase activity. Several chromogens can be used, the most popular being Fast Red TR salt which yields a bright red alcohol soluble reaction product (34).

Protocol 8

The Fast Red TR salt chromogen for use with alkaline phosphatase

Reagents

- Naphthol-AS-MX phosphate, disodium salt (Sigma Chemical Co.)
- N,N-dimethylformamide (BDH Ltd.)
- Tris: 0.1 M Tris–HCl pH 8.2
- Levamisole (Sigma Chemical Co.)
- Fast Red TR salt (Sigma Chemical Co.)
- Mayer's Haematoxylin (Diachem International Ltd.)
- Scott's tap-water substitute (see *Protocol 6*)

Method

1. Immediately before use, dissolve 20 mg of naphthol AS-MX phosphate salt in 1 ml of N,N-dimethylformamide in a glass container. Add 20 ml of Tris buffer gradually with mixing, then add 12 mg of levamisole and 50 mg of Fast Red TR salt, mix well.

2. Filter onto slides and incubate for 10–30 min, control microscopically.

3. Rinse off with deionized water, then wash the slides in running tap-water.

4. Counterstain the nuclei in the Mayer's Haematoxylin, the time will vary according to the age of the solution.[a]

5. Rinse in running tap-water, and 'blue' in the Scott's tap-water substitute for 1 min, after which rinse well in running tap-water to remove all traces of the Scott's, and mount from deionized water using an aqueous mountant.

[a] Do not differentiate with acid–alcohol, as the reaction product is alcohol soluble.

The use of the commercially available Vector Red Substrate kit (Vector Laboratories Ltd.) produces a reaction product that is alcohol and xylene resistant (some workers have reported fading of the colour on storage), and most interestingly the reaction product is intensely red-orange fluorescent when viewed using the rhodamine filter system, which greatly improves its ability to demonstrate very low levels of bound product.

8 Enhancement

It is possible to enhance the DAB reaction product in several ways (35). The simplest one that is routinely employed by many workers is the immersion of immunostained slides into a trough of copper sulfate solution for a few minutes, this darkens the reaction product to a dark brown-black without affecting the background.

Protocol 9

Enhancement of the DAB–HRP reaction product

Reagent

- 0.5% (w/v) copper sulfate, 0.9% (w/v) NaCl in deionized water
- Scott's tap-water substitute (see *Protocol 6*)
- Mayer's Haematoxylin solution (Diachem International Ltd.)

Method

1. Wash slides well in running tap-water after the standard DAB method (see *Protocol 6*).

2. Immerse them in a trough of the copper sulfate solution for 2–10 min.[a]

3. Wash the slides well in running tap-water.

4. Counterstain the nuclei in the Mayer's Haematoxylin, the time will vary according to the age of the solution.

5. Rinse in running tap-water, and 'blue' in the Scott's tap-water substitute for 1 min, after which rinse well in running tap-water to remove all traces of the Scott's.

6. Dehydrate through graded alcohols, clear in xylene, and mount in a resinous mountant.

[a] The DAB reaction product will progressively darken to a brown-black colour, monitor it microscopically.

Heavy metal salts can also be added to the DAB solution itself to alter the reaction product colour (e.g. 0.01% cobalt chloride yields a black to blue-black product). Staining times may need to be shortened to prevent build-up of product in the background due to non-specific precipitation.

8.1 Biotinylated tyramine

Originally described several years ago for use in immunoassay systems (36), biotinylated tyramine amplification has recently been adapted for immunocytochemical use (37). The system is based on the ABC method. H_2O_2 is catalysed to H_2O and oxygen-free radicals by HRP as usual. This then reacts with biotinylated tyramine to produce 'activated' biotinylated tyramine, which binds to nearby proteins in a highly localized manner. The high concentrations of biotin thus produced can then be detected by the application of one additional layer of avidin–reporter conjugate. The system has been used to demonstrate antigens previously only detectable in frozen sections, and to increase primary antibody dilution. Several manufacturers produce kits that are based on the system (CSA from Dako Ltd. and TSA from NEN Life Science Products), or 'home-made' biotinylated tyramine can be produced.

Protocol 10

The production and use of biotinylated tyramine (38)

Equipment and reagents

- 0.45 μm syringe filter (e.g. Nalgene, BDH Ltd.)
- Sulfosuccinimidyl 6-(biotinamido) hexanoate, also called Sulfo-NHS-LC-biotin (Pierce and Warriner (UK) Ltd.)
- Tyramine–HCl (Sigma Chemical Co.)

- 50 mM borate buffer pH 8.0
- TBS: 0.1 M Tris–HCl pH 7.6, 0.15 M NaCl
- TNT: 0.05% (v/v) Tween 20 in TBS
- Tris: 0.05 M Tris–HCl pH 7.6

A. Production of biotinylated tyramine

1. Add 100 mg of sulfosuccinimidyl 6-(biotinamido) hexanoate and 30 mg of tyramine–HCl to 40 ml of borate buffer, then stir gently overnight.

2. Filter through paper, then through a 0.45 μm syringe filter, aliquot, and store frozen at $-20\,°C$.

B. Use of biotinylated tyramine

1. Carry out standard ABC-based method (*Protocol 4*) up to and including step 9.[a]

2. Wash the slides with TNT buffer three times for 5 min each.

3. Apply avidin biotin–HRP complex made in TBS buffer for 30 min (see *Protocol 4*).

4. Wash the slides with TNT buffer three times for 5 min each.

5. Apply the biotinylated tyramine solution diluted 1:500 in Tris buffer containing 0.03% (v/v) H_2O_2. It may be necessary to process a number of slides with different incubation times between 30 sec and 10 min, as the resulting amplification can be quite variable.[b]

6. Wash the slides with TNT buffer three times for 5 min each.

7. Apply avidin biotin–HRP complex made in TBS buffer for 30 min.

Protocol 10 continued

8. Wash the slides with TNT buffer three times for 5 min each.

9. Develop with DAB as in *Protocol 6*.

[a] Primary antibody dilutions will have to be increased significantly (between 10- and 40-fold).

[b] Perry Maxwell, personal communication.

9 Controls

In routine diagnostic practice positive and negative controls are needed.

9.1 Positive control

A section of tissue known to contain epitopes reactive with the primary antibody should be stained alongside the test sections. The expected positive staining pattern will indicate that the method was carried out correctly. The section should have been prepared in exactly the same manner as the test.

9.2 Negative control

Sections from the test are treated in exactly the same way as the positive tests (this includes the use of any antigen retrieval method which the positive has been subjected to), except that the primary antibody is omitted (usually the carrier solution is used in place). Staining in these slides would indicate a cross-reaction of the detection system with a tissue element. Negative control slides are also useful for indicating the presence of any unblocked endogenous enzyme activity, and for showing any endogenous pigments.

Other controls that may be carried out (but not on a day-to-day basis) include the use of Western blotting and other biochemical tests.

9.3 External quality assurance scheme (EQA)

Any laboratory that undertakes diagnostic work should be a member of an appropriate EQA scheme. No one can be an expert in all aspects of such a diverse field. EQA scheme results can highlight areas of weakness that need addressing. Comparing results with a wide range of other laboratories can also provide reassurance and confidence that standards are good.

Acknowledgements

Our thanks go to Steve Bradburn for his invaluable advice and suggestions, and to Jill Gosney for her secretarial assistance.

Suggested further reading

1. Polak, J. M. and Van Noorden, S. (1997). *Introduction to immunocytochemistry* (2nd edn). BIOS Scientific Publishers, Oxford.
2. Polak, J. M. and Van Noorden, S. (ed.) (1986). *Immunocytochemistry: modern methods and applications* (2nd edn). John Wright & Sons Ltd., Bristol.

3. Miller, K. (1996). In *Theory and practice of histological technique* (ed. J. D. Bancroft and A. Stevens), p. 435. Churchill Livingstone, Edinburgh.

4. Jasani, B. and Schmid, K. W. (1993). *Immunocytochemistry in diagnostic pathology.* Churchill Livingstone, Edinburgh.

References

1. Coons, A. H., Creech, H. J., and Jones, R. N. (1941). *Proc. Soc. Exp. Biol. Med.*, **47**, 200.

2. Brandtzaeg, P. (1982). In *Techniques in immunocytochemistry* (ed. G. R. Bullock and P. Petrusz), pp. 1–75. Academic Press, London.

3. Williams, J. H., Mepham, B. L., and Wright, D. H. (1997). *J. Clin. Pathol.*, **50**, 359.

4. Athansou, N. A., Quinn, J., Heryet, A., Woods, C. G., and McGee, J. O. D. (1987). *J. Clin. Pathol.*, **40**, 874.

5. Dodson, A., Davies, E., and Waring, J. (1991). *Histopathology*, **19**, 484 (Letter).

6. Williams, J. (1996). *J. Cell. Pathol.*, **1**, 179 (Letter).

7. Dash, R. C., Ballo, M. S., and Layfield, L. J. (1998). *Appl. Immunohistochem.*, **6**, 145.

8. Shin, H. J. C., Kalapurakal, S. K., Lee, J. J., Ro, J. Y., Hons, W. K., and Lee, J. (1997). *Mod. Pathol.*, **10**, 224.

9. Liau, D. F. and Lai, C. L. (1997). *LabLeader*, **12**, (1), 6 (Shandon Lipshaw Newsletter).

10. Bertheau, P., Cazals-Hatem, D., Meignin, D., deRoquancourt, A., Verola, O., Lesourd, A., *et al.* (1998). *J. Clin. Pathol.*, **51**, 370.

11. Kohler, G. and Milstein, C. (1975). *Nature*, **256**, 495.

12. Mepham, B. L., Frater, W., and Mitchell, B. L. (1979). *Histochemistry*, **11**, 345.

13. Curran, R. C. and Gregory, J. (1978). *J. Clin. Pathol.*, **31**, 974.

14. Shi, S. R., Key, M. E., and Kalra, K. L. (1991). *J. Histochem. Cytochem.*, **39**, 741.

15. Cattoretti, G., Peleri, S., and Parravicini, C. (1993). *J. Pathol.*, **171**, 83.

16. Norton, A. J., Jordon, S., and Yeomans, P. (1994). *J. Pathol.*, **173**, 371.

17. Shi, S. R., Inman, A., Young, L., Cote, R. J., and Taylor, C. R. (1993). *J. Histochem. Cytochem.*, **43**, 193.

18. Taylor, C. R., Shi, S. R., and Cote, R. J. (1996). *Appl. Immunohistochem.*, **4**, 144.

19. Dodson, A. (1996). *J. Cell. Pathol.*, **1**, 180 (Letter).

20. Morgan, J. M., Navabi, H., Schmid, K. W., and Jasani, B. (1994). *J. Pathol.*, **174**, 301.

21. Hsu, S. M., Raine, L., and Fanger, H. (1981). *J. Histochem. Cytochem.*, **29**, 577.

22. Guesdon, J. I., Ternynck, T., and Avrameus, S. (1979). *J. Histochem. Cytochem.*, **27**, 1131.

23. Bussolati, G., Gugliotta, P., Volante, M., Pace, M., and Papotti, M. (1997). *Histopathology*, **31**, 400.

24. Dodson, A. and Campbell, F. (1999). *Histopathology*, **34**, 178 (Letter).

25. Bisgaard, K., Lihme, A., Rolsted, H., and Pluzek, K.-J. (1993). *Scand. Soc. Immunol.* (Abstract). XXIVth Annual Meeting. University of Aarhus, Aarhus, Denmark.

26. Vyberg, M. and Nielsen, S. (1998). *Appl. Immunohistchem.*, **6**, 3.

27. Sabattini, E., Bisgaard, K., Ascani, S., Poggi, S., Piccioli, M., Ceccarelli, C., *et al.* (1998). *J. Clin. Pathol.*, **51**, 506.

28. Cordell, J. L., Falini, B., Erber, W. N., Ghosh, A. K., Abdulaziz, Z., MacDonald, S., *et al.* (1984). *J. Histochem. Cytochem.*, **32**, 219.

29. Avrameas, S. and Uriel, J. (1966). *C. R. Acad. Sci. Paris Ser. D.*, **262**, 2543.

30. Nakane, P. K. and Pierce, G. B. (1966). *J. Histochem. Cytochem.*, **14**, 929.

31. Mason, D. Y. and Sammons, R. E. (1978). *J. Clin. Pathol.*, **31**, 454.

32. Graham, R. C. and Karnovsky, M. J. (1966). *J. Histochem. Cytochem.*, **14**, 291.

33. Pelliniemi, L. J., Dym, M., and Karnovsky, J. (1980). *J. Histochem. Cytochem.*, **28**, 191.

34. Burstone, M. S. (1961). *J. Histochem. Cytochem.*, **9**, 146.

35. Scopsi, L. and Larsson, L. I. (1986). *Histochemistry*, **84**, 221.

36. Bobrow, M. N., Harris, T. D., Shaughnessy, K. J., and Litt, G. J. (1989). *J. Immunol. Methods*, **125**, 279.

37. Berghorn, K. A., Bonnett, J. H., and Hoffman, G. E. (1994). *J. Histochem. Cytochem.*, **42**, 1635.

38. King, G., Payne, S., Walker, F., and Murray, G. I. (1997). *J. Pathol.*, **183**, 237.

Chapter 19

Detection of chemically modified DNA in lymphocytes of patients undergoing chemotherapy

Michael J. Tilby

Paediatric Oncology Laboratory, Cancer Research Unit, Medical School, University of Newcastle upon Tyne, Newcastle upon Tyne NE2 4HH, U.K.

1 Introduction

Antibodies that recognize DNA occur spontaneously in people and animals exhibiting certain types of 'autoimmune' diseases. In rodents not exhibiting autoimmune symptoms, DNA is a weak immunogen, possibly because B cells exhibiting specificity for DNA are deleted (1). However, it was discovered that alteration of the normal structure of DNA by irradiation with UV light greatly increased its immunogenicity and the antibodies, that were then produced, specifically recognized the UV-induced thymine dimers (2). In the 1950s it was also reported that the immunogenicity of DNA could be enhanced by forming an electrostatic complex with methylated bovine serum albumin (2). The mechanisms by which antibodies are made against DNA are not understood. They show the conventional features of oligoclonality, somatic mutation, gene rearrangements, and class switching, but there is no known way that DNA molecules can be processed and presented to T cells in the context of HLA class II molecules (1, 3). Since the 1960s a large number of antisera and monoclonal antibodies have been produced which recognize various DNA structures or modifications. These can be divided pragmatically into three broad groups:

(a) Antibodies raised against atypical DNA conformations (e.g. Z-DNA and cruciform structures) (4–6).

(b) The largest group; antibodies which recognize chemical modifications to the structures of DNA bases (or sugars) (7). Certain antibodies span both these groups since they appear to recognize conformational changes associated with the presence of certain chemical modifications on DNA. (e.g. 8, 9).

(c) Antibodies raised, not against DNA, but against proteins that interact with DNA. Although the epitope recognized by such antibodies may not include a nucleic acid moiety, it is relevant to include these antibodies in the present context because xenobiotic compounds can crosslink proteins to DNA. Also covalent bonds between DNA and enzymes exist in transient reaction intermediates and can be stabilized as in the case of drugs acting on DNA topo-isomerases. Such DNA-linked proteins might be considered to be large DNA adducts. A technique has recently been described which permits immuno-logical detection of proteins crosslinked to DNA in individual cells (10).

In this chapter, production of antibodies against DNA adducts is taken to refer to immunization with nucleic acids modified with low molecular weight adducts.

1.1 Types of modifications to which antibodies can be produced

As indicated above, antibodies can be raised against a wide range of chemically-induced modifications of DNA structure including a wide range of carcinogen adducts and radiation-induced modifications. The examples cited in this chapter concern antibodies raised against specific DNA adducts induced by anticancer chemotherapy agents. During high-dose chemotherapy these modifications are induced at relatively high frequencies, and it is therefore more likely that im-munological techniques will be sufficiently sensitive to quantify these adducts. DNA adducts resulting from, for example, most environmental exposures to carcinogens, will hopefully occur at much lower levels.

1.2 Applications of antibodies against modified DNA

Antibodies that recognize chemically modified DNA have been applied using most of the available immunological approaches, i.e. radioimmunoassay, ELISA, immunohistochemistry, and immunofluorescence. In addition, immunoaffinity isolation of DNA fragments carrying modified bases has been performed. This can be used as a basis for studying DNA damage formation and repair in specific sequences, and as a way of improving the detection of adducts present at very low frequencies by enriching the level of base modification in DNA samples prior to application of chemical, physical, or immunological characterization and detection steps (11, 12).

Quantification of levels of drug-induced DNA modification in patients under-going chemotherapy has a number of objectives. The type of cellular/biochemical responses that occur to DNA damage is dependent upon the level of damage. Therefore knowledge of the level of damage achieved during therapy is relevant to experimental studies of such responses. Also, it is not understood to what extent inter-individual variation in response to drug or carcinogen exposure is determined by variation in access of the chemicals to DNA or to variation in DNA repair.

1.3 Merits of raising antibodies against modified mono/dinucleotides versus modified polymeric DNA

Two approaches have been used to generate antibodies against DNA adducts:

(a) The chemically modified base, nucleoside, or nucleotide is synthesized and then chemically coupled to a carrier protein (BSA, KLH, etc.) for use as the immunogen (13, 14).

(b) Chemically modified polymeric DNA is used as the immunogen. The most widely used immunization protocol involves the use of electrostatic complexes between DNA and a protein.

The first approach generally results in antibodies against a specific DNA adduct present as a free base or a monomeric nucleoside/tide. The resulting antibodies should be useful for quantification of HPLC fractions etc., however, they may not efficiently recognize the same adducts when they are present in polymeric DNA (14, 15). This could be a severe disadvantage if it is planned to study adducts in individual cells by cytological techniques, or to isolate fragments of DNA that carry the modifications. The second approach is more likely to generate antibodies suitable for the latter applications as well as being of use in ELISA techniques to measure adduct levels in polymeric DNA extracted from cells or tissues. Immunization with polymeric DNA or a defined oligonucleotide may be also required in order to produce an antibody that recognizes a conformational change in the polymeric/oligomeric DNA.

Depending on the modification, physical separation and detection methods for modified bases/nucleosides/nucleotides in DNA hydrolysates can already or may eventually provide greater specificity and sensitivity than immunological methods. However, it seems that antibodies recognizing adducts in polymeric DNA provide the only foreseeable basis for specific quantification of certain types of DNA modifications in individual cells and for isolation of DNA fragments carrying those modifications.

Immunization with polymeric DNA has the disadvantage that the modifications present will be heterogeneous (this shortcoming might be overcome by immunization with a specific oligonucleotide carrying a defined modification). Furthermore, even when using DNA modified to a relatively high level, only a minority of the bases will be chemically modified. Therefore there is a significant risk of producing antibodies that recognize normal DNA and success probably depends upon the greater immunogenicity of chemically modified DNA compared to normal DNA together with the power of monoclonal antibody technology to eliminate unwanted immunoreactivities. To maximize the chance of obtaining antibodies against the desired adduct rather than normal DNA it is sensible to immunize with DNA modified to a very high level. However, this increases the chance of obtaining another undesirable type of antibody, namely one which preferentially recognizes closely spaced DNA modifications (16, 17). Assuming that closely spaced modifications of DNA occur by chance, their frequency will tend to be related to the square of the overall adduct level (16). Therefore they will not form a reliable basis for accurate quantification of overall adduct level and, since they will be very rare in DNA modified to low levels, they will not permit sensitive detection of low levels of modifications (17).

The methods addressed in this chapter are specifically concerned with the generation and application of antibodies that recognize adducts in polymeric DNA.

1.4 Requirements for detection of low frequencies of DNA modifications

Even in patients treated with high-dose chemotherapy, the drug-induced DNA modifications are present at very low levels (a few per million bases) and therefore, to be useful, the immunological techniques must be capable first, of detecting very small quantities of modified DNA and secondly, of showing sufficient specificity as to distinguish the modified bases from a million-fold excess of normal bases. The first requirement is usually associated with the need for a high affinity constant for the interaction of antibody with antigen. The latter requirement demands a very low level of cross-reaction with normal DNA. This can be tested initially in screening monoclonal antibodies in direct binding assays. However, the impact of weak cross-reactions can only be assessed fully in the context of the ultimate application of the antibody.

2 Production of appropriate antibodies

2.1 Immunization

The methods for preparation of the chemically modified DNA will depend upon the type of modification in question. The optimal level of modification is probably also dependent upon the modification in question, but in general will be a compromise. As discussed above, a high level of modification will favour production of antibodies against modified compared to normal bases but may also favour production of antibodies that preferentially recognize sites carrying multiple adducts. For cisplatin and melphalan, base modification frequencies of 0.008 and 0.05 respectively, proved to be satisfactory (9, 18). However, no other levels of modification were tested. A similar level of modification has been used to generate antisera against DNA modified by benzo(a)pyrene-deoxyguanosine (19).

Protocol 1

Preparation of immunogen

1. Prepare a solution of DNA carrying modified bases at a frequency of approx. 0.01–0.05 at 150 μg DNA/ml in PBS.

2. On the day of immunization, add methylated BSA so that its concentration by weight is equal to that of the DNA, i.e. add 15 μl of methylated BSA solution (1% (w/v) in water) per ml of DNA solution. This usually results in a fine precipitate of DNA–protein complex.[a]

3. Mix the resulting suspension with an equal volume of Freund's adjuvant—complete for first, incomplete for subsequent immunizations.

[a] The precipitate results from electrostatic molecular aggregation of methylated bovine serum albumin (which has an overall positive charge) with DNA (polyanionic).

Alkylation of guanine at the N7 position is the major outcome of exposure of DNA to many clinically used alkylating agents, however, these products are chemically unstable since the base–sugar bond is labile at low pH and at neutral pH at elevated temperature. This results in loss of the alkylated base from the DNA. On the other hand, at high pH, the alkylated base undergoes a ring open-ing reaction. In the cases that have been studied, the ring-opened products were sufficiently altered in structure as to be no longer recognized by antibodies that recognized the original adducts. (18, 20, 21). However, since the ring-opened adducts have the advantage of being more chemically stable, they have been used to raise antibodies (21, 22).

Volumes and routes of administration of immunogen will probably depend upon immunization protocols already in use. Protocol 1 is satisfactory for rats.

2.2 Screening hybridomas

Hybridoma supernatants should be screened using the direct binding ELISA (Section 3.2). The screen should include wells coated with DNA carrying modifi-cation at a frequency of about 0.01–0.05 modifications per base. In addition, the screen should include wells coated with native and heat denatured control DNA to eliminate unwanted antibodies that recognize the normal DNA structures present in the modified DNA or which show a high degree of cross-reaction for normal DNA. It may be advantageous to also screen on DNA modified to a lower level (about 0.001 per base). This will help to distinguish those antibodies that efficiently detect lower levels of modification from any which recognize only highly modified DNA.

3 ELISA techniques for quantification of DNA modifications

The two ELISA techniques described below have complementary applications. The direct binding assay is used for screening hybridoma supernatants and for initial characterization of antibodies. The only DNA antigen involved in this method is that which becomes absorbed to the surface of the wells. The quan-tity of this DNA is not known accurately but is probably only a few nanograms per well. Although this technique has been used to estimate levels of modifi-cations in samples of DNA it suffers several drawbacks as a method for determining clinically relevant adduct levels.

(a) The amount of DNA bound is not known.

(b) The amount of DNA bound may vary between samples dependent upon differences in purity or molecular weight of the extracted DNA.

(c) The conformation of the DNA molecules absorbed to the plastic is not known, may not be typical of the total DNA applied to the well, and may affect the ability of antibody to interact with adducts.

(d) Since the amount of DNA bound is very small, if adducts are present at the low levels typical of samples exposed to clinical levels of drugs, the quantity of adducts bound per well may be too low to detect. This problem is overcome in the immuno-slot-blot method (23, 24) which is equivalent in principle to the direct binding ELISA but allows the analysis of a larger quantity of DNA by immobilizing it on a blotting membrane.

The second ELISA technique described here is the competitive assay. In this assay, the wells are coated with a uniform quantity of modified DNA and a dilute solution of the antibody is mixed with various quantities of the DNA samples that are to be analysed. When these mixtures are added to the coated assay wells, there is a competition for a limited amount of antibody between two classes of antigen. The more adducts on the DNA in solution, the greater is the tendency for the antibody to bind to this DNA at the expense of its binding to the DNA absorbed to the plastic surface. The assay is calibrated by use of standard concentrations of dissolved competing DNA carrying known levels of adducts. In this method:

(a) The amount of sample DNA in each assay well is accurately known.

(b) Since there is no selective absorption of the sample DNA to the plastic, the conformation of the DNA involved in the assay is well controlled.

(c) Large amounts of DNA can be included in each well so that, if the frequency of modified DNA bases is low, the total quantity of modified bases added to each well can be made high to facilitate detection. However, the amount of DNA that can be added per well is limited by the cross-reaction of antibody with normal DNA (see below).

3.1 Coating wells with DNA

3.1.1 Choice of methods

There are several methods by which DNA can be immobilized on the surface of plastic wells. Covalent crosslinking of DNA to wells requires plates with a specially modified plastic surface. This method is too expensive and time-consuming for routine assays. Streptavidin coated wells can be used in conjunction with biotinylated DNA. Again, these plates are expensive and also, the DNA must be specially modified. Furthermore, the use of biotin/streptavidin-based detection reagents would be precluded. Therefore, as with most ELISA techniques, DNA is simply adsorbed to the plastic surface. Previously described methods have involved drying the DNA solution onto the plastic or UV crosslinking the DNA to protein that had been coated onto the plastic (18). In our experience, the best current method involves adsorption of DNA from a buffered solution in 1 M NaCl. It seems that it is important to incubate at 37 °C overnight in order to ensure uniformity of well coating. This is true even for DNA alkylated by drugs such as melphalan in which the antibody recognizes the thermolabile N7 alkylguanine. We have found that overnight incubation at 4 °C resulted in a high degree of inter-well variation in binding level, but this type of detail may be dependent upon the make of plates and the nature of the DNA preparation.

3.1.2 Level of DNA modification

For use in screening of hybridomas, the level of modification of the DNA used to coat the plates will depend upon the specificity of the antibodies sought, the heterogeneity of modifications present on the DNA, and the relative abundance of the modification in question. For screening hybridomas to obtain antibodies recognizing melphalan and cisplatin adducts, base modification frequencies of 0.01 and 0.06 respectively proved to be satisfactory (9, 18).

Protocol 2

Coating ELISA plates with DNA

Equipment and reagents

- ELISA grade 96-well microplates, high binding type[a] (e.g. from Greiner)
- Coating buffer: 1 M NaCl, 50 mM sodium phosphate, 0.02% (w/v) NaN_3 pH 7.0
- Blocking buffer: PBS containing 1% (w/v) BSA
- Wash buffer: PBS containing 0.1% (v/v) Tween 20

- DNA modified with drug to a relatively high level: this should be pure DNA (e.g. highly purified calf thymus DNA from Merck) with a base modification frequency of approx. 0.01–0.06, stored at −80 °C

Method

1. Dilute a suitable quantity of stock solution of drug modified DNA into coating buffer to give a concentration of about 50 ng DNA/ml.[b]

2. Place into each internal well[c] of the ELISA plates, 50 μl of the solution of highly modified DNA in coating buffer.

3. Tap the plate gently from different directions to ensure that the solution completely covers the bottom of each well.

4. Place the plates in an airtight box to prevent desiccation and incubate overnight at 37 °C.

5. Next day, remove most of the liquid.[d]

6. Add to each coated well, 150 μl of blocking buffer.

7. Leave the plates, with lids on, on the bench for as long as possible (at least 60 min), to allow the BSA to block sites of non-specific binding on the wells.

8. Before use, remove all unbound DNA from the wells by washing twice with wash buffer.

[a] We routinely use ordinary clear ELISA plates for fluorescence assays. These give a significant level of background fluorescence, but this is generally of less a problem than the background resulting from non-specific binding of immunological reagents to the wells. Black or white 96-well plates specifically designed for fluorescence assay are available, however, these are more expensive and the contents of the wells are less visible.

[b] The concentration may have to be optimized for the particular DNA preparation being used.

[c] The outermost wells of ELISA plates are commonly found to give less reproducible results than the inner wells.

[d] Invert and shake the plates over a sink and then tap the inverted plates on paper towels.

3.2 Direct binding assays

3.1.1 Choice of detection system

The following protocol uses the relatively uncommon ELISA detection system of the enzyme β-galactosidase plus the substrate, 4-methylumbelliferyl β-galactoside, which is enzymatically hydrolysed to the highly fluorescent product 4-methylumelliferone. This system has several favourable features:

(a) Detection of 4-methylumelliferone by fluorescence is one or two orders of magnitude more sensitive than detection of typical coloured reaction products by optical density measurements.

(b) Reliable quantification of a substance by fluorescence is linear over a wider range of concentration than is normally possible for quantification by optical density.

(c) β-Galactosidase is a very robust enzyme with a high turnover number.

(d) 4-Methylumbelliferyl β-galactoside has a very low level of fluorescence and exhibits a very low rate of spontaneous hydrolysis at neutral pH.

To fully exploit the high sensitivity of this detection system, low non-specific binding of conjugates to the assay wells is essential and, in the past, only certain commercial conjugates between β-galactosidase and species-specific secondary antibodies were found to be suitable. In recent years the range of commercially available direct conjugates for this enzyme has declined. However, the biotin–streptavidin system permits combination of a secondary antibody exhibiting low non-specific binding, with the desired enzyme and may also give an enhancement in sensitivity.

Protocol 3

Direct binding ELISA

Equipment and reagents

- Plates that have been coated, blocked, and washed as described in *Protocol 2*

- Fluorescence plate reader: wavelengths, excitation = *c*. 355 nm, emission = *c*. 460 nm

- Monoclonal antibody: undiluted culture supernatant when screening hybridomas, or culture supernatant diluted in PBS containing 1% (w/v) BSA and 0.1% (v/v) Tween 20

- Wash buffer (see *Protocol 2*)

- Biotinylated anti-primary antibody second antibody (e.g. from Sigma), typically diluted × 1000 in PBS containing 1% (w/v) BSA and 0.1% (v/v) Tween 20

- Streptavidin–β-galactosidase conjugate (Boehringer Mannheim) typically diluted × 10000 in PBS containing 1% (w/v) BSA and 0.1% (v/v) Tween 20.

- Substrate solution: 80 μg 4-methylumbelliferyl β-galactoside per ml

Method

1. To the prepared plates add neat or appropriately diluted hybridoma culture supernatant (50 μl per well).

Protocol 3 continued

2. Incubate at 37 °C for 1 h.

3. Wash wells with wash buffer five times (see *Protocol 2*).

4. Add biotinylated second antibody (50 μl per well).

5. Incubate at 37 °C for 30 min.[b]

6. Wash wells with wash buffer three times.

7. Add streptavidin–β-galactosidase conjugate (50 μl per well).

8. Incubate at 37 °C for 30 min.[b]

9. Wash wells with wash buffer five times.

10. Add 50 μl of substrate solution to each well.

11. Incubate before reading plates in a fluorescent plate reader. The time and temperature of incubation depends on the application. For high levels of antibodies, 5 min at room temperature can be sufficient. For maximum sensitivity, such as in the competitive ELISA, incubation at 37 °C for between 1–18 h can be used.[b]

[a] Mix solid 4-methylumbelliferyl β-galactoside to a paste with a drop of PBS, add rest of PBS, warm until dissolved, pass through filter paper to remove any undissolved particles of substrate. Add $MgCl_2$ as 1 M stock solution.

[b] Do not stack plates in the incubator unless the incubation time is to be several hours because the rate of warming of the wells of single plates is more uniform. For overnight incubation, the plates should be stacked in an airtight box to prevent desiccation.

3.3 Competitive assays

3.3.1 Isolation of DNA

There are certain requirements regarding the quality of the final DNA preparation for it to be useful in immunoassay of adducts.

(a) It must be free of ethanol as relatively low concentrations of this can affect the antibody–DNA interaction, although the magnitude of this interference will probably be antibody-specific.

(b) The final DNA solution should be in a maximum volume of about 400 μl and in a buffer compatible with the immunoassay.

(c) The DNA must be completely dissolved to ensure proper recognition of adducts by the antibody.

(d) Unless the DNA is to be digested with DNAse (see below), it should consist of relatively short fragments (< approx. 30 kb) because high concentrations of DNA may be used in the assay. The high viscosity of solutions of high molecular weight DNA would prevent accurate pipetting of small volumes and could interfere with mixing and diffusion of antibody molecules.

(e) If labile adducts are being studied, the extraction procedure should not expose the DNA to high temperatures or high pH.

In order to reduce the molecular weight we routinely sonicate the mixture of cells and lysis solution. This also promotes efficient extraction of DNA. Many procedures employ ethanol to concentrate and desalt the DNA. In order to fully remove the ethanol this DNA needs to be thoroughly dried. However, the resulting pellet may then be difficult to redissolve without extended incubation in buffer during which time labile adducts may be lost. Furthermore, in the case of DNA exposed to bifunctional crosslinking agents, there is the possibility of delayed reactions which can result in the formation of crosslinks between DNA fragments and this will impede the redissolution.

To avoid the above problems we routinely use an extraction method in which DNA fragments become bound to hyroxyapatite powder (9). This is handled in the form of a centrifuged chromatography column for the purposes of washing with various buffers and final elution of pure DNA in 0.5 M potassium phosphate. This DNA is concentrated and desalted in a centrifugal ultrafiltration device (Centricon-10 from Amicon). Many commercial kits are available for the isolation of pure DNA from biological samples therefore a detailed description of the above method seems inappropriate, however, choice of the method should take into account the above considerations.

The concentration of the final DNA solution is determined from its OD at 260 nm (for native DNA, concentration = $OD_{260} \times 50$ μg/ml). As a rough check on the purity of the DNA, the ratio of $OD_{260} \div OD_{280}$ should be determined. For pure DNA this ratio is 1.8 and we routinely accept values between 1.75–1.9. Since the DNA concentration often needs to be > 100 μg/ml, and since the photometric accuracy of most spectrophotometers is unreliable above about OD values of 1.5, the solution should be analysed in a 1 mm path-length cuvette. This is more convenient than and avoids errors associated with the alternative approach in which samples of the DNA solution are diluted prior to the photometric analysis. (The samples can be conveniently added to 1 mm path-length cuvettes by using disposable insulin-type syringes and needles. To recover the samples, the same syringe is used whilst holding the cuvette in a partially inverted position.)

3.3.2 Maximization of DNA immunoreactivity

Recognition of adducts in polymeric DNA generally seems to be highly dependent upon whether or not the DNA has been denatured. For antibodies raised against melphalan (18) and cisplatin (9), immunoreactivity and hence assay sensitivity was 10- to 100-fold higher for adducts on denatured DNA despite the fact that the antibodies resulted from immunization with drug modified native DNA. In both these examples, in order to attain assay sensitivity sufficient to detect adducts in samples from patients, it was essential to denature the extracted DNA. In general, DNA can be denatured by heating or by exposure to high pH. Reproducible change of pH is difficult to achieve in small samples because any added alkali must be accurately neutralized before proceeding with the immunoassay. Furthermore, the neutralized solution will then contain a high concentration of salt. In addition, certain types of adducts are labile under alkaline conditions, namely the N7 alkyl adducts of guanine, which undergo a ring-opening reaction (25). However, when heated at neutral pH these adducts undergo a depurination reaction (26). It has been found that, the immunoreactivity of melphalan adducts on native DNA can be increased by hydrolysis of the DNA with DNase I. The suitability of this approach will be dependent upon the specificity of the particular antibody but an ability to recognize adducts in short oligonucleotides rather than in mononucleotides (27) is probably satisfied by the fact that DNase I hydrolyses DNA to fragments three to four nucleotides long and not to monomers. Immunoreactivity of heat stable adducts, such as those formed by cisplatin, is most readily maximized by heat denaturation of DNA.

Protocol 4

Maximization of immunoreactivity of heat stable adducts by thermal denaturation of DNA

Reagent
- Buffer A: 50 mM sodium phosphate, 50 mM NaCl pH 7.0

Method
1. The DNA to be analysed is diluted to a suitable starting concentration[a] in a total volume of 260 μl of buffer A in a screw-capped 1.5 ml microcentrifuge tube.

2. The tube is placed in a boiling water-bath for 5 min and then transferred to ice until required (for 5–60 min).

3. The tube is mixed and briefly centrifuged before removal of its contents.

[a] Since the optimum range of DNA concentrations depends on the characteristics of the assay and the frequency of modified bases that it carries, the starting concentration must be determined empirically.

Protocol 5

Maximization of immunoreactivity of heat labile adducts by digestion of DNA

Reagents
- Buffer A (see *Protocol 4*)
- DNase I (e.g. from Sigma): dissolved at 10 000 Kunitz units/ml in buffer A supplemented with 10 mM $MgCl_2$

Method
1. The DNA dissolved in buffer A at a known concentration is supplemented with $MgCl_2$ by addition of 1 M solution to give a final concentration of 10 mM.

2. DNase I solution is added to the DNA to give 5 U of enzyme/μg DNA.

3. The solution is incubated at 37 °C for 1 h.

4. The hydrolysate is stored at −80 °C until required.

5. DNase I could, in principle, remove DNA coating the wells of assay plates. Therefore it must be inactivated by addition of EDTA (10 mM) in subsequent stages of the competitive ELISA.

3.3.3 Standards

The basic data given by the competitive ELISA can be analysed to determine the concentration of each sample of DNA that is necessary to cause 50% reduction in assay signal. In order to calculate from this the absolute level of adducts in the samples, it is necessary to include in the assay a sample of DNA carrying a known level of modified bases. This should be treated to maximize immunoreactivity in the same manner as DNA samples being analysed. Preparation of this standard and the method of determining its level of base modification will depend on the modification being studied.

3.3.4 Concentration of monoclonal antibody

In principle, the lower the concentration of antibody in the wells of the competitive assay, the smaller the quantity of competing antigen that should be required to inhibit binding of the antibody to the immobilized DNA and hence, the more sensitive the assay. The lowest antibody concentration is determined by the need for the assay signal in the absence of competing antigen to be reliably measurable above the background signal. To determine this concentration it is necessary to perform initial direct binding ELISA experiments in which the hybridoma supernatant is serially diluted × 2 and placed in wells coated as for the competitive assay. The highest dilution that gives a signal at least tenfold higher than wells from which the monoclonal had been omitted should be used for the initial competitive assays. This may need to be adjusted in the light of initial competitive ELISA experiments.

3.3.5 Performing the competitive assay

The following plan for the layout of assay plates and handling of samples has been found to be satisfactory but obviously many alternative arrangements are possible. The assay plate layout shown in *Figure 1* includes numerous wells in which no competing DNA is present as these constitute the reference point for all other wells. It is also necessary to include wells from which monoclonal antibody is omitted. These background wells indicate the lowest signal that can be obtained even when all the monoclonal antibody has become bound to competing antigen. A fixed level of background will be due to the fluorescence of the plastic of the assay well and the remainder will result from non-specific binding of immunological reagents to the well. This part of the background signal will increase with increased periods of incubation of wells with the substrate.

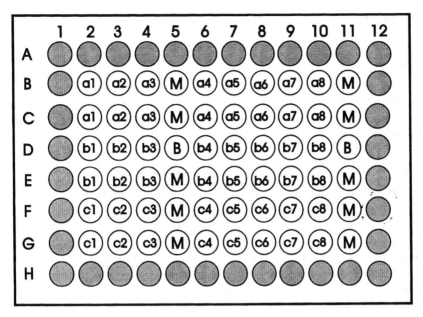

Figure 1 Layout of a 96-well plate suitable for competitive ELISA. M, maximum signal (no competing antigen); B, background (no monoclonal antibody); a1 to a8, b1 to b8, c1 to c8, antibody mixed with serial dilutions of samples a, b, c. Outer shaded wells are not used.

After the step to maximize immunoreactivity of adducts, the experimental samples and the sample of standard DNA are serially diluted in flat-bottomed 96-well plates. These plates are set out to match the layout to be attained in the assay plate.

Protocol 6

Competitive ELISA

Equipment and reagents

- 96-well plates, flat and V-bottomed wells: these only need to be cheap quality microtitration plates rather than the more expensive ELISA grade plates

- Multiway pipetters, ideally including an electronic multidispensing multiway pipetter

- Dilute solution of monoclonal antibody in buffer B[c]

- Buffer B: PBS containing 1% BSA, 0.1% Tween 20, 20 μg/ml phenol red,[a] NaCl concentration increased to 190 mM[b] (NB: if the DNA samples have been digested with DNase I, then 10 mM sodium EDTA should also be included to inactivate the enzyme)

- Wash buffer[d]

- Buffer A (see *Protocol 4*)

Method

1. For each sample designate a row of a flat-bottomed 96-well plate.

2. Place 120 μl of buffer A in wells 3–11.

Protocol 6 continued

3. Place at least 250 μl of starting DNA solution in well 2.

4. Using a multiway pipetter serially dilute several samples simultaneously by transferring 120 μl from wells 2 to wells 3, mixing by pipetting up and down several times. Repeat this transfer from wells 3 to 4 but then transfer from wells 4 to 6, thereby leaving wells 5 containing just buffer. Continue serially diluting from 6 to 7, 7 to 8, 8 to 9, and 9 to 10. Leave wells 11 with just buffer.

5. Transfer 55 μl from each well in the first row of diluted samples to each of rows B and C of a V-bottomed 96-well plate. Then transfer from the second row to rows D and E and so on.

6. To wells D5 and D11 of the V-bottomed plate, add 55 μl of buffer B. To the remaining wells add 55 μl of the monoclonal antibody that has been diluted in buffer B as described above.[e]

7. Briefly mix the contents of the wells on a plate shaker.

8. Incubate plate at 37 °C for 30 min.[f]

9. Using a multichannel pipetter, from each well of the V-bottomed plate, transfer 50 μl to the equivalent wells in each of two flat-bottomed ELISA plates that have been coated with DNA.

10. Incubate the ELISA plates at 37 °C for 1 h.[f]

11. Wash the plates[d] five times.

12. The final steps of the assay are to quantify the monoclonal antibody bound to the wells and are exactly the same as for the direct binding ELISA, i.e. *Protocol 3*, steps 4–11.

[a] Phenol red is added to improve visibility of the solutions—particularly useful when attempting to distinguish wells to which antibody has been added from those only containing solutions of sample.

[b] To give a final concentration of 140 mM when mixed with an equal volume of DNA solution in buffer A.

[c] The dilution must be determined for each combination of individual antibody and plate coating by serial dilution of the antibody or hybridoma supernatant in a direct binding ELISA. The optimal dilution is the highest that gives a satisfactory signal and this is generally at least ten times higher than is obtained for wells from which monoclonal antibody has been omitted.

[d] See *Protocol 2*.

[e] Since buffer B contains protein and detergent it wets the surface of plastic pipette tips and consequently it is not possible to dispense accurate volumes by the normal pipetting with disposable tips. The most convenient way to accurately dispense this solution is to use an electronic multichannel pipetter operating in multidispense mode. Alternatively, a manual multichannel pipetter must be used in 'reverse pipetting' mode, i.e. (i) Depress plunger of pipetter to the second stop. (ii) Pick up solution from reagent reservoir (i.e. > 55 μl). (iii) Dispense by depressing plunger to first stop (dispensing 55 μl and leaving excess solution in the tips). (iv) Keeping plunger depressed to first stop immerse tips in reservoir, then pick up more solution.

[f] See *Protocol 3*, footnote c.

3.3.6 Analysis of data

The following protocol defines the steps necessary for analysis of the plate readings. The actual method of implementing these steps will depend on the software available with the plate reader. If this software cannot perform the necessary analysis, then the values can be entered into a spread sheet program such as *Excel* for initial manipulation and then pasted into a curve-fitting/graph plotting program such as *Prism* (Graph-Pad Software Inc.).

Protocol 7

Analysis of data

All the data from each pair of replicate assay plates is analysed together as follows.

1. Average the four background wells (D5, D11).
2. Subtract this average from the readings for each of the other wells.
3. Calculate the average maximum reading (i.e. the wells marked M in *Figure 1*; total of 20 wells on two plates).
4. For the standard DNA solution calculate the concentration of adducts (moles/well) at each of the serial dilutions. For the other samples calculate the DNA concentration (g/well) at each dilution.
5. Express the readings for each serial dilution of each sample as a percentage of the maximum wells.

$$Fc = \frac{(Sc - BG)}{(MAX - BG)} \times 100\%$$

where Fc = percent value for well containing sample at concentration c;
BG = mean reading for wells without primary antibody;
MAX = mean reading for wells with no competing antigen;
Sc = mean reading for sample wells containing a given concentration (c) of competing antigen.

6. Plot a graph of the values calculated in step 5 (linear ordinate) against the concentration of competing antigen from step 4 (logarithmic abscissa). The data points lie on a sigmoid curve.
7. Fit an equation to the data to calculate the concentration of competing antigen that causes 50% reduction in signal. The simplest fitting procedure is to use the log-logit equation but this is not satisfactory for full sigmoid curves and it is most appropriate to use a logistic equation because these naturally generate full sigmoid lines. The logistic equation of the following form is used:

$$Fc = \frac{M \times C^P}{C^P + K^P}$$

where F_c = assay signal at a given value of C;
C = the concentration of competing antigen;
M = fitted maximum fluorescence value;
K = fitted competitor concentration for which $F_c = M \div 2$;
P = fitted slope value.

Protocol 7 continued

8. The level of adducts in the sample DNA is calculated from the determined concentration of the sample DNA and the determined concentration of adducts in the standard that each cause 50% reduction of assay signal.[a]

$$\text{adduct level in sample DNA} = \frac{\text{STD}}{\text{SA}} \text{ moles adduct per g DNA}$$

where SA = g of sample DNA per assay well causing 50% reduction in assay signal;
STD = moles of standard adduct per assay well causing 50% reduction in assay signal.

[a] Having plotted such a curve for a standard antigen it is possible to read off single values to determine the antigen concentration in specific concentrations of sample. However, we find it preferable to generate, for each sample, data from a dilution series that can be fitted to a curve to determine the dilution required to give 50% inhibition. Not only should this be more accurate, but also the shape of the curve can show up spurious results and reveal information about the quality of the antigen.

4 Staining for adducts in individual cells

4.1 Staining techniques: conventional versus agarose embedded DNA

It is generally important that staining intensity quantitatively reflects the level of adducts present on the DNA, however, immunological quantification of drug–DNA adducts can, in principle, be adversely affected by the following two factors. First, access of antibody to DNA may be hindered to various degrees by cellular proteins, particularly after fixation. Secondly, the antibody–adduct interaction is often dependent upon DNA conformation (e.g. extent of denaturation), but the conformation of DNA in a fixed preparation is difficult to control and/or understand. The use of DNA trapped in agarose in the so-called TARDIS technique for staining (Trapped in Agarose DNA Immunostaining) (28) was developed to minimize the effects of these factors with regard to the staining of melphalan–DNA adducts, but also led directly to the possibility of detecting the presence of specific protein molecules, such as topoisomerases, covalently linked to DNA in individual cells (10). In this technique, living cells are embedded in a thin layer of low melting point agarose spread over a microscope slide. The cells are lysed and their DNA remains trapped (largely in the spaces originally occupied by the intact cells) by virtue of its very high molecular weight. It can then be easily denatured by immersion of the slides in alkaline solutions, and stained by application of antibody solutions to the surface of the agarose.

4.2 Staining agarose embedded DNA

4.2.1 Embedding and lysing cells in agarose on microscope slides

Protocol 8

Embedding cells in agarose on microscope slides

Equipment and reagents

- Clean glass microscope slides[a]
- Microscope slide staining tanks and slide carriers
- Low melting point agarose:[b] 0.5% (w/v) in water

- Low melting point agarose:[b] 2% (w/v) in PBS
- A metal surface at 0 °C[c]
- Lysis solution: 1% (w/v) Sarkosyl, 10 mM EDTA, 10 mM phosphate buffer pH 6.8[d]

Method

1. Coat the microscope slides with a thin layer of agarose. Melt an aliquot of 0.5% agarose in a boiling water-bath and spread a drop of the solution along each slide using the end of a second slide inclined at an angle of about 45°.

2. Leave slides to dry in a 37 °C non-humidified incubator.

3. Melt an aliquot of 2% agarose gel in a boiling water-bath and equilibrate to 37 °C.

4. Place 100–200 μl of cell suspension in a 1.5 ml tube and equilibrate briefly to 37 °C in a water-bath. Then, using a cut-off pipette tip, add an equal volume of the 2% agarose solution at 37 °C, mix briefly by pipetting. Place aliquots of about 100 μl at one end of the coated area of pre-coated slides and immediately spread by drawing the drop along the coated area *behind* the end of a second inclined slide.

5. Immediately place the slide on a refrigerated surface.[c]

6. After a few minutes to allow the agarose to set, gently immerse[e] the slide in the lysis solution in a slide staining tank and leave for at least 5 min at room temperature.

[a] It is essential that the slides permit the solution of agarose to cover uniformly the majority of the surface over which it has been spread. We have found that only certain makes of slide permit the uniform and stable spreading of the agarose solution. On unsuitable slides the thinly spread agarose solution retracts and forms rounded patches. Slides that perform satisfactorily are Premium Microscope Slides (Western Laboratory Service Ltd.). Even with these, the occasional slide has to be rejected as the agarose layer does not remain spread properly before it dries.

[b] Sea-Plaque agarose from FMC Bioproducts. Add the required weight of agarose to water or PBS. Weigh the container and its contents, heat and boil for a few minutes, cool, reweigh, and add water to regain the original weight. Divide the solutions into small bottles and autoclave before storage.

[c] This is to set the agarose and minimize loss of water by evaporation. It is convenient to place a flat metal object such as a retort stand base in a tray of ice, ensuring that the surface does not become immersed in water.

[d] The composition is probably not critical. SDS can be used instead of Sarkosyl. The EDTA is to improve cell disruption and to minimize DNase activities.

[e] It is important to handle the slides gently as the agarose can easily become detached from the slide.

4.2.2 Staining embedded DNA

This is described in *Protocol 9*.

4.2.3 Analysis of immunofluorescence

Methods for quantitative analysis of fluorescence images are beyond the scope of this chapter. Methods for analysis of images obtained by conventional fluorescence microscopy and a cooled slow scan CCD camera are described in detail elsewhere (28, 29). Analysis by laser scan cytometry will be dependent upon the software provided with the instrument.

Protocol 9

Staining embedded DNA

Equipment and reagents

- Fluorescence microscope which, ideally, should be equipped with a sensitive and accurate imaging device[a]
- PBSTB: PBS containing 0.1% (v/v) Tween 20 and 1% (w/v) BSA
- Solution of primary antibody in PBSTB (for melphalan adducts we routinely use hybridoma supernatant diluted × 100)

- Solution of appropriate second antibody conjugated to a suitable fluorochrome diluted according to the manufacturer's recommendations in PBSTB
- 10 μM Hoechst stain (Hoechst No. 33258) in PBS: this is conveniently prepared from a stock solution of the dye (1 mM in water, stored at −80 °C)

Method

1. Initial treatment of trapped DNA will depend on the particular application. For example, staining for melphalan–DNA adducts requires denaturation of DNA by immersion of the slides in 0.1 M NaOH for 5 min, followed by neutralization in PBS (28). Alternatively, for detection of crosslinked proteins, it may be necessary to maximize the removal of non-covalently attached proteins by exposure to a high salt concentration (10).

2. Remove lysis buffer, NaOH, etc. by thorough washing. Leave the slides in a staining tank of wash buffer[b] for at least 10 min.[c] Repeat this step at least once more.

3. For application of immunological reagents drain the slides, and place them on a horizontal staining rack, in a humidified box.

4. Add primary antibody solution—about 250 μl per slide.

5. After 1 h, wash the slides in three changes of wash buffer, as in step 2.

6. Stain the slides with second antibody conjugated to a suitable fluorochrome, as in step 3.

7. Wash the slides, as in step 5.

8. Finally, wash the slides in PBS without detergent.[d]

9. Drain the slides, add a solution of Hoechst stain,[e] and apply a large coverslip.

10. View the slides and analyse using a fluorescence microscope which, ideally, should be equipped with a sensitive and accurate imaging device.[a]

[a] A cooled slow scan CCD camera can provide excellent sensitivity and the accuracy of 16-bit digitization. However, the ideal method for analysing the slides is probably a The LSC® Laser Scanning Cytometer (CompuCyte Corporation, 12 Emily Street, Cambridge, MA 02139). This not only scans slides automatically, but the photometric accuracy is probably superior to what can be attained using conventional microscopy where there are significant problems of non-uniformity of illumination and light transmission which are difficult to correct for.

[b] See Protocol 2.

[c] It is particularly important to soak for extended periods because the significant thickness of the agarose means that removal of reagents is diffusion limited.

[d] The detergent tends to increase background fluorescence of the Hoechst dye.

[e] Other DNA fluorescent dyes are also suitable. Hoechst has the advantage of using a different excitation wavelength to fluorescein. This means that the DNA fluorescence can be viewed and used to select fields without photobleaching the fluorescein. Also, with suitable optical filters, there is no significant cross-over of signal between the DNA fluorescence and the fluorescein fluorescence. This is important when analysing low levels of immunofluorescence where instruments using, for example, propidium iodide and fluorescein need to correct for the cross-over signal resulting from incomplete exclusion of the red DNA fluorescence from the green immunofluorescence.

References

1. Rekvig, O. P., Andreassen, K., and Moens, U. (1998). *Scand. J. Rheumatol.*, **27**, 1.
2. Van Vunakis, H. (1980). *Photochem. Photobiol. Rev.*, **5**, 293.
3. Mond, J. J., Lees, A., and Snapper, C. M. (1995). *Annu. Rev. Immunol.*, **13**, 655.
4. Bell, D., Sabloff, M., Zannis-Hadjopoulos, M., and Price, G. (1991). *Biochim. Biophys. Acta*, **1089**, 299.
5. Brown, B. A., 2nd, Li, Y., Brown, J. C., Hardin, C. C., Roberts, J. F., Pelsue, S. C., *et al.* (1998). *Biochemistry*, **37**, 16325.
6. Sanford, D. G. and Stollar, B. D. (1990). *J. Biol. Chem.*, **265**, 18608.
7. Strickland, P. T. and Boyle, J. M. (1984). *Prog. Nucleic Acids Res. Mol. Biol.*, **31**, 1.
8. Hasan, R., Ali, A., and Ali, R. (1991). *Biochim. Biophys. Acta*, **1073**, 509.
9. Tilby, M. J., Johnson, C., Knox, R., Cordell, J., Roberts, J. J., and Dean, C. J. (1991). *Cancer Res.*, **51**, 123.
10. Willmore, E., Frank, A. J., Padget, K., Tilby, M. J., and Austin, C. A. (1998). *Mol. Pharmacol.*, **53**, 78.
11. Kang, H., Konishi, C., Eberle, G., Rajewsky, M. F., Kuroki, T., and Huh, N. (1992). *Cancer Res.*, **52**, 5307.
12. Hochleitner, K., Thomale, J., Yu. Nikitin, A., and Rajewsky, M. F. (1991). *Nucleic Acids Res.*, **19**, 4467.
13. Erlanger, B. F. and Beiser, S. M. (1964). *Proc. Natl. Acad. Sci. USA*, **52**, 68.
14. Müller, R. and Rajewsky, M. (1980). *Cancer Res.*, **40**, 887.
15. Poirier, M. C., Yuspa, S. H., Weinstein, I. B., and Blobstein, S. (1977). *Nature*, **270**, 186.
16. West, G. J., West, I., and Ward, J. F. (1982). *Radiat. Res.*, **90**, 595.
17. Fichtinger-Schepman, A. J., Baan, R. A., and Berends, F. (1989). *Carcinogenesis*, **10**, 2367.

18. Tilby, M. J., Styles, J. M., and Dean, C. J. (1987). *Cancer Res.*, **47**, 1542.
19. Poirier, M. C., Santella, R., Weinstein, I. B., Grunberger, D., and Yuspa, S. H. (1980). *Cancer Res.*, **40**, 412.
20. Meredith, R. D. and Erlanger, B. F. (1979). *Nucleic Acids Res.*, **6**, 2179.
21. van Delft, J. H. M., van Weert, E. J. M., Schellekens, M. M., Claassen, E., and Baan, R. A. (1991). *Carcinogenesis*, **12**, 1041.
22. Tilby, M. J., McCartney, H., Cordell, J., Frank, A. J., and Dean, C. J. (1995). *Carcinogenesis*, **16**, 1895.
23. Nehls, P., Adamkiewicz, J., and Rajewsky, M. (1984). *J. Cancer Res. Clin. Oncol.*, **108**, 23.
24. Wani, A. A., D'Ambrosio, S. M., and Alvi, N. K. (1987). *Photochem. Photobiol.*, **46**, 477.
25. Lawley, P. D. and Brookes, P. (1963). *Biochem. J.*, **89**, 127.
26. Lawley, P. D. (1957). *Proc. Chem. Soc. Lond.*, 290.
27. Tilby, M. J., Lawley, P. D., and Farmer, P. B. (1990). *Chem. Biol. Interact.*, **73**, 183.
28. Frank, A. J., Proctor, S. J., and Tilby, M. J. (1996). *Blood*, **88**, 977.
29. Frank, A. J. (1999). In *Cytotoxic drug resistance mechanisms* (ed. R. Brown and U. Böger-Brown). Humana Press Inc., Totowa, NJ.

Chapter 20

Monoclonal antibody therapy in organ transplantation

Matt Wise and Diana Zelenika

Sir William Dunn School of Pathology, University of Oxford, South Parks Road, Oxford OX1 3RE, U.K.

1 Introduction

In the two decades since they were first described by Kohler and Milstein monoclonal antibodies have proved to be invaluable tools for research, not only in immunology, but also in many aspects of biomedical research. Monoclonal antibodies revolutionized lymphocyte biology by permitting the precise characterization of cell surface molecules which until that time were ill-defined by polyclonal sera. To date numerous mAbs have been used in animal models as either immunosuppressants or agents to induce tolerance. It may therefore come as some surprise to the reader that since their inception only two mAbs have been licensed for this purpose in humans.

Organ transplantation can no longer be viewed as an experimental procedure but as an acceptable form of therapy for renal, heart, lung, and liver end-organ failure. Triple therapy with azathioprine, cyclosporin A, and steroids has set a standard for immunosuppression, dramatically reducing the incidence of rejection episodes observed in the early days of transplantation. In recent years new immunosuppressive drugs, such as FK506, rapamycin, and mycophenolate mofetil, have been introduced though to what extent these offer additional benefit has yet to be fully evaluated. Whilst newer immunosuppressive drugs become more selective in the cells they target it is still the case that the majority of lymphocytes are penalized because of a minority of errant cells which participate in graft rejection. Patients tread a precarious path between too little immunosuppression facilitating episodes of rejection and too much subjecting them to increased risk of infection and malignancy. The ideal immunosuppressant would therefore be one which could be given for a short period and selectively impair lymphocytes which are aggressive towards the graft for the duration of the recipients life. This process of immunological tolerance occurs naturally towards self-antigens during development of the immune system and it is the goal of transplant immunologists to try and recreate this phenomenon in the adult immune system following transplantation.

2 The challenge of antibody therapy in humans

2.1 Immunological tolerance versus immunosuppression

Later we shall review the extensive literature that exists for monoclonal antibody therapy in animal models both as immunosuppressants and agents to induce antigen-specific immunological tolerance and yet, as already stated, only mAbs to CD3 and CD25 (IL-2 receptor) are licensed for transplantation in humans. These two targets for therapy may not even be the best choices based on experimental data, so why have they had such relative success?

The answer to some extent is that transplantation has been the victim of its own success. If one were to design the ideal drug it would be given as a short course at the time of transplantation and induce tolerance to the graft for an indefinite period. Such a treatment would avoid the current practice of the long-term administration of immunosuppression which has global effects on the immune system exposing patients to increased morbidity and mortality from infection and malignancy. Drugs such as steroids, azathioprine, and cyclosporin have other undesirable adverse effects including osteoporosis, hyperglycaemia, hyperlipidaemia, hepatotoxicity, anaemia, and nephrotoxicity which further limit their usefulness. However, despite these aforementioned problems conventional triple therapy does achieve acceptable levels of graft and patient survival such that one can expect one year graft and patient survival rates of 92% and 97% for renal allografts and 82% and 83% for heart transplants (1). It is against this background of achievement that any new therapy must be judged. The difficulty of introducing any new drug treatment is that it is ethically unacceptable to substitute triple therapy with something which may have only been used previously in animal studies. Typically this means any new treatment is used as an adjunct to conventional immunosuppression. Since survival rates are already quite good it may take large numbers of patients and a considerable period of time before any additional benefit is observed, inevitably this escalates the costs of any clinical trial.

However far more importantly, conventional immunosuppressants may have adverse or synergistic interactions with the therapeutic antibody. An elegant example of this is the use of mAbs to CD3 in NOD mice which spontaneously develop diabetes. Following the onset of glycosuria a five day course of anti-murine CD3 mAb reverses the disease process and re-establishes tolerance to pancreatic autoantigens, but this was prevented by co-administration of cyclosporin A (2). Retrospective analysis by Opelz (3) for the collaborative transplant study also showed that the beneficial effects of using the murine anti-human CD3 mAb OKT3 were abolished unless cyclosporin A had been delayed for several days. Diamentstein *et al.* (4) using a rat cardiac allograft model demonstrated that subtherapeutic doses of cyclosporin A could synergize with an immunosuppressive antibody to the IL-2 receptor leading to prolonged or indefinite graft survival in the majority of animals. The data concerning the interaction of cyclosporin A and co-stimulation blockade through CTLA-4Ig and anti-CD40L is most interesting and instructive. In mice the long-term graft acceptance that is achieved with this strategy is blocked by cyclosporin A (5), but in primates there is synergy for immunosuppression (6). Herein lies the crux of the problem for antibody therapy, tolerance or immunosuppression?

In small animal models, such as rodents, tolerance strategies with mAb are developed in the absence of other drugs and often require high doses (mg per kg body weight) to be administered. In large animal or clinical studies antibody is only ever used in small doses which are immunosuppressive rather than tolerogenic and in combination with other drugs. It should come as no surprise that any antibody successfully entering the clinical arena will be used as an immunosuppressant rather than a tolerance-inducing treatment and show synergy with conventional drugs. This is exemplified by mAb to CD25 which is the only example of a mAb being licensed for transplantation since OKT3 was first used in 1981 (7). Results of animal studies with mAbs to CD25 were far less impressive than many other specificities in achieving immunological tolerance and yet this is a useful target for adjunct immunosuppression in human transplantation.

2.2 Depletion versus non-depletion

Monoclonal antibodies may exert their effects through functional blockade or lysis of the cell expressing the target molecule. Immunoglobulin isotypes vary enormously in their capacity to deplete target cells. Studies using class switch variants of the same antibody idiotype have clearly demonstrated that within a species there is a hierarchy in terms of effector function (8). Different immunoglobulin subclasses fix complement or bind to Fc receptors with different efficiencies, although it is probably the latter which is most relevant in terms of cell depletion *in vivo*. Monoclonal antibodies can be class switched *in vitro* (9) or re-engineered by recombinant DNA technology (10) so that the effector function is tailored to give the desired profile *in vivo*. Whether one should be attempting cell depletion or functional blockade remains controversial. If a mAb is developed for cancer immunotherapy then clearly a lytic mAb would be the best choice, however when treating autoimmune disease or organ rejection it is less clear. Until quite recently the major consensus within both fields was that T cell depletion was undesirable because this represented over-immunosuppression and patients would be more at risk from infection and in the long-term malignancy. Adult patients also tend to be relatively poor and slow at reconstituting lymphocytes following ablative therapy, giving concern about long-term safety. However, follow-up of patients with rheumatoid arthritis treated with the depleting pan-lymphocyte mAb Campath-1 suggests this view may be over-cautious (11). Infective complications were observed only in the first few months after mAb administration and responded to conventional treatment. Infections such as reactivation of Herpes zoster showed normal distribution over one or two dermatomes rather than the widespread involvement which one might have expected if patients were over-immunosuppressed. Malignancy has not been observed at a higher frequency than one would expect for this group despite prolonged CD4 T cell cytopaenia. This suggests that what one measures in peripheral blood long-term following mAb treatment may be not the most relevant marker of immune status. The demonstration by Knechtle and colleagues (12) that prolonged renal allograft survival could be induced in primates with T depletion despite the return to normal of immune responses to third party antigens has stimulated interest amongst transplanters.

2.3 Avoiding the antiglobulin response

One of the conundrums of mAb therapy is that even though an antibody may be immunosuppressive it can evoke strong antiglobulin responses leading to rapid clearance and reduced efficacy (13). The presence of antiglobulin renders repeated courses ineffective and serum sickness may ensue following the formation of immune complexes. Many mAbs are xenogenic which contributes to their immunogenicity, but this is not the whole story. It appears for reasons yet unknown that mAbs which bind to bone marrow-derived cells through their idiotype evoke potent antiglobulin responses. One route to reducing the immunogenicity of a mAb is to 'graft' the CDRs from the parental antibody onto a human framework (10). However, this process of 'humanization' is not sufficient to eliminate the idiotype response even though the individual is tolerant of the majority of epitopes in the protein (14). Administration of additional immunosuppression in the form of steroids can reduce the formation of neutralizing antibody, but does not eliminate the problem. A novel approach to eliminating this adverse effect is based on classical experiments by Dresser (15) who observed that immunoglobulin was immunogenic when given in the form of aggregates but tolerogenic if administered as a monomer. The humanized mAb Campath-1H was re-engineered further by mutating one or two amino acids which were critical for binding to the cell surface target CD52. Administration of these monomeric *non-cell binding* mutants to mice transgenic for human CD52 tolerized to the later injection of 'wild-type' cell binding mAb (16).

2.4 Can monoclonal antibodies ever be used to induce tolerance in humans?

On the basis of the preceding argument it is difficult to ever envisage monoclonal antibody therapy being used to induce transplantation tolerance in patients. However, if an antibody has been shown to be effective in combination with conventional treatment the dose may be increased further whilst reducing or omitting other drugs. Providing there are no deleterious effects one may arrive at a situation where antibody alone or perhaps in combination with one other drug may generate tolerance and allow the patient to be free of all long-term immunosuppression. An obvious caveat, as discussed above, is that the mAbs that make it into the clinical arena in the first place may not be the ones best suited to tolerance strategies.

Another route whereby tolerance regimens for mAb therapy may gain access into transplantation is by first establishing them in the treatment of autoimmune disease. This diverse group of illnesses constitutes a much larger patient pool than organ transplantation. On the whole less aggressive immuno-suppression is employed which is usually stopped once mAb therapy is instituted and only recommenced if relapse is observed. Initial clinical studies tend to involve patients with progressive poor prognosis disease or following failure of conventional therapy. If a treatment is beneficial in this context patients with earlier disease can be treated with the hope of re-establishing tolerance and preventing disease sequelae from developing.

3 Animal models of transplantation and mAb therapy: defining the problem

In order to make rational choices about which mAbs are likely to be useful targets for organ transplantation one must first define which components of the immune system normally participate in the rejection process. Although rejection of an allograft involves diverse effector mechanisms the T lymphocyte plays a central orchestrating role. It became apparent early on that nude mice which lacked T cells were unable to reject even the most immunogenic allograft (17). To what extent different T cell subsets contribute to rejection has been an intense area of research for many years (18, 19) and was heavily dependent on mAbs being available to define cell populations. In general, studies have involved the *in vivo* depletion of T cell subsets using mAbs or by the adoptive transfer of purified T cells into empty 'test-tube' animals. Analysis in these various models has gone into great depth such that one can define which cells are involved in the rejection of isolated MHC disparities or single minor antigens. Though of interest from a scientific viewpoint such detailed study is of little relevance to organ transplantation in an outbred human population where at best matching is only achieved for some MHC alleles.

Cobbold and Waldmann (20) were able to demonstrate in T cell depleted mice that both CD4 and CD8 T cells could independently reject skin grafts which differed from the recipient in the expression of minor or major MHC in-compatibilities. In primary allograft rejection of minors CD8 T cells appeared heavily dependent on help from CD4 T cells, though once primed CD8 T lymphocytes rejected second set minor skin grafts as quickly as intact animals. In some rodent models of vascularized organ graft rejection CD8 T cells also appear critically dependent on CD4 lymphocytes for help and in these cases targeting the CD4 molecule alone with mAb is sufficient to prevent rejection (21). However, any strategy aiming to disrupt transplant rejection should target T cells and in particular both CD4 and CD8 subsets in order to be successful. This is exemplified by experimental studies of transplantation in rodents, primates, and humans, where common specificities include mAbs to CD2, CD3, CD4 (and CD8), CD25, CD45, CD80 and CD86 (B7), CD154 (CD40L), ICAM-1 (CD54), LFA-1 (CD11a and CD18), transferrin receptor, and $\alpha\beta$-TCR. The difficulty in interpreting these numerous experimental studies is that the effects observed after mAb treatment are dependent on several factors which can not be easily compared between experiments.

(a) Different species.

(b) Different strains within a species. Evidence from several experimental systems suggests some strains are easy to tolerize whilst others are relatively resistant (22). The mechanism for susceptibility or resistance is unknown.

(c) Extent of mismatching for minor or major MHC loci. In general tolerance can be achieved more easily to minor mismatched grafts than for those involving major MHC loci. Increasing the number of loci for which there is donor/recipient disparity usually makes tolerance induction more difficult.

(d) Tissue transplanted. Within any donor/recipient combination tissues behave very differently in relation to graft acceptance, despite being equally rejectable in the absence of immunomodulation. Induction of tolerance to vascularized organ grafts is a relatively easy procedure, whilst skin grafts tend to be the most immunogenic. The reason why heart or renal transplants may be accepted indefinitely with relatively small amounts of mAb treatment whilst the same treatment may not even delay skin grafts from the same donor are not clearly understood. There are probably several mechanisms operating which can provide some degree of protection for certain tissues. Thus fully allogeneic testicle transplants are not rejected because of the presence of Fas Ligand (CD95L) on sertoli cells (23), corneal transplants enjoy a degree of immune privilege because they express CD95L (24) and have few donor-derived dendritic cells (25).

(e) Dose of mAb administered. Depending on the factors listed above the amount of mAb used is critical for whether one observes immunosuppression or tolerance induction. This principle was first observed for mAbs to murine CD4 inducing tolerance to their own idiotypes (26), where low doses would suppress responses to third party antigens whilst eliciting antiglobulins to the mAbs own idiotype. However, high doses induced antigen-specific tolerance to the injected monoclonal. One observes the same principles in transplant models. CBA/Ca mice can be tolerized to minor mismatched B10.BR skin grafts (27) with three injections of mAb to CD4 and CD8 over the period of a week (2 mg per injection). However, the same treatment will only prolong fully allogeneic C57/BL10 skin grafts which require three times as much mAb to achieve tolerance (28). Many of the 'failures' of mAb therapy in primate models can be explained by inadequate dosing.

4 Which mAbs for organ transplantation

As outlined above many mAbs have been employed in modulating graft rejection and we shall not attempt to review them all because some have only been used in 'weak' rodent models and do not really warrant further discussion. Instead we shall consider those which at the time of writing are the most relevant either because they are used clinically (anti-CD3 and CD25), display powerful effects in rodent models (anti-CD4 and CD8), or provoking enthusiasm as tools in the near future (anti-CD40L and CTLA4-Ig, and T cell depleting mAbs).

4.1 Anti-CD3 mAb treatment

OKT3 is a mouse mAb to the human epsilon chain of the CD3 complex which is associated with the T cell receptor and is thus present on all T lymphocytes. Perhaps the most astonishing aspect of this mAb is that it was used for the treatment of acute renal allograft rejection in patients only two years after being raised, on the basis of its ability to block CTLs *in vitro*, at a time when the structure of the T cell receptor complex was unknown (7). Although, even now it remains as an effective treatment for the reversal of steroid resistant acute allograft rejection, many of the complications associated with therapy were either underestimated or not anticipated. Nowadays extensive pre-clinical data is required, often involving primate studies, before a mAb can be used in clinical trials. It is likely that if mAbs to CD3 were introduced today rather than in 1981 that they would not have gained a licence for human use. Monoclonal antibodies to CD3 in both human and animal studies have broadly similar effects and are of interest because they elicit many of the complications associated with mAb therapy. How these problems have been circumvented serves a framework for many other mAbs under development.

4.1.1 Experimental use of mAbs to CD3 in animal studies

Probably the best studied mAb to CD3 in experimental models is the hamster anti-murine CD3 mAb 145-2C11 which recognizes an epitope of the epsilon chain in the CD3 complex (29). This mAb is a potent immunosuppressive agent *in vivo* capable of delaying skin allograft rejection to a mean of 32–34 days compared to untreated controls (30). Administration of the first dose of mAb triggers a 'cytokine release syndrome' characterized by the presence of TNF-α, IFN-γ, IL-1, IL-2, IL-4, IL-6, and GM-CSF (31). This massive but self-limiting cytokine storm is dose related and induces a syndrome characterized by somnolence, hypomotility, diarrhoea, hypothermia, piloerection, and pulmonary oedema in mice. Escalating the amount of mAb given increases mortality and so cytokine release is dose limiting. TNF-α is an important mediator of this syndrome which can be blocked by co-administering an anti-TNF mAb (32), however the situation is somewhat more complex since mAbs against the $\alpha\beta$ chains of the TCR provoke isolated but equally high levels of TNF-α without inducing the same pathophysiology (33). 145-2C11 is mitogenic when added *in vitro* to cultures of murine T cells and represents a correlate of the *in vivo* cytokine release syndrome. Both phenomena are dependent on efficient crosslinking of Fc receptors on monocytes to trigger cell activation, apoptosis, and cytokine production since these effects can be blocked by Fc receptor fusion proteins, using variants which are incapable of binding Fc receptors or F(ab)$_2$ fragments of 145-2C11 (30, 34–37). Alegre *et al.* (30) have cloned the variable region of 145-2C11 and made a number of chimeric mAbs between various mouse IgG subclasses and the variable region. Chimeric mAbs with low affinity for Fc receptor binding are not mitogenic *in vitro* unless first immobilized onto plastic and fail to trigger cytokine release in mice whilst suppressing allograft rejection to the same extent as 'wild-type' mAb. Routledge *et al.* (38) humanized the rat anti-human CD3 mAb YTH 12.5 and mutated the Fc region removing a glycosylation site critical for both Fc receptor binding and complement fixation. This 'humanized aglycosyl' mAb retained its' immunosuppressive effects in mice transgenic for human CD3 whilst failing to trigger cytokine release. The immunosuppressive property of CD3 mAbs is therefore independent of cytokine production.

Like the use of OKT3 in humans (39), 145-2C11 being of hamster origin is xenogenic to the mouse and inevitably evokes an antiglobulin response which limits the length of treatment or re-treatment. Combination with the immunosuppressant deoxyspergualin (DSG) eliminated this response and showed synergy for delayed skin transplant rejection (40). However, it is unclear whether synergy was the result of greater efficacy of 145-2C11 in the absence of neutralizing antibody or some other interaction between DSG and the anti-CD3 mAb. In their study with low affinity Fc receptor binding chimeric mAbs, Alegre *et al.* were unable to detect an antiglobulin response to these variants whilst wild-type 145-2C11 induced a strong response. Animals were not tested to see whether the mAb had induced tolerance to itself. An additional feature of non-mitogenic variants of CD3 mAbs is that they are not associated with extensive T cell depletion which occurs by apoptosis which in part is Fas (CD95) dependent. This would be perceived as another desirable characteristic of the antibody if tolerance could be achieved to a graft without any killing of peripheral T cells. However, recent studies in both mice and primates have found that when mAbs to CD3 are made more lytic by conjugating an immunotoxin tolerance or prolonged graft survival was more successful (41–44). One has to be a little cautious with these results because it is unclear how much of the effect is due to T cell depletion compared with the specificity of the anti-CD3 mAb. In other words would a mAb against a different T cell specificity but with the same lytic capacity have given a similar result?

4.1.2 Tolerance and mechanism of action

Most of the experimental studies involving 145-2C11 or similar mAbs have looked at these agents as immunosuppressants rather than tolerance inducing agents. The profound immunosuppression observed in these models correlating well with the use of OKT3 in humans to reverse steroid resistant rejection episodes. Tolerance induction has been achieved in rodent models to bone marrow, heart, and kidney allografts (41, 45, 46). However, tolerance can be obtained to skin only if a heart graft has been accepted first, primary skin allografts being rejected following a significant delay. Failure to tolerize skin may reflect an inadequate dose of mAb being administered since the cytokine release syndrome and antiglobulins limit the amount of mitogenic mAb that can be given. To date there is no published data on attempts to use larger amounts of non-mitogenic anti-CD3 mAbs for tolerance induction to skin.

Perhaps one of the most instructive uses of 145-2C11 to facilitate tolerance is not in a transplant model but in a model of autoimmune diabetes (2). NOD mice develop a spontaneous form of diabetes which is characterized by T cell-mediated destruction of the endocrine pancreas. Overtly diabetic mice can be rendered normoglycaemic by a five day course of 145-2C11. This may also be accomplished with $F(ab')_2$ fragments which fail to deplete T cells peripherally. Tolerance in these animals is lifelong and characterized by the presence of a population of $CD4^+$ T cells which can regulate diabetogenic T cells. Mitogenic anti-CD3 mAbs induce T cell proliferation and apoptosis but lymphocytes which survive show evidence of anergy (47). Recently the group of Lechler has analysed the function of T cell clones anergized by mitogenic CD3 *in vitro*, by injecting these cells into mice transplanted with grafts recognized by the anergic T cells. These CD4 lymphocytes rather than being incompetent *in vivo* were capable of suppressing transplant rejection (Lechler, personal communication, European Journal of Immunology 1999, **29**: 686). There has been a suggestion from some workers that non-mitogenic anti-CD3 mAbs by delivering incomplete signals for T cell activation favour the development of Th2 cells over Th1 cells and this is essential feature of their mechanism of action (48, 49). However, to date nobody has identified Th2 cells as mediating anti-CD3 induced tolerance to allografts. Moreover, there is good evidence that Th2 T cells themselves can mediate rejection (50). Importantly, non-mitogenic mAbs do not have improved efficacy in terms of graft survival over mitogenic variants (30).

4.1.3 Clinical applications of anti-CD3 mAb

OKT3, a murine IgG2a mAb recognizing human CD3, was first used in 1981 to reverse steroid resistant allograft rejection and is still used for this purpose today (7). Subsequently it has been used for the induction of immuno-suppression following transplantation. The Ortho multicentre study (51) compared OKT3 with high dose steroids as the first line treatment of renal rejection in patients maintained on azathioprine and low dose steroids. Reversal was observed in 94% of those receiving OKT3 compared to 75% of the high dose steroid group (p = 0.009). In a retrospective study of patients on triple therapy with histologically confirmed renal rejection reversal was achieved in 98% with one year graft survival of 87.5% (52). However, current protocols of OKT3 administration probably do not offer any additional advantage in terms of efficacy over anti-thymocyte globulin (ATG) (53).

Randomized prospective trials of triple therapy of azathioprine, steroids, and either OKT3 or cyclosporin A induction have demonstrated that time to first rejection episode and number of rejections are reduced with OKT3 induction (54, 55). Retrospective analysis by Opelz for the Collaborative Transplant Study (3) confirmed that OKT3 induction improves survival with high-risk patients gaining the most benefit. However OKT3 has not been proven to be superior to ATG for induction and some studies even suggest it may not be as good (56, 57). In one case at least this correlated with ATG inducing more profound long-term T cell depletion (57).

The adverse effects associated with OKT3 are broadly the same as those described in animal models. In the short-term these constitute a cytokine release syndrome (58–60), human anti-mouse antiglobulins (39), and opportunistic infection. The major long-term complication is increased risk of malignancy particularly EBV-associated lymphoproliferative disease (61). The severity of the cytokine release syndrome can be abrogated by co-administration of high dose steroid or neutralizing TNF-α (62, 63). Many transplant patients are more susceptible to first dose effects because of coexisting disease and abnormal fluid balance. Several groups have now developed humanized non-mitogenic anti-CD3 mAbs for human use which lack this adverse side-effect (64, 65). One of these aglycosyl humanized YTH 12.5 efficiently reverses acute rejection without cytokine release (66). Moreover in the nine patients studied there was no evidence of T cell depletion or formation of antiglobulin. These 'tailored' mAbs offer a great opportunity to use anti-CD3 mAbs in strategies which are more likely to induce T cell tolerance. Larger doses may be given without risk of a first dose effect, neutralizing antiglobulin, or profound T cell depletion, and should allow a reduction of conventional immunosuppression.

A major concern of anti-T cell therapies is the increased risk of malignancy which appears to be related to the total immunosuppression which increases when mAb is used. Whether non-mitogenic anti-CD3 mAbs will reduce the incidence of neoplastic disease in transplant recipients remains to be seen.

4.2 Monoclonal antibodies to CD4 and CD8

These two targets have been the most extensively studied and powerful in experimental studies, but this has not been manifested in clinical success. One can identify two reasons for this 'failure'. The first is that clinical studies use inadequate amounts of mAb to obtain any prolonged effect. The second is that no clinical study has combined mAbs to CD4 and CD8 together and so only some of the T cells participating in rejection are targeted. This form of therapy is at an inherent disadvantage because regulatory issues often prevent the combination of mAbs unless they have shown single agent efficacy. In this respect mAbs to CD3, CD25, or CD52 have an advantage as they target both CD4 and CD8 T cells.

4.2.1 Experimental models

The *first evidence* that mAbs could be used to induce tolerance in the adult immune system came with the discovery that anti-CD4 mAb could induce tolerance to foreign protein antigens whilst immune responses to third party antigens were preserved (26). This led to the concept of a short course of treatment leading to long-term antigen-specific tolerance without the need for maintenance immunosuppression. Although initial studies used depleting mAbs the same results were obtained without depletion (67). Subsequently combination with mAbs to CD8 generated tolerance to bone marrow, heart, and skin across donor/recipient disparities involving not only minors but also minors plus major histocompatibility loci (27, 28, 68–71). Tolerance could even be induced to multiple minor antigens in primed animals that would normally show second set rejection (71).

Mice which had become tolerant following mAb therapy contained a population of CD4 T cells which could regulate other T cells (CD4 and CD8) which may otherwise damage the graft (27, 68). Elimination of CD4 T cells from primed tolerant animals enabled rejection to occur through the CD8 population which had been held 'in check' by regulatory T cells (71). Naive CD4 T cells could also become regulatory if they coexisted with tolerant CD4 T cells for a period of time, a phenomenon coined 'infectious tolerance' (27). Regulatory CD4 T cells can suppress rejection on adoptive transfer (suppression), resist rejection from naive T cells injected into a tolerant recipient (resistance), subsequently guiding these naive T cells to become tolerant too (infectious tolerance), and suppress rejection to third party antigens only if they are present on the same graft as the tolerated antigens (linked suppression). All these phenomena have been described as a mechanism of dominant tolerance and represents the kind of therapy which one would like to be able to introduce into the clinic. The phenotype of these cells has not yet been characterized further, but there is no evidence to suggest that there is deviation towards Th2 (72).

The observation of linked suppression has important implications for organ transplantation. In these experiments (70) mice tolerant of minor mismatched skin grafts (A) following a short course of mAb to CD4 and CD8 will accept the same minor mismatched grafts (A), but reject MHC class I mismatched skin (B) showing that they display antigen-specific tolerance. However, if instead animals are transplanted with an F1 of the two grafts (A × B) about 50% undergo delayed rejection and 50% are accepted. Mice which show prolonged survival can then be transplanted with MHC class I incompatible grafts (B) which are also accepted. Tolerance is therefore achieved to the 'linked antigen'. The implication for organ transplantation is that one does not require tolerance to be induced to all antigens from the outset. Linked suppression can occur solely through the indirect route of antigen presentation and this enables one to establish a two-step strategy for tolerance induction (73). If one can identify common antigens (MHC molecules or dominant minors) which can be given in any form which allows processing through the indirect route undercover of mAb to CD4 (and CD8) then some regulatory cells can be initiated prior to organ transplantation. The advantages are several-fold. Regulatory cells can be induced electively when the patients clinical condition is optimal, this may also be applied to xenoantigens. Peritransplant immunosuppression with conventional drugs may be decreased reducing patient morbidity and mortality. Linked suppression through indirect recognition may be the explanation for the well documented beneficial effects of blood transfusion on allograft survival (74). The group of Wood has studied this two-step strategy of blood transfusion and anti-CD4 therapy extensively in rodents (75) and is now investigating it clinically.

Two questions are pertinent to these studies, are regulatory CD4 T cells a peculiarity to anti-CD4 and CD8 mAb therapy? and do they exists in humans? Regulatory CD4 T cells have also been described as occurring naturally or following treatment with MAb to CD3, CTLA-4Ig, and a small molecule drug (LF 08-0299) and therefore appear to be a general phenomenon (2, 76, 77). The group of Lechler has also described linked suppression *in vitro* using resting and memory human T cells (78).

4.2.2 Clinical trials of mAbs to human CD4

The results of clinical trials with mAbs to human CD4 are few in number and generally disappointing, but none the less predictable on the basis of animal studies. In all studies where these mAbs have been used they have not been combined with anti-CD8 treatment and therefore only part of the T cell repertoire has been targeted. The largest study to date enrolled 30 renal transplant patients and combined mAb induction with triple therapy (79). The murine IgG2a mAb OKT4A was well tolerated without any first dose side-effects. No T cell depletion was observed and the majority (84%) made an antiglobulin response, within three months 37% had experienced a rejection episode. Although this mAb appeared safe there appeared to be little therapeutic benefit. Despite showing that the majority of CD4$^+$ T cells were coated during treatment the presence of an anti-murine antibody response suggests that the doses administered may have been too low in any case.

More encouraging was a small trial using a chimeric mAb cM-T412 which has a murine variable region grafted onto a human IgG1 constant region (80). Induction with this preparation and triple therapy was compared to a control group receiving ATG in place of the anti-CD4 mAb. Each arm of the study included only 11 cardiac transplant patients, nevertheless the anti-CD4 treated group experienced less rejection episodes, fewer infections, and improved graft survival at one year. However, unlike OKT4A this chimeric preparation depleted CD4 T cells from peripheral blood, indeed the dosing schedule was chosen so that the level of depletion was similar to ATG. The apparent discordance in efficacy between these two trials is almost certainly related to this difference in T cell depletion.

Ideally one would like to use large doses of anti-CD4 combined with anti-CD8 in patients free from other immunosuppressants in a similar way to rodent studies, for the reasons outlined in the introduction it is unlikely that any such trial would take place.

4.3 Monoclonal antibodies to CD25 (IL-2 receptor)

The receptor for interleukin-2 consists of a molecular complex of three chains termed α (CD25), β (CD122), and γ (CD132). The α chain is specific for the IL-2 receptor complex whilst the β and γ chains are also found in the receptors for other cytokines. The IL-2Rα subunit is only expressed on activated and not resting T cells. Monoclonal antibodies targeting the α subunit can therefore selectively impair T lymphocytes participating in allograft rejection.

4.3.1 Experimental models

The literature for describing the use of mAbs to CD25 in rodent studies is relatively small when one considers this specificity is a target for human transplantation. Predominantly models of heart or kidney transplantation are described where significant delays in allograft rejection were observed (81–83). The few examples of skin transplantation show only modest effects on survival at best (84). Unlike mAb to CD4 and CD8 the induction of tolerance as a therapeutic procedure could not be routinely accomplished, though long-term survival was demonstrated in a proportion of rats receiving kidney allografts (83). However, animals with intact transplants were not tested for tolerance by re-grafting. Several studies in rodents demonstrated synergy with sub-therapeutic doses of cyclosporin and this is probably the most important feature when one considers that mAb to CD25 is used as an adjunct to triple therapy (4, 82).

4.3.2 Clinical experience

Early clinical trials used mouse or rat mAbs for the induction of immuno-suppression or reversal of rejection in recipients of renal or liver allografts and found them to be safe and as effective as ATG (85). Antiglobulins were inevitably evoked and so humanized versions of anti-CD25 were generated. The re-engineered antibody demonstrated enhanced effector function and depletion. It was less immunogenic *in vivo* and with a longer half-life than the murine counterpart. Importantly it was also more effective at prolonging cardiac allo-graft survival in primates. A randomized, placebo-controlled phase III study has since been conducted using humanized anti-CD25 in combination with triple therapy to prevent renal allograft rejection (86). A significant reduction in acute graft rejection was observed at six months. On the basis of this trial a licence was granted for the prevention of renal transplant rejection. At no time have mAbs to CD25 fulfilled the transplant immunologists goal of inducing tolerance in the adult immune system. However, they adequately fit the criteria discussed earlier of a safe drug which can synergize for immunosuppression (4, 82). Indeed if one does ever evoke regulatory T cells in a clinical setting such a treatment may well be counterproductive since such cells are almost certainly CD25$^+$. A much longer period of time will be necessary to determine if current use of these mAbs will lead to improved graft and patient survival at five years.

4.4 Blockade of co-stimulation through CD40 and CD28 pathways

In 1996 Larsen *et al.* demonstrated that simultaneous blockade of CD40 and CD28 pathways permitted long-term survival of fully allogeneic cardiac and skin grafts in mice (5). Transplant immunologists were excited by this new finding because it offered a potentially powerful tool for interrupting the rejection process. The receptor ligand pairs CD28–B7 (CD80 and CD86) and CD40–CD40L (CD154) are essential for the initiation and amplification of T cell responses. CD28–B7 provides essential co-stimulatory signals for T cell activation, whilst CD40–CD40L are critical for dendritic cell and T cell maturation. CD40L is selectively expressed on activated T cells, probably CD4$^+$ lymphocytes more than CD8$^+$. Importantly longstanding cardiac allografts were free of chronic vascular rejection which has become a major cause of graft dysfunction clinically since prevention and treatment of acute rejection has improved. Long-term acceptance of skin allografts could be inhibited by concomitant administration of cyclosporin A (5).

In a small number of Rhesus renal allografts Kirk et al. (87) reported that blockade of CD28 and CD40 pathways for 28 days permitted survival for an impressive greater than 180 day survival. Cyclosporin A appears to show synergy for graft survival in primate studies (6), but whether as discussed earlier synergy for long-term acceptance will also be demonstrated given the conflicting mouse data remains to be seen. To date no clinical data is available and one must therefore wait to see if this initial promise is fulfilled. One cautionary note is that in mice with long-term acceptance of skin allografts rejection was provoked by re-transplantation through CD8$^+$ T cells (6).

4.5 T cell depletion

Strategies involving depletion of T cells have enjoyed renewed optimism because of encouraging studies involving primates. This approach had been unpopular because the whole T cell compartment was compromised in an attempt to disrupt the minority of lymphocytes participating in rejection. Concerns also existed that individuals would be over-immunosuppressed and fail to reconstitute immune function to third party antigens.

Knechtle and colleagues (42) developed an immunotoxin to study renal allograft survival in rhesus monkeys. A diphtheria toxin binding site mutant (a mutant unable to bind cells) was conjugated to an anti-CD3 rhesus mAb to create an immunotoxin which would only kill CD3$^+$ T cells. Administration of this conjugate one week prior to transplantation reduced T cells to 1% of pretreatment levels in peripheral blood and lymph nodes and permitted renal allograft survival for more than 100 days in the majority of recipients without further immunosuppression despite the return to normal of T cell numbers and third party immune responses. Administration of antibody at the time of transplantation seemed to be less efficient at prolonging graft survival (43). Contreras et al. (44) investigated this further by combining the anti-CD3 immunotoxin with DSG or steroids. It appeared that in this model the cytokine release syndrome triggered by the anti-CD3 immunotoxin prevented long-term allograft survival. Combination with DSG restored prolonged transplant survival even when the immunotoxin was given at the time of transplantation. The authors contributed the beneficial effect of DSG to inhibition of IFN-γ, rather than any independent effect of this drug. Importantly, these workers showed that T cell depletion was an essential feature as the 'wild-type' mAb which depletes by only 30–50% could not generate long-term graft survival.

Encouraged by these results Calne et al. (88) transplanted 13 patients with cadaveric renal allografts under cover of the pan-lymphocyte depleting mAb Campath-1H (anti-CD52) and 500 mg of methylprednisolone to minimize any cytokine release which is sometimes observed with this mAb. Post-transplant immunosuppression was with 'half' dose cyclosporin only which was delayed for 48 hours after T cell depletion. During follow-up (6–11 months) after surgery only one episode of rejection (reversed with steroids) was observed despite recovery of lymphocyte counts and the reduced level of monotherapy. A randomized trial is proposed to compare this protocol with standard immuno-suppression. If these initial results are repeated in a larger group of patients the impact on transplantation will be enormous. Although this strategy may not induce tolerance in the way we have discussed earlier, a reduced dose of mono-therapy decreases drug costs. More importantly the overall reduction of main-tenance immunosuppression will reduce patient morbidity and mortality whilst graft function is sustained.

4.6 Summary

Monoclonal antibodies have revolutionized our understanding of immunology but their impact on human transplantation remains limited. We have outlined many reasons for the low clinical impact that exists at present, however tolerance induction following a short course of mAb therapy still remains a realistic goal. Transplantation is still a relatively new field and it is perhaps salient that in the 1997 United Network for Organ Sharing report (1), which analysed the results of 97 000 transplants in the United States from 1988–1994, one of the major factors influencing survival was the centre where transplanta-tion was undertaken.

References

1. Lin, H.-M., Kauffman, M., McBride, M. A., Davies, D. B., Rosendale, J. D., Smith, C. M., et al. (1998). *J. Am. Med. Assoc.*, **280**, 1153.
2. Chatenoud, L., Primo, J., and Bach, J.-F. (1997). *J. Immunol.*, **158**, 2947.
3. Opelz, G. (For the collaborative transplant study.) (1995). *Transplantation*, **60**, 1220.
4. Diamantstein, T., Volk, H.-D., Tilney, N. L., and Kupiec-Weglinski, J. W. (1986). *Immunobiology*, **172**, 391.
5. Larsen, C. P., Elwood, E. T., Alexander, D. Z., Ritchie, S. C., Hendrix, R., Tucker-Burden, C., et al. (1996). *Nature*, **381**, 434.
6. Larsen, C. P. (1998). *Imperial College School of Medicine Transplantation Symposium*, 19–20th November 1998.
7. Cosimi, A. B., Burton, R. C., and Colvin, B. (1981). *Transplantation*, **32**, 535.
8. Bruggemann, M., Teale, C., Clark, M., Bindon, C., and Waldmann, H. (1989). *J. Immunol.*, **142**, 3145.
9. Hale, G., Cobbold, S. P., Waldmann, H., Easter, G., Matejtschuk, P., and Coombs, R. R. (1987). *J. Immunol. Methods*, **103**, 59.
10. Riechmann, L., Clark, M., Waldmann, H., and Winter, G. (1988). *Nature*, **332**, 323.
11. Isaacs, J., Greer, S., Symmons, D., Sharma, S., Hale, G., Waldmann, H., et al. (1998). *Arthritis Rheum.*, **41**, S56.
12. Fechner, J. H., Vargo, D. J., Geissler, E. K., Graeb, C., Wang, J., Hanaway, M. J., et al. (1997). *Transplantation*, **63**, 1339.

13. Benjamin, R. J., Cobbold, S. P., Clark, M. R., and Waldmann, H. (1986). *J. Exp. Med.*, **163**, 1539.

14. Lockwood, C. M., Thiru, S., Isaacs, J. D., Hale, G., and Waldmann, H. (1993). *Lancet*, **341**, 1620.

15. Dresser, D. W. (1962). *Immunology*, **5**, 378.

16. Gilliland, L., Walsh, L., Frewin, M., Wise, M., Tone, M., Hale, G., *et al.* (1999). *J. Immunol*, 3663.

17. Miller, J. F. A. P. (1961). *Lancet*, **ii**, 748.

18. Hall, B. M. (1991). *Transplantation*, **51**, 1141.

19. Rosenberg, A. S. and Singer, A. (1992). *Annu. Rev. Immunol.*, **10**, 333.

20. Cobbold, S. P. and Waldmann, H. (1986). *Transplantation*, **41**, 634.

21. Darby, C. R., Morris, P. J., and Wood, K. J. (1992). *Transplantation*, **54**, 483.

22. Davies, J. D., Cobbold, S. P., and Waldmann, H. (1997). *Transplantation*, **63**, 1570.

23. Bellgrau, D., Gold, D., Selawry, H., Moore, J., Franzusoff, A., and Duke, R. C. (1995). *Nature*, **377**, 630.

24. Yamagami, S., Kawashima, H., Tsuru, T., Yamagami, H., Kayagaki, N., Yagita, H., *et al.* (1997). *Transplantation*, **64**, 1107.

25. Sano, Y., Ksander, B. R., and Streilein, J. W. (1996). *Transplant Immunol.*, **4**, 53.

26. Benjamin, R. J. and Waldmann, H. (1986). *Nature*, **320**, 449.

27. Qin, S., Cobbold, S. P., Pope, H., Elliot, J., Kioussis, D., Davies, J., *et al.* (1993). *Science*, **259**, 974.

28. Cobbold, S. P., Martin, G., and Waldmann, H. (1990). *Eur. J. Immunol.*, **20**, 2747.

29. Leo, O., Foo, M., Sachs, D. H., Samelson, L. E., and Bluestone, J. A. (1987). *Proc. Natl. Acad. Sci. USA*, **84**, 1374.

30. Alegre, M.-L., Tso, J. Y., Sattar, H., Smith, J., Desalle, F., Cole, M., *et al.* (1995). *J. Immunol.*, **155**, 1544.

31. Ferran, C., Dy, M., Merite, S., Sheehan, K., Schreiber, R., Leboulenger, F., *et al.* (1990). *Transplantation*, **50**, 642.

32. Ferran, C., Dy, M., Sheehan, K., Schreiber, R., Grau, G., Bluestone, J. A., *et al.* (1991). *Eur. J. Immunol.*, **21**, 2349.

33. Chatenoud, L., Legendre, C., Kurrle, R., Kreis, H., and Bach, J.-F. (1993). *Transplantation*, **55**, 443.

34. Debets, J. M., Van de Winkel, J. G., Cueppens, I. E., Dieteren, M., and Buurman, W. A. (1990). *J. Immunol.*, **144**, 1304.

35. Hirsch, R., Bluestone, J. A., DeNenno, L., and Gress, R. E. (1990). *Transplantation*, **49**, 1117.

36. Krutmann, J., Kirnbauer, R., Kock, A., Schwarz, T., Schopf, E., May, T., *et al.* (1990). *J. Immunol.*, **145**, 1337.

37. Vossen, A. C., Tibbe, G. J., Kroos, M. J., van de Winkel, J. G., Benner, R., and Savelkoul, H. F. (1995). *Eur. J. Immunol.*, **25**, 1492.

38. Routledge, E. G., Falconer, M. E., Pope, H., Lloyd, I., and Waldmann, H. (1995). *Transplantation*, **60**, 847.

39. Jaffers, G. J., Colvin, R. B., Cosimi, A. B., Giorgi, J. V., Goldstein, G., Fuller, T. C., *et al.* (1983). *Transplant. Proc.*, **15**, 646.

40. Alegre, M.-L., Sattar, H. A., Herold, K. C., Smith, J., Tepper, M. A., and Bluestone, J. A. (1994). *Transplantation*, **57**, 1786.

41. Mottram, P. L., Han, W. R., Murray-Segal, L. J., Mandel, T. E., Pietersz, G. A., and McKenzie, I. F. C. (1997). *Transplantation*, **64**, 684.

42. Knechtle, S., Vargo, D., Fechner, J., Zhai, Y., Wang, J., Hanaway, M., *et al.* (1997). *Transplantation*, **63**, 1.

43. Armstrong, N., Buckley, P., Oberley, T., Fechner, J., Dong, Y., Hong, X., *et al.* (1998). *Transplantation*, **66**, 5.

44. Contreras, J. L., Wang, P. X., Eckhoff, D. E., Lobashevsky, A. L., Asiedu, C., Frenette, L., *et al.* (1998). *Transplantation*, **65**, 1159.

45. Mackie, J. D., Pankewycz, O. G., Bastos, M. G., Kelley, V. E., and Strom, T. B. (1990). *Transplantation*, **49**, 1150.

46. Nicolls, M. R., Aversa, G. G., Pearce, N. W., Spinelli, A., Berger, M. F., Gurley, K. E., *et al.* (1993). *Transplantation*, **55**, 459.

47. Chai, J.-G. and Lechler, R. I. (1997). *Int. Immunol.*, **9**, 935.

48. Smith, A. S., Tso, J. Y., Clark, M. R., Cole, M. S., and Bluestone, J. A. (1997). *J. Exp. Med.*, **185**, 1413.

49. Smith, A. S. and Bluestone, J. A. (1997). *Curr. Opin. Immunol.*, **9**, 64.

50. Zelenika, D., Adams, E., Melloe, A., Simpson, E., Channdler, P., Stockinger, B., *et al.* (1998). *J. Immunol.*, **161**, 1868.

51. Ortho Multicenter Transplant Study Group. (1985). *N. Engl. J. Med.*, **313**, 337.

52. Kamath, S., Dean, D., Peddi, V. R., Schroeder, T. J., Alexander, J. W., Cavallo, T., *et al.* (1997). *Transplantation*, **64**, 1428.

53. Cantarovich, M., Latter, D. A., and Loertscher, R. (1997). *Clin. Transplant.*, **11**, 316.

54. Abramowicz, D., Goldman, M., De Pauw, L., Vanherweghem, J. L., Kinnaert, P., and Vereerstraeten, P. (1992). *Transplantation*, **54**, 433.

55. Norman, D. J., Kahana, L., Stuart, F. J., Thistlethwaite, J. J., Shield, C. E., Monaco, A., *et al.* (1993). *Transplantation*, **55**, 44.

56. Malinow, L., Walker, J., Klassen, D., Oldach, D., Schweitzer, E., Bartlett, S. T., *et al.* (1996). *Clin. Transplant.*, **10**, 237.

57. Bock, H. A., Gallati, H., Zurcher, R. M., Bachofen, M., Mihatsch, M. J., Landmann, J., *et al.* (1995). *Transplantation*, **59**, 830.

58. Chatenoud, L., Ferran, C., Reuter, A., Legendre, C., Gevaert, Y., Kreis, H., *et al.* (1989). *N. Engl. J. Med.*, **320**, 1420.

59. Chatenoud, L., Ferran, C., Legendre, C., Thouard, I., Merite, S., Reuter, A., *et al.* (1990). *Transplantation*, **49**, 697.

60. Kan, E. A. R., Wright, S. D., Welte, K., and Wang, C. Y. (1986). *Cell. Immunol.*, **98**, 181.

61. Hanto, D. W., Frizzera, G., Purtilo, D. T., Sakamoto, K., Sullivan, J. L., Saemundsen, A. K., *et al.* (1981). *Cancer Res.*, **41**, 4253.

62. Charpentier, B., Hiesse, C., Lantz, O., Ferran, C., Stephens, S., O'Shaugnessy, D., *et al.* (1992). *Transplantation*, **54**, 997.

63. Eason, J. D., Pascual, M., Wee, S., Farrell, M., Phelan, J., Boskovic, S., *et al.* (1996). *Transplantation*, **61**, 224.

64. Bolt, S., Routledge, E., Lloyd, I., Chatenoud, L., Pope, H., Gorman, S. D., *et al.* (1993). *Eur. J. Immunol.*, **23**, 403.

65. Alegre, M.-L., Peterson, L. J., Xu, D., Sattar, H. A., Jeyarajah, D. R., Kowalkowski, K., *et al.* (1994). *Transplantation*, **57**, 1537.

66. Friend, P., Rubello, P., Chatenoud, L., Hale, G., Phillips, J., and Waldmann, H. (1999). Transplantation (In press).

67. Qin, S., Wise, M., Cobbold, S. P., Leong, L., Kong, Y.-C., Parnes, J. R., *et al.* (1990). *Eur. J. Immunol.*, **20**, 2737.

68. Chen, Z. K., Cobbold, S. P., Waldmann, H., and Metcalfe, S. (1996). *Transplantation*, **62**, 1200.

69. Davies, J. D., Leong, L., Mellor, A., Cobbold, S. P., and Waldmann, H. (1996). *J. Immunol.*, **156**, 3602.

70. Davies, J. D., Martin, G., Phillips, J., Marshall, S. E., Cobbold, S. P., and Waldmann, H. (1996). *J. Immunol.*, **157**, 529.

71. Marshall, S. E., Cobbold, S. P., Davies, J. D., Martin, G. M., Phillips, J. M., and Waldmann, H. (1996). *Transplantation*, **62**, 1614.

72. Cobbold, S. P., Adams, E., Marshall, S. E., Davies, J. D., and Waldmann, H. (1996). *Immunol. Rev.*, **149**, 5.

73. Wise, M. P., Bemelman, F., Cobbold, S. P., and Waldmann, H. (1998). *J. Immunol.*, **161**, 5813.

74. Opelz, G., Sengar, D. P. S., Mickey, M. R., and Terasaki, P. I. (1973). *Transplant. Proc.*, **5**, 253.

75. Bushell, A., Morris, P. J., and Wood, K. J. (1995). *Eur. J. Immunol.*, **25**, 2643.

76. Tran, H. M., Nickerson, P. W., Restifo, A. C., Ivis-Woodward, M. A., Patel, A., Strom, T. B., *et al.* (1997). *J. Immunol.*, **159**, 2232.

77. Andoins, C., Fornel, D., Annat, J., and Dutartre, P. (1996). *Transplantation*, **62**, 1543.

78. Frasca, L., Carmichael, P., Lechler, R., and Lombardi, G. (1997). *Eur. J. Immunol.*, **27**, 3191.

79. Cooperative Clinical Trials in Transplantation Research Group. (1997). *Transplantation*, **63**, 1087.

80. Meiser, B. M., Reiter, C., Reichenspurner, H., Uberfuhr, P., Kreuzer, E., Rieber, E. P., *et al.* (1994). *Transplantation*, **58**, 419.

81. Kirkman, R., Barrett, L., Gaulton, G., Kelley, V., Ythier, Y., and Strom T. B. (1985). *J. Exp. Med.*, **162**, 358.

82. Kupiec-Weglinski, J. W., Diamantstein, T., and Tilney, N. J. (1988). *Transplantation*, **46**, 785.

83. Tellides, G., Dallman, M., and Morris, P. (1987). *Br. J. Surg.*, **74**, 1145.

84. Kirkman, R., Barrett, L., Gaulton, G., Kelley, V., Koltun, W., Schoen, F., *et al.* (1985). *Transplantation*, **40**, 719.

85. Kirkman, R., Shapiro, M., Carpenter, C., McKay, D., Milford, E., Ramos, E., *et al.* (1991). *Transplantation*, **51**, 107.

86. Vincenti, F., Kirkman, R., Light, S., Bumgardner, G., Pecovitz, M., Halloran, P., *et al.* (1998). *N. Engl. J. Med.*, **338**, 161.

87. Kirk, A., Harlan, D., Armstrong, N., Davis, T., Dong, Y., Gray, G., *et al.* (1997). *Proc. Natl. Acad. Sci. USA*, **94**, 8780.

88. Calne, R., Friend, P., Moffat, S., Bradley, A., Hale, G., Firth, J., *et al.* (1998). *Lancet*, **351**, 1701.

Chapter 21

Monoclonal antibody therapy in rheumatoid arthritis

Ernest H. S. Choy, Gabrielle H. Kingsley, and
Gabriel S. Panayi

Department of Rheumatology, King's College and St. Thomas' Hospitals' Medical
and Dental Schools, Guy's Hospital, London SE1 9RT, U.K.

1 Introduction

Rheumatoid arthritis (RA) is the commonest inflammatory and destructive
arthropathy in the world with a prevalence of 1000 per 100 000. It is characterized by a chronic symmetrical polyarthritis that leads to joint damage in most
patients. The commonest affected joints are the small joints of the hands and
feet but it can involve any synovial joint. Rheumatoid factor is present in over
70% of the patients. The long-term prognosis of RA is poor as most patients
develop significant disability despite treatment (1). Hence it is a major socio-
economical burden. In the UK, the cost of RA was estimated at £1.3 billion/year
in 1992 (2).

Disease modifying anti-rheumatic drugs (DMARDs), the main treatment in
RA, suppress inflammation and ameliorate symptoms. However they do not
improve the long-term disease outcome (1) since they have limited efficacy but
high toxicity. DMARDs rarely induce complete disease remission (3) and few
reduce progression of joint damage (4). The high incidence of side-effect leads to
50% of the patients discontinuing treatment after two years (5). Therefore there
is a need for more effective and less toxic treatments in RA.

2 Pathogenesis of rheumatoid arthritis

Histologically, the rheumatoid synovium is characterized by hyperplasia of the
synovial lining layer and lymphocytic infiltration of the interstitial area (6).
Most of these lymphocytes are CD4$^+$ T cells. As RA is associated with specific
major histocompatibility complex (MHC) class II molecules, HLA-DRB1*0101,
HLA-DRB1*0401, and HLA-DRB1*0404, most researchers agree that disease is
initiated by an antigenic or autoantigenic peptide complexed to MHC molecules
on the surface of an antigen presenting cell. This peptide–MHC complex activates antigen-specific CD4$^+$ lymphocytes resulting in increased cell surface
expression of activation markers such as interleukin (IL)-2 receptor, MHC class II

molecules, and CD69. Activation also leads to increased production of lympho-kines including interferon gamma (IFNγ) and IL-2. Activated CD4$^+$ lymphocytes stimulate monocytes/macrophages and synovial fibroblasts to release mono-kines and matrix metalloproteinases respectively through direct cell surface contact (7). Matrix metalloproteinases especially collagenase and stromelysin, can degrade connective tissue matrix and are thought to be the main mediators of joint damage in RA. Monokines, in particular IL-1 and tumour necrosis factor alpha (TNFα), are potent pro-inflammatory cytokines that have a wide range of actions on mesenchymal and endothelial cells (8). They up-regulate the ex-pression of adhesion molecules on endothelial cells, adhesion molecule-1 (ICAM-1), and vascular cellular adhesion molecule-1 (VCAM-1), which facilitate the recruitment of inflammatory cells into the synovial joints and therefore perpetuate inflammation.

Unlike the initiation of RA, the mechanisms that maintain synovitis have remained controversial. Currently, there are two opposing theories about the perpetuation of chronic RA known as the mesenchymal and the T cell hypo-theses. The proponents of the former base their argument on the scarcity of T cell cytokines such as IL-2 and IFNγ (9) in the rheumatoid synovium in contrast with the abundance of monokines such as IL-1, TNFα, and IL-6 (10, 11). Con-sequently, they argue that, in chronic RA, T cells are suppressed in the rheuma-toid synovium and inflammation is perpetuated by an autocrine and paracrine feedback loop involving monocytes, synoviocytes, and various monokines (12). However, the proponents of the T cell hypothesis argue that since the in-flammatory cascade leads to amplification of the inflammatory response, only a small quantity of lymphokines is necessary to sustain synovitis (13). In addition, lymphokines are short-lived, act mainly in the local environment, and cell to cell contact may be essential for their function (14); therefore the level of lym-phokines is no guide to their involvement in pathogenesis. It is probable that RA is initiated and maintained by CD4 lymphocytes but that the mesenchymal response is responsible for joint damage.

3 Treatment strategies in rheumatoid arthritis

There are many therapeutic targets in RA including the MHC molecule, CD4$^+$ lymphocytes, and cytokines. Clearly the ideal therapeutic target is the arthrito-genic peptide but this remains unknown. Therefore current attempts have focused on mediators further down the inflammatory cascade. Indeed, medi-ators at the base of the inflammatory cascade are more directly responsible for inflammation and joint damage. Agents inhibiting these are more likely to be effective anti-inflammatory therapies although they will not be disease specific (*Figure 1*). Furthermore, repeated treatments are probably necessary to maintain therapeutic benefit since disease is sustained by proximal mediators, their effects are likely to be transient, and they may lead to general immuno-suppression.

450

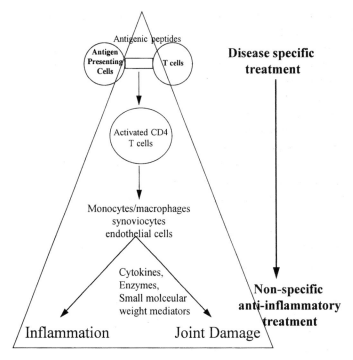

Figure 1 The immunopathogenesis of rheumatoid arthritis.

4 Therapeutic monoclonal antibodies

The advent of hybridoma technology allows generation of monoclonal antibodies (mAb) in large quantities (15). However, most of these mAbs were mainly murine in origin and when these were administered to patients, virtually all the recipients developed human anti-mouse antibody responses (16). This hindered their use in therapy because repeated treatments were less effective and carried the risk of anaphylaxis; as exemplified by the use of a murine anti-ICAM-1 mAb in RA (17). Recent biotechnological advances enabled us to develop less antigenic 'man-made' chimeric and humanized mAbs (18) that could be used in therapy. In the former only the murine Fc is replaced by a human sequence. In the latter, not only the Fc but even the framework region of the antigen binding site has been replaced by human sequences. Consequently, these mAbs are less immunogenic and therefore more suitable as therapeutic agents. Moreover, the half-lives of these 'man-made' mAbs are five to six times longer than their murine equivalents. Consequently, dosing frequency and the total dose of 'man-made' mAbs needed for a particular effect are usually less than murine mAbs.

The use of phage display libraries technique makes generating completely human mAbs feasible in the future (19). A large and diverse repertoire of Fab fragments or single chain variable region genes are generated by polymerase chain reaction. These are inserted into bacterial phages and fused with a gene that encodes a protein expressed on the surface of the phage. The resultant phage will have the antibody Fab anchored on its surface mimicking a B cell. The phage with the desired antibody specificity can be selected by passing the library through a solid phase antigen column. The bound phages are then eluted and used to infect *E. coli*, which will produce more phages for selection. Repeated passages will lead to isolation of phages expressing antibodies with high binding affinities.

5 Monoclonal antibodies in rheumatoid arthritis

A large number of mAbs targeting different mediators have been tested in RA. They inhibit either cytokines, adhesion molecules, MHC, or T cells.

5.1 Anti-cytokine mAbs in rheumatoid arthritis

5.1.1 Anti-TNFα mAbs

TNFα is a potent pro-inflammatory cytokine that has been implicated in the pathogenesis of RA (20). In particular, the TNFα transgenic mice, who express human TNFα constitutively in the joint, develop a spontaneous inflammatory arthritis resembling RA (21). Treatment with anti-TNFα mAb suppresses inflammation in several animal models of RA (22). In RA patients, TNFα can be found both in the blood and the joint particularly at the cartilage–pannus junction (23, 24).

Both chimeric (cA2, Centocor) and humanized (CDP571, Celltech Therapeutics Ltd., also known as Bay103356, Bayer Corp.) anti-TNFα mAbs have been used in RA. In placebo-controlled trials (25, 26), intravenous high doses of both mAbs (10 mg/kg) produced rapid clinical improvement lasting four to eight weeks. The clinical response was apparent within three days. The duration of clinical improvement was related to plasma antibody concentration. Clinical improvement was accompanied by parallel reductions in acute phase reactants and serum IL-6 level (27). In the high dose group, plasma stromelysin as well as urine pyridinoline and deoxypyridinoline levels (27) were also reduced. The latter could be considered as surrogate markers of bone and cartilage turnover, therefore these data suggest that inhibition of TNFα reduces joint damage. However, there was no effect on T cell activation markers such as soluble CD4 and IL-2R levels. Perhaps this explains why repeated treatment with anti-TNFα mAbs is necessary to sustain clinical improvement. Interestingly, cA2 but not CDP571 treatment reduced circulating monocyte/macrophage numbers (28). This difference may be explained by their different Fc portions. cA2 has a human IgG1 Fc while CDP571 has a human IgG4. Hence, cA2 is complement fixing and capable of killing TNFα-bearing monocytes and macrophages while CDP571 is not.

Anti-TNFα mAbs are highly effective anti-inflammatory treatments although repeated treatments are necessary to maintain disease improvement. However, the duration of clinical improvement may progressively shorten with repeated treatments (29, 30) which may be the result of the anti-antibody response.

Interestingly, concomitant low dose methotrexate seems to suppress antibody response to cA2 and therefore may enhance the efficacy of the latter (31) although larger clinical trials are necessary to assess the long-term safety of this combination. Side-effects of repeated anti-TNFα mAbs include skin rashes, infections and, surprisingly, the development of autoantibodies including anti-dsDNA and anti-cardiolipin antibodies (29, 32). In a recent study, repeated cA2 treatment combined with methotrexate administered eight weekly was efficacious with sustained disease improvement and no severe side-effects. Since TNFα also has an important role in tumour surveillance, lymphoproliferative disease is a potent long-term risk that needs to be assessed.

Chimeric anti-TNFα mAbs are antigenic and many patients developed anti-antibody response. This can be circumvented by using soluble TNFα receptor conjugated to human Ig Fc (sTNFR-hIg), completely human anti-TNFα mAbs, or pegylating soluble TNF receptor. Etanercept (Immunex) is a sTNFR-hIg that is administered as subcutaneous injection twice weekly. It is effective and well tolerated in RA (33). Infection risk appears lower than anti-TNF mAbs and the only apparent side-effect is mild injection site reaction. Antibody to the construct was not detected after repeated treatments. It has been recently licensed by the Food and Drug Administration in the USA for the treatment of RA. Completely human anti-TNFα mAbs for example D2E7 (Knoll) can be generated by phage display library technology (19). Recent placebo-controlled trial of such an antibody in RA showed promising results (34).

However both mAbs and sTNFR-hIg are expensive to manufacture which is a major disadvantage. Pegylating proteins can reduce their antigenicity and furthermore, the cost of manufacturing such conjugates is less than mAbs and sTNFR-hIg constructs. Pegylated soluble TNF receptor has shown positive results in an early phase I/II study in RA (35).

Anti-TNFα mAbs are powerful anti-inflammatory agents with proven efficacy in RA. High dose intravenous treatment produces rapid disease improvement but repeated therapies can lead to significant immunosuppression with the risk of infection. Frequent subcutaneous low dose treatment may be a safer alternative since the immune response can recover quickly once anti-TNFα mAb therapy is stopped.

5.1.2 Anti-IL-6 and anti-IL-6R mAbs

In RA, IL-6 is produced by monocytes and synoviocytes in the synovium. It induces B cells to mature into plasma cells and stimulates hepatocytes to produce acute phase reactants such as C reactive protein (CRP) and serum amyloid A protein. High levels of IL-6 are found in the blood and synovial fluid of RA patients. Plasma IL-6 levels correlated with disease activity, erythrocyte sedimentation rate (ESR), and CRP (36).

A murine anti-IL-6 mAb has been tested in RA. Wendling *et al.* administered 10 mg of B-E8 intravenously to RA patients daily for ten days in an open study (37). Clinical improvement was accompanied by a reduction in ESR and CRP lasting six months. Clearly the clinical efficacy of anti-IL-6 mAb needs to be confirmed in randomized placebo-controlled trial. Interestingly, serum IL-6 measured by enzyme-linked immunoassay increased but no biologically active IL-6 was detectable. This may be explained by the formation of IL-6-anti-IL-6 mAb complexes that have a longer half-life than free IL-6. An alternative way to inhibit the IL-6 pathway is to use a mAb against the IL-6 receptor, which was efficacious in collagen, induced arthritis in monkeys (38). A clinical trial of a humanized anti-IL-6 receptor mAb is currently in progress.

5.2 Anti-adhesion molecule mAbs

Leucocyte migration into the synovial joint is an important mechanism that perpetuates chronic synovitis in RA. Trafficking of leucocytes into different tissues is controlled by adhesion molecules expressed on the surface of leuco-cytes and endothelial cells. Initially leucocytes roll on the surface of endothelial cells; a process mediated by the adhesion molecules sialyl-Lewis-X and E-selectin expressed by leucocytes and endothelial cells respectively. Leucocytes then attach themselves firmly onto the endothelial cells which is mediated by integrins such as leucocyte function associated antigen-1 (LFA-1) and ICAM-1 (39). This interaction then initiates a sequence of events, including Very Late Activation Antigen-4 (VLA-4) binding to VCAM-1, which eventually leads to the transmigration of the leucocytes into the joint. LFA-1/ICAM-1 and VLA-4/VCAM-1 interactions also play crucial roles in thymic selection and T cell activation. ICAM-1 is a 90 kDa cell surface glycoprotein expressed by endothelial cells, monocytes, and antigen presenting cells. LFA-1 is one of its natural ligands. The interruption of the interaction between ICAM-1 and LFA-1 should theoretically inhibit leucocyte migration into the synovial joint and thereby suppresses inflammation in RA.

In an open trial, 32 refractory RA patients were treated with three different doses of the murine anti-ICAM-1 mAb, BIRR-1 (Boehringer Ingelheim) daily for five days (16). 13 patients, particularly those in the high dose group, showed a good response lasting one month. In nine patients improvement lasted two months. In the high dose group, furthermore, plasma concentration achieved *in vivo* was greater than the concentration needed to inhibit leucocyte adhesion *in vitro*. Most patients developed a peripheral blood leucocytosis after treatment, which may be the result of reduced leucocyte migration into the joints (40). Cutaneous delayed hypersensitivity reaction and *ex vivo* lymphocyte prolifer-ation to antigens was also reduced. This impairment of T cell activation supports the *in vitro* evidence suggesting that ICAM-1 is an important co-stimulatory molecule in T cell activation. Thus anti-ICAM-1 mAb may induce its effect not only by inhibiting lymphocyte migration but also by an effect on T cell function.

All the patients developed human anti-mouse antibodies after treatment. In eight patients whose disease relapsed, repeated treatments were given but these were less effective. Eight patients developed allergic reactions, six had urticaria, and two had angioedema (17) which highlights the limitation of wholly murine mAbs as therapeutic agents. Chimeric or humanized BIRR-1 should be used in future trials.

5.3 Anti-T cell mAbs in RA

The controversy surrounding the role of T cells in the perpetuation of rheumatoid synovitis has been discussed previously. The effect of anti-T cell mAbs has been perceived as a test of the truth of T cell hypothesis.

5.3.1 Depleting anti-T cell mAbs

Once T cells were recognized to be pathogenic in RA, a number of strategies aiming to deplete T cells were tested in clinical trials. These included total lymphoid irradiation (41–44), lymphocytapheresis (45), and thoracic duct drainage (46). These produced promising results but are impractical for everyday clinical use. Therefore a number of depleting anti-lymphocyte mAbs, targeting all or a subset of T cells, were tested in RA. Unfortunately, the clinical effects of these depleting mAbs have been disappointing. Three major factors limited the use of depleting anti-lymphocyte mAbs in RA but are important in other aspects of immune function. First, some of the T cells targeted are not pathogenic in RA. Secondly, treatment produced dose-related severe and prolonged lymphopenia with the risk of long-term immunosuppression. Thirdly, depleting antibodies are more efficient in eliminating naive, circulating T cells than removing the activated T cells present in the joint which are likely to be pathogenic.

(i) Anti-CD7 mAbs

CD7 is expressed by T cells although its function is unknown. Anti-CD7 mAb acts synergistically with a suboptimal level of mitogen or antigen to stimulate T cells to secrete IL-2 (47). *In vitro* by blocking anti-CD7 mAb inhibits mitogen stimulated lymphocyte proliferation (48). Anti-CD7 mAb is an effective treatment for transplant rejection (49). Based on this, both murine and chimeric anti-CD7 mAbs have been used in refractory rheumatoid patients (50, 51). However, neither produced significant clinical improvement although circulating $CD7^+$ T cells were depleted (50, 51). Lazarovits *et al.* later showed that in the RA synovial joint, lymphocytes are predominantly $CD7^-$ (52) suggesting $CD7^+$ lymphocytes are unlikely to have a major pathogenic role in RA. The development of anti-CD7 mAbs for the treatment of RA has therefore been abandoned.

(ii) CD5-PLUS

CD5-PLUS is an immunotoxin composed of the murine anti-CD5 mAb conjugated to the A chain of the toxin, ricin. CD5 is expressed by all mature T cells and a subset of mature B cells. Ligation of CD5 and CD28 activates T cells to proliferate and express CD69 on their cell surface (53, 54). In the initial open study in established RA (55), a 50% reduction in disease activity was seen. A subsequent study in early RA showed a 70% clinical improvement. Both studies showed a marked reduction in the number of circulating $CD3^+CD5^+$ T cells. In common with other murine mAbs, all but one patient developed a human anti-murine antibody response. Unfortunately, a double-blind placebo-controlled trial of CD5-PLUS failed to show any significant clinical improvement despite inducing significant CD5 lymphopenia (14). However, in this study the CD5-

PLUS dose was lower than that used in the open studies. Indeed, the degree of CD5 lymphopenia was less than that obtained in the open studies raising doubts regarding the biological equivalence of the open and controlled studies. The placebo-control trial was further compounded by an uncommonly high placebo response of approximately 40%. CD5-PLUS has been abandoned for the treatment of RA.

(iii) Campath-1H

Campath-1H is a humanized anti-CDw52 mAb with a human IgG1 Fc. The CDw52 molecule is expressed by T cells, monocytes, and B cells. Although the precise function of the CDw52 molecule is unknown; it is an excellent target for complement-mediated lysis (56). Indeed Campath-1H is extremely potent in activating complement. It has been tested in lymphoma (57) and multiple sclerosis (58). In open studies in RA, Campath-1H reduced tender and swollen joint scores by 50% without significant changes in ESR or CRP (59, 60). All the patients developed profound and protracted dose-related lymphopenia after treatment. In the high dose groups, there was prolonged absence of circulating lymphocytes although disease improvement was unrelated to lymphopenia. Indeed disease was seen to relapse even in the presence of severe lymphopenia. Synovial tissues in some of these patients, who subsequently underwent arthroplasty, showed diffuse mononuclear infiltrates with both $CD4^+$ and $CD8^+$ lymphocytes still present (61). This confirms that the T cell depletion seen in peripheral blood was not reflected in the synovium and suggests that Campath-1H is more efficient in eliminating naive and circulating T cells than the activated, memory pathogenic T cells in the joint. The experience with the chimeric depleting anti-CD4 mAb, cM-T412 (Centocor Inc.) discussed below, throws further light on this problem. Although the development of Campath-1H in RA has been stopped, it is being used in the treatment of refractory lymphoproliferative diseases.

(iv) cM-T412

cM-T412 is a chimeric anti-CD4 mAb with a human IgG1. Treatment with cM-T412 in RA patients produced a dose-dependent reduction in circulating $CD4^+$ lymphocyte numbers that was protracted especially when given with con-comitant methotrexate (62, 63). In open studies, clinical improvement was variable and, as with Campath-1H, did not correlate with peripheral blood CD4 lymphopenia. Three placebo-controlled trials in RA (62, 64, 65) failed to show significant clinical benefit although, similar to CD5-PLUS, the doses used in the placebo-controlled trials were lower than those used in open studies.

Sequential analysis of paired peripheral blood and synovial fluid samples showed that after a single 50 mg dose of cM-T412, there was a severe CD4 lymphopenia and over 90% of the peripheral blood $CD4^+$ lymphocytes were coated with cM-T412. In contrast, there was no change in synovial fluid $CD4^+$ lymphocyte numbers and only 11% of synovial $CD4^+$ cells were coated with cM-T412. After five daily treatments with cM-T412, there was a small but statistically significant reduction in the number of synovial fluid CD4$^+$ lymphocytes. The percentage of synovial fluid lymphocytes coated with cM-T412 varied greatly among patients and the concentration of free cM-T412 was uniformly low (63). Interestingly, the percentage of cM-T412 coated synovial lymphocytes correlated with the percentage of clinical improvement (63).

The likely explanation for these results is that cM-T412 when given intravenously, binds to CD4$^+$ targets in the circulatory system. These include not only CD4$^+$ lymphocytes but also CD4$^+$ monocytes as well as CD4$^+$ reticulo-endothelial cells (66). Most of these cells are not involved in the pathogenesis of rheumatoid synovitis. Only when the peripheral blood CD4 targets are saturated, does cM-T412 enter the joints to bind to synovial CD4$^+$ lymphocytes. Therefore cM-T412 must be given in a dose sufficient to saturate the circulation and penetrate into the joint. Moreover, among the synovial CD4$^+$ lymphocytes, most are recruited non-specifically to the joint and only a small proportion are the disease driving arthritogenic lymphocytes. Therefore, if one aims to improve arthritis by depleting synovial arthritogenic CD4$^+$ lymphocytes, high doses must be given to ensure removal of all synovial CD4$^+$ lymphocytes. In support of this, there is evidence that if a large proportion of synovial fluid CD4$^+$ lymphocytes is bound by cM-T412, the severity of synovitis is reduced (63). However, these high doses of depleting mAbs may produce a severe and prolonged depletion of peripheral CD4$^+$ lymphocytes resulting in an unacceptable level of immunosuppression. This problem is likely to arise with all depleting anti-T cell mAbs. Therefore, the T cell depletion strategy has been abandoned in favour of a strategy aiming to tolerize T cells.

5.3.2 Non-depleting anti-T cell mAbs

In streptococcal cell wall arthritis, an animal model of RA, an inflammatory polyarthritis is induced by injecting streptococcal cell wall into susceptible animals. Treatment with a single course of a non-depleting anti-CD4 mAb, at the time of streptococcal cell wall injection, prevents the development of arthritis. Furthermore, the treated animals acquire a resistance to further attempts at disease induction without repeated treatment with anti-CD4 mAbs. This refractory state appears to be due to the induction of immunological tolerance (67). Three non-depleting anti-CD4 mAbs have been studied in RA. The initial results are encouraging. In an open study using a humanized non-depleting anti-CD4 mAb, 4162W94 (Glaxo Wellcome), disease activity scores and CRP levels decreased in patients treated with either 100 mg or 300 mg daily for five days (68). Interestingly, there were also reductions in the synovial fluid TNFα and IL-1 levels. A recent placebo-controlled trial confirms the anti-inflammatory effect of 4162W94 although repeated treatment was associated with a high incidence of skin rash and, in some patients, significant CD4 lymphopenia (69). These adverse events prevent the use of repeated treatment with 4162W94. A placebo-controlled trial of a privatized non-depleting anti-CD4 mAb (IDEC-CE9.1/SB-210396, Smith Kline and Beecham) has also been carried out in RA (70). It is known as a privatized mAb because it was raised initially in macaques. Patients with refractory RA were treated either with placebo or three different doses (40 mg, 80 mg, and 140 mg) of this antibody twice weekly for four weeks. In the two high dose groups, there was a statistically significant clinical improvement although some patients in this group developed leukocytoclastic vasculitis; dosing in this cohort was discontinued. A placebo-controlled trial using a humanized non-deleting ani-CD4 mAb, OKT4 has recently been completed (71) but the full results of the study are not yet available.

Treatment with these relatively non-depleting anti-CD4 mAbs showed that they are effective anti-inflammatory treatment and support the hypothesis that CD4 lymphocyte continues to have an important role in sustaining synovitis. It is also clear that high doses of both antibodies with human IgG1 Fc can produce significant CD4 lymphopenia after high dose repeated treatments. New IgG4 antibodies, which should be less depleting, are now being tested in clinical trials. Randomized controlled studies with repeated treatments will be necessary to assess whether they can induce sustained clinical improvement.

6 Conclusion

The use of mAbs in RA has given us further insights in the roles of various inflammatory mediators in disease pathogenesis although ultimately mAbs will be judged by their ability to be used as a therapy for RA in routine clinical practice. The ideal treatment for RA should be safe, produce sustained disease remission, stop radiological damage, it should have a permanent and prolonged effect after a short course of treatment, and should be immunomodulatory rather than immunosuppressive or cytotoxic. If repeated treatments are necessary, they must be safe and economical to use in the long-term. It is against these criteria that new treatments for RA, including mAb, should and will be assessed.

At present, mAbs are administered as intravenous infusions that are costly, and frequent day case treatments will be a significant practical burden. However, if the dose of mAb treatment is small, antibody could be administered either subcutaneously or intramuscularly although such routes of administration can theoretically increase the likelihood of the development of antiglobulin antibodies (72).

Monoclonal antibodies, in particular, anti-TNFα mAbs, have been shown to be clinically effective in suppressing inflammation in RA. However, they are expensive and their role in clinical practice remains to be determined. Nevertheless, they may be used to treat disease flares and in combination with conventional disease modifying anti-rheumatic drugs to achieve better disease control. Some mAbs, such as anti-CD4, improve disease for a prolonged period in animal models in RA although this has yet to be demonstrated in human disease. Nevertheless, it holds exciting possibilities for the future. Long-term clinical trials with or without disease modifying anti-rheumatic drugs in the next decade will define better their precise role in clinical practice.

Acknowledgements

Studies with cM-T412, CDP571, and 4162W94 were funded by Centocor Inc., Celltech Therapeutics Ltd., and Glaxo Wellcome respectively. Dr E. H. S. Choy is the Arthritis Research Campaign Senior Lecturer in Rheumatology at Guy's, King's College, and St Thomas Hospitals Medical and Dental Schools of King's College London. The Academic Rheumatology Unit is supported by an ICAC grant (PO526) from the Arthritis Research Campaign of Great Britain.

References

1. Scott, D. L., Symmons, D. P., Coulton, B. L., and Popert, A. J. (1987). *Lancet*, **1**, 1108.
2. McIntosh, E. (1996). *Br. J. Rheumatol.*, **35**, 781.
3. Wolfe, F. and Hawley, D. J. (1985). *J. Rheumatol.*, **12**, 245.
4. Iannuzzi, L., Dawson, N., Zein, N., and Kushner, I. (1983). *N. Engl. J. Med.*, **309**, 1023.
5. Pincus, T., Marcum, S. B., and Callahan, L. F. (1992). *J. Rheumatol.*, **19**, 1885.
6. Hedfors, E. (1989). *Br. J. Rheumatol.*, **28**, 278.
7. Isler, P., Vey, E., Zhang, J. H., and Dayer, J. M. (1993). *Eur. Cytokine Network*, **4**, 15.
8. Brennan, F. M. and Feldmann, M. (1992). *Curr. Opin. Immunol.*, **4**, 754.
9. Firestein, G. S., Xu, W. D., Townsend, K., Broide, D., Alvar-Garcia, J. M., Glasebrook, A., *et al.* (1988). *J. Exp. Med.*, **168**, 1573.
10. Buchan, G., Barrett, K., Turner, M., Chantry, D., Maini, R. N., and Feldmann, M. (1988). *Clin. Exp. Immunol.*, **73**, 449.
11. Houssiau, F., Devogelaer, J. P., Van Damme, J., Nagant de Deuxchaisnes, C., and Van Snick, J. (1988). *Arthritis Rheum.*, **31**, 784.
12. Firestein, G. S. and Zvaifler, N. J. (1990). *Arthritis Rheum.*, **33**, 768.
13. Panayi, G. S., Lanchbury, J. S., and Kingsley, G. H. (1992). *Arthritis Rheum.*, **35**, 729.
14. Olsen, N. J., Brooks, R. H., Cush, J. J., Lipsky, P. E., St.Clair, E. W., Matteson, E. L., *et al.* (1996). *Arthritis Rheum.*, **39**, 1102.
15. Kohler, G. and Milstein, C. (1975). *Nature*, **256**, 495.
16. Kavanaugh, A. F., Davis, L. S., Nichols, L. A., Norris, S. H., Rothlein, R., Scharschmidt, L. A., *et al.* (1994). *Arthritis Rheum.*, **37**, 992.
17. Kavanaugh, A. F., Schulze-Koops, H., Davis, L. S., and Lipsky, P. E. (1997). *Arthritis Rheum.*, **40**, 849.
18. Winter, G. and Milstein, C. (1991). *Nature*, **349**, 293.
19. Marks, C. and Marks, J. D. (1996). *N. Engl. J. Med.*, **335**, 730.
20. Brennan, F. M., Maini, R. N., and Feldmann, M. (1992). *Br. J. Rheumatol.*, **31**, 293.
21. Keffer, J., Probert, L., Cazlaris, H., Georgopoulos, S., Kaslaris, E., Kioussis, D., *et al.* (1991). *EMBO J.*, **10**, 4025.
22. Williams, R. O., Feldmann, M., and Maini, R. N. (1992). *Proc. Natl. Acad. Sci. USA*, **89**, 9784.
23. Brennan, F. M., Maini, R. N., and Feldmann, M. (1992). *Br. J. Rheumatol.*, **31**, 293.
24. Chu, C. Q., Field, M., Allard, S., Abney, E., Feldmann, M., and Maini, R. N. (1992). *Br. J. Rheumatol.*, **31**, 653.
25. Elliott, M. J., Maini, R. N., Feldmann, M., Kalden, J. R., Antoni, C., Smolen, J. S., *et al.* (1994). *Lancet*, **344**, 1105.
26. Rankin, E. C., Choy, E. H., Kassimos, D., Kingsley, G. H., Sopwith, A. M., Isenberg, D. A., *et al.* (1995). *Br. J. Rheumatol.*, **34**, 334.
27. Choy, E. H. S., Kassimos, D., Kingsley, G. H., Panayi, G. S., Rankin, E. C. C., Isenberg, D. A., *et al.* (1995). *Arthritis Rheum.*, **38** (**Suppl**), S185.
28. Lorenz, H. M., Antoni, C., Valerius, T., Repp, R., Grunke, M., Schwerdtner, N., *et al.* (1996). *J. Immunol.*, **156**, 1646.

29. Elliott, M. J., Maini, R. N., Feldmann, M., Long-Fox, A., Charles, P., Bijl, H., *et al.* (1994). *Lancet*, **344**, 1125.

30. Rankin, E. C. C., Choy, E. H. S., Sopwith, M., Vetterlein, O., Panayi, G. S., and Isenberg, D. A. (1995). *Arthritis Rheum.*, **38** (**Suppl**), S185.

31. Maini, R. N., Breedveld, F. C., Kalden, J. R., Smolen, J. S., Davis, D., Macfarlane, J. D., *et al.* (1998). *Arthritis Rheum.*, **41**, 1552.

32. Rankin, E. C. C., Ravirajan, C. T., Ehrenstein, M. R., Bodman, K. B., Choy, E. H. S., Panayi, G. S., *et al.* (1995). *Br. J. Rheumatol.*, **34** (**Suppl 1**), 101.

33. Moreland, L. W., Baumgartner, S. W., Schiff, M. H., Tindall, E. A., Fleischmann, R. M., Weaver, A. L., *et al.* (1997). *N. Engl. J. Med.*, **337**, 141.

34. van de Putte, L. B., van Riel, P. L. C. M., Den Broeder, A., Sander, O., Rau, R., Binder, C., *et al.* (1998). *Arthritis Rheum.*, **41** (**Suppl**), S57.

35. McCabe, D., Moreland, L., Caldwell, J., Sack, M., Weisman, M., and Edward III, C. K. (1998). *Arthritis Rheum.*, **41** (**Suppl**), S58.

36. Dasgupta, B., Corkill, M., Kirkham, B., Gibson, T., and Panayi, G. (1992). *J. Rheumatol.*, **19**, 22.

37. Wendling, D., Racadot, E., and Wijdenes, J. (1993). *J. Rheumatol.*, **20**, 259.

38. Masahiko, M., Kotoh, M., Oda, Y., Kumagai, E., Takagi, N., Tsunemi, K., *et al.* (1997). *Arthritis Rheum.*, **40** (**Suppl**), S133.

39. Pitzalis, C., Kingsley, G., and Panayi, G. (1994). *Ann. Rheum. Dis.*, **53**, 287.

40. Davis, L. S., Kavanaugh, A. F., Nichols, L. A., and Lipsky, P. E. (1992). *Arthritis Rheum.*, **35**, S43.

41. Kotzin, B. L., Strober, S., Engleman, E. G., Calin, A., Hoppe, R. T., Kansas, G. S., *et al.* (1981). *N. Engl. J. Med.*, **305**, 969.

42. Strober, S., Tanay, A., Field, E., Hoppe, R. T., Calin, A., Engleman, E. G., *et al.* (1985). *Ann. Intern. Med.*, **102**, 441.

43. Field, E. H., Strober, S., Hoppe, R. T., Calin, A., Engleman, E. G., Kotzin, B. L., *et al.* (1983). *Arthritis Rheum.*, **26**, 937.

44. Strober, S., Kotzin, B. L., Hoppe, R. T., Slavin, S., Gottlieb, M., Calin, A., *et al.* (1981). *Int. J. Radiat. Oncol. Biol. Phys.*, **7**, 1.

45. Karsh, J., Klippel, J. H., Plotz, P. H., Decker, J. L., Wright, D. G., and Flye, M. W. (1981). *Arthritis Rheum.*, **24**, 867.

46. Paulus, H. E., Machleder, H. I., Levine, S., Yu, D. T., and MacDonald, N. S. (1977). *Arthritis Rheum.*, **20**, 1249.

47. Jung, L. K., Roy, A. K., and Chakkalath, H. R. (1992). *Cell. Immunol.*, **141**, 189.

48. Costantinides, Y., Kingsley, G. H., Pitzalis, C., and Panayi, G. S. (1991). *Clin. Exp. Immunol.*, **85**, 164.

49. Lazarovits, A. I., Rochon, J., Banks, L., Hollomby, D. J., Muirhead, N., Jevnikar, A. M., *et al.* (1993). *J. Immunol.*, **150**, 5163.

50. Kirkham, B. W., Pitzalis, C., Kingsley, G. H., Chikanza, I. C., Sabharwal, S., Barbatis, C., *et al.* (1991). *Br. J. Rheumatol.*, **30**, 459.

51. Kirkham, B. W., Thien, F., Pelton, B. K., Pitzalis, C., Amlot, P., Denman, A. M., *et al.* (1992). *J. Rheumatol.*, **19**, 1348.

52. Lazarovits, A. I., White, M. J., and Karsh, J. (1992). *Arthritis Rheum.*, **35**, 615.

53. Verwilghen, J., Vand enberghe, P., Wallays, G., Anthony, N., Panayi, G. S., and Cueppens, J. L. (1992). *J. Immunol.*, **150**, 835.

54. Vand enberghe, P., Verwilghen, J., Van Vaeck, F., and Cueppens, J. L. (1993). *Immunology*, **78**, 210.

55. Strand, V., Lipsky, P. E., Cannon, G. W., Calabrese, L. H., Wiesenhutter, C., Cohen, S. B., *et al.* (1993). *Arthritis Rheum.*, **36**, 620.

56. Greenwood, J., Clark, M., and Waldmann, H. (1993). *Eur. J. Immunol.*, **23**, 1098.

57. Hale, G., Dyer, M. J., Clark, M. R., Phillips, J. M., Marcus, R., Riechmann, L., *et al.* (1988). *Lancet*, **2**, 1394.

58. Moreau, T., Thorpe, J., Miller, D., Moseley, I., Hale, G., Waldmann, H., *et al.* (1994). *Lancet*, **344**, 298.

59. Isaacs, J. D., Watts, R. A., Hazelman, B. L., Hale, G., Keogan, M. T., Cobbold, S. P., *et al.* (1992). *Lancet*, **340**, 748.

60. Isaacs, J. D., Manna, V. K., Rapson, N., Bulpitt, K. J., Hazleman, B. L., Matteson, E. L., *et al.* (1996). *Br. J. Rheumatol.*, **35**, 231.

61. Ruderman, E. M., Weinblatt, M. E., Thurmond, L. M., Pinkus, G. S., and Gravallese, E. M. (1995). *Arthritis Rheum.*, **38**, 254.

62. Choy, E. H. S., Chikanza, I. C., Kingsley, G. H., Corrigall, V., and Panayi, G. S. (1992). *Scand. J. Immunol.*, **36**, 291.

63. Choy, E. H., Pitzalis, C., Cauli, A., Bijl, J. A., Schantz, A., Woody, J., *et al.* (1996). *Arthritis Rheum.*, **39**, 52.

64. Moreland, L. W., Pratt, P. W., Mayes, M. D., Postlethwaite, A., Weisman, M. H., Schnitzer, T., *et al.* (1995). *Arthritis Rheum.*, **38**, 1581.

65. van der Lubbe, P. A., Dijkmans, B. A., Markusse, H. M., Nassander, U., and Breedveld, F. C. (1995). *Arthritis Rheum.*, **38**, 1097.

66. Scoazec, J. Y. and Feldmann, G. (1990). *Hepatology*, **12**, 505.

67. Van den Broek, M. F., Van de Langerijt, L. G., Van Bruggen, M. C., Billingham, M. E., and Van den Berg, W. B. (1992). *Eur. J. Immunol.*, **22**, 57.

68. Panayi, G. S., Choy, E. H. S., Connolly, D. J. A., Manna, V. K., Regan, T., Rapson, N., *et al.* (1996). *Arthritis Rheum.*, **39** (**Suppl**), S244.

69. Panayi, G. S., Choy, E. H. S., Emery, P., Madden, S., Breedveld, F. C., Kraan, M. C., *et al.* (1998). *Arthritis Rheum.*, **41**, S56.

70. Levy, R., Weisman, M., Wiesenhutter, C., Yocum, D., Schnitzer, T., Goldman, A., *et al.* (1996). *Arthritis Rheum.*, **39** (**suppl**), S122.

71. Schulze-Koops, H., Davis, L. S., Haverty, P., Wacholtz, M. C., and Lipsky, P. (1997). *Arthritis Rheum.*, **40** (**Suppl**), S191.

72. Matteson, E. L., Yocum, D. E., St Clair, E. W., Achkar, A. A., Thakor, M. S., Jacobs, M. R., *et al.* (1995). *Arthritis Rheum.*, **38**, 1187.

List of suppliers

Accurate Chemical and Scientific Co., 300 Shames Drive, Westbury, NY 11590, USA.
Web site: www.accurate-assi-leeches.com

Agar Scientific Ltd., 66a Cambridge Road, Stanstead, Essex CM24 8DA, UK.
Tel: 01279 813519

Althin Medical Ltd., Unit 25, Science Park, Milton Road, Cambridge CB4 4FW, UK.

A. Menarini Diagnostics Ltd., Wharfedale Road, Winnersh, Wokingham, Berkshire RG41 5RA, U.K.
Tel: 0118 944 4100   Fax: 0118 944 4111

Amersham Pharmacia BioTech

Amersham Pharmacia Biotech, Amersham Place, Little Chalfont, Buckinghamshire HP7 9NA, UK.
Tel: 0800 515 313   Web site: www.apbiotech.com

Pharmacia Biotech (Biochrom) Ltd., Unit 22, Cambridge Science Park, Milton Road, Cambridge CB4 0FJ, UK.
Tel: 01223 423723   Fax: 01223 420164   Web site: www.biochrom.co.uk

Pharmacia and Upjohn Ltd., Davy Avenue, Knowlhill, Milton Keynes, Buckinghamshire MK5 8PH, UK.
Tel: 01908 661101   Fax: 01908 690091   Web site: www.eu.pnu.com

Pharmacia, 23 Grosvenor Road, St. Albans, Hertfordshire AL1 3AW, UK.

Amersham Pharmacia Biotech, 800 Centennial Avenue, PO Box 1327, Piscataway, NJ 08855, USA.
                          Web site: www.apbiotech.com

Ambion, Inc., 2130 Woodward, Ste 200, Austin, TX 78744, USA
                Fax: 1 512 651 0201   Web site: www.ambion.com

Amicon, Millipore Corporation, 80 Ashby Road, Bedford, MA 01730, USA.
                          Web site: www.millipore.com/analytical/amicon/index.html

Anderman and Co. Ltd., 145 London Road, Kingston-upon-Thames, Surrey KT2 6NH, UK.
Tel: 0181 541 0035   Fax: 0181 541 0623

Baxter Health Care Corp., Thetford, Norfolk.

Beckman Coulter Inc.

Beckman Coulter Inc., 4300 N Harbor Boulevard, PO Box 3100, Fullerton, CA 92834-3100, USA.
Tel: 001 714 871 4848   Fax: 001 714 773 8283   Web site: www.beckman.com

Beckman Coulter (UK) Ltd., Oakley Court, Kingsmead Business Park, London Road, High Wycombe, Buckinghamshire HP11 1JU, UK.
Tel: 01494 441181   Fax: 01494 447558   Web site: www.beckman.com

Becton Dickinson and Co.

Becton Dickinson UK Ltd., Between Towns Road, Cowley, Oxford OX4 3LY, UK.
Tel: 01865 748844   Fax: 01865 781627   Web site: www.bd.com

Becton Dickinson and Co., 1 Becton Drive, Franklin Lakes, NJ 07417-1883, USA.
Tel: 001 201 847 6800   Web site: www.bd.com

The Binding Site Ltd., PO Box 4073, Birmingham B29 6AT, UK.

Bio 101 Inc.

Bio 101 Inc., c/o Anachem Ltd., Anachem House, 20 Charles Street, Luton, Bedfordshire LU2 0EB, UK.
Tel: 01582 456666   Fax: 01582 391768   Web site: www.anachem.co.uk

Bio 101 Inc., PO Box 2284, La Jolla, CA 92038-2284, USA.
Tel: 001 760 598 7299   Fax: 001 760 598 0116   Web site: www.bio101.com

Biogenex, 4600 Norris Canyon Road, San Ramon, CA 94583, USA.

Bio-Rad Laboratories Ltd.

Bio-Rad Laboratories Ltd., Bio-Rad House, Maylands Avenue, Hemel Hempstead, Hertfordshire HP2 7TD, UK.
Tel: 0181 328 2000   Fax: 0181 328 2550   Web site: www.bio-rad.com

Bio-Rad Laboratories Ltd., Division Headquarters, 1000 Alfred Noble Drive, Hercules, CA 94547, USA.
Tel: 001 510 724 7000   Fax: 001 510 741 5817   Web site: www.bio-rad.com

Bio-Whittaker UK Ltd., 1 Ashville Way, Wokingham, Berkshire RG41 2PL, UK.

Boehringer

Boehringer, Bell Lane, Lewes, East Sussex BN7 1LG, UK.

Boehringer, 9115 Hague Road, Indianapolis, IN 46250, USA.

BPL Bio Products, Dagger Lane, Elstree, Hertfordshire WD6 3BX, UK.

British BioCell International Ltd., Golden Gate, Ty Glas Avenue, Cardiff CF4 5DX, UK.
Tel: +44 (0) 1222 747232

Cambridge Bioscience, 24-25 Signet Court, Newmarket Road, Cambridge CB5 8LA, UK.

CellPro, St-Pietersplein 11/12, B-1970 Wezembeek-Oppem, Belgium.

Clontech Laboratories, Inc., 1020 East Meadow Circle, Palo Alto, CA 94303–4230, USA
                    Fax: 1 650 424 8222   Web site: www.clontech.com

Cobe International, Blood Component Technology, Mercuriusstraat 30, 1930 Zaventum, Belgium.

Corning Inc., Science Products Division, 45 Nagog Park, Acton, MA 01720, USA.
                              Web site: www.corningcostar.com

CP Instrument Co. Ltd., PO Box 22, Bishop Stortford, Hertfordshire CM23 3DX, UK.
Tel: 01279 757711   Fax: 01279 755785   Web site: www.cpinstrument.co.uk

Dako

Dako Ltd., Denmark House, Angel Drove, Ely, Cambridge CB7 4ET, UK.

Dako Corp., 6392 Via Road, Carpinteria, CA 93013, USA.

Diachem International Ltd., Unit 5, Gardiners Place, West Gillibrands, Skelmersdale, Lancashire WN8 9SP, UK.

Diatome Ltd., 2501 Bienne, PO Box 551, Switzerland.

Drukker International, Beverstraat 20, 5431 SH Cuijk, The Netherlands.
Tel: +31(0) 485 39 57 00

Dupont

Dupont (UK) Ltd., Industrial Products Division, Wedgwood Way, Stevenage, Hertfordshire SG1 4QN, UK.
Tel: 01438 734000   Fax: 01438 734382   Web site: www.dupont.com

Dupont Co. (Biotechnology Systems Division), PO Box 80024, Wilmington, DE 19880-002, USA.
Tel: 001 302 774 1000   Fax: 001 302 774 7321   Web site: www.dupont.com

Eastman Chemical Co., 100 North Eastman Road, PO Box 511, Kingsport, TN 37662-5075, USA.
Tel: 001 423 229 2000   Web site: www.eastman.com

Elga Ltd., Lane End, High Wycombe, Buckinghamshire HP14 3JH, UK.

Fisher Scientific

Fisher Scientific UK Ltd., Bishop Meadow Road, Loughborough, Leicestershire LE11 5RG, UK.
Tel: 01509 231166   Fax: 01509 231893   Web site: www.fisher.co.uk

Fisher Scientific, Fisher Research, 2761 Walnut Avenue, Tustin, CA 92780, USA.
Tel: 001 714 669 4600   Fax: 001 714 669 1613   Web site: www.fishersci.com

Fluka

Fluka, PO Box 2060, Milwaukee, WI 53201, USA.
Tel: 001 414 273 5013   Fax: 001 414 2734979   Web site: www.sigma-aldrich.com

Fluka Chemical Co. Ltd., PO Box 260, CH-9471, Buchs, Switzerland.
Tel: 0041 81 745 2828   Fax: 0041 81 756 5449   Web site: www.sigma-aldrich.com

FMC BioProducts, 191 Thomaston Street, Rockland, ME 04841, USA.
Tel: 207 584 3400   Fax: 1 207 594 3491   Web site: www.bioproducts.com

Greiner Labortechnik Ltd., Brunel Way, Stroudwater Business Park, Stonehouse, Gloucester GL10 3SX, UK.
Tel: 01453 825255

Hybaid

Hybaid Ltd., Action Court, Ashford Road, Ashford, Middlesex TW15 1XB, UK.
Tel: 01784 425000   Fax: 01784 248085   Web site: www.hybaid.com

Hybaid US, 8 East Forge Parkway, Franklin, MA 02038, USA.
Tel: 001 508 541 6918   Fax: 001 508 541 3041   Web site: www.hybaid.com

HyClone Laboratories, 1725 South HyClone Road, Logan, UT 84321, USA.
Tel: 001 435 753 4584   Fax: 001 435 753 4589   Web site: www.hyclone.com

Invitrogen

Invitrogen BV, PO Box 2312, 9704 CH Groningen, The Netherlands.
Tel: 00800 5345 5345   Fax: 00800 7890 7890   Web site: www.invitrogen.com

Invitrogen Corp., 1600 Faraday Avenue, Carlsbad, CA 92008, USA.
Tel: 001 760 603 7200   Fax: 001 760 603 7201   Web site: www.invitrogen.com

Kirkegaard and Perry Laboratories, Inc., (KPL) 2 Cessna Court, Gaithersburg, MD 20897, USA.
                    Web site: www.kpl.com

Leica, Davy Avenue, Knowlhill, Milton Keynes MK5 8LB, UK.
Tel: 01908 666663   Web site: www.leica.com

Life Technologies

Life Technologies Ltd., PO Box 35, Free Fountain Drive, Incsinnan Business Park, Paisley PA4 9RF, UK.

Tel: 0800 269210   Fax: 0800 838380   Web site: www.lifetech.com

Life Technologies Inc., 9800 Medical Center Drive, Rockville, MD 20850, USA.

Tel: 001 301 610 8000   Web site: www.lifetech.com

Merck Sharp & Dohme

Merck Sharp & Dohme Research Laboratories, Neuroscience Research Centre, Terlings Park, Harlow, Essex CM20 2QR, UK.

Web site: www.msd-nrc.co.uk

MSD Sharp and Dohme GmbH, Lindenplatz 1, D-85540, Haar, Germany.

Web site: www.msd-deutschland.com

Millipore

Millipore (UK) Ltd., The Boulevard, Blackmoor Lane, Watford, Hertfordshire WD1 8YW, UK.

Tel: 01923 816375   Fax: 01923 818297

Web site: www.millipore.com/local/UK.htm

Millipore Corp., 80 Ashby Road, Bedford, MA 01730, USA.

Tel: 001 800 645 5476   Fax: 001 800 645 5439   Web site: www.millipore.com

Nanoprobes Inc., 25 E Loop Road, Sye. 124, Stony Brook, NY 11790-3350 USA.

NEN™, Life Science Products, 549-3 Albany Street, Boston, MA 02118, USA.

Web site: www.nenlifesci.com

New England Biolabs, 32 Tozer Road, Beverley, MA 01915-5510, USA.

Tel: 001 978 927 5054

Nikon

Nikon Corp., Fuji Building, 2-3, 3-chome, Marunouchi, Chiyoda-ku, Tokyo 100, Japan.

Tel: 00813 3214 5311   Fax: 00813 3201 5856

Web site: www.nikon.co.jp/main/index_e.htm

Nikon Inc., 1300 Walt Whitman Road, Melville, NY 11747-3064, USA.

Tel: 001 516 547 4200   Fax: 001 516 547 0299   Web site: www.nikonusa.com

Nycomed

Nycomed Amersham plc, Amersham Place, Little Chalfont, Buckinghamshire HP7 9NA, UK.

Tel: 01494 544000   Fax: 01494 542266   Web site: www.amersham.co.uk

Nycomed Amersham, 101 Carnegie Center, Princeton, NJ 08540, USA.

Tel: 001 609 514 6000   Web site: www.amersham.co.uk

Nycomed AS Pharma, Diagnostic Division, PO Box 4284 Torshov, N-0401 Oslo, Norway.

Ortho Diagnostic Systems, PO Box 653, Enterprise House, Station Road, Loudwater, Buckinghamshire HP10 9XH, UK.

Oxoid Ltd., Basingstoke, Hampshire, UK.

PE Applied Biosystems, 850 Lincoln Centre Drive, Foster City, CA 94404, USA.

Fax: 1 203 761 2542   Web site: www.perkin-elmer.com

Perkin Elmer Ltd., Post Office Lane, Beaconsfield, Buckinghamshire HP9 1QA, UK.

Tel: 01494 676161   Web site: www.perkin-elmer.com

PerSeptive Biosystems Inc., 500 Old Connecticut Path, Framingham, MA 01701, USA.

Web site: www.pbio.com

Pharmacia [please see Amersham Pharmacia BioTech]

Pharmingen: distributed by Cambridge Bioscience

Pierce Chemical Co., 3747 N Meridian Road, Rockford, IL 61105, USA.

Web site: www.piercenet.com

Pierce and Warriner (UK) Ltd., 44 Upper Northgate Street, Chester CH1 4EF, UK.

Tel: 01244 382525   Fax: 01244 373212

Promega

Promega UK Ltd., Delta House, Chilworth Research Centre, Southampton SO16 7NS, UK.

Tel: 0800 378994   Fax: 0800 181037   Web site: www.promega.com

Promega Corp., 2800 Woods Hollow Road, Madison, WI 53711-5399, USA.

Tel: 001 608 274 4330   Fax: 001 608 277 2516   Web site: www.promega.com

Qiagen

Qiagen UK Ltd., Boundary Court, Gatwick Road, Crawley, West Sussex RH10 2AX, UK.

Tel: 01293 422911   Fax: 01293 422922   Web site: www.qiagen.com

Qiagen Inc., 28159 Avenue Stanford, Valencia, CA 91355, USA.

Tel: 001 800 426 8157   Fax: 001 800 718 2056   Web site: www.qiagen.com

Roche Diagnostics

Roche Diagnostics Ltd., Bell Lane, Lewes, East Sussex BN7 1LG, UK.

Tel: 01273 484644   Fax: 01273 480266   Web site: www.roche.com

Roche Diagnostics Corp., 9115 Hague Road, PO Box 50457, Indianapolis, IN 46256, USA.

Tel: 001 317 845 2358   Fax: 001 317 576 2126   Web site: www.roche.com

Roche Diagnostics GmbH, Sandhoferstrasse 116, 68305 Mannheim, Germany.

Tel: 0049 621 759 4747   Fax: 0049 621 759 4002   Web site: www.roche.com

Schleicher and Schuell Inc., Keene, NH 03431A, USA.

Tel: 001 603 357 2398

Serotec Ltd., 22 Bankside, Station Approach, Kidlington, Oxford OX5 1JE, UK.

Shandon Scientific Ltd., 93-96 Chadwick Road, Astmoor, Runcorn, Cheshire WA7 1PR, UK.

Tel: 01928 566611   Web site: www.shandon.com

Sigma-Aldrich

Sigma-Aldrich Co. Ltd., The Old Brickyard, New Road, Gillingham, Dorset XP8 4XT, UK.

Tel: 01747 822211   Fax: 01747 823779   Web site: www.sigma-aldrich.com

Sigma-Aldrich Co. Ltd., Fancy Road, Poole, Dorset BH12 4QH, UK.

Tel: 01202 722114   Fax: 01202 715460   Web site: www.sigma-aldrich.com

Sigma Chemical Co., PO Box 14508, St Louis, MO 63178, USA.

Tel: 001 314 771 5765   Fax: 001 314 771 5757   Web site: www.sigma-aldrich.com

Sorvall Centrifuges (distributors): Medi-Tech International, Inc., 2924 NW 109th Avenue, Miami, FL 33172, USA.

Web site: www.sorvall.com

Stedim, Z.I. des Paluds, BP1051-13781, Aubagne, France.

Stratagene

Stratagene Europe, Gebouw California, Hogehilweg 15, 1101 CB Amsterdam Zuidoost, The Netherlands.

Tel: 00800 9100 9100   Web site: www.stratagene.com

Stratagene Inc., 11011 North Torrey Pines Road, La Jolla, CA 92037, USA.

Tel: 001 858 535 5400   Web site: www.stratagene.com

TAAB Laboratories Equipment Ltd., 3 Minerva House, Calleva Park, Aldermaston, Berkshire RG7 8NA, UK.

Tel: 0118 981 7775

Therapeutic A. L. Centre, Oxford University, Old Road, Headington, Oxford OX3 7JT, UK.

Tropix, 47 Wiggins Avenue, Bedford, MA 01730, USA.

Web site: www.tropix.com

United States Biochemical, PO Box 22400, Cleveland, OH 44122, USA.

Tel: 001 216 464 9277

Vector

Vector, 30 Ingold Road, Burlingame, CA 94010, USA.

Vector, 16 Wulfric Square, Bretton, Peterborough PE3 8RF, UK.

Wallac Inc., 9238 Gaither Road, Gaithersburg, MD 20877, USA.

Web site: www.wallac.com

Western Laboratory Service Ltd., Unit 8, Redan Hill Estate, Redan Road, Aldershot, Hampshire, UK.

Tel: 01252 312128

Index